Geophysical Monograph Series

Including

IUGG Volumes

Maurice Ewing Volumes
Mineral Physics Volumes

GEOPHYSICAL MONOGRAPH SERIES

Geophysical Monograph Volumes

1. Antarctica in the International Geophysical Year *A. P. Crary, L. M. Gould, E. O. Hulburt, Hugh Odishaw, and Waldo E. Smith (Eds.)*
2. Geophysics and the IGY *Hugh Odishaw and Stanley Ruttenberg (Eds.)*
3. Atmospheric Chemistry of Chlorine and Sulfur Compounds *James P. Lodge, Jr. (Ed.)*
4. Contemporary Geodesy *Charles A. Whitten and Kenneth H. Drummond (Eds.)*
5. Physics of Precipitation *Helmut Weickmann (Ed.)*
6. The Crust of the Pacific Basin *Gordon A. Macdonald and Hisashi Kuno (Eds.)*
7. Antarctic Research: The Matthew Fontaine Maury Memorial Symposium *H. Wexler, M. J. Rubin, and J. E. Caskey, Jr. (Eds.)*
8. Terrestrial Heat Flow *William H. K. Lee (Ed.)*
9. Gravity Anomalies: Unsurveyed Areas *Hyman Orlin (Ed.)*
10. The Earth Beneath the Continents: A Volume of Geophysical Studies in Honor of Merle A. Tuve *John S. Steinhart and T. Jefferson Smith (Eds.)*
11. Isotope Techniques in the Hydrologic Cycle *Glenn E. Stout (Ed.)*
12. The Crust and Upper Mantle of the Pacific Area *Leon Knopoff, Charles L. Drake, and Pembroke J. Hart (Eds.)*
13. The Earth's Crust and Upper Mantle *Pembroke J. Hart (Ed.)*
14. The Structure and Physical Properties of the Earth's Crust *John G. Heacock (Ed.)*
15. The Use of Artificial Satellites for Geodesy *Soren W. Henricksen, Armando Mancini, and Bernard H. Chovitz (Eds.)*
16. Flow and Fracture of Rocks *H. C. Heard, I. Y. Borg, N. L. Carter, and C. B. Raleigh (Eds.)*
17. Man-Made Lakes: Their Problems and Environmental Effects *William C. Ackermann, Gilbert F. White, and E. B. Worthington (Eds.)*
18. The Upper Atmosphere in Motion: A Selection of Papers With Annotation *C. O. Hines and Colleagues*
19. The Geophysics of the Pacific Ocean Basin and Its Margin: A Volume in Honor of George P. Woollard *George H. Sutton, Murli H. Manghnani, and Ralph Moberly (Eds.)*
20. The Earth's Crust: Its Nature and Physical Properties *John C. Heacock (Ed.)*
21. Quantitative Modeling of Magnetospheric Processes *W. P. Olson (Ed.)*
22. Derivation, Meaning, and Use of Geomagnetic Indices *P. N. Mayaud*
23. The Tectonic and Geologic Evolution of Southeast Asian Seas and Islands *Dennis E. Hayes (Ed.)*
24. Mechanical Behavior of Crustal Rocks: The Handin Volume *N. L. Carter, M. Friedman, J. M. Logan, and D. W. Stearns (Eds.)*
25. Physics of Auroral Arc Formation *S.-I. Akasofu and J. R. Kan (Eds.)*
26. Heterogeneous Atmospheric Chemistry *David R. Schryer (Ed.)*
27. The Tectonic and Geologic Evolution of Southeast Asian Seas and Islands: Part 2 *Dennis E. Hayes (Ed.)*
28. Magnetospheric Currents *Thomas A. Potemra (Ed.)*
29. Climate Processes and Climate Sensitivity (Maurice Ewing Volume 5) *James E. Hansen and Taro Takahashi (Eds.)*
30. Magnetic Reconnection in Space and Laboratory Plasmas *Edward W. Hones, Jr. (Ed.)*
31. Point Defects in Minerals (Mineral Physics Volume 1) *Robert N. Schock (Ed.)*
32. The Carbon Cycle and Atmospheric CO_2: Natural Variations Archean to Present *E. T. Sundquist and W. S. Broecker (Eds.)*
33. Greenland Ice Core: Geophysics, Geochemistry, and the Environment *C. C. Langway, Jr., H. Oeschger, and W. Dansgaard (Eds.)*
34. Collisionless Shocks in the Heliosphere: A Tutorial Review *Robert G. Stone and Bruce T. Tsurutani (Eds.)*
35. Collisionless Shocks in the Heliosphere: Reviews of Current Research *Bruce T. Tsurutani and Robert G. Stone (Eds.)*
36. Mineral and Rock Deformation: Laboratory Studies—The Paterson Volume *B. E. Hobbs and H. C. Heard (Eds.)*
37. Earthquake Source Mechanics (Maurice Ewing Volume 6) *Shamita Das, John Boatwright, and Christopher H. Scholz (Eds.)*
38. Ion Acceleration in the Magnetosphere and Ionosphere *Tom Chang (Ed.)*
39. High Pressure Research in Mineral Physics (Mineral Physics Volume 2) *Murli H. Manghnani and Yasuhiko Syono (Eds.)*
40. Gondwana Six: Structure, Tectonics, and Geophysics *Gary D. McKenzie (Ed.)*
41. Gondwana Six: Stratigraphy, Sedimentology, and Paleontology *Garry D. McKenzie (Ed.)*
42. Flow and Transport Through Unsaturated Fractured Rock *Daniel D. Evans and Thomas J. Nicholson (Eds.)*
43. Seamounts, Islands, and Atolls *Barbara H. Keating, Patricia Fryer, Rodey Batiza, and George W. Boehlert (Eds.)*

44 **Modeling Magnetospheric Plasma** *T. E. Moore and J. H. Waite, Jr. (Eds.)*

45 **Perovskite: A Structure of Great Interest to Geophysics and Materials Science** *Alexandra Navrotsky and Donald J. Weidner (Eds.)*

46 **Structure and Dynamics of Earth's Deep Interior (IUGG Volume 1)** *D. E. Smylie and Raymond Hide (Eds.)*

47 **Hydrological Regimes and Their Subsurface Thermal Effects (IUGG Volume 2)** *Alan E. Beck, Grant Garven, and Lajos Stegena (Eds.)*

48 **Origin and Evolution of Sedimentary Basins and Their Energy and Mineral Resources (IUGG Volume 3)** *Raymond A. Price (Ed.)*

49 **Slow Deformation and Transmission of Stress in the Earth (IUGG Volume 4)** *Steven C. Cohen and Petr Vaníček (Eds.)*

50 **Deep Structure and Past Kinematics of Accreted Terranes (IUGG Volume 5)** *John W. Hillhouse (Ed.)*

51 **Properties and Processes of Earth's Lower Crust (IUGG Volume 6)** *Robert F. Mereu, Stephan Mueller, and David M. Fountain (Eds.)*

52 **Understanding Climate Change (IUGG Volume 7)** *Andre L. Berger, Robert E. Dickinson, and J. Kidson (Eds.)*

53 **Plasma Waves and Instabilities at Comets and in Magnetospheres** *Bruce T. Tsurutani and Hiroshi Oya (Eds.)*

54 **Solar System Plasma Physics** *J. H. Waite, Jr., J. L. Burch, and R. L. Moore (Eds.)*

55 **Aspects of Climate Variability in the Pacific and Western Americas** *David H. Peterson (Ed.)*

56 **The Brittle-Ductile Transition in Rocks** *A. G. Duba, W. B. Durham, J. W. Handin, and H. F. Wang (Eds.)*

57 **Evolution of Mid Ocean Ridges (IUGG Volume 8)** *John M. Sinton (Ed.)*

58 **Physics of Magnetic Flux Ropes** *C. T. Russell, E. R. Priest, and L. C. Lee (Eds.)*

59 **Variations in Earth Rotation (IUGG Volume 9)** *Dennis D. McCarthy and Williams E. Carter (Eds.)*

60 **Quo Vadimus Geophysics for the Next Generation (IUGG Volume 10)** *George D. Garland and John R. Apel (Eds.)*

61 **Cometary Plasma Processes** *Alan D. Johnstone (Ed.)*

62 **Modeling Magnetospheric Plasma Processes** *Gordon R. Wilson (Ed.)*

63 **Marine Particles: Analysis and Characterization** *David C. Hurd and Derek W. Spencer (Eds.)*

64 **Magnetospheric Substorms** *Joseph R. Kan, Thomas A. Potemra, Susumu Kokubun, and Takesi Iijima (Eds.)*

65 **Explosion Source Phenomenology** *Steven R. Taylor, Howard J. Patton, and Paul G. Richards (Eds.)*

66 **Venus and Mars: Atmospheres, Ionospheres, and Solar Wind Interactions** *Janet G. Luhmann, Mariella Tatrallyay, and Robert O. Pepin (Eds.)*

67 **High-Pressure Research: Application to Earth and Planetary Sciences (Mineral Physics Volume 3)** *Yasuhiko Syono and Murli H. Manghnani (Eds.)*

68 **Microwave Remote Sensing of Sea Ice** *Frank Carsey, Roger Barry, Josefino Comiso, D. Andrew Rothrock, Robert Shuchman, W. Terry Tucker, Wilford Weeks, and Dale Winebrenner*

69 **Sea Level Changes: Determination and Effects (IUGG Volume 11)** *P. L. Woodworth, D. T. Pugh, J. G. DeRonde, R. G. Warrick, and J. Hannah*

70 **Synthesis of Results from Scientific Drilling in the Indian Ocean** *Robert A. Duncan, David K. Rea, Robert B. Kidd, Ulrich von Rad, and Jeffrey K. Weissel (Eds.)*

71 **Mantle Flow and Melt Generation at Mid-Ocean Ridges** *Jason Phipps Morgan, Donna K. Blackman, and John M. Sinton (Eds.)*

72 **Dynamics of Earth's Deep Interior and Earth Rotation (IUGG Volume 12)** *Jean-Louis Le Mouël, D.E. Smylie, and Thomas Herring (Eds.)*

73 **Environmental Effects on Spacecraft Positioning and Trajectories (IUGG Volume 13)** *A. Vallance Jones (Ed.)*

74 **Evolution of the Earth and Planets (IUGG Volume 14)** *E. Takahashi, Raymond Jeanloz, and David Rubie (Eds.)*

75 **Interactions Between Global Climate Subsystems: The Legacy of Hann (IUGG Volume 15)** *G. A. McBean and M. Hantel (Eds.)*

76 **Relating Geophysical Structures and Processes: The Jeffreys Volume (IUGG Volume 16)** *K. Aki and R. Dmowska (Eds.)*

77 **The Mesozoic Pacific: Geology, Tectonics, and Volcanism—A Volume in Memory of Sy Schlanger** *Malcolm S. Pringle, William W. Sager, William V. Sliter, and Seth Stein (Eds.)*

78 **Climate Change in Continental Isotopic Records** *P. K. Swart, K. C. Lohmann, J. McKenzie, and S. Savin (Eds.)*

79 **The Tornado: Its Structure, Dynamics, Prediction, and Hazards** *C. Church, D. Burgess, C. Doswell, R. Davies-Jones (Eds.)*

80 **Auroral Plasma Dynamics** *R. L. Lysak (Ed.)*

81 **Solar Wind Sources of Magnetospheric Ultra-Low Frequency Waves** *M. J. Engebretson, K. Takahashi, and M. Scholer (Eds.)*

82 **Gravimetry and Space Techniques Applied to Geodynamics and Ocean Dynamics (IUGG Volume 17)** *Bob E. Schutz, Allen Anderson, Claude Froidevaux, and Michael Parke (Eds.)*

83 **Nonlinear Dynamics and Predictability of Geophysical Phenomena (IUGG Volume 18)** *William I. Newman, Andrei Gabrielov, and Donald L. Turcotte (Eds.)*

84 **Solar System Plasmas in Space and Time** *J. Burch, J. H. Waite, Jr. (Eds.)*

85 **The Polar Oceans and Their Role in Shaping the Global Environment** *O. M. Johannessen, R. D. Muench, and J. E. Overland (Eds.)*

86 **Space Plasmas: Coupling Between Small and Medium Scale Processes** *Maha Ashour-Abdalla, Tom Chang, and Paul Dusenbery (Eds.)*

87 **The Upper Mesosphere and Lower Thermosphere: A Review of Experiment and Theory** *R. M. Johnson and T. L. Killeen (Eds.)*

88 **Active Margins and Marginal Basins of the Western Pacific** *Brian Taylor and James Natland (Eds.)*

89 **Natural and Anthropogenic Influences in Fluvial Geomorphology** *John E. Costa, Andrew J. Miller, Kenneth W. Potter, and Peter R. Wilcock (Eds.)*

90 **Physics of the Magnetopause** *Paul Song, B.U.Ö. Sonnerup, and M.F. Thomsen (Eds.)*

91 **Seafloor Hydrothermal Systems: Physical, Chemical, Biological, and Geological Interactions** *Susan E. Humphris, Robert A. Zierenberg, Lauren S. Mullineaux, and Richard E. Thomson (Eds.)*

92 **Mauna Loa Revealed: Structure, Composition, History, and Hazards** *J. M. Rhodes and John P. Lockwood (Eds.)*

93 **Cross-Scale Coupling in Space Plasmas** *James L. Horwitz, Nagendra Singh, and James L. Burch (Eds.)*

94 **Double-Diffusive Convection** *Alan Brandt and H.J.S. Fernando (Eds.)*

95 **Earth Processes: Reading the Isotopic Code** *Asish Basu and Stan Hart (Eds.)*

96 **Subduction Top to Bottom** *Gray E. Bebout, David Scholl, Stephen Kirby, and John Platt (Eds.)*

97 **Radiation Belts—Models and Standards** *J. F. Lemaire, D. Heynderickx, and D. N. Baker (Eds.)*

98 **Magnetic Storms** *Bruce T. Tsurutani, Walter D. Gonzalez, Yohsuke Kamide, and John K. Arballo (Eds.)*

Maurice Ewing Volumes

1 **Island Arcs, Deep Sea Trenches, and Back-Arc Basins** *Manik Talwani and Walter C. Pitman III (Eds.)*

2 **Deep Drilling Results in the Atlantic Ocean: Ocean Crust** *Manik Talwani, Christopher G. Harrison, and Dennis E. Hayes (Eds.)*

3 **Deep Drilling Results in the Atlantic Ocean: Continental Margins and Paleoenvironment** *Manik Talwani, William Hay, and William B. F. Ryan (Eds.)*

4 **Earthquake Prediction—An International Review** *David W. Simpson and Paul G. Richards (Eds.)*

5 **Climate Processes and Climate Sensitivity** *James E. Hansen and Taro Takahashi (Eds.)*

6 **Earthquake Source Mechanics** *Shamita Das, John Boatwright, and Christopher H. Scholz (Eds.)*

IUGG Volumes

1 **Structure and Dynamics of Earth's Deep Interior** *D. E. Smylie and Raymond Hide (Eds.)*

2 **Hydrological Regimes and Their Subsurface Thermal Effects** *Alan E. Beck, Grant Garven, and Lajos Stegena (Eds.)*

3 **Origin and Evolution of Sedimentary Basins and Their Energy and Mineral Resources** *Raymond A. Price (Ed.)*

4 **Slow Deformation and Transmission of Stress in the Earth** *Steven C. Cohen and Petr Vaníček (Eds.)*

5 **Deep Structure and Past Kinematics of Accreted Terranes** *John W. Hillhouse (Ed.)*

6 **Properties and Processes of Earth's Lower Crust** *Robert F. Mereu, Stephan Mueller, and David M. Fountain (Eds.)*

7 **Understanding Climate Change** *Andre L. Berger, Robert E. Dickinson, and J. Kidson (Eds.)*

8 **Evolution of Mid Ocean Ridges** *John M. Sinton (Ed.)*

9 **Variations in Earth Rotation** *Dennis D. McCarthy and William E. Carter (Eds.)*

10 **Quo Vadimus Geophysics for the Next Generation** *George D. Garland and John R. Apel (Eds.)*

11 **Sea Level Changes: Determinations and Effects** *Philip L. Woodworth, David T. Pugh, John G. DeRonde, Richard G. Warrick, and John Hannah (Eds.)*

12 **Dynamics of Earth's Deep Interior and Earth Rotation** *Jean-Louis Le Mouël, D.E. Smylie, and Thomas Herring (Eds.)*

13 **Environmental Effects on Spacecraft Positioning and Trajectories** *A. Vallance Jones (Ed.)*

14 **Evolution of the Earth and Planets** *E. Takahashi, Raymond Jeanloz, and David Rubie (Eds.)*

15 **Interactions Between Global Climate Subsystems: The Legacy of Hann** *G. A. McBean and M. Hantel (Eds.)*

16 **Relating Geophysical Structures and Processes: The Jeffreys Volume** *K. Aki and R. Dmowska (Eds.)*

17 **Gravimetry and Space Techniques Applied to Geodynamics and Ocean Dynamics** *Bob E. Schutz, Allen Anderson, Claude Froidevaux, and Michael Parke (Eds.)*

18 **Nonlinear Dynamics and Predictability of Geophysical Phenomena** *William I. Newman, Andrei Gabrielov, and Donald L. Turcotte (Eds.)*

Mineral Physics Volumes

1 **Point Defects in Minerals** *Robert N. Schock (Ed.)*

2 **High Pressure Research in Mineral Physics** *Murli H. Manghnani and Yasuhiko Syona (Eds.)*

3 **High Pressure Research: Application to Earth and Planetary Sciences** *Yasuhiko Syono and Murli H. Manghnani (Eds.)*

Geophysical Monograph 99

Coronal Mass Ejections

Nancy Crooker
Jo Ann Joselyn
Joan Feynman
Editors

American Geophysical Union

Published under the aegis of the AGU Books Board

Library of Congress Cataloging-in-Publication Data
Coronal mass ejections / Nancy Crooker, Jo Ann Joselyn, Joan Feynman, editors
        p.   cm. -- (Geophysical Monograph ; 99)
    Includes bibliographical references.
    ISBN 0-87590-081-X
    1. Sun--Corona. I. Crooker, Nancy, 1944- . II. Joselyn, J. A.
III. Feynman, Joan. IV. Series
QB529.C68.1997
523.7'5--DC21   97-26277
                                                                                   CIP

ISBN 0-87590-081-X
ISSN 0065-8448

Copyright 1997 by the American Geophysical Union
2000 Florida Avenue, N.W.
Washington, DC 20009

Figures, tables, and short excerpts may be reprinted in scientific books and journals if the source is properly cited.

Authorization to photocopy items for internal or personal use, or the internal or personal use of specific clients, is granted by the American Geophysical Union for libraries and other users registered with the Copyright Clearance Center (CCC) Transactional Reporting Service, provided that the base fee of $1.50 per copy plus $0.35 per page is paid directly to CCC, 222 Rosewood Dr., Danvers, MA 01923. 0065-8448/97/$01.50+0.35.
   This consent does not extend to other kinds of copying, such as copying for creating new collective works or for resale. The reproduction of multiple copies and the use of full articles or the use of extracts, including figures and tables, for commercial purposes requires permission from AGU.

Printed in the United States of America.

CONTENTS

Preface
Nancy Crooker, Jo Ann Joselyn, Joan Feynman   ix

An Introduction
A. J. Hundhausen   1

Coronal Mass Ejections: An Overview
J. T. Gosling   9

Solar Views

Observations of CMEs from SOHO/LASCO
R. A. Howard, G.E. Brueckner, O. C. St. Cyr, D. A. Biesecker, K.P. Dere, M. J. Koomen, C. M. Korendyke, P. L. Lamy, A. Llebaria, M. V. Bout, D. J. Michels, J. D. Moses, S. E. Paswaters, S. P. Plunkett, R. Schwenn, G. M. Simnett, D. G. Socker, S. J. Tappin, and D. Wang   17

Soft X-Ray Signatures of Coronal Ejections
Hugh S. Hudson and David F. Webb   27

Initiation

The Role of Coronal Mass Ejections in Solar Activity
B. C. Low   39

Evolving Magnetic Structures and Their Relation to Coronal Mass Ejections
Joan Feynman   49

The Initiation of Coronal Mass Ejections by Magnetic Shear
Zoran Mikić and Jon A. Linker   57

Coronal Mass Ejections: Causes and Consequences—A Theoretical View
James Chen   65

A Self-Consistent Numerical Magnetohydrodynamic (MHD) Model of Helmet Streamer and Flux-Rope Interactions: Initiation and Propagation of Coronal Mass Ejections (CMEs)
S. T. Wu and W. P. Guo   83

Solar and Interplanetary Topology

Solar Magnetic Topologies and Reconnection
P. Démoulin   91

Opening Solar Magnetic Fields: Some Analytical and Numerical MHD Aspects
T. Amari, J. F. Luciani, J. J. Aly, and Z. Mikić   101

The Topology and Instability of Complex Magnetic Fields
Aaron William Longbottom   111

Helicity Conservation
D. M. Rust   119

Predicting the Sign of Magnetic Helicity in Erupting Filaments and Coronal Mass Ejections
S. F. Martin and A. H. McAllister   127

CONTENTS

The Field Configuration of Magnetic Clouds and the Solar Cycle
V. Bothmer and D. M. Rust   137

Interplanetary Magnetic Flux Ropes and Solar Filaments
K. Marubashi   147

Magnetic Clouds
Vladimir Osherovich and L. F. Burlaga   157

Flux Ropes and Spheromaks: A Numerical Study
M. Vandas, S. Fischer, D. Odstrčil, M. Dryer, Z. Smith, and T. Detman   169

Recent Work on Modelling the Global Field Line Topology of Interplanetary Magnetic Clouds
C. J. Farrugia   177

Using Energetic Particles to Probe the Magnetic Topology of Ejecta
I. G. Richardson   189

Using Charged Particles to Trace Interplanetary Magnetic Field Topology
S. W. Kahler   197

Energetic Particle Acceleration

The Current Status in Our Understanding of Energetic Particles, Coronal Mass Ejections, and Flares
H. V. Cane   205

Energetic Particles and the Structure of Coronal Mass Ejections
Donald V. Reames   217

Particle Acceleration and Transport at CME-Driven Shocks
Martin A. Lee   227

Interplanetary Views and Effects at Earth

Mass Ejections Observed in Radio Propagation Measurements Through the Solar Corona
Richard Woo   235

Particle and Field Signatures of Coronal Mass Ejections in the Solar Wind
Marcia Neugebauer and Raymond Goldstein   245

Minor Ion Composition in CME-Related Solar Wind
A. B. Galvin   253

Global Modeling of CME Propagation in the Solar Wind
V. J. Pizzo   261

Extending Coronal Models to Earth Orbit
Jon A. Linker and Zoran Mikić   269

Coronal Mass Ejections, Corotating Interaction Regions, and Geomagnetic Storms
A. H. McAllister and N. U. Crooker   279

CMEs and Space Weather
J. G. Luhman   291

PREFACE

The early 1970's can be said to mark the beginning of The Enlightenment in the history of the Space Age, literally as well as by analogy to European history. Instruments blinded by Earth's atmosphere were lifted above and, for the first time, saw clearly and continuously the ethereal white light and sparkling x-rays from the solar corona. From these two bands of the light spectrum came images of coronal mass ejections and coronal holes, respectively. But whereas coronal holes were immediately identified as the source of high-speed solar wind streams, at first coronal mass ejections were greeted only by a sense of wonder. It took years of research to identify their signatures in the solar wind before the fastest ones could be identified with the well-known shock disturbances that cause the most violent space storms.

This volume marks a turning point in coronal mass ejection research. It marks the culmination of that initial research, based on data from OSO-7, Skylab, Solwind, SMM, ISEE 3, Helios 1 and 2, Pioneer-Venus Orbiter, and IMP 8, and the beginning of intensive studies with data from the more recent Yohkoh, SOHO, Ulysses, and Wind missions. With these newer data, we are making major advances in understanding the physics of coronal mass ejection formation on the Sun, propagation in the interplanetary medium, and impact at Earth. This is the first volume ever that is wholly devoted to the subject, thus celebrating its maturation as an exciting new subfield of space physics, spanning the areas from solar to heliospheric to magnetospheric physics.

Nearly all aspects of observational and theoretical research on coronal mass ejections are covered, primarily in review format. These include their x-ray, white-light, interplanetary scintillation, and solar wind signatures and the relationships between them, their ion composition, their magnetic topology, the particles energized by their shocks, mechanisms for lift-off from the Sun, models of their form and propagation in the solar wind, and their impact on Earth's magnetosphere. The papers have been sorted into five topics ordered by distance from the Sun. The sorting and ordering could only be done in an approximate way, however, since many papers cover several topics and wide ranges of space.

The papers are based on material presented at the 1996 Chapman Conference in Bozeman, Montana, entitled "Coronal Mass Ejections: Causes and Consequences." The Editors are indebted to Loren Acton at Montana State University, whose local arrangements helped foster lively scientific exchange, and to keynote speaker Art Hundhausen and program committee member Jack Gosling, both outstanding leaders in the field since its inception, who kindly provided an introduction and overview for this volume.

The Editors thank the following referees for their constructive and timely reviews: D. Alexander, S. Antiochos, T. Armstrong, A. Bhattacharjee, J. Bieber, V. Bothmer, S. Bravo, J. Burkepile, L. Burlaga, R. Canfield, P. Cargill, J. Chen, E. Cliver, R. Dahlburg, P. Demoulin, T. Detman, T. Forbes, K. Harvey, M. Heinemann, A. Hewish, S. Kahler, J. Karpen, J. Klimchuk, M. Kojima, J. Kota, R. Lepping, P. Liewer, R. Lin, J. Linker, R. Marsden, K. Marubashi, P. Martens, S. Martin, J. Mazur, A. McAllister, Z. Mikic, M. Moldwin, M. Neugebauer, A. Otto, S. Plunkett, P. Riley, C. Russell, D. Rust, A. Ruzmaikin, J. Ryan, B. Sanahuja, T. Sanderson, J. Scudder, R. Sheldon, J. Steinberg, S. Suess, M. Vandas, D. Webb, L. Weiss, R. Wolfson, and R. Zwickl.

Nancy Crooker
Center for Space Physics
Boston University
Boston, Massachusetts

Jo Ann Joselyn
Space Environment Center
National Oceanographic and Atmospheric Administration
Boulder, Colorado

Joan Feynman
Jet Propulsion Laboratory
California Institute of Technology
Pasadena, California

An Introduction

A. J. Hundhausen

High Altitude Observatory, NCAR[1], P.O. Box 3000, Boulder, Colorado

Coronal mass ejections have been a topic of extensive study since they were first identified in observations made with space-borne coronagraphs in the 1970s. The mass ejection phenomenon has proven scientifically interesting from many different points of view; for example, as a form of solar activity unrecognized by most solar physicists until just 25 years ago, as a massive expulsion of plasma from an atmosphere that is, in the main, gravitationally and magnetically bound, or as a major source of transient interplanetary disturbances that, in turn, have significant terrestrial effects. This interest from a broad spectrum of the solar, interplanetary, and geomagnetic research communities led to a workshop on the causes and consequences of coronal mass ejections. This monograph is an outgrowth of that conference.

Research on coronal mass ejections has been reviewed many times since the 1970s; e.g., [*Gosling*, 1975; *Hildner*, 1977; *MacQueen*, 1980; *Rust and Hildner*, 1980; *Dryer*, 1982; *Fisher*, 1984; *Hundhausen et al.*, 1984a; *Wagner*, 1984; *Hildner*, 1986; *Kahler*, 1987; *Hundhausen*, 1987; *Webb*, 1992; *Gosling*, 1990, 1992, 1993a; *Dryer*, 1994; *Hundhausen*, 1997a, 1997b]. Another review by this author is neither appropriate or necessary. Rather, I will limit myself here to a few introductory remarks. These remarks are intended as a reminder of the sources of our empirical knowledge of coronal mass ejections. They will also reveal a few personal opinions on the high (and low) points attained in our attempts to understand this interesting but complex phenomenon.

Coronal Mass Ejections
Geophysical Monograph 99
Copyright 1997 by the American Geophysical Union

1. WHAT "WHITE LIGHT" OBSERVATIONS TELL US

The observations made with conventional "white light" coronagraphs are the historical and empirical foundations of our present knowledge of coronal mass ejections. These instruments record the photospheric radiation (or white light) scattered by electrons in the ionized coronal plasma; they thus offer a direct diagnostic for coronal density that is independent of other physical characteristics of that plasma (such as its temperature). Coronagraph images reveal the density structure of the corona. Time sequences of such images reveal temporal changes in that structure, and in particular the transient expulsion of coronal plasma from the gravitational field of the Sun that is the essence of the mass ejection phenomenon.

This essence is illustrated in a time sequence of images in Plate 1. A coronal mass ejection observed on 14 April 1980, with the coronagraph on the Solar Maximum Mission Spacecraft has been displayed many times in research papers and the review papers cited above. In Plate 1 a pre-event coronagraph image has been subtracted from a sequence of four images from that day to show the temporal development of the mass ejection. In this color display of the resulting "difference images," the hot colors (red and orange) denote an increase in the scattered radiation (and therefore an increase in the density) with respect to the pre-event corona, while cool colors (blues) denote a decrease in that radiation (and in the density). The outward passage of two "loop-like" regions of enhanced brightness (red in color) in this time sequence demonstrates the expulsion of material from the Sun. The outermost of these features has a diffuse appearance and is interpreted as a shell of dense coronal

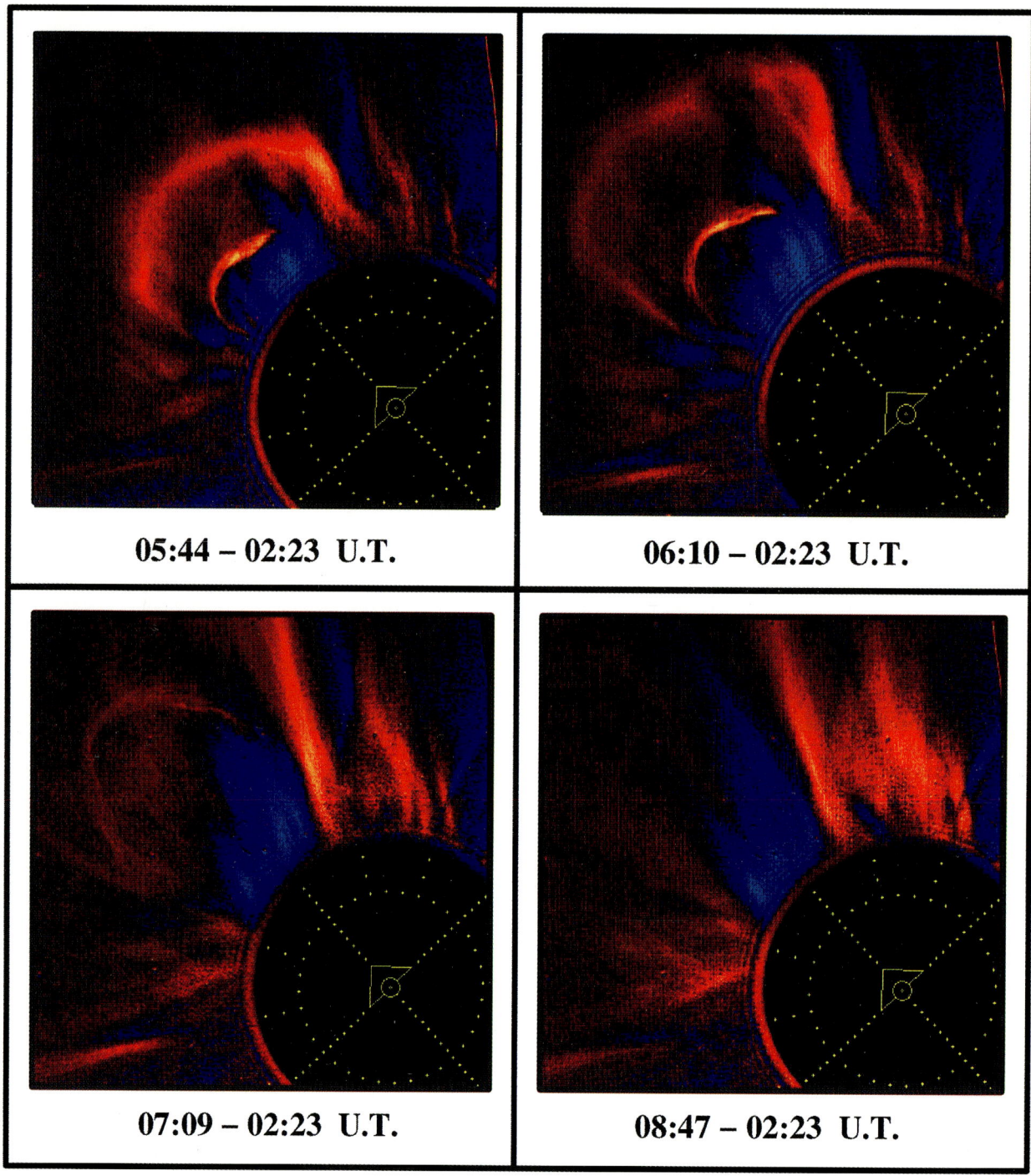

Plate 1. A time sequence of differences between four images taken with the Solar Maximum Mission coronagraph during a coronal mass ejection on 14 April 1980 and a single "pre-event" image. Positive differences (brightenings of the corona since the pre-event image) are shown in red and orange, negative differences in blue. A pair of bright (red) loops moved outward through the corona between 0544 and 0709 UT on this date, leaving a wedge of depleted (blue) corona behind them (as at 0847 UT). Coronal features to the sides of the loops were progressively pushed away from the ejection during its passage through the coronagraph field of view, and are thus visible on these difference images.

Table 1: Some Average Characteristics of Coronal Mass Ejections

	Skylab (1973-74)	Solwind (1979-80 & 1984-85)	SMM (1980, 1984-89)
Angular Size	42°	43°	47°
Speed	470 km sec^{-1}	460 km sec^{-1}	350 km sec^{-1}
Mass	—	4.0×10^{15} gm	3.3×10^{15} gm
Kinetic Energy	—	3.4×10^{30} ergs	6.7×10^{30} ergs
Potential Energy	—	—	7.1×10^{30} ergs
Mechanical Energy	—	—	1.38×10^{31} ergs

(10^{6}°K) plasma, rimming a three-dimensional mound of tenuous plasma reflected in the reduced brightness (blue in color) beneath the "loop." The inner-most bright feature has a more structured appearance and is observed to be emitting in the Hα line of neutral hydrogen, indicating presence of much cooler plasma. This inner loop is interpreted as a solar prominence that erupted, from beneath the field of view of the coronagraph, in conjunction with the *coronal* mass ejection.

Integration over the brightness enhancements in these two loops yields estimates of $2.1 \times 10^{15} gm$ of ionized coronal plasma in the outer feature and $1 \times 10^{14} gm$ of ionized plasma in the inner "prominence" feature. The ejection of the coronal material from the Sun is confirmed by the last image in this sequence; a (blue) wedge of diminished coronal brightness is left in the "wake" of the mass ejection and often remains visible for many hours after the event. A prominence underlying the site of a coronal mass ejection is often seen to erupt above the solar limb (or disappear from view over the solar disc) as part of a mass ejection. The essence of the mass ejection process is illustrated by the outward motion of solar material and its resulting removal from the corona and chromosphere (the prominence) in observations such as this. Further effects of the mass ejection can be seen outside of the region spanned by the loop-like features. Outlying coronal rays are commonly seen to be progressively pushed away from the mass ejection site, as indicated by the nearly radial blue and red structures in the difference images of Plate 1 These "deflections" of pre-existing coronal structure can be seen to extend well away from the ejection, sometimes all of the way around the Sun. They can be interpreted as the effect of compressive waves propagating at the fast MHD wave speed away from the ejected mass.

Several thousand coronal mass ejections have been observed in white light by both space-borne and ground-based coronagraphs. These observations provide both most of the simple, classic examples of the phenomenon (as in Plate 1) and form the broadest basis for statistical description of mass ejection characteristics. For the record, Table 1 displays the average values of six important mass ejection properties that have been tabulated from three extensive sets of spacecraft observations. This tabulation is an update of that in *Hundhausen* [1997]; the reader is referred to that source and to *Gosling et al.* [1976], *Howard et al.* [1985, 1986], *Hundhausen* [1993], and *Hundhausen et al.* [1994a and b] for further details and discussion of some physical implications of these measurements.

2. WHAT SHORT WAVELENGTH OBSERVATIONS TELL US

The major thermal emission from the million degree coronal plasma occurs at wavelengths well short of the visible portion of the solar spectrum. Hence X-ray and extreme ultraviolet observations of the corona provide the most direct indication of its thermal state (to which the white-light observations described above are not sensitive). Soft X-ray observations made from Skylab gave the first direct indications of hot coronal structures related to coronal mass ejections. Although both SMM and GOES X-ray observations have added to our knowledge of that relationship, all controversies have not been resolved and, more important, many scientific ambiguities remain. The availability of high quality soft X-ray images from the Yohkoh spacecraft since late 1991 offers the best opportunity since the Skylab mission of 1973 – 1974 to settle these controversies and resolve these ambiguities.

The analysis of these Yohkoh observations in the context of coronal mass ejections and particularly as evidence of the physical causes of mass ejections remains in its infancy; one might hope that this workshop (and these proceedings) will produce a wealth of new results. This author has participated in comparisons of Yohkoh SXT observations with the coronal white light and chromospheric Hα observations from the Mauna Loa Observation. In collaboration with colleagues J. T. Burkepile, E. Hiei, A. H. McAllister, and D. G. Sime, some tentative (in that they are based on about a dozen examples) conclusions have been drawn. There are several clear examples (see *Hiei et al.* [1993], *Sime et al.* [1994], *Hundhausen* [1997]) of the formation of bright X-ray loops or arcades of loops spanning a magnetic neutral line beneath the site of a mass ejection. These loops are first seen after the onset of the mass ejection, spread laterally and rise vertically with time, and remain visible long after the departure of the mass ejection from the corona. The intensity of soft X-ray emission from such loops is extremely variable and seems to be poorly-correlated with mass ejection properties such as size, speed, mass, or energy. These results are in basic accord with the results of Skylab era studies of such post-ejection X-ray emission and the interpretation of these results in terms of magnetic reconnection along a vertical current sheet formed in the mass ejection. The Yohkoh observations extend this concept to very weak large-scale arcades of loops over high latitude neutral lines, as illustrated in Fig 1. The emission from these features is much weaker than commonly identified as post-ejection loops in Skylab data and is not discernible in the integrated X-ray flux from the Sun, as routinely provided from the GOES satellites.

In addition to these published results, I would like to offer here three additional and very tentative conclusions, largely as a challenge to others at this workshop who may be engaged in similar analyses.

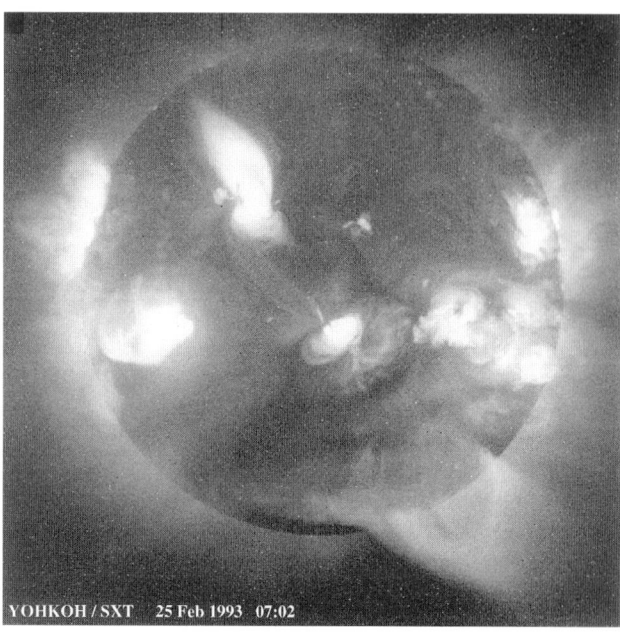

Figure 1. An image obtained by the Yohkoh Soft X-ray Telescope on 25 February 1993. The system of bright loops just above and to the left of sun center is an X-ray flare seen (with an earlier peak intensity at the C 1.8 level) in the GOES integrated solar X-ray flux. This flare was typical of those associated with coronal mass ejections seen in Skylab or Solar Maximum Mission observations. The much fainter arcade of bright loops near the south pole of the sun (extending above the solar limb at the lower right) was not discernible above the background level ($\sim$ B2) in the GOES X-ray flux. It was probably related to a high latitude mass ejection, as in the case described by *Hiei et al.* [1993].

1. Our examinations of Yohkoh images at times when mass ejections were underway and visible in the Mauna Loa data have yet to reveal an X-ray counterpart to the bright frontal features seen in white light (e.g., the outer loop in Plate 1). This result may be mitigated by the limited range of heliocentric distance over which soft X-ray emission is seen. However, there is also little trace in the Yohkoh images of the bright sides of mass ejections seen as they form in the Mauna Loa field of view.

2. As in *Hudson et al.* [1997] and *Webb and Hudson* [poster papers at the Montana meeting], *decreases* in X-ray intensity are sometimes seen near the sites of, and in approximate temporal coincidence with coronal mass ejections seen in white light. Some of these are seen extending outward from the limb of the Sun and may be the result of a coronal density depletion in a mass ejection, as illustrated in Plate 1. Distinct regions of decreased X-ray intensity are sometimes seen on the solar disc (see *Hiei et al.* [1997, and at this meeting]), spanning a magnetic neutral line from which a prominence has erupted and over which the typical post-ejection formation of loops or arcades subsequently occurs. A very conspicuous example of such an event is seen in Fig. 2. It is tempting to interpret these disc regions of diminished X-ray emission as the *bases* of the three-dimensional coronal regions depleted by mass ejections.

3. Outward moving bright X-ray features are visible in Yohkoh images obtained during a few mass ejections. In those cases where detailed comparisons with Mauna Loa data were possible (as in *Sime et al.* [1994]), these could be identified with an interior structure (such as a prominence) rather than the outer or frontal bright feature of the mass ejection.

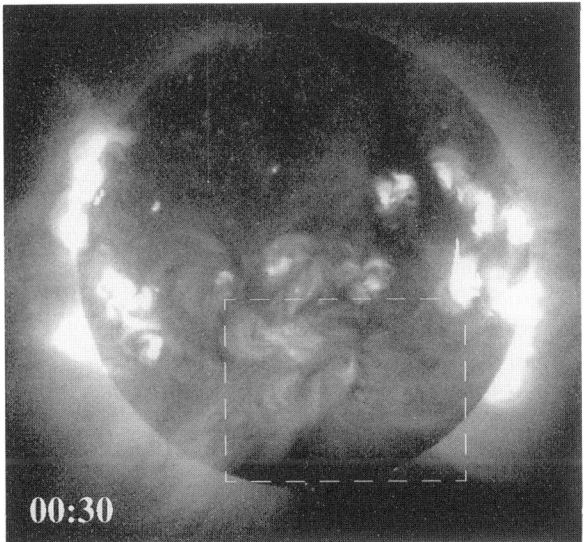

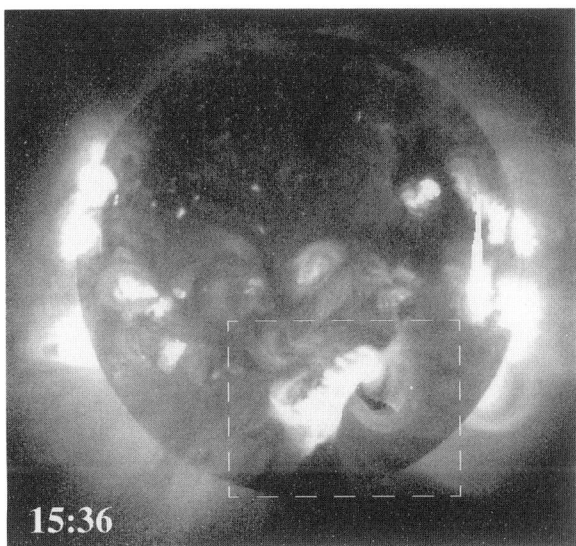

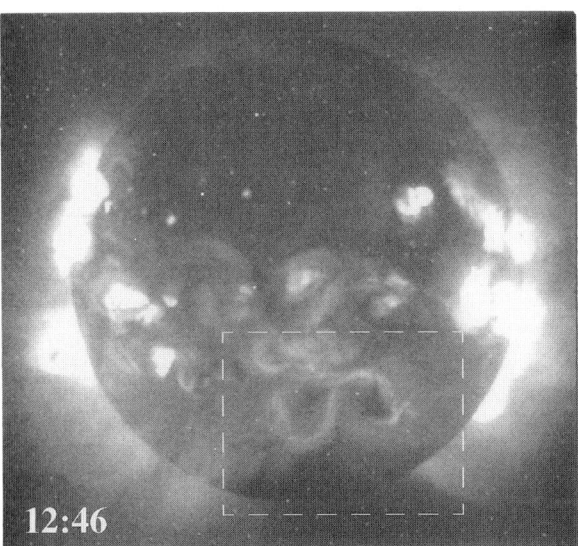

Figure 2. A time sequence of Yohkoh soft X-ray images from 16 January 1993 showing the "dimming" of a region on the solar disc that may be the base of a coronal mass ejection. Between the first and second of these images, the "figure-8" shaped region in the dashed box below (south of) sun center dims by a factor of 2 to 3; the two lobes of the figure are separated by a magnetic neutral line from which a filament (prominence) disappeared. An arcade of bright X-ray emitting loops had formed across the neutral line by the 1536 UT time of the third image. These changes can be interpreted (as in *Hiei et al.* [1997]) as the disc signature of the coronal depletion (seen at the limb in Plate 1) produced by a mass ejection (indicated by the filament disappearance) followed by the reformation of closed (bright) magnetic loops.

Despite the tentative nature of these empirical conclusions based on Yohkoh soft X-ray data, this author cannot resist two comments regarding their physical implications. The absence of any distinct X-ray emission from the fronts (or sides) of the coronal material in mass ejections is not consistent with a thermal driving of the ejection phenomenon. The major heating of the corona detected in the Yohkoh data is qualitatively consistent with the Skylab observations of post-ejection emission and the interpretation of this emission in terms of the reformation (likely through the physical process of magnetic reconnection) of closed magnetic structures in regions "blown open" by coronal mass ejections.

3. WHAT NO OBSERVATIONS TELL US

The reader is referred to *Gosling* [1993], *Hudson et al.* [1994], and *Svestka* [1994] for a sample of the remaining controversy regarding the relationship of mass ejections to the enhanced X-ray emission (or flares) indicative of the presence of hot coronal plasma, and to *Hundhausen* [1997] for this author's views on some remaining scientific ambiguities. In any case, it is clear that many puzzles remain concerning the physical causes of coronal mass ejections and that those puzzles raise serious difficulties in any attempt to predict earth-directed mass ejections (and their interplanetary and geomag-

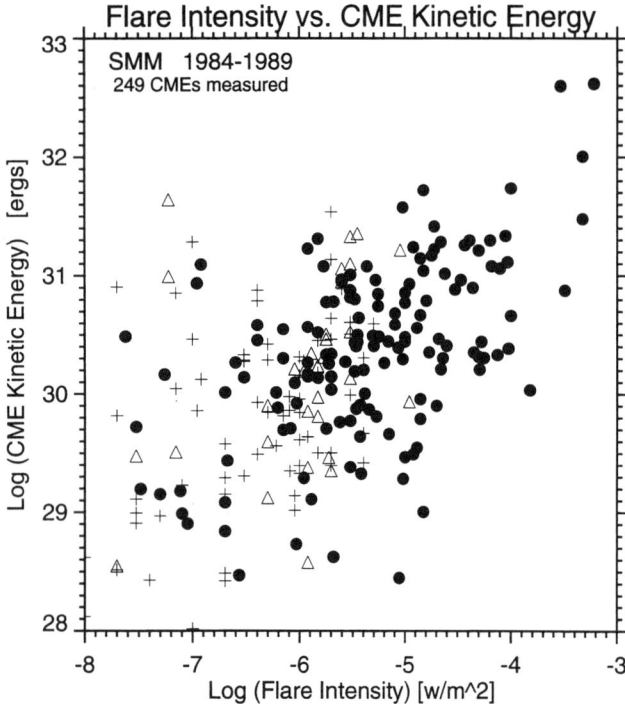

Figure 3. A scatter plot of the logarithms of coronal mass ejection (CME) kinetic energies and of the peak intensities of associated X-ray flares seen in the GOES integrated soft X-ray flux. Mass ejection onset times were deduced from the trajectories of ejections seen in the field of view of the Solar Maximum Mission coronagraph; X-ray flares were regarded as associated if they occurred within the conventional ±90 minutes of the onset time. The dots indicate peak intensities that were a factor of two above the GOES background level and the triangles indicate discernible peaks that were less than a factor of two above background. The plus signs indicate background levels when no elevation of the X-ray flux were discernible; they are thus *upper limits* for the intensity of any associated X-ray flare.

netic consequences) on the basis of presently available solar observations. Figure 3 illustrates these difficulties with respect to a predictor that has been widely used in the past, the intensity of X-ray flares observed with the GOES spacecraft. The figure is a scatter plot of the logarithms of the estimated kinetic energy in 249 mass ejections observed with the Solar Maximum Mission Coronagraph versus the logarithms of the peak intensity of the brightest GOES X-ray flare that occurred within a ±90 minute temporal window of the estimated start time of each mass ejection. Both the mass ejection kinetic energies and the X-ray peak intensities are spread over five orders of magnitude. The cross correlation of the *logarithms* of these quantities is 0.53, indicating some relationship. But this relationship is poor in the sense that, given the intensity of a flare associated with a mass ejection, the kinetic energy of the ejection can still be spread over a range of at least three orders of magnitude.

It is this author's opinion that our remaining difficulties in understanding the physical origins of mass ejections and constructing a reasonable scheme for predicting the occurrence and characteristics of mass ejections stem from the lack of knowledge of a crucial physical parameter. It is now widely recognized that mass ejections are large-scale disruptions of a quasi-equilibrium between the coronal plasma and magnetic field. Individual mass ejections are often preceded by a slow evolution of the coronal structures that eventually erupt. The view that mass ejections are magnetically driven can be advocated both from theoretical models and from empirical evidence.

But how well do we know the magnetic field in the corona? The Answer is simple. We have only a smattering of observational information on the strengths and directions of the fields in mass ejections or in the coronal structures that are involved in their formation. Most of the empirical conclusions we draw about those fields are qualitative, based on the assumption that the coronal features observed in white light or X-rays trace out magnetic field lines. Theoretical models can be used to extend the fields from the solar photosphere, where they can be measured, outward into the corona. However these models often involve extreme idealizations concerning the nature of electric currents in the corona or include more appropriate physics at the expense of geometric idealization (or spatial resolution in numerical models). It is here that I believe that we come face-to-face with our true ignorance of the solar corona. Magnetic fields play a crucial role in forming the structure of the solar corona, and we have only the poorest knowledge of their nature. Our progress toward both understanding and predictability may well hinge on gaining that knowledge.

Acknowledgments. The author thanks C. Worster for producing this manuscript and T. Manyik for assistance with word processing equipment. J. T. Burkepile and A. L. Stanger produced the figures in this paper. The author has greatly benefited from discussion of mass ejection physics with numerous colleagues; in particular I would like to acknowledge J. T. Burkepile, E. Hiei, A. H. McAllister, and D. G. Sime.

REFERENCES

Dryer, M., Coronal transient phenomenon, in *Space Sci. Rev.*, *33*, 233-275, 1982.

Fisher, R. R., Coronal mass ejection events, in *Adv. Space Res.*, *4*, 163-174, 1984.

Gosling, J. T., Coronal mass ejections and magnetic flux ropes in interplanetary space, in *Physics of Magnetic Flux Ropes, Geophys. Monogr. Series*, vol. 58, edited by C. T. Russell, E. R. Priest, and L. C. Lee, (Washington, DC: AGU), 343-364, 1990.

Gosling, J. T., In situ observations of coronal mass ejections in interplanetary space, in *Eruptive Solar Flares*, edited by Z. Svestka, B. V. Jackson, and M. E. Machado, (New York, NY: Springer-Verlag), 258-267, 1992.

Gosling, J. T., Coronal mass ejections: the link between solar and geomagnetic activity, in *Phys. Fluids B 5*, (7), 2638. 1993a.

Gosling, J. T., The solar flare myth, in *J. Geophys. Res.*, *98*, 18,937-18,949, 1993b.

Gosling, J. T., E. Hildner, R. M. MacQueen, R. H. Munro, A. I. Poland, and C. L. Ross, The speeds of coronal mass ejection events, in *Solar Phys.*, *48*, 389-397, 1976.

Hiei, E., A. J. Hundhausen, and D. G. Sime, Reformation of a coronal helmet streamer by magnetic reconnection after a coronal mass ejection, in *J. Geophys. Res.*, *20*, 2785-2788, 1993.

Hildner, E., Mass ejections from the corona into interplanetary space, in *Study of Travelling Interplanetary Phenomena*, edited by M. A. Shea *et al.*, (Hingham, MA: D. Reidel), 3-21, 1977.

Hildner, E. Do we understand coronal mass ejections yet?, in *Adv. Space Res.*, *6*, 297-306, 1986.

Howard, R. A., N. R. Sheeley, Jr., M. J. Koomen, and D. J. Michels, Coronal mass ejections, 1979-1981, in *J. Geophys. Res.*, *90*, 8,173, 1985.

Howard, R. A., N. R. Sheeley, Jr., M. J. Koomen, and D. J. Michels, The solar cycle dependence of coronal mass ejections, in *The Sun and the Heliosphere in Three Dimensions*, edited by R. G. Marsden, (Dordrecht: D. Reidel), 107, 1986.

Hudson, H., B. Haisch, and K. T. Strong, Comments on "The solar flare myth," *J. Geophys. Res.*, *100*, 3473-3477, 1995.

Hundhausen, A. J., The origin and propagation of coronal mass ejections, in *Proceedings of the Sixth International Solar Wind Conference*, edited by V. J. Pizzo, T. E. Holzer, and D. G. Sime, NCAR Technical Note NCAR/TN-306+Proc, Boulder, 181-214, 1987.

Hundhausen, A. J., Coronal mass ejections: A summary of SMM observations from 1980 and 1984-1989, to be published in *The Many Faces of the Sun; Scientific Highlights of the Solar Maximum Mission*, edited by K. T. Strong *et al.*, (New York, NY: Springer-Verlag), 1997a?.

Hundhausen, A. J. Coronal mass ejections, to be published in *Cosmic Winds and the Heliosphere*, edited by J. R. Jokipii *et al.*, (Tucson, AZ: Univ. Arizona Press), 1997b.

Hundhausen, A. J. *et al.*, Coronal transients and their interplanetary effects, in *Solar Terrestrial Physics: Present and Future*, edited by D. M. Butler and K. Papadopoulos, NASA Reference Publication 1120, Washington, 6-1 - 6-32, 1984a.

Hundhausen, A. J., J. T. Burkepile, and O. C. St.Cyr, The speeds of coronal mass ejections: SMM observations from 1980 and 1984-1989. in *J. Geophys. Res.*, *99*, 6543-6552, 1994a.

Hundhausen, A. J., A. L. Stanger, and S. A. Serbicki, Mass and energy contents of coronal mass ejections: SMM results from 1980 and 1984-1988, in *Proc. of the Third SOHO Workshop — Solar Dynamical Phenomena and Solar Wind Consequences* (ESA SP-373), 409-412, 1994b.

Kahler, S., Observations of coronal mass ejections near the sun, in *Proc. of the Sixth International Solar Wind Conference*, edited by V. J. Pizzo, T. E. Holzer, and D. G. Sime, NCAR Technical Note NCAR/TN-306+Proc, Boulder, 181-214, 1987.

Kahler, S., Solar flares and coronal mass ejections, in *Ann. Rev. Astron. Astrophys.*, *30*, 113-141, 1992.

MacQueen, R. M., Coronal transients: A summary, in *Phil Trans. R. Soc. Lond.*, *A297*, 605-620, 1980.

Rust, D. M. and E. Hildner *et al.*, Mass ejections, in *Solar Flares*, edited by P. A. Sturrock, (Boulder, CO: Colorado Assoc. Univ. Press), 273-339, 1980.

Sime, D. G., E. Hiei, and A. J. Hundhausen, Coronal eruptive events on April 4 and May 4, 1992, in *X-ray Solar Physics from Yohkoh*, edited by Uchida *et al.*, (Tokyo, Japan: Universal Academy Press), 197-200, 1994.

Svestka, Z., On "the solar flare myth" postulated by Gosling, in *Solar Phys.*, *160*, 153-156, 1995.

Wagner, W. J., Coronal mass ejections, in *Ann. Rev. Astron. Astrophys.*, *22*, 267-289, 1984.

Webb, D. F., The solar sources of coronal mass ejections, in *Eruptive Solar Flares*, edited by Z. Svestka, B. V. Jackson, and M. Machado, (New York, NY: Springer-Verlag), 234-247, 1992.

A. Hundhausen, High Altitude Observatory, National Center for Atmospheric Research, P.O. Box 3000, Boulder, CO 80307-3000

---

[1] The High Altitude Observatory is part of the National Center for Atmospheric Research which is Sponsored by the National Science Foundation under the management of the University Corporation for Atmospheric Research

Coronal Mass Ejections: An Overview

J. T. Gosling

Los Alamos National Laboratory, Los Alamos, New Mexico

This paper provides a brief overview of present understanding of coronal mass ejections, or CMEs, as observed both in the corona and in the solar wind far from the Sun. The overview provides historical background material on ideas about mass ejections prior to their direct detection by orbiting coronagraphs and outlines significant results and insights obtained from the early OSO 7 and Skylab coronagraph experiments. This is followed by summaries of current understanding about the CME phenomenon. The important role of CMEs in producing transient shock wave disturbances in the solar wind, large solar energetic particle events, and large nonrecurrent geomagnetic storms is noted. The overview concludes with a series of questions concerning the origin of CMEs and their evolution in interplanetary space.

HISTORICAL BACKGROUND

Ideas about coronal mass ejections, or CMEs, predate our knowledge of the existence of the solar wind or actual observations of CMEs with satellite-borne coronagraphs. Commenting upon observed associations between various forms of solar activity and large geomagnetic storms, *Lindemann* [1919] appears to have been the first to suggest that geomagnetic storms result from transient ejections of plasma from the Sun that impact the Earth's magnetic field several days following the eruption. It was later recognized that such transient ejections of solar material should drag magnetic loops out into interplanetary space [e.g., *Cocconi et al.*, 1958; *Piddington*, 1958; *Gold*, 1962], owing to the high electrical conductivity of the coronal plasma. Initially such loops would be connected to the Sun at both ends. The magnetic "bottles" so formed would effectively channel energetic particles produced at the Sun into interplanetary space and would serve to exclude some galactic cosmic rays from their interiors [e.g., *Obayashi*, 1962]. Piddington considered the possibility that some of these field lines might pinch off (i.e., reconnect) and sever completely their magnetic connection to the Sun; however, he concluded it was unlikely that reconnection would be very effective in severing field lines. It was widely thought that these ejections were a consequence of solar flares [e.g., *Hale*, 1931; *Chapman*, 1950; *Piddington*, 1958]. Commenting upon the abrupt onsets of many geomagnetic storms, *Gold* [1955] suggested that the transient plasma ejections must often produce shock disturbances in the interplanetary gas as they force their way outward from the Sun.

Parker developed an alternative view of interplanetary disturbances associated with solar activity [e.g., *Parker*, 1963]. His early models of the solar wind expansion suggested that solar wind speed scaled directly as the temperature in the low corona. Reasoning that solar flares produced a sudden local heating of the corona, he suggested that solar wind shock disturbances were produced by solar flares and resulted from a sudden increase in the rate of expansion of a part of the corona already participating in the solar wind expansion, i.e., a region already magnetically open to interplanetary space. In his model of solar wind disturbances no additional magnetic flux is transported outward by disturbances associated with solar activity, and all field lines in the solar wind remain open to the outer boundary of the heliosphere.

Coronal Mass Ejections
Geophysical Monograph 99
Copyright 1997 by the American Geophysical Union

Early *in situ* observations showed that transient shock disturbances were relatively common in the solar wind near Earth [e.g., *Gosling et al.*, 1968; *Hundhausen et al.*, 1970] and that they were driven outward from the Sun by plasma having a different character from that of the normal wind. The material was often rich in He^{++} [e.g., *Hirshberg et al.*, 1972] and, in contrast to expectations from Parker's model, usually had anomalously low (rather than high) proton and electron temperatures [*Gosling et al.*, 1973; *Montgomery et al.*, 1974]. At least one example was noted where the plasma driving a shock contained suprathermal electrons counterstreaming along the interplanetary magnetic field [*Montgomery et al.*, 1974], suggesting that field lines threading the driving plasma were closed, either connected to the Sun at both ends as proposed by Cocconi et al. and others, or disconnected from it entirely. By making reasonable assumptions about the Sun-centered solid angle filled by these disturbances, one could estimate the total mass ejected from the Sun as well as the total energy content in these events. Estimates of the order of 10^{16} g and 10^{32} ergs were obtained for some of the larger disturbances [*Hundhausen et al.*, 1970], in remarkably good agreement with estimates later obtained from coronagraph observations. These early *in situ* observations indicated that Parker's model of solar wind disturbances was incorrect; rather, the disturbances were associated with eruptions from previously closed field regions in the solar atmosphere.

EARLY RESULTS FROM ORBITING CORONAGRAPHS

The nearly continuous monitoring of the solar corona provided by coronagraphs flown on OSO 7 [*Tousey*, 1973] and, particularly, Skylab [*MacQueen et al.*, 1974] in the early 1970's conclusively established that the transient ejections of material from the solar atmosphere that we now call coronal mass ejections were a common occurrence. The ejections appeared to arise preferentially from closed magnetic field regions in the solar atmosphere where the field normally was sufficiently strong to constrain the coronal plasma from expanding outward into the heliosphere. Moreover, many of the ejections had the appearance of closed magnetic loops attached to the Sun at both ends. It was estimated that a typical ejection contained ~4 x 10^{15} g [*Gosling et al.*, 1974]. Motions observed in adjacent coronal regions during the outward progression of some ejections indicated that the ejections often drove pressure waves into the surrounding corona and solar wind. When the outward speed was sufficiently high, the ejections produced shock wave disturbances in the solar wind far from the Sun [*Gosling et al.*, 1975].

There were, however, a number of surprises. First, far more ejections were observed than would have been expected on the basis of the frequency of occurrence of shock wave disturbances in the solar wind or large transient geomagnetic storms. Second, the ejections were more commonly observed in association with eruptive prominences than with impulsive solar flares [*Gosling et al.*, 1974; *Munro et al.*, 1979]. Many of the ejecta were, however, followed by long-duration (hours), soft x-ray events associated with magnetic loop formation in the corona [*Sheeley et al.*, 1975; *Webb et al.*, 1976; *Kahler*, 1977]; such x-ray events have since come to be called gradual flares. Third, most of the material in the ejections was of coronal origin rather than being prominence material or flare ejecta [*Hildner et al.*, 1975]. Fourth, individual ejecta exhibited a wide range of outward speeds ranging from less than 100 km s^{-1} to greater than 1200 km s^{-1} [*Gosling et al.*, 1976]; it was thus obvious that many ejections did not produce significant solar wind or geomagnetic disturbances because their speeds were too low. Fifth, it was difficult, if not impossible, to detect magnetic disconnection of the ejecta from the Sun even though it seemed that such disconnection was required to maintain a rough balance of magnetic flux in the solar wind far from the Sun [*Gosling*, 1975].

MORE RECENT RESULTS FROM ORBITING CORONAGRAPHS

Subsequent observations obtained from the SOLWIND, Solar Maximum Mission (SMM) and, more recently, SOHO coronagraphs have established the essential characteristics of coronal mass ejections close to the Sun. CMEs have a frequency of occurrence that varies by more than an order of magnitude over the course of the 11-year solar activity cycle; near solar activity minimum they occur at a whole-Sun rate of ~0.2 events day^{-1}, while near solar activity maximum they occur at a rate of ~3.5 events day^{-1} [*Webb and Howard*, 1994]. Ejection masses typically are in the range from 10^{+15} to 10^{+16} g [e.g., *Hundhausen*, 1988], while leading edge speeds (within about 5 solar radii of the surface) range from less than 50 to greater than 2000 km s^{-1} [*Howard et al.*, 1985; *Hundhausen et al.*, 1994]. Approximately half of all CMEs have speeds close to the surface of the Sun that are less than that of the the minimum solar wind speed near 1 AU (~280 km s^{-1}), indicating that these CMEs receive substantial further acceleration at heliocentric distances greater than 5 solar radii. The distribution of CME speeds close to the Sun appears to be independent of solar latitude.

Substantial evidence has been gathered that indicates that CMEs are not, in general, produced by impulsive solar flares [*Gosling*, 1993]. On those occasions when CMEs and flares do occur in close temporal association with one another, the CMEs usually begin to lift off from the Sun before any substantial flaring activity has occurred [*Harrison*, 1986; *Hundhausen*, 1988; *Harrison et al.*, 1990; *Hundhausen*, 1997]. CMEs tend to occur near magnetic neutral lines in the solar atmosphere [e.g., *Hundhausen*, 1988]. Near solar activity maximum such neutral lines can be found over a wide range of solar latitudes, and CMEs at this phase of the solar cycle thus commonly occur at both high and low latitudes. Near solar activity minimum, however, most CMEs occur at low heliographic latitudes near the magnetic equator.

Many CMEs begin as a slow swelling of a coronal streamer ("bugles") on a time scale of several days [e.g., *Hundhausen*, 1997]. Such CMEs often appear to have a 3-part structure consisting of an outer loop, an inner cavity relatively devoid of material, and an embedded prominence, mirroring the structure of their place of origin lower in the solar atmosphere [*Hundhausen*, 1988]. CMEs represent the opening up to interplanetary space of previously closed coronal magnetic field lines. A reformation of the corona is commonly observed after an ejection that is visible in both white light (on the limb) and in soft x-rays (long-duration events) [e.g., *Sheeley et al.*, 1983; *Hiei et al.*, 1993]. This reformation is commonly believed to be associated with reconnection that occurs close to the Sun within the magnetic "legs" of the departing CMEs [e.g., *Kopp and Pneuman*, 1976]. Typically CMEs span about 45 degs in latitude [*Hundhausen*, 1993]; CME longitudinal extents are less well established due to the fact that CMEs are observed projected against the plane perpendicular to the line of sight, i.e., in longitude. Observations of reformation behind CMEs suggest that CMEs can often have substantial longitudinal extents, covering as much as 180 deg in longitude, or more, on occasion [e.g., *McAllister et al.*, 1996].

CORONAL MASS EJECTIONS IN THE SOLAR WIND: RESULTS FROM *IN SITU* OBSERVATIONS

A great deal has also been learned about the CME phenomenon in interplanetary space in the last 25 years, although the identification of the ejected material in the solar wind is still something of an art [e.g., *Neugebauer and Goldstein*, this volume]. Common signatures of CMEs in the solar wind include counterstreaming suprathermal electrons [e.g., *Gosling et al.*, 1987] and energetic protons [e.g., *Marsden et al.*, 1987], helium abundance enhancements [e.g., *Borrini et al.*, 1982], anomalously low proton [e.g., *Richardson and Cane*, 1995] and electron [e.g., *Gosling et al.*, 1987] temperatures, strong magnetic fields [e.g., *Burlaga and King*, 1979], low plasma beta, smooth field rotations (flux ropes) [e.g., *Burlaga*, 1991], and low magnetic field variance [e.g., *Pudovkin et al.*, 1979]. Occasionally unusual ionization states are also observed within CMEs in the solar wind [e.g., *Bame et al.*, 1979; *Schwenn et al.*, 1980; *Galvin*, this volume]. However, few CMEs produce all of these signatures. It is notable that high plasma number density and/or high flow speed are not commonly distinguishing traits of CMEs in the solar wind far from the Sun. Indeed, CME speeds in the solar wind generally fall within the range of speeds of the ambient wind at both low and high heliographic latitudes, and beyond ~2 AU CMEs are commonly distinguished by low plasma density (caused by expansion). Single point occurrence frequencies in the ecliptic plane as derived from counterstreaming suprathermal electron events are ~72 events year^{-1} near solar activity maximum and ~8 events year^{-1} near solar activity minimum. Averaged over the solar cycle CMEs account for about 7% of the solar wind flow in the ecliptic plane near 1 AU [*Gosling et al.*, 1992]; their contribution at high latitudes is substantially less than this.

Suprathermal electron measurements indicate that magnetic field lines threading CMEs in the solar wind commonly are attached to the Sun at both ends [e.g., *Gosling*, 1996], although it is becoming increasingly clear that many CMEs in interplanetary space also contain some open and (less often) disconnected field lines as well [e.g., *Kahler*, this volume]. At least 1/3 of all CMEs in the solar wind appear to be nearly force-free magnetic flux ropes [cf. *Gosling*, 1990; *Marubashi*, this volume], consisting of series of helical field lines of ever increasing pitch wrapped about a central axis [*Goldstein*, 1983; *Farrugia*, this volume]. When the field strength is high and the plasma beta low, such CMEs are known as "magnetic clouds" [e.g., *Burlaga*, 1991; *Osherovich and Burlaga*, this volume]. It has been suggested that magnetic clouds may be interplanetary extensions of solar filaments [e.g., *Burlaga*, 1991; *Rust*, 1994]. An alternative suggestion is that the flux rope field topology of magnetic clouds is a consequence of 3-dimensional reconnection within the magnetic legs of CMEs [*Gosling*, 1990; *Martin and McAllister*, this volume]; when the plasma beta is low the helical field lines resulting from such reconnection relax to the nearly force-free flux rope configuration observed in interplanetary space. Sustained 3-dimensional reconnection is also capable of producing open and disconnected field lines within CMEs [*Gosling et al.*, 1995], as is sometimes observed. Consistent with CME formation near neutral lines, some magnetic clouds have been observed to carry the polarity change between opposite magnetic sectors in the form of a large-scale rotation in the field rather than as a current sheet [*Crooker and Intriligator*, 1996].

At low latitudes only occasionally do CMEs near 1 AU have speeds greater that 750 km s^{-1}, and they never have speeds less than the minimum solar wind speed of ~280 km s^{-1} [e.g., *Gosling et al.*, 1987]. Thus the fastest CMEs at low latitudes are decelerated in interplanetary space as they interact with slower plasma ahead, while the slowest CMEs observed in the corona are accelerated further by the time they reach 1 AU. Usually it appears that the slow CMEs are accelerated outward by the same forces that accelerate the normal solar wind. About 1/3 of all CMEs observed in the ecliptic plane have speeds high enough relative to the ambient wind to produce shock wave disturbances [e.g., *Sheeley et al.*, 1985; *Gosling et al.*, 1991]. As might be expected, these tend to be the faster CMEs, but relative speed between a CME and the ambient wind ahead is the crucial factor in determining whether a shock is produced. At high heliographic latitudes all CMEs observed in the solar wind to date have had high speeds, comparable to that of the ambient wind at such latitudes [*Gosling et al.*, 1994a].

CMEs evolve considerably as they interact with the ambient wind. The strong fields and high plasma densities commonly found in the leading portions of disturbances driven by fast CMEs are largely a consequence of compression that occurs in interplanetary space as the CMEs overtake slower plasma ahead [e.g., *Hundhausen and Gentry*, 1969]. The interaction causes a transfer of momentum and energy from the CME to the ambient wind and also causes the ambient interplanetary magnetic field to drape about the CME. Such draping can play an important role in reorienting the ambient magnetic field in interplanetary space [e.g., *Gosling and McComas*, 1987]. In contrast, when a CME has a lower speed than the ambient wind ahead a rarefaction forms at the interface between the two plasmas. The spreading of this rarefaction produces an expansion and acceleration of the CME and a deceleration of the ambient wind [e.g., *Gosling and Riley*, 1996]. Similar types of effects occur on the trailing edges of CMEs when substantial speed differentials between the CMEs and the trailing ambient wind are present.

Many CMEs near 1 AU and beyond are expanding as they propagate out through the heliosphere [e.g., *Burlaga et al.*, 1981; *Klein and Burlaga*, 1982]. Such expansion can be a result of the CME's interaction with the ambient wind, as noted above, or it can be a result of an initial high internal pressure or an initial front-to-rear speed gradient. The range of observed CME radial widths is quite large, but the average event has a width of ~0.2 AU near Earth's orbit [*Gosling et al.*, 1987] Near 5 AU CME radial widths as large as 2.5 AU have been observed on occasion. At high heliographic latitudes CME expansion driven by a high internal pressure commonly produces a forward-reverse shock pair surrounding the CME and a deep rarefaction within the CME itself [*Gosling et al.*, 1994b]; however, relative motion between the CME and the ambient wind can effectively prevent the formation of one or both of these expansion shocks.

CORONAL MASS EJECTIONS AND LARGE GEOMAGNETIC STORMS

It is now well-documented that large, nonrecurrent geomagnetic storms are caused by interplanetary disturbances driven by fast CMEs [*Gosling et al.*, 1990, 1991]. Solar wind disturbances associated with magnetic clouds are particularly effective in this regard [e.g., *Burlaga, et al.*, 1987]. Even some recurrent storms are enhanced by solar wind disturbances produced by CMEs [*Crooker and Cliver*, 1994]. This result is much as first suggested by *Lindemann* [1919] and others [e.g., *Hale*, 1931; *Chapman*, 1950], although we now know that solar flares are not fundamentally responsible for either CMEs or large geomagnetic storms. Moreover, many CMEs do not produce large disturbances either in the solar wind or the Earth's magnetosphere. Present evidence indicates that only about 1 in 6 CME-driven disturbances striking the Earth's magnetosphere is effective in stimulating a large geomagnetic storm. A strong southward interplanetary magnetic field is crucial for stimulating geomagnetic activity [e.g., *Burton et al.*, 1975; *Gonzalez and Tsurutani*, 1987]. Since strong fields in the solar wind are primarily a result of compression in interplanetary space, the initial speed of a CME close to the Sun appears to be the most crucial factor in determining if an earthward-directed event will be effective in exciting a large geomagnetic disturbance. Compression produces high field strengths both within the ambient wind ahead of a fast CME as well as within the leading portion of the CME itself; thus the main phases of large nonrecurrent geomagnetic storms often begin before any CME actually encounters the magnetosphere.

CORONAL MASS EJECTIONS AND LARGE ENERGETIC PARTICLE EVENTS

The strong shocks driven by the fastest CMEs are effective in accelerating a small fraction of the coronal and solar wind particles they intercept to very high energies [e.g., *Lee and Ryan*, 1986; *Lee*, this volume]. The gradual, but intense, energetic particle events produced by CME-driven shocks typically last for several days or longer and are found in association with CMEs originating from virtually anywhere on the visible solar disk. The detailed temporal profiles that are observed depend sensitively on the longitude where the CMEs originate relative to the observer [e.g., *Cane et al.*, 1988]. Nevertheless, it is now well established that most major energetic proton events (i.e., long lasting events with sustained high particle fluxes) observed in the vicinity of Earth are the result of particle acceleration occurring in the outer corona and in interplanetary space at shocks driven by fast CMEs [e.g., *Kahler et al.*, 1984; *Kahler*, 1992; *Reames* 1992, this volume; *Reames et al.*, 1996; *Cane*, this volume].

CONCLUDING COMMENTS

Despite the wealth of observational knowledge gained during the last 25 years concerning the CME phenomenon both close to and far from the Sun, it is remarkable that we still do not understand the chain of physical processes that produce these most spectacular of all solar and solar wind events. It is, however, widely appreciated that CMEs play a fundamental role in the long-term evolution of the solar corona [e.g., *Low*, this volume]. They seem to be an essential part of the way the corona responds to changes in the solar magnetic field associated with the advance of the ~11-year solar activity cycle. In fact, they may be the primary way by which the corona is reconfigured in response to those changes. Some outstanding questions concerning CME origins are as follows: What is the physics of initiation and what are its signatures? What determines when and where a CME will occur and how fast the ejection of material will be? What are the processes by which CMEs are accelerated? How are CMEs specifically related to the long-term evolution of the corona and the solar magnetic field? Is magnetic reconnection responsible for CME initiation or is it just an aftereffect of the CME process? Where and when does magnetic reconnection occur and what is its role in determining the overall field topology of CMEs? What role does magnetic helicity and its conservation play in the CME phenomenon? How does shear of the magnetic footpoints of the loops within coronal streamers affect CME initiation? What role does newly emerging flux from beneath the solar surface play in the initiation of CMEs? Are large-scale instabilities the trigger mechanism for CMEs and, if so, which instabilities?

Our understanding of the evolution of CMEs in interplanetary space appears to be on a firmer footing. Far from the Sun the most interesting questions seem to be associated with the interpretation of observed phenomena. Any list of such questions would probably include the following: How can we best identify CMEs in the solar wind? Why is the helium abundance often high within CMEs? Why are kinetic temperatures typically low within CMEs? Why aren't all CMEs in the solar wind low beta objects? What produces the nearly force-free magnetic flux rope configuration common to so many CMEs in the solar wind? Why do mixtures of field topologies occur within some CMEs? How is magnetic flux balance maintained in interplanetary space in the presence of CMEs? Do CMEs eventually disconnect completely from the Sun? How long does it take for such complete disconnection to occur? What are the primary reasons why most CMEs expand as they propagate out through the heliosphere? How are slow CMEs accelerated up to the speed of the ambient wind? What is the relative importance of magnetic forces and ordinary fluid forces in effecting CME evolution? Why are CME disturbances at high heliographic latitudes so different from such disturbances at low latitudes? What is the fate of CMEs in the outer heliosphere? How can we determine which CMEs are going to be geomagnetically effective and/or produce major energetic particle events? Is initial CME speed close to the Sun the most important factor? Can we use existing numerical models of CME evolution to relate observations of CMEs close to the Sun to observations of the related disturbances in interplanetary space?

Clearly there is much interesting and important work yet to be done on the CME phenomenon. This volume is a testament to current interest in this topic.

Acknowledgements. I thank Nancy Crooker for her patience and persistence in pursuing this overview - it would not otherwise have been completed. This work was performed under the auspices of the U.S. Department of Energy with support from an internal Los Alamos National Laboratory research grant.

REFERENCES

Bame, S. J., J. R. Asbridge, W. C. Feldman, E. E. Fenimore, and J. T. Gosling, Solar wind heavy ions from flare heated coronal plasma, *Solar Phys., 62*, 179, 1979.

Borrini, G., J. T. Gosling, S. J. Bame, and W. C. Feldman, Helium abundance enhancements in the solar wind, *J. Geophys. Res., 87,* 7370, 1982.

Burlaga, L. F., Magnetic clouds, in *Physics of the Inner Heliosphere II,* edited by R. Schwenn and E. Marsch, Springer-Verlag, Berlin, pp. 1-22, 1991.

Burlaga, L. F., and J. H. King, Intense interplanetary magnetic fields observed by geocentric spacecraft during 1963-1975, *J. Geophys. Res., 84,* 6633, 1979.

Burlaga, L. F., E. Sittler, F. Mariani, and R. Schwenn, Magnetic loop behind an interplanetary shock: Voyager, Helios, and IMP 8 observations, *J. Geophys. Res., 86,* 6673, 1981.

Burlaga, L. F., K. W. Behannon, and L.W. Klein, Compound streams, magnetic clouds and major magnetic storms, *J. Geophys. Res., 92,* 5727, 1987.

Burton, R. K., R. L. McPherron, and C. T. Russell, An empirical relationship between interplanetary conditions and Dst, *J. Geophys. Res., 80,* 4204, 1975.

Cane, H. V., The current status in our understanding of energetic particles, coronal mass ejections, and flares, this volume.

Cane, H. V., D. V. Reames, and T. T. von Rosenvinge, The role of interplanetary shocks in the longitude distribution of solar energetic particles, *J. Geophys. Res., 93,* 9555, 1988.

Chapman, S., Corpuscular influences upon the upper atmosphere, *J. Geophys. Res., 55,* 361, 1950.

Cocconi, G., T. Gold, K. Greisen, S. Hayakawa, J. P. Morrison, The cosmic ray flare effect, *Nuovo Cimento, 8,* 161, 1958. 1958.

Crooker, N. U., and E. W. Cliver, Postmodern view of M-regions, *J. Geophys. Res., 99,* 23,383, 1994.

Crooker, N. U., and D. S. Intriligator, A magnetic cloud as a distended flux rope occlusion in the heliospheric current sheet, *J. Geophys. Res., 101,* 24,343, 1996.

Farrugia, C. F., Recent work on modeling the global field line topology of interplanetary magnetic clouds, this volume.

Galvin, A. B., Ion composition and charge states in solar transient related solar wind, this volume.

Gold, T. Discussion of shock waves and rarefied gases, in *Gas Dynamics of Cosmic Clouds,* edited by J. C. van de Hulst and J. M. Burgers, North-Holland Publishing Co., Amsterdam, p.103, 1955.

Gold, T., Magnetic storms, *Space Sci. Rev., 1,* 100, 1962.

Goldstein, H., On the field configuration in magnetic clouds, in *Solar Wind Five,* NASA Conference Publ. 2280, edited by M. Neugebauer, pp. 731-733, 1983.

Gonzalez, W. D., and B. T. Tsurutani, Criteria of interplanetary parameters causing intense magnetic storms (Dst < -100 nT), *Planet. Space Sci., 35,* 1101, 1987.

Gosling, J. T., Large scale inhomogeneities in the solar wind of solar origin, *Rev. Geophys. and Space Phys., 13,* 1053, 1975.

Gosling, J. T., Coronal mass ejections and magnetic flux ropes in interplanetary space, in *Physics of Magnetic Flux Ropes,* edited by C. T. Russell, E. R. Priest, and L. C. Lee, Geophys. Monogr. 58, Amer. Geophys. Union, pp.343-364, 1990.

Gosling, J. T., The solar flare myth, *J. Geophys. Res., 98,* 18,937, 1993.

Gosling, J. T., Magnetic topologies of coronal mass ejection events: Effects of 3-dimensional reconnection, in *Solar Wind Eight,* edited by D. Winterhalter, J. T. Gosling, S. R. Habbal, W. S. Kurth, and M. Neugebauer, American Institute of Physics, Conference Proceedings 383, New York, pp. 438-441, 1996.

Gosling, J. T., and D. J. McComas, Field line draping about fast coronal mass ejecta: A source of strong out-of-the-ecliptic magnetic fields, *Geophys. Res. Lett., 14,* 355, 1987.

Gosling, J. T., and P. Riley, The acceleration of slow coronal mass ejections in the high-speed solar wind, *Geophys. Res. Lett., 23,* 2867, 1996.

Gosling, J. T., J. R. Asbridge, S. J. Bame, A. J. Hundhausen, and I. B. Strong, Satellite observations of interplanetary shock waves, *J. Geophys. Res., 73,* 43, 1968.

Gosling, J. T., V. Pizzo, and S. J. Bame, Anomalously low proton temperatures in the solar wind following interplanetary shock waves: Evidence for magnetic bottles?, *J. Geophys. Res., 78,* 2001, 1973.

Gosling, J. T., E. Hildner, R. M. MacQueen, R. H. Munro, A. I. Poland, and C. L. Ross, Mass ejections from the Sun: A view from Skylab, *J. Geophys. Res., 79,* 4581, 1974.

Gosling, J. T., E. Hildner, R. M. MacQueen, R. H. Munro, A. I. Poland, and C. L. Ross, Direct observations of a flare related coronal and solar wind disturbance, *Solar Phys., 40,* 439, 1975.

Gosling, J. T., E. Hildner, R. M. MacQueen, R. H. Munro, A. I. Poland, and C. L. Ross, The speeds of coronal mass ejection events, *Solar Phys., 48,* 389, 1976.

Gosling, J. T., D. N. Baker, S. J. Bame, W. C. Feldman, R. D. Zwickl, and E. J. Smith, Bidirectional solar wind electron heat flux events, *J. Geophys. Res., 92,* 8519, 1987.

Gosling, J. T., S. J. Bame, D. J. McComas, and J. L. Phillips, Coronal mass ejections and large geomagnetic storms, *Geophys. Res. Lett., 17,* 901, 1990.

Gosling, J. T., D. J. McComas, J. L. Phillips, and S. J. Bame, Geomagnetic activity associated with Earth passage of interplanetary shock disturbances and coronal mass ejections, *J. Geophys. Res., 96,* 7831, 1991.

Gosling, J. T., D. J. McComas, J. L. Phillips, and S. J. Bame, Counterstreaming solar wind halo electron events: Solar cycle variations, *J. Geophys. Res., 97,* 6531, 1992.

Gosling, J. T., S. J. Bame, D. J. McComas, J. L. Phillips, B. E. Goldstein, and M. Neugebauer, The speeds of coronal mass ejections in the solar wind at mid heliographic latitudes: Ulysses, *Geophys. Res. Lett., 21,* 1109, 1994a.

Gosling, J. T., D. J. McComas, J. L. Phillips, L. A. Weiss, V. J. Pizzo, B. E. Goldstein, and R.J. Forsyth, A new class of forward-reverse shock pairs in the solar wind, *Geophys. Res. Lett., 21*, 2271, 1994b.

Gosling, J. T., J. Birn, and M. Hesse, Three-dimensional magnetic reconnection and the magnetic topology of coronal mass ejection events, *Geophys. Res. Lett., 22*, 869, 1995.

Hale, G. E., The spectrohelioscope and its work, Part III. Solar eruptions and their apparent terrestrial effects, *Astrophys. J., 73*, 379, 1931.

Harrison, R. A., Solar coronal mass ejections and flares, *Astron. Astrophys., 162*, 283, 1986.

Harrison, R. A., E. Hildner, A. J. Hundhausen, D. G. Sime, and G. M. Simnett, The launch of solar coronal mass ejections: Results from the coronal mass ejection onset program, *J. Geophys. Res., 95*, 917, 1990.

Hiei, E., A. J. Hundhausen, and D. G. Sime, Reformation of a coronal helmet streamer by magnetic reconnection after a coronal mass ejection, *Geophys. Res. Lett., 20*, 2785, 1993.

Hildner, E., J. T. Gosling, R. T. Hansen, and J. D. Bohlin, The sources of material comprising a mass ejection coronal transient, *Solar Phys., 45*, 363, 1975.

Hirshberg, J., S. J. Bame, and D. E. Robbins, Solar flares and solar wind helium enrichments: July 1965-July 1967, *Solar Phys., 23*, 467, 1972.

Howard, R., N. R. Sheeley, M. J. Koomen, and D. J. Michels, Coronal mass ejections: 1979-1981, *J. Geophys. Res., 90*, 8173, 1985.

Hundhausen, A. J., The origin and propagation of coronal mass ejections, in *Proceedings of the Sixth International Solar Wind Conference*, TN306+Proc, edited by V. Pizzo, T. E. Holzer, and D. G. Sime, National Center for Atmospheric Research, Boulder, pp. 181-214, 1988.

Hundhausen, A. J., Sizes and locations of coronal mass ejections: SMM observations from 1980 and 1984-1989, *J. Geophys. Res., 98*, 13,177, 1993.

Hundhausen, A. J., A summary of SMM observations from 1980 and 1984-1989, in *The Many Faces of the Sun*, edited by K. Strong, J. Saba, and B. Haisch, Springer-Verlag, New York, in press, 1997.

Hundhausen, A. J., and R. A. Gentry, Numerical simulation of flare-generated disturbances in the solar wind, *J. Geophys. Res., 74*, 2908, 1969.

Hundhausen, A. J., S. J. Bame, and M. D. Montgomery, The large scale characteristics of flare-associated solar wind disturbances, *J. Geophys. Res., 75*, 4631, 1970.

Hundhausen, A. J., J. T. Burkepile, and O. C. St. Cyr, The speeds of coronal mass ejections: SMM observations from 1980 and 1984-1989, *J. Geophys. Res., 99*, 6543, 1994.

Kahler, S. W., The morphological and statistical properties of solar x-ray events with long decay times, *Astrophys. J., 214*, 891, 1977.

Kahler, S. W., Solar flares and coronal mass ejections, *Ann. Rev. Astron. Astrophys, 30*, 113, 1992.

Kahler, S. W., Using charged particles to trace interplanetary magnetic field topology, this volume.

Kahler, S. W., N. R. Sheeley, R. A. Howard, M. J. Koomen, D. J. Michels, R. E. McGuire, T. T. von Rosenvinge, and D. V. Reames, Associations between coronal mass ejections and solar energetic proton events, *J. Geophys. Res., 89*, 9683, 1984.

Klein, L. W., and L. F. Burlaga, Magnetic clouds at 1 AU, *J. Geophys. Res., 87*, 613, 1982.

Kopp, R. A., and G. W. Pneuman, Magnetic reconnection in the corona and loop prominence phenomenon, *Solar Phys., 50*, 85, 1976.

Lee, M. A., Particle acceleration and transport at CME-driven shocks, this volume.

Lee, M. A., and J. M. Ryan, Time-dependent coronal shock acceleration of energetic solar flare particles, *Astrophys. J., 303*, 829, 1986.

Lindemann, F. A., Note on the theory of magnetic storms, *Phil. Mag., 38*, 669, 1919.

Low, B. C., The role of coronal mass ejections in solar activity, this volume.

MacQueen, R. M., J. A. Eddy, J. T. Gosling, E. Hildner, R. H. Munro, G. A. Newkirk, A. I. Poland, and C. L. Ross, The outer solar corona as observed from Skylab: Preliminary results, *Astrophys. J. Lett., 187*, L85, 1974.

Marsden, R. G., T. R. Sanderson, C. Tranquille, K.-P. Wenzel, and E. J. Smith, ISEE 3 observations of low-energy proton bidirectional events and their relation to isolated interplanetary magnetic structures, *J. Geophys. Res., 92*, 11,009, 1987.

Martin, S., and A. H. McAllister, Predicting the sign of helicity in erupting filaments and coronal mass ejections, this volume.

Marubashi, K., Interplanetary flux ropes and solar filaments, this volume.

McAllister, A. H., M. Dryer, P. McIntosh, H. Singer, and L. Weiss, A large polar crown coronal mass ejection and a "problem" geomagnetic storm: April 14-23, 1994, *J. Geophys. Res., 101*, 13,497, 1996.

Montgomery, M. D., J. R. Asbridge, S. J. Bame, and W. C. Feldman, Solar wind electron temperature depressions following some interplanetary shock waves: Evidence for magnetic merging? *J. Geophys. Res., 79*, 3103, 1974.

Munro, R. H., J. T. Gosling, E. Hildner, R. M. MacQueen, A. I. Poland, and C. L. Ross, The association of coronal mass ejection transients with other forms of solar activity, *Solar Phys., 61*, 201, 1979.

Neugebauer, M., and R. Goldstein, Particle and field signatures of mass ejections in the solar wind, this volume.

Obayashi, T., Propagation of solar corpuscles and interplanetary magnetic fields, *J. Geophys. Res., 67*, 1717, 1962.

Osherovich, V. A., and L. F. Burlaga, Magnetic clouds, this volume.

Parker, E. N., *Interplanetary Dynamical Processes*, John Wiley and Sons, New York, 1963.

Piddington, J. H., Interplanetary magnetic field and its

control of cosmic-ray variations, *Phys. Rev., 112*, 589, 1958.

Pudovkin, M. I., S. A. Zaitseva, and E. E. Benevslenska, The structure and parameters of flare streams, *J. Geophys. Res., 84*, 6649, 1979.

Reames, D. V., Trapping and escape of the high energy particles responsible for major proton events, in *Eruptive Solar Flares, Lecture Notes in Physics 399*, edited by Z. Svestka, B. V. Jackson, and M. E. Machado, Springer-Verlag, Berlin, pp. 180-185, 1992.

Reames, D. V., Energetic particles and the structure of coronal mass ejections, this volume.

Reames, D. V., L. M Barbier, and C. K. Ng, The spatial distribution of particles accelerated by coronal mass ejection-driven shocks, *Astrophys. J., 466*, 473, 1996.

Richardson, I. G., and H. V. Cane, Regions of abnormally low proton temperature in the solar wind (1965-1991) and their association with ejecta, *J. Geophys. Res., 100*, 23,397, 1995.

Rust, D. M., Spawning and shedding helical magnetic fields in the solar atmosphere, *Geophys. Res. Lett., 21*, 241, 1994.

Schwenn, R., H. Rosenbauer, and K.-H. Muhlhauser, Singly-ionized helium in the driver gas of an interplanetary shock wave, *Geophys. Res. Lett., 7*, 201, 1980.

Sheeley, N. R., et al., Coronal changes associated with a disappearing filament, *Solar Phys., 45*, 377, 1975.

Sheeley, N. R., R. A. Howard, M. J. Koomen, and D. J. Michels, Associations between coronal mass ejection events and soft x-ray events, *Astrophys. J., 272*, 349, 1983.

Sheeley, N. R., R. A. Howard, M. J. Koomen, D. J. Michels, R. Schwenn, K.-H. Muhlhauser, and H. Rosenbauer, Coronal mass ejections and interplanetary shocks, J. *Geophys. Res., 90*, 163, 1985.

Tousey, R., The solar corona, *Adv. Space Res., 13*, 713, 1973.

Webb, D. F., A. S. Krieger, and D. M. Rust, Coronal x-ray enhancements associated with Hα filament disappearances, *Solar Phys., 48*, 159, 1976.

Webb, D. F., and R. A. Howard, The solar cycle variation of coronal mass ejections and the solar wind mass flux, *J. Geophys. Res., 99*, 4201, 1994.

J. T. Gosling, Los Alamos National Laboratory, MS D466, Los Alamos, NM 87545

Observations of CMEs from SOHO/LASCO

R.A. Howard[1], G.E. Brueckner[1], O.C. St. Cyr[2], D.A. Biesecker[6], K.P. Dere[1],
M.J. Koomen[3], C.M. Korendyke[1], P.L. Lamy[4], A. Llebaria[4], M.V. Bout[4], D.J. Michels[1], J.D. Moses[1],
S.E. Paswaters[5], S.P. Plunkett[6], R. Schwenn[7], G.M. Simnett[6], D.G. Socker[1], S.J. Tappin[6], D. Wang[5]

The LASCO experiment on board the SOHO satellite has been making observations of the solar corona out to 30 R☉ since first light on 29 December 1995, and routinely since 15 May 1996. Over 120 coronal mass ejections have been observed to date. LASCO represents a significant advance over previous coronagraphs in many ways, but principally in an expanded field of view, increased sensitivity and increased dynamic range. The satellite, orbiting about the Lagrangian point, L1, is in continual sunlight, providing the opportunity to view the corona continuously, uninterrupted by orbit night as is common with near-Earth orbits. While the CMEs observed by LASCO are similar to those observed with previous coronagraphs, there are several new aspects: (1) Many are accompanied by a global response of the solar corona, (2) Many show acceleration to the edge of the field, (3) Disconnection is a frequent occurrence, (4) CMEs are occurring more frequently than had been expected at this minimum phase of the solar activity cycle, and (5) CMEs undergo extensive internal evolution as they move outward.

1. INTRODUCTION

The LASCO (Large Angle Spectrometric Coronagraph) coronagraph experiment [*Brueckner et al.*, 1995] on the joint NASA and ESA Solar and Heliospheric Observatory (SOHO) mission is the latest in a series of spacecraft carrying white light coronagraphs: OSO-7, Skylab, P78-1 (Solwind), and SMM. Within a few months of the launch of the first coronagraph on the seventh of NASA's Orbiting Solar Observatory series, the first coronal mass ejection (CME) was observed on 14 December 1971 [*Tousey*, 1973]. OSO-7 had a very low data rate and duty cycle for observing transient phenomena such as CMEs, but a total of 20 CMEs [*Howard et al.*, 1975] were observed before the last tape recorder failed in 1973. The Skylab coronagraph [*MacQueen et al.*, 1974], launched in 1973, then observed many beautiful events during its 9 month mission. The coronagraphs on P78-1, launched in 1979 [*Michels et al.*, 1980], and SMM, launched in 1980 [*MacQueen et al.*, 1980], together operated for almost a solar cycle and observed over 2000 CMEs. This long set of observations enabled detailed analyses of the properties of CMEs [e.g., *Hundhausen et al.*, 1984; *Howard et al.*, 1985]. (For reviews of the CME phenomenon, see *Kahler* [1992], *Webb* [1995], Hundhausen [1997] and Gosling [this volume].)

The term CME originally referred to the observation of material being ejected from the sun and traveling through the

[1]*E.O. Hulburt Center for Space Research, Naval Research Laboratory, Washington, DC*
[2]*Computational Physics, Inc, Fairfax, VA*
[3]*Sachs-Freeman Associates, Landover, MD*
[4]*Laboratoire d'Astronomie Spatiale, Marseille, France*
[5]*Interferometrics, Inc, Chantilly, VA*
[6]*School of Physics and Space Research, University of Birmingham, Birmingham, UK*
[7]*Max-Planck-Institut für Aeronomie, Katlenburg-Lindau, Germany*

Coronal Mass Ejections
Geophysical Monograph 99
Copyright 1997 by the American Geophysical Union

field of view of a white light coronagraph. CME was used to distinguish ejecta events from other types of coronal transients for which material is not seen leaving the solar corona, such as a rearrangement of the global coronal structure. However, an interesting question is whether such rearrangements are always accompanied by a CME. The CME is a complex phenomenon, but in its simplest form, has a three-part structure [Hundhausen et al. 1984; Illing and Hundhausen, 1985], consisting of a curved, loop-like density enhancement at the front, followed by a density depletion, followed by a second, interior density enhancement.

In the trailing phase of a CME, a disconnection signature [*Illing and Hundhausen*, 1983] is occasionally observed. In such an event a concave outward structure is seen to propagate outward. The concave outward structure is suggestive of a topology in which previously open magnetic field lines have become disconnected. This aspect has been studied in detail [*Burkepile and St Cyr*, 1993; *Webb and Cliver*, 1995]. *Burkepile and St. Cyr* [1993] estimated that disconnections were relatively infrequent, occurring about 3% of the time in the SMM data, whereas *Webb and Cliver* [1995], examining all space-borne coronagraphs, suggested that they may be occurring in at least 10% of the CMEs.

The SOHO satellite [*Domingo et al.*, 1995] was launched on 2 December 1995 and cruised for 4 months to reach a halo orbit about the L1 Lagrangian point, which is located along the sun-earth line at about 1.6×10^6 km from the earth toward the sun. A very careful checkout phase of the spacecraft itself as well as the instrument complement was conducted during the cruise phase, lasting until May 1996. In a fixed sun-earth reference frame, the projection of the halo orbit onto the plane of the sky is an ellipse, with semi-major and semi-minor axes of 650,000 and 200,000 km, respectively. A distinct advantage of this orbit is that the satellite is able to observe the sun continuously, without the interruptions usually associated with near-earth orbits. Furthermore, the L1 orbit provides an extremely stable thermal environment which results in very stable sun-centered pointing.

2. LASCO INSTRUMENT DESCRIPTION

The LASCO instrument [*Brueckner et al.*, 1995] consists of three optical systems: an internally occulted coronagraph and two traditional externally occulted coronagraphs. Table 1 presents some characteristics of the three coronagraphs. Together, the LASCO telescope complement observes the solar corona from 1.1 to 30 $R_\odot$. The field of view of the C3 telescope extends to 32 $R_\odot$, but at the equator and at the poles, the Charge Coupled Device (CCD) detector limits the field to 30 $R_\odot$. Polarization and color filters aide the separation of the observed brightness distribution into the electron (K) and dust (F) coronal components. Table 2 shows a comparison between LASCO and the first orbiting coronagraph on NASA's Orbiting Solar Observatory, OSO-7. The only parameter for the P78-1/Solwind coronagraph different from the OSO-7 was the inner field limit, which was 2.5 $R_\odot$, rather than 3.0 $R_\odot$.

LASCO represents a major advance over previous instruments in three aspects. The first is the increased field of view, both closer to and farther from the sun. Secondly, the stray light level in the C2 and C3 telescopes is extremely low. For example, the stray light level in Solwind, which observed the corona to 10 $R_\odot$, was about 5×10^{-11} $B/B_\odot$, compared to about 1×10^{-12} $B/B_\odot$ for C3. Finally, the dynamic range of the CCD detectors coupled with the low system noise of the cameras provide an unprecedented sensitivity over film and vidicon detectors. The CCD detectors are cooled to about -80 C, which eliminates the effect of thermal dark noise. However, the CCDs are sensitive to energetic particles, which create streaks in the images of varying lengths and orientations.

In the C1 telescope, a Fabry-Perot interferometer (FPI) permits spectrally resolved images to be taken. The FPI transmits a very narrow, 0.07 nm, pass band in many spectral orders. A broadband blocking filter is used to select one of the 5 spectral regions shown in Table 1 and thus isolate one of the spectral orders of the FPI. The FPI can be commanded to any wavelength in the 5 nm spectral bandpass of a blocking filter. Since a separate image is taken at each Fabry-Perot wavelength step, about 10-15 images are required to obtain a complete spectral description of an emission line such as the green line of FeXIV at 530.3nm. In addition to the green line, blocking filters were included for three other emission lines: Fe X, Ca XV, and Hα. The electron corona is obtained in two different ways. The first method is the spectral Grotrian technique [*Billings*, 1966], which uses the difference between the depth of a Fraunhofer line (Na I) in the corona to the depth on the solar disk to deduce the K-corona. The second method is a "broadband" technique, which measures the intensity in the continuum to obtain the "white-light" coronal brightness.

In the C2 and C3 telescopes, the total coronal brightness is obtained through the orange filter, a 100 nm passband centered at 590 nm. In addition, C3 has a clear filter with about five times the transmittance of the orange filter. This clear filter is the one normally used to obtain the total brightness because the exposure times are shorter, reducing the effect of the cosmic ray impacts (about 4 cm^{-2} sec^{-1}) on the CCD. All of the measurements to obtain pB are made through the orange filters in both C2 and C3. The F-corona has significant polarization in the range observed by C3, but the color of the F-corona is redder than is the K-corona. Thus, the color filters aide in the separation of the K- and F-coronae.

All of the LASCO cameras are identical and use a 1024 x 1024 CCD with 21 μm square pixels. The electron full well

Table 1. Characteristics of LASCO Coronagraphs

Telescope Parameter	C1	C2	C3
Field of View	1.1 - 3.0 $R_\odot$	1.7 - 6.0 $R_\odot$	3.7 - 32.0 $R_\odot$
Occulter Type	Internal	External	External
Spectral Bandpass	Fabry-Perot Interferometer (0.07 nm)	Broadband	Broadband
Color Filters	Fe XIV 530.3 nm Fe X 637.4 nm Ca XV 564.9 nm Na I 589.0 nm Hα 656.2 nm	Orange Blue Red Hα	Clear Orange Blue Red Infrared Hα
Polarization Analysers	0°, 60°, -60° linear polarizers	0°, 60°, -60° linear polarizers	0°, 60°, -60° linear polarizers
Stray Light	1×10^{-8} B/B$_\odot$	5×10^{-11} B/B$_\odot$	1×10^{-12} B/B$_\odot$
CCD Size	1024 x 1024	1024 x 1024	1024 x 1024
Pixel Size	5.6 arc sec	11.2 arc sec	56.0 arc sec

capacity of the CCD is approximately 200,000 electrons which is digitized to 14 bits (16384). The quantization step has been set such that one quantization step is equivalent to about 12 electrons. Since the quantum efficiency is 0.4, the quantization step is equivalent to 30 photons at the detector and the full well capacity is 5×10^5 photons. Since the total system noise is on the order of 30 photons per pixel, the dynamic range of the cameras is ~16000. The maximum signal to noise ratio is also outstanding at 700:1, being limited by the photon noise, not by the system noise.

LASCO is capable of taking a full image about every 6 minutes. An image sequence consists of configuring the telescope mechanisms and camera, exposing the CCD, reading out the camera, processing the image, and transferring the image to a telemetry buffer. The majority of the time is involved in processing the image to reduce the amount of data transmitted. Although several compression schemes are available [Brueckner et al., 1995], the lossless compression algorithm is used almost exclusively. This is because the lossy technique, an adaptive discrete cosine transform, does not handle the very high or low contrast features properly, rendering the new detector capabilities useless. A new, lossy technique based on wavelets appears to be very promising, but is not yet available. The present telemetry allocation is such that it takes about 22 minutes to downlink a full image losslessly. Images can be acquired at the 6 minute rate until the buffer fills, at which time the observation sequence pauses until the spacecraft has emptied the buffer sufficiently. Thus, a sustained cadence is limited by the telemetry downlink rate.

3. OBSERVATIONS OF CMES

From the initial first light of LASCO (C3) on 29 December 1995 through September, 1996, over 120 CMEs have been observed. Generally, the CMEs have been confined to equatorial latitudes. However, several large CMEs have been seen over the solar poles, one of which (20 July 1996) would be classified as a halo CME, since it surrounded the occulting disk by 270°. In this section we describe some of the observations from LASCO and highlight new aspects to the CME phenomenon.

3.1 *The CME on January 15, 1996*

Figure 1 shows the development of a complex CME that occurred on January 15, 1996. In these images, the C3 field of view that was transmitted to the ground was restricted to be

Table 2. Comparison of LASCO/C3 and OSO-7 Coronagraphs

Telescope Parameter	LASCO/C3	OSO-7
Field of View	3.7 - 32 $R_\odot$	2.5 - 10 $R_\odot$
Resolution	56 arc sec/pixel	75 arc sec/pixel
Image Array	1024 x 1024 1 x 10^6 pixels	256 x 256 64 x 10^3 pixels
Stray Light Level	1 x 10^{-12} $B_\odot$	5 x 10^{-11} $B_\odot$
Dynamic Range	16,000:1	5:1
Signal to Noise Ratio	700:1	70:1
Quantization	14 bits (16384)	7 bits (128)

±15 $R_\odot$ from sun center in the equatorial (E-W) direction and ±7 $R_\odot$ in the polar (N-S) direction. By restricting the number of pixels in this way, the cadence of observation was increased to one image about every 4 minutes. The figure shows a series of difference images in which a base image taken prior to the event was subtracted from the images at the times that are indicated in the figure. Using this type of differencing scheme clearly shows the increases (white) and decreases (black) of column brightness, hence column electron density, from the pre-event time.

There are several phases to this CME, which would have been classified as a complex event in the Solwind classification scheme [*Howard et al.*, 1985]. In the first phase, a narrow, bright loop extends outward just ahead of the main loop. At 06:54 UT it is just extending past the occulting disk. At 10:59 UT it is the narrow, bright feature just north of the equator. The second phase has a large, flat-topped loop, similar to some seen during the Skylab mission [*Gosling et al.*, 1974] and which have been associated with slow mode shocks [*Steinolfson and Hundhausen*, 1988]. A third phase, a system of fine arcs, can be seen toward the southern boundary of the event. Finally, a fourth phase, an oval bubble, can be seen to travel outward at the northern boundary of the event. The trailing part of the bubble is concave outward and is interpreted as an example of a disconnection event [*Illing and Hundhausen*, 1983]. Such disconnection events are seen more frequently in LASCO than expected from the SMM observations [*Burkepile and St Cyr*, 1993].

The amount of ejected mass is computed in a manner similar to *Poland et al.* [1981]. The excess number of electrons is computed from the Thomson scattering function for a volume element located at a single distance from the plane of the sky. Since polarization measurements were not made, the distance from the sky plane cannot be determined, and a value of 20° was used. The mass is obtained by assuming a completely ionized mix of hydrogen and 10% helium. The total ejected mass over the west limb increases as the leading edge extends outward and more of the event is in the field of view. In order to compute the mass, the front of the CME must be in the field. Thus, previous mass estimates have been computed only when the leading edge was less than 5-8 $R_\odot$. At a height of 6 $R_\odot$, the mass of the January 15 CME is found to be 3 x 10^{15} g, which is a typical value for an average CME. However, the maximum mass in this event is found to be 3 x 10^{16} g, when the leading edge is at 13 $R_\odot$.

The east limb also was undergoing transient activity. The deflection of a streamer is evident due to the alternating black and white structure. In addition, a faint outward expulsion of material can be seen traveling along the deflected streamer on the east limb. The motion is especially evident in a time-lapse movie. This near-simultaneous ejection of mass occurring on the opposite limb is one example of the global nature of some CMEs [*Brueckner*, 1996]. The east limb density enhancement is traveling outward at about the same speed (600 km s-1) as the west limb ejection. A simple deflection of the streamer cannot explain the observations, because an integration over all pixels on the east limb confirms that a net mass increase occurs at this time, and the rate of increase is the same as the main west limb CME.

3.2 *CME Speeds and Accelerations*

To date, the speeds of the CMEs observed by LASCO are not unusual, ranging from 20-900 km s^{-1}, which is well within

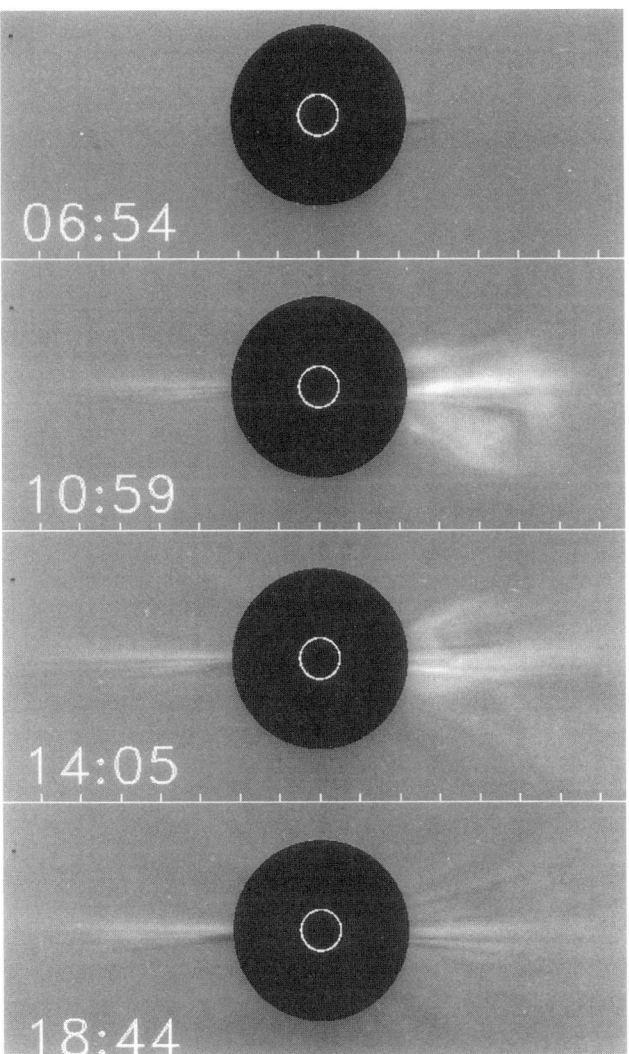

Figure 1. January 15, 1996 CME. The field of view of the LASCO/C3 image is 30 x 15 $R_\odot$ with a pixel size of 56 arc sec. The C3 occulting disk has a radius of 3.7 $R_\odot$. Each image has been differenced from the same pre-event image. The narrow bright loop in the center is moving outward at a speed of 800 km s^{-1}. The large loop-like event with a flat top has a speed of about 600 km s^{-1}. Extensive internal structure can be seen which evolves as the event travels through the field. In this and in all subsequent figures, Solar north is up and west is to the right.

the range seen earlier [*Gosling et al.*, 1976; *Howard et al.*, 1985; *Hundhausen et al.*, 1994]. The most unusual aspect is that acceleration is frequently observed to the edge of the field of view of the C3 coronagraph at 30 $R_\odot$. During the previous CME studies, acceleration was occasionally observed. However, to observe acceleration, the range of heights over which the event is measured is a significant factor.

Figure 2 gives the height-time diagram for the January 15, 1996 CME. In this event six features are tracked: the leading edge of each of the four phases, the trailing disconnection feature and one internal feature. A constant speed is seen for all of the structures, with no convincing evidence for acceleration in any of them. However, in this case the height range available was less than the full range of the C3 coronagraph.

Figure 3 shows the CME that occurred on February 3, 1996 in the C2 coronagraph on the east limb. An empirical background model has been subtracted from each image to remove the F-coronal background leaving the pre-existing streamers in the east and west (as well as polar plumes). In this event a bubble propagates out along a pre-existing streamer. The rear of the bubble has the characteristic disconnection appearance. The positions of the leading edge at each time have been measured and are shown in the upper panel of Figure 4 as asterisks. The solid line is the result of a least squares fit to a second degree polynomial. As can be seen, a constant acceleration gives a very good fit to the observed points. Using the coefficients to the fit, we compute a velocity vs height curve which is shown in the lower panel of Figure 4. A constant acceleration of 7.6 m s^{-2} results in the velocity increasing from 10 km s^{-1} to 400 km s^{-1}. The large height range enables a definite determination of acceleration over this range. There is, of course, no implication that the speed continues to increase beyond the field of view with the same constant acceleration. A linear fit to the heights above 6 $R_\odot$ yields a speed of 300 km s^{-1}.

Figure 5 shows a large loop ejection that occurred on May 1, 1996 in the C3 coronagraph on the west limb. At this time the LASCO observations were designed to observe comet Hyakutake, which entered the C3 field of view on April 29, 1996. The comet observation program was at the expense of observations using the C1 and C2 telescopes, preventing the observation of the leading edge of the CME in those telescopes. In this event, the span was nearly 90 degrees. Figure 5 is a difference image in which a pre-event image has been subtracted from the image during the event. This technique shows the changes to the corona since the earlier time. Additional mass is shown as white and less mass is shown as black. The speed profile is very similar to that shown in Figure 4 for the much smaller event on February 3, except for a slightly smaller acceleration constant of 5.8 m s^{-2}. During the several days around this event, many small density enhancements are ejected from the sun on both limbs.

A class of events has been discovered which invariably show acceleration. They have not been consistently included in the preliminary count of CMEs given above, especially when they are very weak or small. The events generally would have been below the sensitivity of the Solwind coronagraph. The Solwind class of "Jet along a streamer" (JS) [*Howard et*

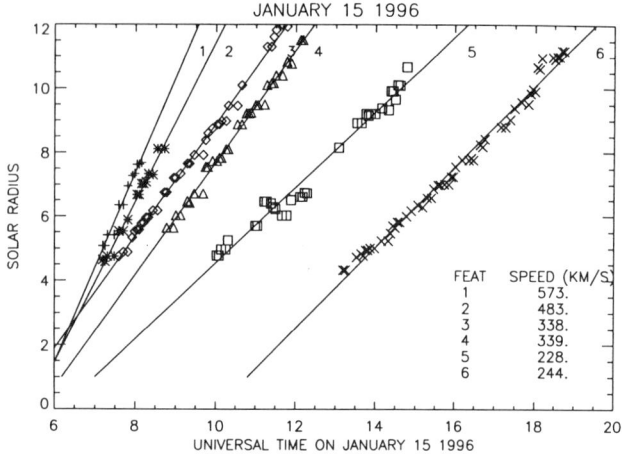

Figure 2. Height-Time Curve for the January 15, 1996 CME. Six features are tracked: (1) The leading edge of the narrow loop in the center, (2) The leading edge of the flat-topped loop, (3) A bright internal filament behind feature 2, (4) The leading edge of the system along the southern boundary, (5) The leading edge of the oval bubble along the northern boundary, (6) The center of the disconnection feature along the northern boundary.

al., 1985] were probably the brighter members of this class. Characteristic of this class is a small, low contrast density enhancement (puff), which propagates through the field of view of C2 and C3 and along a pre-existing streamer [*Sheeley et al.*, 1997]. There are no surface manifestations. In the upper panel of Figure 6, the asterisks give the measured position of the density enhancement as a function of time. The solid line is a least squares quadratic fit to the points with an acceleration constant of 3.5 m s^{-2}. The lower panel of Figure 6 is a plot of the speed calculated from the fit against the observed solar radius. The speed of this puff increased from 20 to 400 km s^{-1}, which is typical for such events [*Sheeley et al.*, 1997]. All of the puffs have acceleration values between 3 and 5 m s^{-2} and appear to be tracking the flow speed of the solar wind.

3.3 *CME Occurrence Rate*

It is premature to present a definitive CME occurrence rate over the entire mission due to the complexity of the operating modes during the spacecraft and instrument checkouts. This complexity necessitates corrections for the duty cycle. With the ±30 R$_\odot$ field of view of C3, a CME traveling at an average rate of about 500 km s^{-1} would take nearly 12 hours to travel through the field of view. Thus observations that are taken 6 hours apart would be more than sufficient to detect the occurrence of a CME in the C3. During a two-month period from May 15, 1996 to July 15, 1996, the instrument was

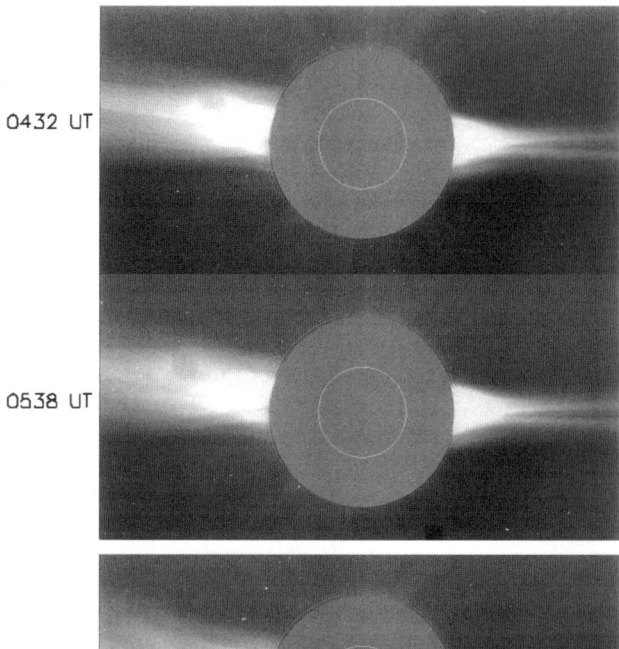

Figure 3. February 3, 1996 CME. The field of view of the LASCO/C2 image is 12 x 12 R$_\odot$ with a pixel size of 11.2 arc sec. The C2 occulting disk has a radius of 1.7 R$_\odot$. Each image has been differenced from the same base image: an empirical background model which removes the static F-corona and stray light. The CME is the small bubble in the pre-existing streamer that can be seen to progress outward. The trailing portion of the bubble has the appearance of a disconnection event. Note that the empirical model also removes the pylon which is in the southeast.

operated almost continuously, thus obviating the need to consider any duty cycle corrections. During this period, the rate of CMEs was 0.8 day^{-1}, considerably higher than the rate of about 0.1 day^{-1} which was what was derived from actual coronagraph CME rates during the previous Solar minimum [*Webb and Howard*, 1994].

3.4 *Fe XIV Transients*

Up to the present time, the C1 observation time has been largely devoted to observing the morphology of the Fe XIV

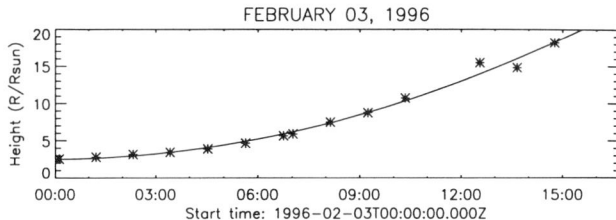

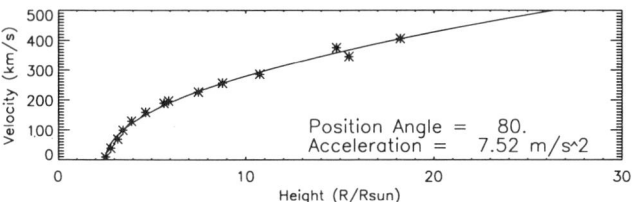

Figure 4. Speed of February 3, 1996 CME. The upper panel shows the Height-Time curve showing the measured position of the leading edge as the asterisks. The solid line is a least-squares fit to a quadratic polynomial. The lower panel shows the Velocity-Height curve giving the computed speed at each of the measured heights as asterisks. The solid line is the computed from the least squares fit to the positions. The acceleration constant is 7.6 m s^{-2}.

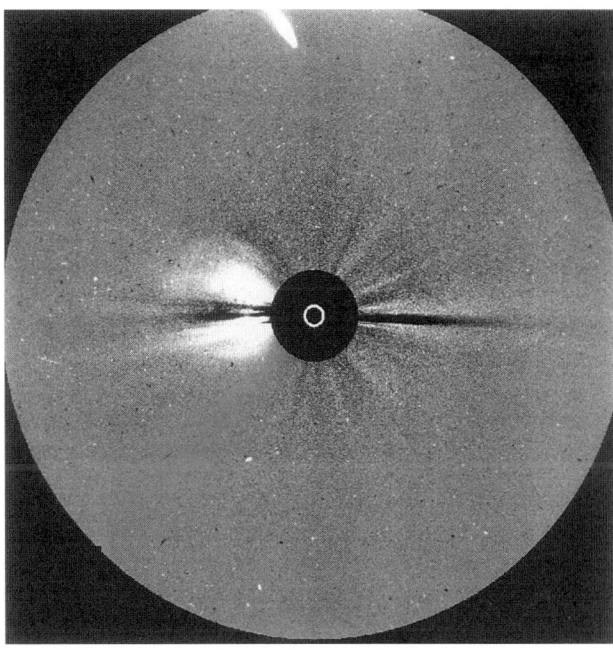

Figure 5. May 1, 1996 CME. The field of view of the LASCO/C3 image is 30 x 30 $R_\odot$, with a pixel size of 56 arc sec. The C3 occulting disk has a radius of 3.7 $R_\odot$. Each image has been differenced from the same image taken at a time prior to the CME. Comet Hyakutake can be seen at its perihelion above the north pole of the sun. This CME has a similar height-time profile as shown in Figure 4 for the 2 February event, but with a smaller acceleration constant of 5.8 m s^{-2}.

green emission line corona and observing line profiles of the emission lines a few times each day. This line is formed at a temperature of about 1.8×10^6 K and appears to trace the low-lying magnetic field. *Schwenn et al.* [1997] have presented some initial observations from C1. In Figure 7 the green line structure during February 1996 is shown. In this figure an off-band image taken close in time has been subtracted from the on-band image to remove the scattered light and the electron corona, leaving only the emission from a 0.07 nm spectral band at the core of the Fe XIV green line.

Very evident is the quadrupolar nature to the structures, with a small system of loops overlying the magnetic neutral lines in the photospheric magnetograms and presumably over the polar crown filament region [*Schwenn et al.*, 1997]. These small loops are quite stable, but show continual small-scale motions. The impression is that the motions are outward, but no loops are actually seen at these latitudes at higher heights. This may be an observational problem, due to the emission line moving out of the narrow passband of the Fabry-Perot interferometer. The helmet streamer that is seen in white light in the C2 telescope encompasses this entire system, so that the small loop systems are underneath the large streamer. Note that this would imply multiple current sheets extending into the corona and solar wind [*Crooker et al.*, 1993; *Dahlberg and Karpen*, 1995].

Figure 8 shows a transient observed on June 17, 1996 in the Fe XIV images. Again, an off-band image has been subtracted from the on-band images. A loop system is seen to emerge at the equator, in between the two stable, higher latitude loop systems. As the loops emerge to the height of the pre-existing loops at about 1.4 $R_\odot$, they immediately expand to the northern and southern boundaries of the large-scale helmet structure that overlies the FeXIV loops. The event was seen in the C2 as a very minor, amorphous density enhancement.

4. DISCUSSION

The CMEs observed from LASCO have properties that are quite similar to those observed previously. However, it is still too soon after the launch to perform detailed statistical analyses to compare with the studies from prior missions. All that we can do is to give a few results and our impressions.

The frequency of occurrence is higher (0.8/day) than expected. The CME occurrence rate at the minima of the previous two solar cycles (1976 and 1985/1986) was about 0.2/day [*Webb and Howard*, 1994]. Some, but not all, of this increase might be due to increased instrumental sensitivity. The small puffs that are occurring frequently were not

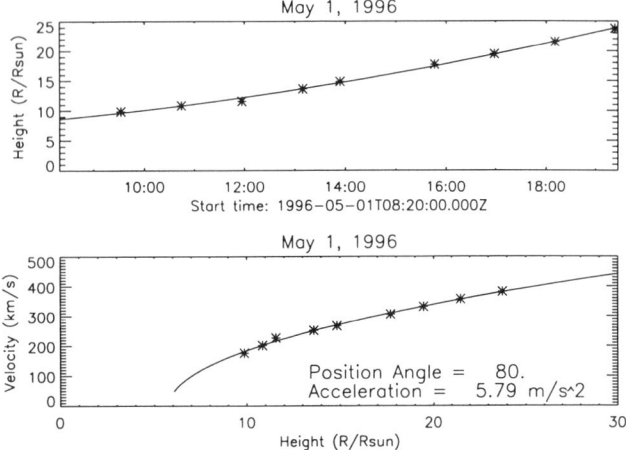

Figure 6. Small Density Enhancement. The upper panel is a plot of the height versus the time of observation. The lower panel gives a plot of the computed velocity versus the height curve. The measured heights were fit to a quadratic polynomial, and then the speeds were computed using this function. A constant acceleration through the entire field of about 3.4 m s^{-2} is obtained from the fit. A first or third degree polynomial does not produce as good a fit.

included in the CME count, but some of the CMEs are indeed very faint, and may not have been detected by Solwind or SMM.

It should be noted that the correlation of CME rate to the sunspot activity cycle [*e.g., Hildner et al.*, 1977; *Howard et al.*, 1986], only holds for long time periods [*Howard et al.*, 1985]. That is, the relation holds when comparing the activity averaged over a year, not when average over shorter time periods, such as over a month. Thus the increased level of CMEs at this time may not be at odds with the correlation to sunspot number. Clearly, a longer interval of LASCO observations will be necessary.

MacQueen and Fisher [1983] showed that CME height time profiles could be divided into two classes, one associated with eruptive events and the other with flares. The eruptive-associated CME underwent large accelerations (50 m s^{-2}) while the flare-associated CMEs started out with the highest speeds and underwent little acceleration. Here we can divide the events into two classes, one that accelerates and one that doesn't. No systematic attempt has been made yet to associate the coronal eruptions with disk sources (eruptive events or flares). In the first class, the events start very slowly (e.g. 10-20 km s^{-1}) and then accelerate with a constant rate, until they leave the field of view of LASCO at a speed of about 300-500 km s^{-1}. In the second class, the events start with a high speed and show no or little acceleration (<0.2 m s^{-2}).

The mass calculated for the January 15, 1996 CME has an unresolved contradiction with previous estimates of mass.

Figure 7. Fe XIV Emission Line Corona (530.3nm) on February 3, 1996. The occulter has a radius of 1.1 R$_\odot$ and the pixel size is 5.6 arc sec. Note the loop systems at high latitudes overlying the polar crown filaments. This structure is indicative of a quadrupolar magnetic field.

Jackson and Howard [1993] and *Webb et al.* [1996] discussed the problem of continued mass expulsion after the front had left the field of view of a coronagraph. These authors concluded that, if flow was considered, the reported mass of a CME was underestimated by up to a factor of 3. They also noted that the masses of CMEs observed by Helios in the solar wind were about three times greater than for the same CMEs as measured by Solwind or SMM. For the January 15 CME, we calculate the mass to be nearly 10 times greater when the leading edge is at 15 R$_\odot$ than when it is at 8 R$_\odot$. Thus, the total mass associated with a CME may be underestimated by as much as a factor of 10, not a factor of 3. This will be investigated in more detail in future papers.

The increased sensitivity of LASCO has revealed some new phenomena. Two such discoveries are the small ejecta that are seen to travel along pre-existing streamers. These ejecta are not the result of a displacement of the pre-existing streamer structure for several reasons. One is that the structure is diffuse and has a form that is unlike the streamer and undergoes proper motion. The second reason is that there is a net mass increase on the order of 10^{14} g. A simple, lateral displacement would not result in any mass increase. Of course, coronal evolution could result in a mass increase (or decrease).

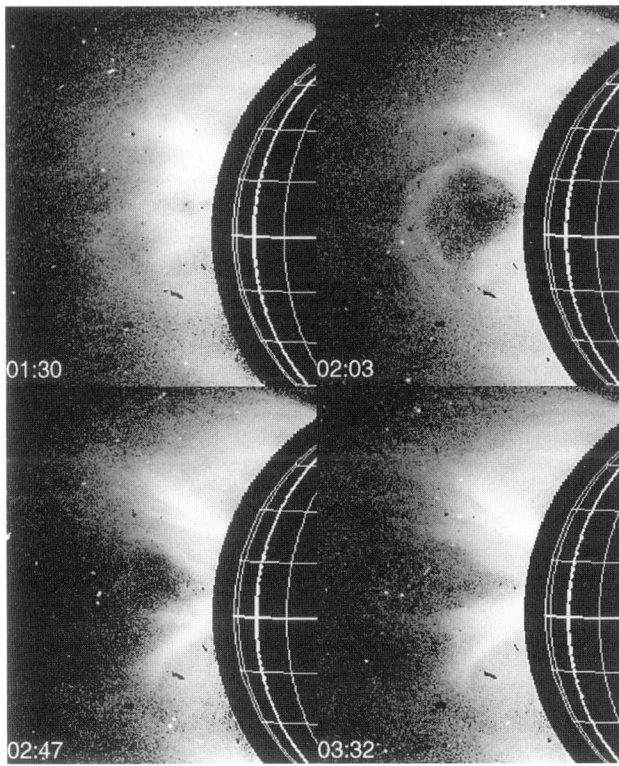

Figure 8. Fe XIV Transient on June 17, 1996. The occulting disk is at 1.1 $R_\odot$. Latitude and longitude lines are shown at the position of the solar disk. A loop system is seen to emerge at the equator on the east limb at 0203 UT in between the two high latitude loop systems, pushing the pre-existing structure seen at 0130 UT aside. After the loop system has passed by, the residual structures no longer show closed loops. A lower limit to the speed of the leading edge is estimated to be about 150 km s^{-1}.

These small ejecta have two speed populations: slow starters and fast starters. The slow appear to be tracking the slow solar wind [*Sheeley et al.*, 1997]. They may be related to the outward expansion of loops seen in the C1 FeXIV emission line. The fast starters are seen in conjunction with CMEs on the opposite limb and have a similar speed profile to the main ejecta. The mechanism for producing the global response is not understood. It could be due an MHD or magnetosonic wave traveling around the corona or an increase in current in the current sheet. Deflections of nearby streamers in a lateral direction in response to a CME is well known. A deflection of a streamer in longitude might explain the observed mass increase. But this would involve a deflection over a much greater range than is seen in latitude.

LASCO's increased sensitivity and increased field of view promises to provide a unique set of observations of the CME phenomena. It will continue the observation of CMEs into the third solar cycle since their discovery in 1971. The expanded field coverage will hopefully answer some of the questions associated with their initiation and propagation into the solar wind.

Acknowledgments. SOHO is an international collaboration between NASA and ESA and is part of the International Solar Terrestrial Physics Program. LASCO was constructed by a consortium of institutions: the Naval Research Laboratory (Washington, DC, USA), the University of Birmingham (Birmingham, UK), the Max-Planck-Institut fur Aeronomie (Katlenburg-Lindau, Germany) and the Laboratoire d'Astronomie Spatiale (Marseille, France). The authors gratefully thank all the members of the teams at the consortium institutions as well as the various project teams who were involved with the management, fabrication, launch and operation of this very successful mission. The NRL effort was supported by NASA under NDPR S-92385-D. The British effort was supported by the Science and Engineering Council in the United Kingdom. The French effort was supported by the Centre National d'Etudes Spatiales (CNES). The German effort was supported under grant 010C88024 by the Deutsche Agentur für Raumfahrtangelegenheiten (DARA).

REFERENCES

Billings, D.E., *A Guide to the Solar Corona*, p. 99, Academic Press, New York, 1966.

Brueckner, G.E., Dynamics of the solar corona as seen from the three LASCO coronagraphs on the SOHO satellite (abstract), *EOS*, 77, S2041, 1996.

Brueckner, G.E., R.A. Howard, M.J. Koomen, C.M. Korendyke, D.J. Michels, J.D. Moses, D.S. Socker, K.P. Dere, P.L. Lamy, A. Lleberia, M.V. Bout, R. Schwenn, G.M. Simnett, D.K. Bedford, and C.J. Eyles, The Large Angle Spectroscopic Coronagraph (LASCO), *Sol. Phys.*, 162, 357, 1995.

Burkepile, J.T. and O.C. St.Cyr, A revised and expanded catalogue of mass ejections observed by the solar maximum mission coronagraph, NCAR/TN-369+STR, Boulder, 1993.

Crooker, N.U., G.L. Siscoe, S. Shodhan, D.F. Wegg, J.T. Gosling, and E.J. Smith, Multiple heliospheric current sheets and coronal streamer belt dynamics, *J. Geophys. Res.*, 98, 9371, 1993.

Dahlburg, R.B., and J.T. Karpen, A triple current sheet model for adjoining coronal helmet streamers, *J. Geophys. Res.*, 100, 23489, 1995.

Domingo, V., B. Fleck, and A.I. Poland, The SOHO Mission: An Overview, *Sol. Phys.*, 162, 1, 1995.

Gosling, J.T., Coronal mass ejections: An overview, this volume.

Gosling, J.T., E. Hildner, R.M. MacQueen, R.H. Munro, A.I. Poland, and C.L. Ross, Mass Ejections from the Sun: a View from Skylab, *J. Geophys. Res.*, 79, 4581, 1974.

Gosling, J.T., E. Hildner, R.M. MacQueen, R.H. Munro, A.I. Poland, and C.L. Ross, The Speeds of CMEs, *Sol. Phys.*, 48, 389, 1976.

Hildner, E., Mass ejections from the corona into interplanetary space, in *Study of Traveling Interplanetary Phenomena*, edited by M.A. Shea et al., p 3, D. Reidel, Hingham, MA, 1977.

Howard, R.A., M.J. Koomen, D.J. Michels, R. Tousey, C.R. Detwiler, D.E. Roberts, R.T. Seal, J.T. Whitney, R.T. Hansen, S.F. Hansen, C.J. Garcia, and E. Yasukawa, Synoptic observations of the solar corona during carrington rotations 1580 - 1596 (11 October 1971 - 15 January 1973), *World Data Center A*, Report UAG 48A, 1975.

Howard, R.A., N.R. Sheeley, Jr., M.J. Koomen, and D.J. Michels, Coronal Mass Ejections 1979-1981, *J. Geophys. Res.*, *90*, 8173, 1985.

Howard, R.A., N.R. Sheeley, Jr., M.J. Koomen, D.J. Michels, The solar cycle dependence of coronal mass ejections, in *The Sun and the Heliosphere in Three Dimensions*, edited by R.G. Marsden, p 107, D. Reidel, Dordrecht, 1986.

Hundhausen, A.J., Coronal Mass Ejections, in *Cosmic Winds and the Heliosphere*, edited by J.R. Jokipi, C.P. Sonett, and M.S. Giampapa, in press, University of Arizona Press, Tucson, AZ, 1997.

Hundhausen, A.J., C.B. Sawyer, L. House, R.M.E. Illing, and W.J. Wagner, Coronal mass ejections observed during the Solar Maximum Mission: Latitude distribution and rate of occurrence, *J. Geophys. Res.*, *89*, 2639, 1984.

Hundhausen, A.J., R.M. MacQueen, and D.G. Sime, The morphology and geometry of coronal mass ejections (abstract), *EOS*, *65*, 1069, 1984.

Hundhausen, A.J., J.T. Burkepile, and O.C. St. Cyr, Speeds of coronal mass ejections, SMM observations from 1980 and 1984-1989, *J. Geophys. Res.*, *90*, 6543, 1994.

Illing, R.M.E. and A.J. Hundhausen, Possible observation of a disconnected magnetic structure in a coronal transient, *J. Geophys. Res.*, *88*, 10210, 1983.

Illing, R.M.E. and A.J. Hundhausen, Observation of a coronal transient from 1.2 to 6 Solar radii, *J. Geophys. Res.*, *90*, 275, 1985.

Jackson, B.V. and R.A. Howard, A CME mass distribution derived from Solwind coronagraph observations, *Solar Phys.*, *148*, 359, 1993.

Kahler, S., Solar flares and coronal mass ejections, *Ann. Rev. Astron. Astrophys.*, *30*, 113, 1992.

MacQueen, R.M. and R.R. Fisher, The kinematics of solar inner coronal transients, *Sol. Phys.*, *89*, 1983.

MacQueen, R.M., J.A. Eddy, J.T. Gosling, E. Hildner, R.H. Munro, G.A. Newkirk, A.I. Poland, and CL. Ross, The outer corona as observed from Skylab, *Astrophys. J. Lett.*, *187*, L85, 1974.

MacQueen, R.M., A. Csoeke-Poecke, E. Hildner, R. Reynolds, A. Stanger, H. Te Poel, and W.J. Wagner, The High Altitude Observatory coronagraph/polarimeter on the Solar Maximum Mission, *Solar Phys.*, *65*, 91, 1980.

Michels, D.J., R.A. Howard, M.J. Koomen, and N.R. Sheeley, Jr., Satellite observations of the outer corona near sunspot maximum, in *Radio Physics of the Sun*, edited by M.R. Kundu and T. Gergely, p 439, D. Reidel, Hingham, MA., 1980.

Poland, A.I., R.A. Howard, M.J. Koomen, D.J. Michels, and N.R. Sheeley, Jr, Coronal transients near sunspot maximum, *Solar Phys.*, *69*, 169, 1981.

Schwenn, R., B. Inhester, S.P. Plunkett, A. Epple, B. Podlipnik, G.E. Brueckner, K.P. Dere, R.A. Howard, P.L Lamy, M.J. Koomen, C.M. Korendyke, D.J. Michels, N.E. Moulton, G.M. Simnett, O.C. St. Cyr, and D.G. Socker, First view on the extended green line emission corona at solar activity minimum using the LASCO-C1 coronagraph on SOHO, *Sol. Phys.*, in press, 1997.

Sheeley, N.R., Jr, Y.-M. Wang, S.H. Hawley, G.E. Brueckner, K.P. Dere, R.A. Howard, M.J. Koomen, C.M. Korendyke, D.J. Michels, S.E. Paswaters, D.G. Socker, O.C. St. Cyr, D. Wang, P.L. Lamy, A. Llebaria, R. Schwenn, G.M Simnett, S. Plunkett, and D.A. Biesecker, Measurements of Flow Speeds in the Corona between 2 and 30 R, *Astrophys. J.*, in press, 1997.

Steinolfson, R.S. and A.J. Hundhausen, Density and white light brightness in looplike coronal mass ejections: temporal evolution, *J. Geophys. Res.*, *93*, 14269, 1988.

Tousey, R., The solar corona, in *Space Research XIII*, edited by M.J.Rycroft and S.K. Runcorn, p 713, Akademie-Verlag, Berlin, 1973.

Webb, D.F., The solar sources of coronal mass ejections, in *Eruptive Solar Flares*, edited by Z. Svestka, B.V. Jackson, and M. Machado, Springer-Verlag, New York, 1992.

Webb, D.F. and E.W. Cliver, Evidence for magnetic disconnection of mass ejections in the corona, *J. Geophys. Res*, *100*, 5853, 1995.

Webb, D.F. and R.A. Howard, The Solar Cycle Variation of Coronal Mass Ejections, *J. Geophys. Res.*, *98*, 9371, 1994.

Webb, D.F., R.A. Howard, and B.V. Jackson, Comparison of CME masses and kinetic energies near the sun and in the inner heliosphere, in *Proc. of Solar Wind 8*, edited by D. Winterhalter et al., NASA Jet Propulsion Lab., Pasadena CA, 1996.

G.E. Brueckner, K.P. Dere, R.A. Howard, M.J. Koomen, C.M. Korendyke, D.J. Michels, J.D. Moses, S.E. Paswaters, S.P. Plunkett, D.G. Socker, O.C. St.Cyr, D. Wang, Code 7660, E.O. Hulburt Center for Space Research, Naval Research Laboratory, Washington, DC 20375.

M.V.Bout, P.L.Lamy, A.Llebaria, Laboratoire d'Astronomie Spatiale, Rue Trois Lucs, Marseille, France

D.A.Biesecker, S.P.Plunkett, G.M Simnett, S.J. Tappin, School of Physics and Space Research, University of Space Research, University of Birmingham, Birmingham, UK

R. Schwenn, Max-Planck-Institut für Aeronomie, Katlenburg-Lindau, Germany

Soft X-Ray Signatures of Coronal Ejections

Hugh S. Hudson

Solar Physics Research Corp. and ISAS

David F. Webb

Institute for Scientific Research, Boston College

We have a new view of the time behavior of the inner solar corona via the extensive soft X-ray observations of *Yohkoh*. These show many forms of expansion of coronal material, ranging from highly collimated structures (jets) to large-scale evacuations ("dimmings") of the diffuse corona. We review these effects, emphasizing those probably related to CMEs, and present a preliminary classification scheme of the large-scale ejections seen in soft X-rays. The new observations bring clarity and focus to the well-established scenario of filament channel → CME → X-ray arcade. The associated X-ray brightening typically occurs in a close temporal relationship with the ejection as inferred from the dimming, but the location of the ejected mass may be displaced from the region of strongest brightening. The soft X-ray observations thus far have shown no clear example of the familiar three-part structure of CMEs as seen in white-light coronal images.

1. INTRODUCTION

The solar wind and disturbances within it originate in the solar corona. This paper reviews the recent increases in our knowledge of this region of space, emphasizing the new views of coronal transient phenomena made possible by the Soft X-ray Telescope (SXT) on *Yohkoh*. The coronal mass ejections (CMEs) constitute the largest and most energetic of these transient events. In our usage "coronal mass ejection" refers to the phenomena seen with a white-light coronal imaging instrument [*e.g., Hundhausen*, 1996].

The soft X-ray view of the solar corona differs substantially from the familiar view provided by white-light imagers. We therefore begin this review by discussing the differences. We then list various kinds of soft X-ray disturbances interpreted as ejecta, and conclude by describing the X-ray phenomena we believe to be most directly associated with the CME phenomenon. The original soft X-ray imaging observations from *Skylab* (see *Sturrock* [1980]) showed some of the same phenomena seen by *Yohkoh*, but not in such detail; see for example *Kahler* [1992] or *Webb* [1992] for reviews of the earlier data on CMEs.

The work of identifying X-ray coronal features with coronal mass ejections remains in its early stages. An initial survey of the *Yohkoh* data [*Klimchuk et al.*, 1994] showed that many X-ray eruptive events could be observed above the limb, like the *Skylab* event of *Rust and Hildner* [1978]. *Hiei et al.* [1993] described an X-ray "streamer re-formation" event that almost certainly involved a CME, and *Sime et al.* [1994] gave preliminary comparisons of two events observed by the HAO Mauna

Loa K-coronameter with *Yohkoh* observations. Beyond this, the current literature presents few examples of approximately simultaneous white-light and X-ray observations. There are several reports of associations between coronal X-ray events and either interplanetary observations or non-simultaneous coronagraph observations [*Lemen et al.*, 1996; *McAllister et al.*, 1996a; *Weiss et al.* 1996]. Data from the K-coronameter at Mauna Loa Solar Observatory and from the coronal instruments on board SOHO are already changing this situation. Because of the rapid pace of observational work, we have tried to include some literature published after 1996 (the Chapman Conference on CMEs), and to incorporate ideas from these observations in the discussion.

Finally, we note that CMEs play an important role in stimulating geomagnetic activity [*Gosling*, 1991; *Webb*, 1995]. *Hudson* [1997] has recently reviewed some of the same material presented here but with an emphasis on the application of the solar observations to geomagnetic storms. That paper presents some additional data, including a list of dimming events (see below) observed prior to 1996.

2. THE X-RAY VIEW OF THE CORONA

We normally image the solar corona in white light or in the coronal emission lines in the visible range. This can be done during eclipses or via coronagraphs, but the brightness of the solar disk restricts the observations to the region above the solar limb. To see the corona against the disk requires going to long (radio) or short (X-ray) wavelengths to reduce the photospheric competition. The lower temperature of the visible photosphere makes it appear dark in X-rays; at radio wavelengths the height of optical depth unity rises far up into the corona, so again the photosphere becomes invisible. Because most of the corona is hot, it has the highest contrast in EUV and soft X-radiation. For general information please refer to standard monographs, *e.g.*, *Kundu* [1964] or *Zirin* [1988]. In this section we briefly summarize coronal morphology as seen in soft X-rays, then make some comments about the specific limitations of the X-ray observations, with emphasis upon the *Yohkoh* soft X-ray telescope.

The X-ray spectrum of hot plasma consists of the emission lines and continua of highly ionized atoms and, at typical coronal temperatures (*i.e.*, 1-3×10^6 K, or below the temperatures achieved by flares), the emission lines dominate energetically. The complicated structure we see in the lower solar atmosphere in an X-ray image is defined by small-scale magnetic fields. These simplify with height and eventually map into the solar-wind flow, which has a predominantly bidirectional magnetic structure (with inward-pointing and outward-pointing sectors) dominated by the flow itself. Most of the magnetic field lines extending into the heliosphere originate in coronal holes, which thus provide the sources of the high-speed solar wind. Other areas of "open" magnetic fields (defined as field lines that extend beyond the Alfvénic critical point of the outward flow) exist elsewhere, for example in the hearts of active regions as evidenced by the occurrence of Type III radio bursts with high starting frequencies. The larger open-field regions are dark as viewed with a soft X-ray telescope. The slow component of the solar wind is less well understood, but probably originates in or near large closed-field regions of the solar atmosphere at lower latitudes [*e.g.*, *Withbroe et al.*, 1991].

In addition to the quasi-steady structure of the X-ray corona, there are transient events of many kinds. These usually take the form of injections of mass into apparently stable magnetic loop structures (microflares, flares, arcade events). Ejective events, well-observed in the sense that successive images actually show the motion of the structure, frequently occur, most often in association with flare-like brightenings. The ejected mass, along with its frozen-in magnetic fields, may become a part of the outward flow into the solar wind [*Uchida et al.*, 1992]. The soft X-ray emission rapidly drops with height, however, so that identifying soft X-ray ejecta with solar-wind features requires some interpolation.

Because in soft X-rays we see the general corona and transient features such as ejections, we have the possibility of observing CMEs directly in emission as they actually start. The *Yohkoh* observations represent the first soft X-ray imaging data suitable for these purposes since the *Skylab* obervations. *Yohkoh* began observations in September, 1991. A comparison of these improved X-ray data with those in white light (or other wavelengths), at the times of CME launches, should have high priority.

There are several substantial differences between soft X-ray observations of the *Yohkoh* type, and white-light observations. Most of our knowledge of CMEs comes from the white-light data, so it is important to understand the differences in the data reported here. The X-ray images show the entire Earth-facing hemisphere, offer a better view near the surface of the Sun, and provide some information on temperatures and densities. The fields of view are typically different, with a coronagraph viewing structures higher up in the corona ($\geq 1.2 R_S$ projected height) compared to those viewed near the surface ($\leq 1.5 R_S$) in X-rays. Another difference between the white light and X-ray observations is that of their line-of-sight dependences. The white light emission arises from Thomson scattering of photospheric radiation, but the X-ray emission arises directly from the emission of the hot coronal gas; X-radiation is isotropic and not concentrated in the plane of the sky. X-rays thus let us see the source of the emission against the disk and so determine the heliographic coordinates of any discrete source.

The white-light brightness (K-corona) varies in proportion to the electron density, and does not depend on the temperature. In contrast, the X-ray emission is a complicated function of density (approximately the square) and the temperature. For the thinnest *Skylab* X-ray filter, for example, *Kahler* [1976] found the signal to depend on the square of the electron pressure. The SXT signal S has a stronger dependence on temperature T ($d(\ln S)/d(\ln T) \sim 2\text{-}6$) than did *Skylab* and is biased towards shorter wavelengths. These different instrumental dependences mean that similar coronal features may look quite different when viewed in X-ray or white-light emissions.

Tsuneta et al. [1991] give a full description of the *Yohkoh* SXT instrument, which uses a grazing-incidence mirror, a set of broad-band filters sensitive in the range 0.3 - 3 keV (4 - 40 Å), and a CCD sensor with 1024×1024 pixels 2.45″ square (see *Acton et al.* [1992] for an survey of the data). The telemetry capacity of *Yohkoh* allows the transmission of about 21 whole-Sun images, with 2×2-pixel summations ($\sim 5''$) per 97-minute orbital period of the spacecraft. The *Yohkoh* SXT provides a global view of the corona within its total field of view, which is a square 0.70° across. A flare mode is normally triggered at about the GOES C2 level, some 2×10^{-3} ergs(cm^{2}sec)$^{-1}$ in the 2–8 Å (1.5-4 keV) band. This results in the loss of full-Sun imaging for an extended interval, in exchange for more telemetry devoted to high-resolution observations of the flare itself, with a maximum field of view 10′ square. The optical axis of the telescope is almost always pointed so that the entire disk is visible, except during special operations.

3. SXT OBSERVATIONS OF SOLAR MASS LOSS

The *Skylab*, *Solwind P78-1*, and *Solar Maximum Mission* observations generally showed a complicated relationship between the white-light CMEs they observed and the soft X-ray corona as inferred from the GOES photometry [*e.g., Kahler et al.*, 1992]. This situation could be attributable to poor temporal data coverage in the coronagraphs, to poor GOES X-ray sensitivity to the main bulk of the ejected coronal material, or of course to a weak physical relationship. With the improved data cadence and sensitivity of the *Yohkoh* X-ray images we hoped to find consistent signatures of CMEs in the low corona [*e.g., Klimchuk et al.*, 1994]. This turned out not to be so simple: some of the CME motions may be observed directly in X-rays, but the morphology can also be different. *Yohkoh* detects some kinds of coronal mass ejection not resembling CMEs at all. We summarize the ejection events in different categories below. While the categories may overlap, there are clearly different physical effects at work, as distinguished by the direction (parallel or perpendicular to the field) and speed of the motion. Parallel flows much slower than the Alfvén speed are common, suggesting hydrodynamic driving, while perpendicular flows provide good evidence for magnetic driving. Following this list we discuss "dimming", which we think of as one of the best soft X-ray signatures of the onset of a CME, in a separate section.

3.1 Expanding active-region loops

One of the first discoveries in the new X-ray data was the tendency for some active-region loops to expand at intermediate speeds (10-50 km s^{-1}), rather than remain static [*Uchida et al.*, 1992]. This observation suggests that magnetically-driven outward flows from active regions may contribute to the global coronal structure or even to the slow, dense component of the solar wind [*e.g., Hick et al.*, 1995]. Such a mechanism may also be of general interest for stellar mass loss.

3.2 Soft X-ray jets

Highly collimated jets, of various types, occur frequently in the *Yohkoh* soft X-ray observations [*Shibata et al.*, 1992; *Strong et al.*, 1992]. These appear to be essentially hydrodynamically driven flows along the large-scale magnetic field, and are strongly linked to flare-like effects near their feet. *Yokoyama and Shibata* [1995] suggest magnetic reconnection in an emerging-flux scenario as the key physical mechanism.

We believe that at least some of the magnetic fields involved are unipolar and open to the heliosphere because of the identification of the jets with meter-wave Type III bursts [*Aurass et al.*, 1994; *Kundu et al.*, 1995; *Raulin et al.*, 1996]. Closed-field structures also support similar behavior, including meter-wave U-bursts [*Pick et al.*, 1994], "two-sided loop jets" [*Shimojo et al.*, 1996], and jets that re-enter the chromosphere at large (0.5 solar radii) distances from their point of origin [*Strong et al.*, 1992].

3.3 Flare ejecta

Direct *Yohkoh* soft X-ray imaging of flares may show motion, mostly outwards. These motions are distinguishable (as image displacements) from the presumably field-aligned flows associated with "evaporation" detected mainly in soft X-ray emission-line blue shifts [*e.g., Acton et al.*, 1982]. Compact flare ejecta range from the relatively slow (<500 km s^{-1}) outward motion of compact blobs, to faster outward motions during some flares (>500 km s^{-1}). The flare-related flows [*Shibata et al.*, 1995] in the soft X-ray observations may take various forms, as do those at lower temperatures (surges and sprays). We distinguish these from jets (above) but do not really know yet whether this distinction is physically justified. Certainly, many flare ejecta are not jet-like from the point of view of collimation and velocity.

3.4 CME-like ejecta at the limb

Klimchuk et al. [1994], in an early study of the *Yohkoh* data based upon the standard movie images, found expanding features at the limb which had similar parameters (widths, speeds and occurrence rates) as those of CMEs observed in coronagraphs (see also *Sime et al.* [1994]; *Hundhausen* [1996]). Recently *Gopalswamy et al.* [1996] have described *Yohkoh* observations of a slow ejection at the limb on 10-11 July 1993 that appeared to incorporate the three elements of a "classic" CME – front, cavity and embedded filament. In this case the filament was well-observed with the Nobeyama 17 GHz radioheliograph, and the front clearly was a slowly-rising magnetic structure. The mass of this event was estimated as 1.2×10^{14} g, at the low end of the range of masses of white-light CMEs [*Hundhausen et al.*, 1994], but not very different from the mass found by *Hudson et al.* [1996b] for a different *Yohkoh* SXT mass-ejection event. It has not yet been possible to do a thorough calibration of the *Yohkoh* data set against coronagraph data, but there are many common event periods with the Mauna Loa K-coronameter observations (A. Hundhausen, personal communication 1995; also see below).

3.5 Filament eruptions

The behavior of Hα filaments has always provided one of the best guides to the occurrence of a solar eruptive event (the *disparition brusque*). *Skylab* data showed the X-ray coronal structure of the channel in which the filament forms and the bright, long-duration loop arcade which appears late in the event. The new X-ray data show many beautiful examples of arcades of this type [*Alexander et al.*, 1994; *Hanaoka et al.*, 1994; *Khan et al.*, 1994; *Lemen et al.*, 1996; *McAllister et al.*, 1992; 1996a; *Tsuneta et al.*, 1992; *Watanabe et al.*, 1992; 1994; *Watari et al.*, 1996]; these observations are bringing the puzzling relationship between the filament and the CME into sharper focus. The sense of chirality of the filament, in such cases, appears to be related to that of the magnetic cloud resulting from the eruption [*Bothmer and Schwenn*, 1994; *Rust*, 1994; *Bothmer and Rust*, this volume]. An important new link in this chain is presented by *Martin and McAllister* [this volume]: the chirality of an erupting filament bears a fixed relationship to the skew of the resulting X-ray arcade. These insights offer encouragement that the pre-event coronal configuration can eventually be linked to the geoeffectiveness of the event, since this depends upon the field orientation and flow properties.

4. THE "DIMMING SIGNATURE"

At times and locations expected for CME launches, *i.e.* near an LDE flare or large arcade event, a large volume of the soft X-ray corona may rapidly become significantly dimmer. Coronal depletions were first described using HAO K-coronameter data by *Hansen et al.*, [1974], and corresponding X-ray effects (in the *Skylab* data) by *Rust and Hildner* [1978]. *Rust* [1983] describes the dimming effect as viewed against the disk as a "transient coronal hole". The *Skylab* observations, however, were limited in sampling and photometry and not well optimized for detecting such effects. The *Yohkoh* SXT data also have sampling limitations; with more frequent images and better image dynamic range the velocity field could have been measured more easily. The new results are nevertheless extensive, and will be described in more detail below.

In some of the cases studied thus far, the dimming appears to be amorphous and unstructured. It results in a decrease of the coronal surface brightness directly above the accompanying brightening. In other cases a structured mass flows outward from the region that dims. The dimming or outward mass flow can occur either above or near the brightening, or can be widespread in the vicinity of the brightening. *Hudson et al.* [1996b] point out that the radiative cooling time for such temperatures and spatial scales greatly exceeds the characteristic time scale of the dimming, consistent with the interpretation in terms of material ejection, assuming unity filling factor in the estimate of source volume or density. The strong temperature dependence of the soft X-ray signal noted above suggests that adiabatic cooling upon expansion will result in a rapid brightness decrease with height, as is observed during the outward motion.

Large-scale clouds adjacent to the sites of solar flares have been directly observed to move outwards and disappear during the flare brightening. Large-scale X-ray clouds were also observed during Skylab, but not to disappear rapidly [*Rust and Webb*, 1978]. In the best-studied SXT event, on 13 November 1994, the coronal cloud moved outward in a direction consistent with radial and a projected (constant) velocity consistent with the range expected from a CME [*Hudson et al.*, 1996b]. Other excellent examples of disappearing clouds are the events on 27 February 1994 and 6 February 1995. Some large-scale flare ejecta [*e.g., Manoharan et al.*, 1996; R. L. Moore *et al.* (poster paper, Chapman Conference on CMEs, 1996)] show two-lobed structures in the pattern described by *Moore and LaBonte* [1979], strongly suggesting non-vertical motions. In many of these cases the outward motion takes the form of a succession of large-scale loops, each physically moving outwards as established by the image continuity from frame to frame.

The first *Yohkoh* SXT event directly associated with white light observations, on 23-24 January 1992, consisted of a streamer disruption followed by a re-formation observed by the Mauna Loa K-coronameter [*Hiei et al.*, 1993]. This event also provides an excellent example of coronal dimming [*Hudson*, 1996], allowing a determination of the probable time of the CME launch (the bulk of the dimming occurred between 08:00 and 11:00 UT, consistent with the non-observation of a CME at Mauna Loa because of local night). The event occurred just at the limb, and the coronal dimming occurred both above the location of the arcade formation and to either side. The dimming region thus appeared to envelop the region that brightened. In some other large arcade events, the standard *Yohkoh* movie clearly shows the dimming to occur on a large scale. The event of 24 February 1993, for example, appears to unmask a large region of the south polar coronal hole and also to be in the "enveloping" category (see below) [*Harvey et al.* 1996]. However, dimming is not obvious in all large X-ray events.

We can measure the mass of such a dimming region, especially if it has the form of a discrete cloud. For the ejected cloud of 13 November 1994, we derive a lower limit of 4×10^{14}g. Only a lower limit is possible because of confusion with the brighter parts of the flare, the lack of complete knowledge of the differential emission-measure distribution, and the theoretical problem of estimating any replenishing mass coming from the deeper atmosphere. This latter difficulty results from the continuous nature of the outward flow during the interval of flare brightening, and from the contrast problem (the difficulty of detecting faint features near bright ones). For the 24 January 1992 large-scale event, a rough mass estimate is about 10^{15}g. These estimates are consistent with the range of white light CME masses, which are 10^{14} to 10^{16} g [*Hundhausen et al.*, 1994].

For events in which we can see the ejected cloud of material actually departing from the low corona (*e.g.*, on 21 February 1992, 28 August 1992, 27 February 1994

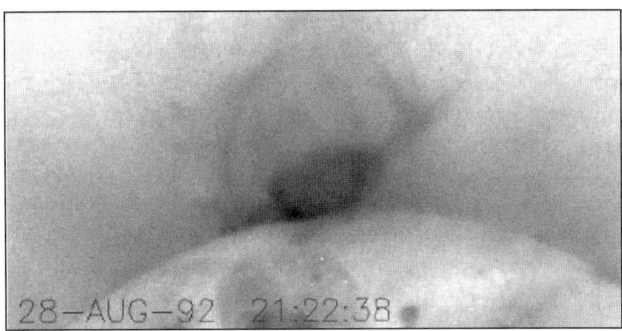
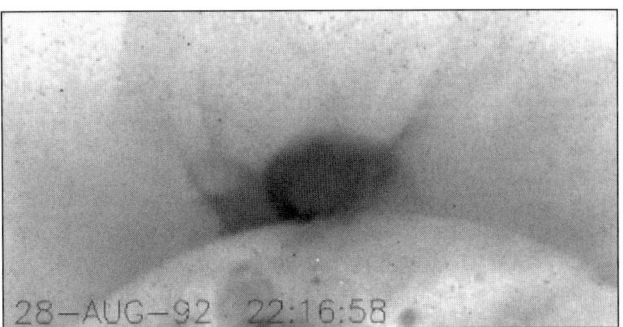
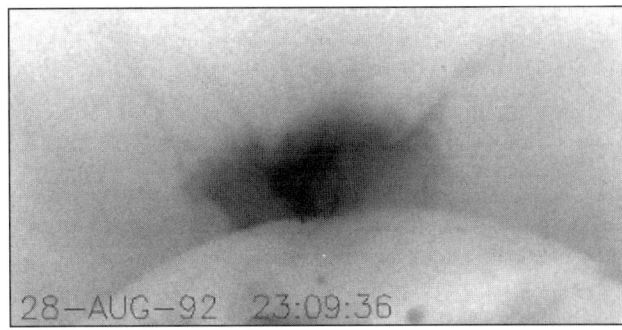

Figure 1. An "LDE flare dimming" in a limb event of 28 August 1992. This and the subsequent figures have reversed colors and have north at the top, west to the right. The top frames show the filamentary structure that had formed above the flare loops prior to the flare. This filamentary structure rises as the diffuse corona in the same region dims, almost simultaneously with the flare brightening. *Hudson and Khan* [1996] show difference images, demonstrating the outward motion of the filamentary structure. As mentioned in the text, these hot filaments appeared to wrap around a cool filament seen in Hα (*Ta. Watanabe*, personal communication).

and 13 November 1994) we can in principle determine the flow field of the ejected material. Difference images show that the flow tends to be everywhere outwards, with no obvious trace of the inward flow that might be expected if large-scale magnetic reconnection were the

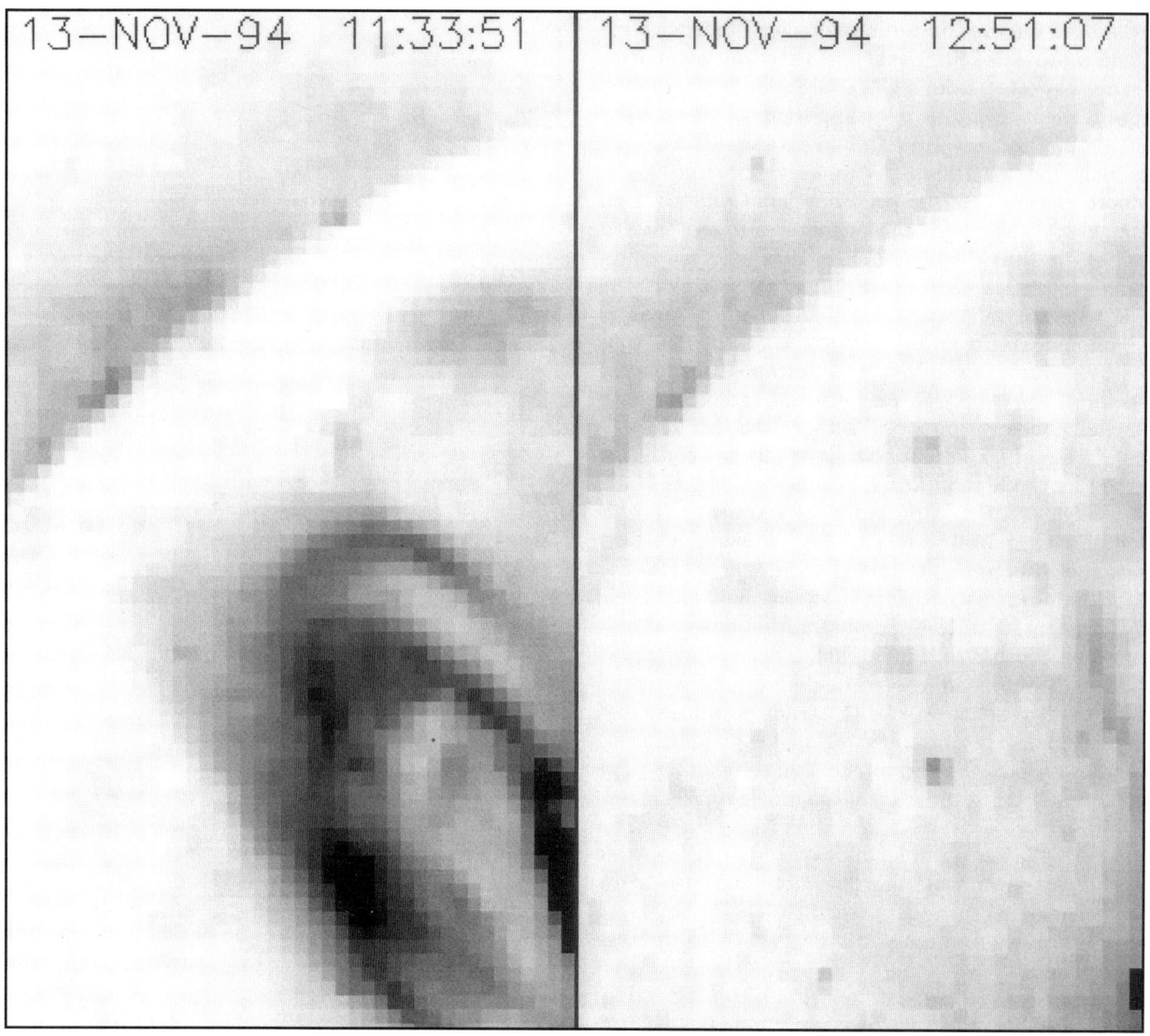

Figure 2. A "cloud dimming" associated with the long-duration flare of 13 November 1994 [*Hudson et al.*, 1996b]. The flare proper is outside the field of view of the two images to the SW (lower right). Prior to the left image, the structure seen had been rising steadily as the flare brightened, and in the interval shown (about an hour) it disappeared almost completely except for the dimly-seen legs.

source of the flare energy [*Hudson et al.*, 1996c]. T. Watanabe *et al.* (poster paper, Chapman Conference on CMEs, 1996) presented an Hα image of the 28 August 1992 event locating an erupting filament within the X-ray cloud whose outward motion constitutes the dimming. This suggests an identification of the X-ray dimming volume with the void component of the classical 3-part visible CME.

5. CLASSIFICATION OF X-RAY DIMMING SIGNATURES

The X-ray dimming phenomena most probably associated with CME launching have a variety of morphological properties. We discuss here a preliminary classification scheme for such events, with main emphasis on the description of the dimming signature. The formation of a long-enduring arcade of hot loops occurs in each case; statistically such long-duration X-ray events are well associated with white-light CME occurrence [*e.g., Sheeley et al.*, 1983]. The dimming signature is of fundamental importance because it probably represents at least some of the expelled mass of the CME itself. Not all arcade events with filament eruptions or other good CME proxy signatures show clear dimming signatures. We do not know at present if this is due to the detection biases intrinsic to the X-ray observations, but we suspect so and suggest that appropriate coronal observations will always show a depletion or dimming of the corona at the time of a CME. Figures 1-4 shows representative examples of each of the four types of dimming event.

Dimming above an LDE flare (Figure 1). There are several examples of events in which the dimming signature appears over a well-defined volume *above* the arcade that is forming: 21 February 1992 [*Tsuneta*, 1996] and 28 August 1992 are the prototype events *Hudson et al.* [1996c]. Such events must be observed at the limb.

Cloud ejections (Figure 2). In other flare-associated events, a well-defined X-ray coronal cloud adjacent to the flare moves away from the flare region and disappears. Events in this category include 27 February 1994 [*Hudson et al.*, 1996a] and 13 November 1994 [*Hudson et al.*, 1996b]. The observations suggest a large-scale twist (approximately one turn) in each cloud. The double-lobed ejecta of the 25 October 1994 event [*Manoharan et al.*, 1996] may also fit in this category.

Enveloping dimmings (Figure 3). We identify these as "streamer blowout" CME events seen in white light [*Howard et al.* 1985; *Hundhausen*, 1993]. The classic example of this in the *Yohkoh* data is the event of 24 January 1992 [*Hiei et al.*, 1993; *Hudson*, 1996].

Transient coronal holes (Figure 4). The diffuse corona near an arcade development on the disk occasionally dims at or near the time of the X-ray brightening, strongly suggesting the formation of a new area of open field lines. These areas are not permanent, gradually filling in after several hours or a day or so. The term "coronal hole" is used because in X-rays the brightness of these areas can decrease to approximately the level of the larger and more permanent coronal holes.

6. ASSOCIATION OF X-RAY AND WHITE-LIGHT SIGNATURES

The comparison of the new X-ray signatures of coronal ejections must be understood in the context of

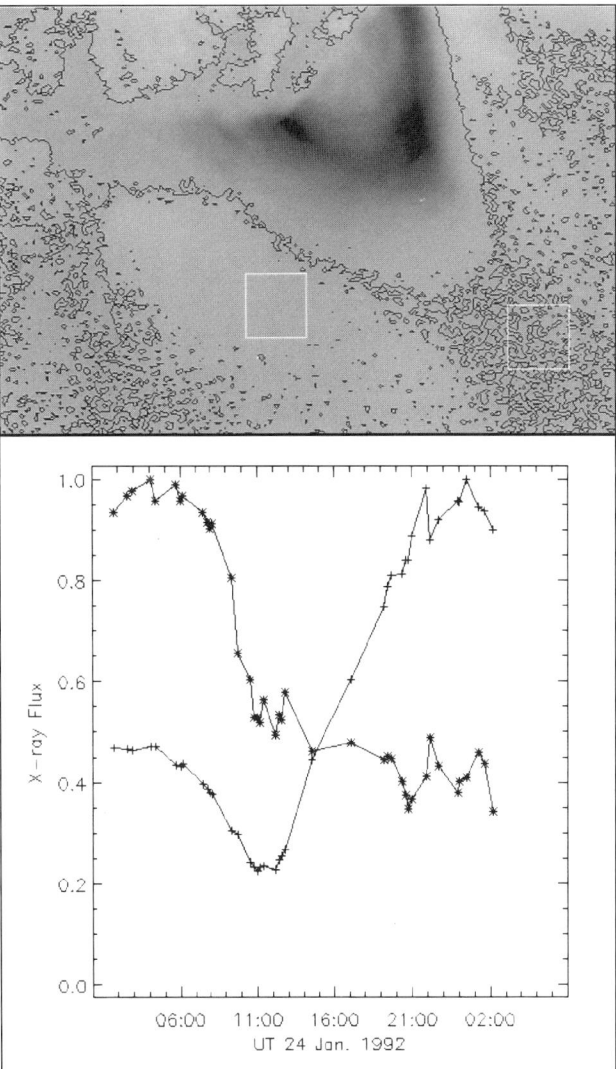

Figure 3. An "enveloping" dimming event, that of 24 January 1992 [*Hiei et al.*, 1993; *Hudson*, 1996]. The plot at the bottom shows light curves from two regions of the corona, one (*) to the S of the developing arcade, and one (+) at its brightest point at about 12:00 UT. The image at the top shows the difference of two images (14:33 minus 06:01), with the zero contour overlaid; the boxes show the locations integrated to generate the light curves. Note that the light curve from the cusp region shows an initial dimming, followed by an increase as the tip of the bright cusp enters the integration box. The voids left by the blowout are the regions S and N of the arcade, plus the areas at the top of the image on the disk.

white-light coronagraph data. Since the demise of SMM in late 1989 there have been no spaceborne coronagraph instruments, until the current era of SOHO. Thus, during most of the lifetime (to date) of *Yohkoh*, white light

34 SOFT X-RAY SIGNATURES

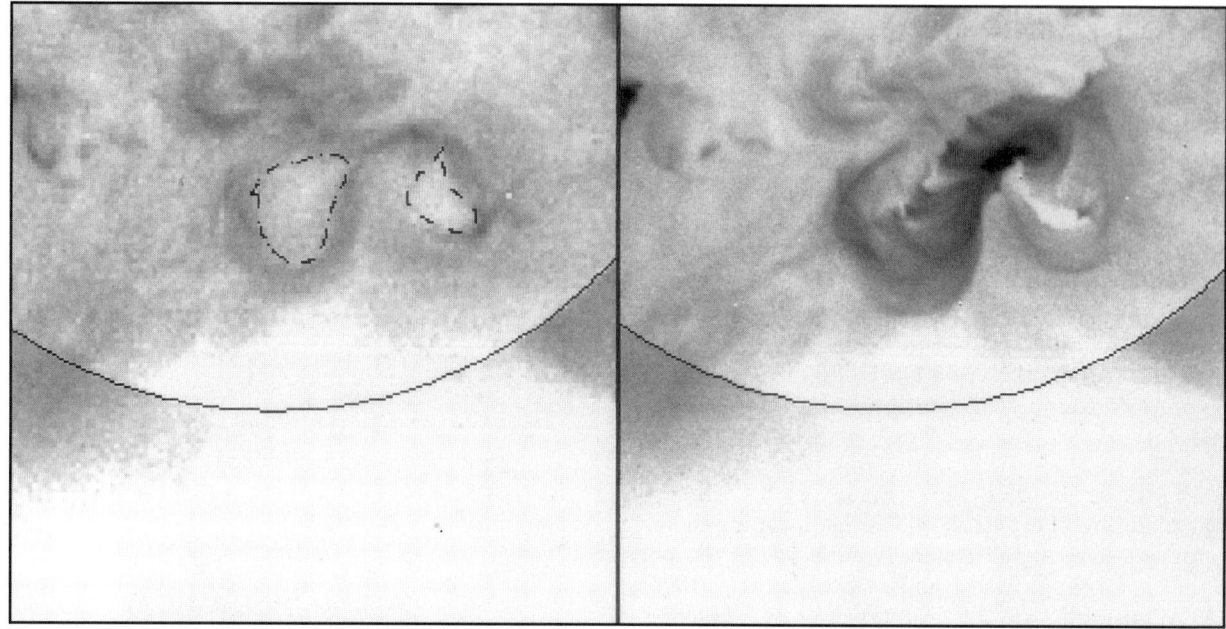

Figure 4. Example of transient coronal holes. The image at left (16 January 1993, 12:46:23 UT) shows two regions (enclosed within dashed lines) that dimmed suddenly. These regions fit within the S-shaped bright structure that evolves with time to become the bright arcade seen at right (14:58:23 UT). The dimmed region does not quite reach the darkness level of the south polar hole.

CME observations were only available from the ground-based Mauna Loa K-coronameter with the usual problems of day/night cycles (a typical observing day of ~5 hr.) and weather (including the airborne dust effects of Mt. Pinatubo).

D. Webb and H. Hudson (poster paper, Chapman Conference on CMEs, 1996) reported on a preliminary examination of the SXT data for transient X-ray features (involving brightenings and outward motion) near the appropriate limb location occurring before and during periods when white light CMEs were observed at Mauna Loa. We found that nearly 2/3 of the CMEs were associated with a transient X-ray structure, usually a loop, and a majority of these loops had at least one foot in a flaring active region. Consistent with previous results, the X-ray feature typically did not lie symmetrically underneath the CME.

The other major new source of white-light data is the LASCO suite of instruments on SOHO. These data represent a significant advancement over previous white-light coronal observations [see *Howard et al.*, this volume]. Although LASCO observes transients resembling the classical events observed by the *Skylab*, SOLWIND and SMM coronagraphs, it has better sensitivity and a larger field of view than these instruments, enabling it to observe other kinds of outward flows that may also be identifiable in the soft X-ray observations. At this time, no study comparing flows or ejections observed by SOHO at visual wavelengths and by *Yohkoh* in soft X-rays has been carried out. However we have made preliminary surveys which show that there have been many events observed in common, both with LASCO (*C. St. Cyr*, private communication 1996) and also with other SOHO instruments.

7. TIMING AND CAUSALITY

The physical processes involved in launching a CME remain poorly understood, so the X-ray view of the behavior of the lower corona during well-defined CMEs is of great interest. In cases where a large-scale flow can actually be observed in soft X-rays, we can learn about the geometry of the ejection process and the origin of the ejected mass. Even in the more common cases where only a dimming of diffuse coronal material can be detected, the relative timing of the X-ray dimming signature and the X-ray brightening might point towards the direction of causality. Standard models of large-scale magnetic reconnection suggest that post-flare loop arcades might occur after some delay relative to the CME onset. This indeed appeared to be the case from earlier soft X-ray observations with less sensitivity than the *Yohkoh* observations, *e.g., Hundhausen* [1996], but the *Yohkoh* observations indicate that the flare-related arcade brightening may occur with little or no delay [*Hudson*, 1997].

The filament eruption is well-known to begin, by activation including turbulent motion and a slow rise, well before the main part of the X-ray arcade development. Recent data confirm this pattern well (*Hanaoka et al.*, 1994; *Khan et al.*, 1994; *McAllister et al.*, 1996b]) and suggest that the main arcade development takes place when the filament has risen to about 10 times the width of the arcade.

The *Yohkoh* observations show a wide variety of soft X-ray loop arcades, extending this type of observation beyond that found with *Skylab* and *Solar Maximum Mission*. They confirm the *SMM* observation of "giant arches", which may be morphologically distinct from the post-flare loops [*Švestka et al.*, 1996]. Either kind of loop system – or perhaps both or neither – may be identifiable with the large-scale reconnection scenario (see *Tsuneta* [1996], for a positive view of such a picture, or *Hudson and Khan* [1997], for a skeptical view). *Klimchuk* [1996] comments on the absence of reconnection signatures in CME development, but *Kahler and Hundhausen* [1992] point out that the "legs" of CMEs may in fact consist of multiple cusp (bipolar) structures.

8. CONCLUSIONS

Soft X-ray imaging of the solar corona reveals several forms of mass ejecta, some of which are new with the *Yohkoh* SXT observations. Compared with the traditional coronagraph or K-coronameter observations, the *Yohkoh* data have better sensitivity and sampling, and view the entire visible hemisphere. Events probably associated with CMEs often show clearly measurable dimmings of the X-ray corona near the site of a flare or arcade brightening. We interpret the X-ray dimmings as the expansion and opening of magnetic field lines during the early phase of a CME. Transient coronal holes, dimmings seen against the disk, usually appear at the same time or later than the first arcade brightening and are skewed relative to its center, suggesting that they mark the evacuated feet of the flux ropes of the rising CME. In general the main arcade brightening follows the mass ejection and the impulsive phase of any associated flare ("impulsive" here means non-thermal energy release as detected in hard X-ray bremsstrahlung; *Hudson et al.* [1994] show that this occurs even in slowly-rising arcade events).

In some of the cases that have been studied, the *Yohkoh* SXT observations show details of the origin of the ejected material. For example, the 13 November 1994 cloud event has the appearance of a large structure with approximately one full twist ($\sim 2\pi$); the structure appears to be anchored at one end in a flaring active region [*Hudson et al.*, 1996b]. The 21 February 1992 event and others, on the contrary, appear to dim only above the developing arcade, even when viewed from apparently different perspectives relative to the arcade axis. Finally some of the large-scale arcade events appear to show large-scale dimming both at and remote from the arcade location. These different dimming signatures suggest that there may be a variety of physical processes involved.

The relationship between CMEs and flares or flare-like brightenings also does not seem now so simple as previously thought; it now appears in many cases that there is no appreciable delay between the launching of mass and the associated flare brightening [*e.g., Hudson* 1997]. This is consistent with recent discussions of this relationship by *Feynman and Hundhausen* [1994] and *Harrison* [1995]. Our comparison of SXT and Mauna Loa data better elucidates the physical interpretation of CME onsets [Webb and Hudson, poster paper, Chapman Conference on CMEs, 1996]. Under the CME one can usually find a brightening and expanding X-ray structure, which is typically looplike with one end embedded in a flaring active region. On the other hand, such expanding loops are not usually associated with the Mauna Loa CMEs, despite having speeds, widths and occurrence rates similar to those of CMEs [*Klimchuk et al.*, 1994]. Thus, we conclude that the dimming effect seems a more consistent X-ray signature of CME *onset*, if not occurrence.

Do flares and CMEs divide naturally into two classes of events? We note that no single parameter of a solar flare or arcade disturbance has been reported to exhibit actual bimodality, *i.e.* a distribution function with two resolved maxima. All parameters seem to have broad or unimodal distributions, suggesting that flares and CMEs form a continuum with the same underlying physics. On the other hand several properties of the interplanetary counterparts of CMEs clearly have bimodality (see *Reames* [1994] and references therein).

Phenomena observable in the low corona, on the solar disk, in principle give us our earliest possible hint that a CME has been launched and might strike the Earth. We can hope that further understanding of these phenomena may even help us to predict the terrestrial consequences ("space weather"). Unfortunately we currently have no firm theoretical basis for such predictions, but we can hope that further empirical understanding will also help us to understand the basic physics.

Acknowledgments. This work was supported under NASA contract NAS 8-37334 (HSH) and under NASA contract NAGW-4578 and Air Force contract AF19628-96-K-0030 (DFW). *Yohkoh* is a mission of the Institute of Space and Astronautical Sciences (Japan), with participation from the U. S. and U. K. We would like to thank A. Hundhausen and E. Hiei for pointing out the 16 January 1993 event, and C. St. Cyr for maintaining a list of CME events observed on SOHO. E. Hildner pointed out the possible role of adiabatic cooling in the soft X-ray observations of CME counterpart events during discussion at the Chapman Conference on CMEs, 1996.

REFERENCES

Acton, L. W., R. C. Canfield, T. A. Gunkler, H. S. Hudson, A. L. Kiplinger, and J. W. Leibacher, Chromospheric evaporation in a well-observed compact flare, *Ap. J.*, **263**, 409-422, 1982.

Acton, L. W., S. Tsuneta, Y. Ogawara, R. Bentley, M. Bruner, R. Canfield, L. Culhane, G. Doschek, E. Hiei, and T. Hirayama, The *Yohkoh* mission for high-energy solar physics, *Science*, **258**, 618-621, 1992.

Alexander, D., H. S. Hudson, G. Slater, A. McAllister, and K. Harvey, The large-scale coronal eruptive event of April 14, 1994, in *Solar Dynamic Phenomena and Solar Wind Consequences: Proceedings of the Third SOHO Workshop*, ESA SP-373, 187-190, 1994.

Aurass, H., K.-L. Klein, P. C. H. Martens, First detection of correlated electron beams and plasma jets in radio and soft X-ray data *Solar Phys. Lett.*, **155**, 203-206, 1994.

Bothmer, V., and D. M. Rust, The field configuration of magnetic clouds and the solar cycle, this volume.

Bothmer, V., and R. Schwenn, Eruptive prominences as sources of magnetic clouds in the solar wind, *Space Sci. Revs.*, **70**, 215-220, 1994.

Feynman, J., and A. J. Hundhausen, Coronal mass ejections and major solar flares: The great active center of March 1989, *J. Geophys. Res.*, **99**, 8451-8464, 1994.

Gopalswamy, N., Y. Hanaoka, M. R. Kundu, S. Enome, J. R. Lemen, and M. Akioka, and A. Lara, Radio and X-ray studies of a CME associated with a very slow prominence eruption, *Ap. J.*, submitted, 1996.

Gosling, J. T., D. J. McComas, J. L. Phillips, and S. J. Bame, Geomagnetic activity associated with Earth passage of interplanetary shock disturbances and coronal mass ejections, *J. Geophys. Res.*, **96**, 7831-7839, 1991.

Gosling, J. T., Corotating and transient solar wind flows in three dimensions, *Annu. Rev. Astron. Astrophys.*, **24**, 35, 1996.

Hanaoka, Y., H. Kurokawa, S. Enome, H. Nakajima, K. Shibasaki, M. Nishio, T. Takano, C. Torii, H. Sekiguchi, and S. Kawashima, Simultaneous observations of a prominence eruption followed by a coronal arcade formation in radio, soft X-rays, and Hα, *Publ. Astr. Soc. Japan*, **46**, 205-216, 1994.

Hansen, R. T., C. G. Garcia,., S. F. Hansen, and E. Yasukawa, Abrupt depletions of the inner corona, *Publ. Astr. Soc. Pacific*, **86**, 500-515, 1974.

Harrison, R., The nature of solar flares associated with coronal mass ejection, *Astr. Astrophys.*, **304**, 585-595, 1995.

Harvey, K., A. McAllister, H. Hudson, D. Alexander, J. R. Lemen, and H. P. Jones, Comparison and relation of HeI 1083 nm two-ribbon flares and large-scale coronal arcades observed by *Yohkoh*, in *Solar Drivers of Interplanetary and Terrestrial Disturbances*, eds. K. S. Balasubramaniam *et al.*, ASP Conference Ser. Vol. 95, pp. 100-107, San Francisco, 1996.

Hiei, E., A., Hundhausen, and D. Sime, Reformation of a coronal helmet streamer by magnetic reconnection after a coronal mass ejection, *Geophys. Res. Lett.*, **20**, 2785-2788, 1993.

Hick, P., B. V. Jackson, G. Woan, G. Slater, K. Strong, and Y. Uchida, Synoptic IPS and Yohkoh soft X-ray observations, *Geophys. Res. Lett.*, **22**, 643-646, 1995.

Howard, R. A., N. R. Sheeley, Jr., M. J. Koomen, and D. J. Michels, Coronal mass ejections: 1979-1981, *J. Geophys. Res.*, **90**, 8173-8191, 1985.

Howard, R. A., *et al.*, Observations of CMEs from SOHO LASCO, this volume.

Hudson, H. S., *Yohkoh* observations of coronal mass ejections, in *Proc. I.A.U. Colloq. 153*, in press, 1996.

Hudson, H. S., The solar antecedents of geomagnetic storms, in *Magnetic Storms*, edited by B. T. Tsurutani, W. D. Gonzales, and Y. Kamide, in press, American Geophysical Union, Washington, D. C., 1997.

Hudson, H. S., L. W. Acton, A. S., Sterling, S. Tsuneta, J. Fishman, C. Meegan, W. Paciesas, and R. Wilson, Nonthermal effects in slow solar flares, in *X-ray Solar Physics from Yohkoh*, eds. Y. Uchida *et al.*, Universal Academy Press, Tokyo, pp. 143-146, 1994.

Hudson, H. S., L. W. Acton, D. Alexander, S. L. Freeland, J. R. Lemen, K. L. Harvey, 1996, *Yohkoh*/SXT observations of sudden mass loss from the solar corona, *Solar Wind Eight*, eds. R. Winterhalter *et al.*, AIP, Woodbury NY, 84-87, 1996a.

Hudson, H. S., L. W. Acton, and S. L. Freeland, A long-duration solar flare with mass ejection and global consequences, *Ap. J.*, **470**, 629-635, 1996b.

Hudson, H. S., J. R. Lemen, and D. F. Webb, Coronal dimming in two limb flares, *Proc. of the Yohkoh Conference on Observations of Magnetic Reconnection in the Solar Atmosphere*, eds. R. D. Bentley and J. T. Mariska, ASP Conference Series, 111, 379-382, 1996c.

Hudson, H. S., and Khan, J. I., Problems with standard reconnection models of solar flares, *Proc. of the Yohkoh Conference on Observations of Magnetic Reconnection in the Solar Atmosphere*, eds. R. D. Bentley and J. T. Mariska, ASP Conference Series, 111, 135-144, 1996.

Hundhausen, A. J., Sizes and locations of coronal mass ejections - SMM observations from 1980 and 1984-1989, *Geophys. Res. Lett.*, **98** 13,177-13,200, 1993.

Hundhausen, A. J., Coronal mass ejections, in *Cosmic Winds and the Heliosphere*, eds. J.R. Jokipii *et al.*, Arizona, Tucson, in press, 1996.

Hundhausen, A. J., A. L. Stanger, and S. A. Serbicki, Mass and energy contents of coronal mass ejections: SMM results from 1980 and 1984-1989, in *Solar Dynamical Phenomena and Solar Wind Consequences*, ESA SP-373, pp. 409-412, 1994.

Kahler, S. W., Determination of the energy or pressure of a solar X-ray structure using X-ray filtergrams from a single filter, *Solar Phys.*, **48**, 255, 1976.

Kahler, S. W., Solar flares and coronal mass ejections, *Annu. Revs. Astron. Astrophys.* **30**, 113-141, 1992.

Kahler, S. W., and A. J. Hundhausen, A. J., The magnetic topology of solar coronal structures following mass ejections, *J. Geophys. Res.*, **97**, 1619, 1992

Khan, J. I., Y. Uchida, A. H. McAllister, and Ta. Watanabe, *Yohkoh* soft X-ray observations related to a prominence eruption and arcade flare on 7 May 1992, in *X-ray Solar Physics from Yohkoh*, eds. Y. Uchida, T. Watanabe, K. Shibata, and H. Hudson, Universal Academy Press, Tokyo, 201-204, 1994.

Klimchuk, J. A., Post-eruption arcades and 3-D magnetic reconnection, in *Magnetic reconnection in the solar atmosphere, Proc. of the Yohkoh Conference on Observations of Magnetic Reconnection in the Solar Atmosphere*, eds. R. D. Bentley and J. T. Mariska, ASP Conference Series, 111, 319-322, 1996.

Klimchuk, J. A., L. W. Acton, K. L. Harvey, H. S. Hudson, K. L. Kluge, D. G. Sime, K. T. Strong, and Ta. Watanabe, Coronal eruptions observed by *Yohkoh*, in Y. Uchida *et al.* (eds.), *X-ray Solar Physics from Yohkoh*, Tokyo, Universal Academy Press, 181-186, 1994.

Kundu, M. R., *Solar Radio Astronomy*, New York, Academic, 1964.

Kundu, M. R., J. P. Raulin, N. Nitta, H. S. Hudson, M. Shimojo, K. Shibata, and A. Raoult, Detection of nonthermal radio emission from coronal X-ray jets, *Ap. J.*, **447**, L135-L139, 1995.

Lemen, J. R., L. W. Acton, D. Alexander, A. B. Galvin, K. L. Harvey, J. T. Hoeksema, X. Zhao, and H. S. Hudson, Solar identification of solar-wind disturbances observed at *Ulysses*, in *Solar Wind Eight*, eds. R. Winterhalter *et al.*, AIP, Woodbury NY, 92-95, 1996.

Manoharan, P. K., L. van Driel-Gesztelyi, M. Pick, and P. Demoulin, Evidence for large-scale magnetic reconnection obtained from radio and X-ray measurements, *Ap. J.*, **468**, L73-L76, 1996.

Martin, S., and A. McAllister, Predicting the sign of helicity in erupting filaments and coronal mass ejections, this volume.

McAllister, A., Y. Uchida, S. Tsuneta, K. Strong, L. Acton, E. Hiei, M. Brunner, Ta. Watanabe, and K. Shibata, The structure of the coronal soft X-ray source associated with the dark filament disappearence of September 28, 1991 using the *Yohkoh* soft X-ray telescope, *Publ. Astr. Soc. Japan*, **44**, L205-L208, 1992.

McAllister, A., M. Dryer, P. McIntosh, H. Singer and L. Weiss, A large polar crown CME and a "problem" geomagnetic storm: April 14-23, 1994, *J. Geophys. Res.*, **101**, 13,497-13,511, 1996a.

McAllister, A. H., H. Kurokawa, K. Shibata, and N. Nitta, A filament eruption on november 5, 1992: comparison of SXT and Hα, *Solar Phys.*, **169**, 123-149, 1996b.

Moore, R. L., and B. LaBonte, The filament eruption in the 3B flare of July 29, 1973 - Onset and magnetic field configuration, in *Proc. Symposium on Solar and Interplanetary Dynamics*, Reidel, Dordrecht, pp. 207-210, 1979.

Pick, M., A. Raoult, G. Trottet, N. Vilmer, K. Strong, and A. Magalhaes, Energetic electrons and magnetic field structures in the corona, in S. Enome and T. Hirayama (eds.), *Proceedings of Kofu Symposium*, NRO Report No. 360, 263-266, 1994.

Raulin, J.-P, M. R. Kundu, H. S. Hudson, N. Nitta, and A. Raoult, Metric type III bursts associated with soft X-ray jets, *Astr. Astrophys.* **306**, 299-307, 1996.

Reames, D. V., Acceleration of energetic particles which accompany coronal mass ejections, *Proc Third SOHO Workshop*, edited by A. Poland, ESA SP-373, 107, 1994.

Rust, D. M., Coronal disturbances and their terrestrial effects, *Space Science Reviews*, **34**, 21-36, 1983.

Rust, D. M., Spawning and shedding helical magnetic fields in the solar atmosphere, *Solar Phys.*, **21**, 241-244, 1994.

Rust, D.M., and E. Hildner, Expansion of an X-ray coronal arch into the outer corona, *Solar Phys.* **48**, 381-387, 1978.

Rust, D. M., and D. F. Webb, Soft X-ray observations of large-scale sctive region brightenings, *Solar Phys.* **54**, 403, 1977.

Sheeley, N. R., Jr., R. A. Howard, M. J. Koomen, and D. J. Michels, Associations between coronal mass ejections and soft X-ray events, *Ap. J.*, **272**, 349, 1983.

Shibata, K., Y. Ishido, L.W. Acton, K.T. Strong, T. Hirayama, Y. Uchida, A.H. McAllister, R. Matsumoto, S. Tsuneta, T. Shimizu, H. Hara, T. Sakurai, K. Ichimoto, Y. Nishino, and Y. Ogawara, Y., Observations of X-ray jets with the *Yohkoh* soft X-ray telescope, *Publ. Astr. Soc. Japan*, **44**, L173–L180, 1992.

Shibata, K., S. Masuda, M. Shimojo, H. Hara, T. Yokoyama, S. Tsuneta, T. Kosugi, and Y. Ogawara, Hot plasma ejections associated with compact-loop solar flares, *Ap. J.*, **451**, L83-L85, 1995.

Shimojo, M., S. Hashimoto, K. Shibata, T. Hirayama, H. S. Hudson, and L. W. Acton, Statistical study of solar X-ray jets observed with the Yohkoh Soft X-ray telescope, *Pub. Astr. Soc. Japan*, **48**, 123-136, 1996.

Sime, D. G., E. Hiei, and A. J. Hundhausen, Coronal eruptive events on April 4, and May 4, 1992, in *X-ray Solar Physics from Yohkoh*, eds. Y. Uchida *et al.*, Universal Academy Press, Tokyo, 197-200, 1994.

Švestka, F. Farnik, H. S. Hudson, Y. Uchida, P. Hick, and J.R. Lemen, Large-Scale Active Coronal Phenomena in Yohkoh SXT Images, I, *Solar Phys.* **161**, 331-343, 1995.

Strong, K. T., K. L. Harvey, T. Hirayama, N. Nitta, T. Shimizu, T., and S. Tsuneta, Observations of the variability of coronal bright points by the soft X-ray telescope on *Yohkoh*, *Publ. Astr. Soc. Japan*, **44**, L161–L166, 1992.

Sturrock, P. A. (ed.), *Solar flares: A monograph from Skylab Solar Workshop II*, Colorado, Boulder, 1980.

Tsuneta, S., Structure and dynamics of magnetic reconnection in a solar flare, *Ap. J.* **456**, 840-849, 1996.

Tsuneta, S., L. Acton, M. Bruner, J. Lemen, W. Brown, R. Caravalho, R. Catura, S. Freeland, B. Jurcevich, M. Morrison, Y. Ogawara, T. Hirayama, and J. Owens, The soft X-ray telescope for the Solar-A mission, *Solar Phys.*, **136**, 37-67, 1991.

Tsuneta, S., H. Hara, T. Shimizu, L. W. Acton, K. T. Strong, H. S. Hudson, and Y. Ogawara, Global restructuring of the coronal magnetic fields observed with the *Yohkoh* Soft X-ray Telescope, *Publ. Astr. Soc. Japan*, **44**, L63–L67, 1992.

Uchida, Y., A. McAllister, K. T. Strong, Y. Ogawara, T. Shimizu, R. Matsumoto, and H. S. Hudson, Continual expansion of the active-region corona observed by the *Yohkoh* Soft X-ray telescope, *Publ. Ast. Soc. Japan* **44**, L155-L158, 1992.

Watanabe, Ta., Y. Kozuka, M. Ohyama, M. Kojima, K. Yamaguchi, S. Watari, S. Tsuneta, J. A. Joselyn, K. Harvey, and L. W. Acton, Coronal/interplanetary disturbances associated with disappearing solar filaments, *Publ. Astr. Soc. Japan*, 44, L193-L197, 1992.

Watanabe, Ta., M. Kojima, Y. Kozuka, S. Tsuneta, J. R. Lemen, H. S. Hudson, J. A. Joselyn, and J. A. Klimchuk, Interplanetary consequences of transient coronal events, in *X-Ray Solar Physics from Yohkoh*, edited by Y. Uchida, T. Watanabe, K. Shibata, and H. S. Hudson, Universal Academy Press, Tokyo, 207-210, 1994.

Watari, S., T. Detman, and J. A. Joselyn, A large arcade along the inversion line, observed on May 19, 1992 by Yohkoh, and enhancement of interplanetary energetic particles, *Solar Phys.*, 169, 167-179, 1996.

Webb, D. F., The solar sources of coronal mass ejections, in *Eruptive Solar Flares*, eds. Z. Švestka, B. V. Jackson, and M. E. Machado, Springer-Verlag, Berlin, 234, 1992.

Webb, D. F., Coronal Mass Ejections: the key to major interplanetary and geomagnetic disturbances, *Rev. of Geophysics (Suppl.)*, 33, 577-583, 1995.

Webb, D. F., and R. A. Howard, The solar cycle variation of coronal mass ejections and the solar wind mass flux, *J. Geophys. Res.* 99, 4201-4220, 1994.

Webb, D., H. Hudson, and R. Howard, X-ray signatures of CMEs observed in white light, *EOS*, 77, F563, 1996.

Weiss, L. A., J. T. Gosling, A. McAllister, A. J. Hundhausen, J. T. Burkepile, J. L. Phillips, K. T. Strong, and R. J. Forsyth, A comparison of interplanetary coronal mass ejections at *Ulysses* with *Yohkoh* soft X-ray coronal events, *J. Geophys. Res.*, 101, 13,497-13,515, 1996.

Withbroe, G., W. C. Feldman, and H. S. Ahluwahlia, The solar wind and its coronal origins, in A. N. Cox, W. C. Livingston, and M. S. Matthews (eds.), *Solar Interior and Atmosphere* (Arizona), 1087-1106, 1991.

Yokoyama, T., and Shibata, K., Magnetic reconnection as the origin of X-ray jets and Hα surges on the Sun, *Nature*, 375, 42-44, 1995.

Zirin, H., *Astrophysics of the Sun*, Cambridge, 1988.

Hugh Hudson, ISAS c/o Ogawara, 3-1-1 Yoshinodai Sagamihara-shi, Kanagawa 229, Japan

The Role of Coronal Mass Ejections in Solar Activity

B. C. Low

High Altitude Observatory, NCAR[1], P.O. Box 3000, Boulder, Colorado

Observation and theory relating to coronal mass ejections are critically reviewed. The suggestion is explored that the coronal mass ejection is the hydromagnetic process by which the corona reconfigures systematically over an eleven-year magnetic cycle to culminate in the reversal of its global magnetic field. Central to this suggestion is the idea that magnetic fields in twisted form rise bodily, over evolutionary time scales, into the corona from below to form the cavity in a helmet streamer. The magnetic field accumulated in the cavity subsequently leaves the corona with the coronal mass ejection when the helmet streamer erupts dynamically. This may be the process which removes the old magnetic flux from the corona for replacement by the flux of the opposite polarity belonging to the new magnetic cycle.

1. INTRODUCTION

Coronal mass ejections (CMEs) and flares are the two major forms of eruptive hydromagnetic phenomena in the solar corona. Flares have been studied for many decades but CMEs are a relatively recent discovery; [e.g., Gosling, this volume]. Here, we take our interest beyond the physical fascination with an individual CME event to the physics of where in the tapestry of solar activity the CMEs fit in. A general synthesis of coronal phenomena and physical processes into a global physical picture has been published recently [Low, 1996], which incorporates a central role for the CMEs. In this paper, we explore the possibility that CMEs are the basic mechanism by which the large-scale corona reconfigures itself systematically to reverse its global magnetic polarity in the course of an eleven-year solar cycle. The mass in a CME, typically of the order of $\sim 10^{15}$ g, takes with it 10^{31-32} erg of kinetic and gravitational potential energy, and the flare of the type often observed to follow a CME liberates a comparable amount of energy in the rapid, intense heating of the coronal plasma. It is unlikely that these coronal eruptions, developing at typical speeds comparable to coronal Alfvenic speeds > 700 km s^{-1}, are driven directly by the massive but slowly moving photosphere (0.5 km s^{-1}). The total energy liberated is likely to have been stored in stressed magnetic fields prior to the eruptions. Accepting this idea, a key question not often asked is: In addition to the liberation of energy, does the CME affect the coronal magnetic field in any permanent way? There is a popular model widely discussed in the solar-physics literature, which proposes that the shearing displacement of magnetic footpoints may build up energy for a catastrophic release by some suitable instability or nonequilibrium process. The magnetic field is returned to essentially the initial state from which it starts, so that the process may repeat. The process is interesting in its own right, but it has no role to play in the long-term evolution of the corona which proceeds from a simple form at activity minimum to a complex form at activity maximum when the global magnetic-field reversal occurs.

---
[1]The National Center for Atmospheric Research is sponsored by the National Science Foundation

Coronal Mass Ejections
Geophysical Monograph 99
Copyright 1997 by the American Geophysical Union

This evolution requires removal of magnetic flux threading through the corona early in the magnetic cycle, to be replaced with the new flux welling up from below the corona. The extensive coronal holes over the poles at activity minimum each contains an enormous amount of unipolar magnetic flux spread out into interplanetary space. Coronal holes persist with stability for an extended period of time and, because their rotations significantly deviate from the differential rotation observed at the photosphere, one simple interpretation — the one adopted in this paper — is that their magnetic fields are rooted deep below the photosphere. Yet that deep rooted connection must relent at some point during an eleven-year cycle to allow for the decay of the coronal hole, so that, at the end of the cycle, it can be replaced with a hole made up of opposite magnetic flux of the new cycle. The high electrical conductivity of the corona does not allow the magnetic flux to be locally destroyed in bulk. A simple possibility of removal would be the physical transport of magnetic flux by its embedded plasma out of the corona. In the rest of this paper, we lay the basis for suggesting that the CME is the end product of a process which accomplishes this transport. This suggestion implies that magnetic fields do not just emerge and thread through the photosphere but, in the course of solar activity, do rise bodily into the corona and are eventually removed out and into interplanetary space. If this suggestion is valid, the significance of CMEs goes beyond coronal and heliospheric concerns because, through them, a distinct physical coupling between the corona and the interior solar dynamo would have been identified. The field reversal is of course driven by the solar-interior dynamo. What is novel here is the implication that a significant fraction of the field (of a given sign) generated in each dynamo cycle is transported out of the solar interior, into the corona, and, ultimately, out into interplanetary space.

2. THE MAGNETIC ORIGIN OF CMEs

Let us review the phenomenology of CMEs to lead to some understanding of their underlying physics. The review is brief and qualitative; an extensive list of references to original sources is given in *Low* [1996] and elsewhere in this monograph.

The observed close association between CMEs and flares naturally suggests the possibility, considered in the late seventies, that CMEs are the dynamical response of the corona to the sudden input of energy liberated by a flare at the coronal base (Dryer 1982). This is not supported by observations which show that the flare associated with a CME generally appears from minutes to an hour after the CME has fully formed and traveled out of the corona at speeds of several 100 km s^{-1} [*Harrison*, 1986; *Hundhausen*, 1995]. Recent comparisons of white-light data from the HAO ground-based coronameter and soft X-ray data from Yohkoh have confirmed this temporal ordering of the CME and its associated flare [*Hiei, Sime, and Hundhausen*, 1994]. These data comparisons also imply that the CME is largely an ideal MHD process with little resistive heating of the corona [*Low*, 1994]. Only when the associated flare has set in do the soft X-ray data show significant plasma heating.

The CME-flare phenomenon can be explained simply in terms of a two-step process [*Hundhausen*, 1995; *Low*, 1990]. First there is the CME which opens up an initially closed coronal magnetic field to eject mass previously trapped in the closed magnetic field. This is followed by the flare which results from the re-closing of the opened field by magnetic reconnection [*Hirayama*, 1974; *Kopp and Pneuman*, 1976]. The first step is an ideal MHD process, and the second step a dissipative MHD turbulence process [*Low*, 1994].

To understand why such a two-step process should occur in the corona, we need to appreciate the forces which structure the large-scale corona. The million-degree corona cannot be entirely confined by solar gravity. The rapid inverse-square fall-off of solar gravity and the high thermal conductivity of the corona give rise to the solar-wind expansion [*Parker*, 1963]. The magnetic field provides the only means by which this expansion can be thwarted. The Lorentz force includes an isotropic magnetic-pressure force which can only enhance the tendency of the corona to expand. It is the other component of the Lorentz force, the magnetic tension force, which may locally trap plasma against the expansion, provided the field is closed and is sufficiently intense. Thus the corona is made up of two distinct types of dynamical regions, the open-field regions where the plasma flows out as the solar wind and the closed-field regions where the magnetic field confines quasi-static pockets of plasma. The latter appear as the conspicuous helmet-streamers outside of which density is low and the magnetic field is open. At times of activity minimum, the open-field regions over the poles produce the extensive polar coronal holes.

Subject to quasi-steady changes at the base of the corona, across which mass, energy and magnetic flux do pass continually into the corona during the course of a solar cycle, some of the initially closed magnetic fields in the corona may open up to produce a CME. In fact, many of the CMEs observed by the SMM Coronagraph and the NRL P-78 Coronagraph originated from the breakup of coronal helmet streamers [*Howard et al.*, 1985; *Hundhausen*, 1995). The dynamical breakup of a pre-existing coronal structure is likely to be the result of a loss of equilibrium in the course of slow quasi-steady evolution [*Low*, 1982; *Wolfson*, 1982]. This scenario has been pointed out early in the study of CMEs from theoretical considerations but gained acceptance after the observational support for it had been established [*Harrison*, 1986; *Hundhausen*, 1995; *Kahler*, 1992; *Low*, 1990].

A loss of equilibrium can result from the stressing of the closed magnetic fields in the helmet by the large-scale magnetic footpoint displacements on the solar surface, such as due to the solar differential rotation (e.g., [*Mikic and Linker*, 1994]). This effect is likely to be a trigger mechanism which, by itself, is not sufficient to explain the CME phenomenon. The limitation can be seen in models using the force-free magnetic field. Figure 1 shows a helmet-streamer identified with a region of closed bipolar magnetic field with the two ends of its lines of force anchored to the coronal base. A magnetic field of this topology in the force-free state cannot have an energy in excess of any state in which the field is fully open with one end of each line of force dragged out to infinity and the other end rigidly anchored to the base [*Aly*, 1991]. This means that the force-free field stressed by surface motion cannot gain enough energy to spontaneously open up fully, let alone deliver the enormous energy above the open state to be carried away by a CME [*Low and Smith*, 1993]. The stressed field may erupt but, because of this energy constraint, the eruption cannot be of the type associated with CMEs.

Subject to quasi-static stressing, force-free fields in infinite space achieve equilibrium by expanding outward. Although its energy density increases with stress, the amount of increase is limited as the result of the expansion adjustment. If the force-free field is confined to a fixed volume of space by rigid walls, its energy can grow without bound by intensive compression when subject to ever increasing stress. In the solar atmosphere there are of course no rigid walls. However, the weight of plasma in a non-force-free magnetic field acts like a rigid wall to confine a magnetic field. This provides a means of putting more magnetic energy in the atmosphere than possible with strictly force-free fields [*Low and Smith*, 1993]. The force-free magnetic field is, of course, not a uniformly valid approximation over the scales of helmet streamers and CMEs. This approximation is good near or at the coronal base, but must break down at increasingly large coronal heights where the nearly isothermal coronal gas pressure always dominates, with the solar wind combing all field lines at these heights out into interplanetary space. Insofar as observation is concerned, this theoretical development away from a pure force-free field model is in the right direction. One of the physically interesting properties of the CME is its large mass and the significant gravitational potential energy associated with that mass. Therefore, to address the physics of the CME, a viable model must ultimately allow for interaction between plasma and magnetic field in the pre-CME equilibrium to account for its large mass and gravitational potential energy.

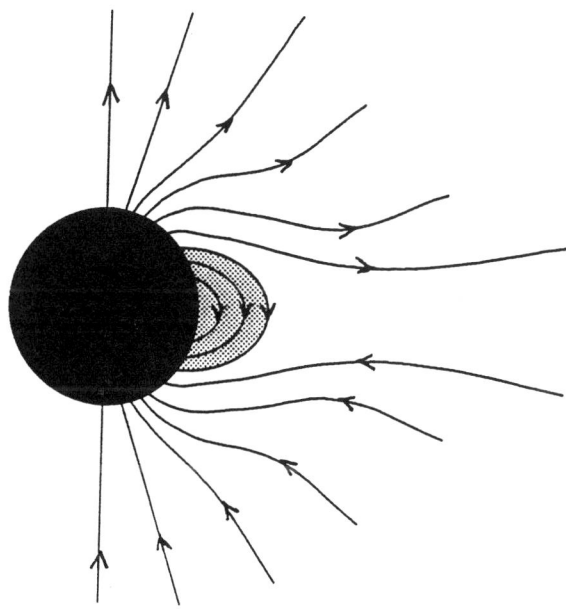

Figure 1. Magnetic lines of force of an axisymmetric corona showing anchored bipolar fields at equator sandwiched between two hemispheres of open fields.

Once we recognize that over the large scale of the helmet streamer, magnetic fields are not force-free but do interact with the plasma atmosphere, Aly's energy constraint ceases to be restrictive. With confinement provided by the weight of the atmosphere, it is conceivable that a highly twisted magnetic field may be trapped with an energy greater than possible for a uniformly force-free field.

An important means of getting more magnetic energy stored in the atmosphere is to have detached magnetic fields in the solar atmosphere [*Low and Smith*, 1993]. A fully detached field can travel bodily out of the corona and, by expansion, give up all its magnetic energy. Figure 2 is a sketch of a helmet-streamer distinct from that in Figure 1. In addition to the bipolar, anchored, field is a region of fields which close upon themselves entirely in the corona. Allowing for a strong magnetic field component out of the meridian plane in this closed-field region, this region contains a magnetic flux rope run-

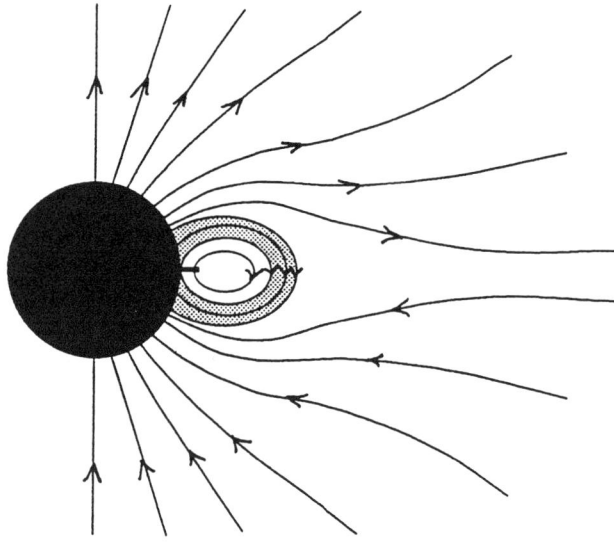

Figure 2. Magnetic lines of force of an axisymmetric corona showing anchored bipolar fields at equator, containing completely closed loops of fields and sandwiched between two hemispheres of open fields. The closed loops of fields carry a significant azimuthal field component so that these fields represent an azimuthal rope of twisted fields running above the equator. The thick line represents a vertical sheet of prominence materials suspended at the base of the closed loop region.

ning in the azimuthal direction, fully detached from the surface of the Sun. In both Figures 1 and 2, we have assumed for simplicity an axisymmetric corona. In the real corona, we do not expect to find the global azimuthal flux rope in Figure 2. A flux rope in the corona would have ends which fold down in realistic geometry to blend into the background magnetic fields. What distinguishes it from a simple bipolar anchored field is that it has lines of force winding once or more times about each other within the corona, whereas, the latter has lines of force which are direct paths from one footpoint to the other. Figure 2 may then be interpreted to be describing the cross-section of some suspended mid-section of the realistic flux rope.

Bearing in mind the more complex situation involving realistic geometry, the magnetic field in Figure 2 explains many observational features of the helmet structure. Many helmets come with a three-part structure: high-density dome, low-density cavity at the base, and the quiescent prominence in the cavity. The high density region is identified with the anchored bipolar fields in Figure 2. Subject to the ubiquitous coronal heating, the region of bipolar fields achieves thermal equilibrium by having its temperature raised to a point where its radiative loss and thermal conductive flux down into the photosphere can balance the heat input. The high temperature attained implies a large hydrostatic density scale height, thus explaining the helmet high density in terms of heated materials having risen from the lower atmosphere along the bipolar field. The flux rope region has a distinctly different magnetic topology. In the axisymmetric atmosphere in Figure 2, this region is completely cut off from the photosphere which is the ultimate source of energy and mass. In the realistic flux rope, with anchored ends, it is the long path lengths of winding field lines which produce a qualitatively similar effect of thermal and mass isolation. Inside this flux rope, if the isolation effect is sufficiently strong, not only is the mass supply from the photosphere restricted, condensation instabilities of various types conceivably can cause evacuation of the local mass and the formation of the prominence in the density cavity produced. In this process, the magnetic pressure of the flux rope holds out against self-collapse. The thermodynamic states are thus controlled by magnetic topology, in terms of suppression of thermal conduction and mass transport across the magnetic field. The abrupt difference between the two field topologies at the boundary of the flux rope explains the sharp boundary of the low-density cavity observable in (white-light) coronagraph images if that boundary is suitably oriented with the line of sight. Finally, many well known properties of prominence structures have recently been explained in terms of the magnetic topology of the flux rope within which the prominence sits [*Low and Hundhausen*, 1995]. The above explanation makes good physical sense and relates a diversity of otherwise disjoint observational results.

If we accept this explanation for the three-part structure of the helmet, several physical issues with the CMEs can be resolved satisfactorily. Many CMEs show a corresponding three-part structure of (i) the leading high-density shell of ejected mass having the appearance of a loop when viewed in projection on the plane of the sky, (ii) a trailing cavity with a sharp boundary, and, (iii) an erupted prominence in the cavity. Since many CMEs erupt from pre-existing helmet streamers, it is reasonable to identify the three-part structure in one with that of the other. Most CMEs do not originate in an impulsive release of energy but do result from a pre-existing structure going out of equilibrium. Therefore, it is not surprising that the coherent form of the magnetized plasma with its three parts is preserved in this transition from equilibrium to fully developed motion. This point is bolstered by the observation that the speeds of CMEs range impressively from as low as 10 km s^{-1} to as high as 1000 km s^{-1}, with a median speed of 350 km s^{-1} in the data set from the SMM Coronagraph/Polarimeter [*Hundhausen, Holzer,* and Low, 1987]. This median speed is supersonic but sub-Alfvenic. The energy source and the trigger for the CME can also be explained. It is the magnetic flux rope which provides the necessary amount of energy to drive the ejection and open up the field. In turn the 10^{15} g of ejected mass plays the role of confining the flux rope to hold it in equilibrium prior to eruption. Finally, when that confinement fails, the CME is initiated by the magnetic buoyancy of the cavity flux rope [*Low*, 1981; *Fisher and Poland*, 1981; *Low, Munro and Fisher*, 1982; *Illing and Hundhausen*, 1986].

A basic physical point needs to be clarified before we proceed. Consider the question of whether a truly detached flux rope in axisymmetry is energetically more favored to open up than a highly twisted, non-axisymmetric flux rope with its ends anchored to the solar surface. For a force-free magnetic field, the former is not limited by the energy of the fully open field, under Aly's constraint, whereas the latter is. The answer to the question is not necessarily yes. For an anchored flux rope in fully three-dimensional geometry to open up to infinity, opening up all the magnetic fields in the entire atmosphere is not required. In the absence of axisymmetry, the flux rope can open by simply pushing its way amidst neighboring magnetic fields which can remain closed. This is not incompatible with the observation that when a CME erupts from a very long magnetic arcade, only a sub-length of the arcade of closed magnetic field opens up with the CME. In contrast, a flux rope in axisymmetry such as the one in Figure 2 needs to open up all lines of force anchored to the Sun; this is just a topological necessity of the axisymmetry assumed. Aly's constraint sets an upper limit to the amount of energy stored in terms of the everywhere fully open field [*Wolfson and Low*, 1992]. This limit is relevant to the axisymmetric flux rope but not relevant to the anchored flux rope which can open up without requiring all fields to also open up [*Low*, 1996].

This interesting point needs to be confirmed theoretically but it is really not very essential to our concern with CMEs. The more important point to be making is that a magnetic field if sufficiently twisted can have an energy exceeding any limit as long as it is possible to keep the highly twisted magnetic field confined in equilibrium by plasma weight in the corona. This point is demonstrated in *Low and Smith* [1993]. We suggest that in this idea lies the relationship between the CME mass and the magnetic energy which fuels the CME expulsion: this mass is found in the pre-eruption helmet and it plays the role of confining the magnetic field in the helmet cavity where the energy for the eruption is stored.

The above brief review provides a basis for us to explore the implications of CMEs for the global coronal magnetic field.

3. THE REMOVAL OF MAGNETIC FLUX BY CMEs

The idea that the corona may remove magnetic flux by ejection out into the interplanetary space has been suggested independently by *Rust* [1994] with the proposal that the prominence is a twisted magnetic flux rope which, upon its ejection, produces the interplanetary magnetic cloud [*Burlaga et al.*, 1981]. Our view is more complete. We recognize that the eruption of a prominence is only a part of the large-scale coronal eruption which is the CME [*Low*, 1994]. In this view, the magnetic flux rope is identified not directly with the prominence but with the helmet cavity, within which lies the prominence. In fact, both in the CME phenomenon and the helmet-streamer before eruption, it is the cavity and not the prominence which is the fundamental hydromagnetic structure [*Low and Hundhausen*, 1995].

What is the origin of the cavity magnetic flux rope? One possibility is the formation of the flux rope in the pre-existing coronal magnetic fields by a process taking place entirely in the corona. Many theoretical calculations have been published to show how suitable shearing displacements of magnetic footpoints on the photosphere combined with magnetic reconnection may convert an initial magnetic field with anchored bipolar lines of force into one with a "magnetic island" to form a magnetic flux rope in the corona (e.g., [*Inhester, Birn, and Hesse*, 1992; *Choe and Lee*, 1992]. An important aspect of these models is that the motion of the footpoints is required to have a significant component which brings pairs of footpoints to converge at the polarity inversion line for the formation of the magnetic island by reconnection. This converging motion is motivated by observations of magnetic elements of the opposite polarities drifting toward each other on the photosphere to meet and annihilate at the polarity inversion line. Such apparent motions are especially associated with the formation of a prominence over the polarity inversion line [*Martin*, 1989].

An alternative mechanism, favored in this paper, is that the cavity magnetic flux rope originates from the interior of the Sun [*van Ballegooijen and Martens*, 1990]. The turbulent dynamo in the solar interior is likely to generate magnetic fields which are highly tangled and twisted so that at the time of emergence of these fields, they are already twisted. These fields may emerge through the photosphere in two forms depending on their field strengths. If the field is extremely intense (kilo-gauss strength) and coherent over a large scale ($\sim 10^4$ km), they form sunspot-like structures. In such a case, although the field may still be deformed strongly by the sub-photospheric plasma, its coherence may persist so as to be observed as the twisted magnetic field seen in flare related H_α structures [*Tanaka*, 1991; *Zirin*, 1988]. Recently, observations of photospheric vector magnetic fields suggest that flux emergence may take place in the form of a coherent rope of twisted magnetic fields in a closed geometry [*Lites et al.*, 1995].

Another form of the emerged magnetic field, probably the more common one, is a relatively weak field, highly tangled by the turbulent motion in the photosphere and below. Although the field has a net twist over its volume, it is too chaotic and spatially spread out to show its global twist. As this magnetic field makes its way into the corona, the different parts of it expand in the tenuous medium of the corona to press into each other. Magnetic reconnection (giving rise to heating) simplifies the global field as it settles to a quiescent state [*Parker*, 1994]. It has been pointed out that in consequence of the extremely high electrical conductivity of the corona, reconnection merely transfers twist among the magnetic lines of force which in the process liberates magnetic energy [*Taylor*, 1986; *Low*, 1996]. The magnetic twist cannot locally vanish as in the case of an electrically resistive medium [*Berger*, 1984]. Therefore, if the magnetic field upon emergence has a net twist, the final relaxed magnetic field in the corona would have that net twist. If the twist is significant, the final state must take on a relatively ordered rope of twisted magnetic fields which, in our interpretation, we identified with the helmet cavity. This process explains the observational fact that active regions upon their decay leave behind long lines of polarity inversion above which a well formed quiescent prominence can be found with a conspicuous helmet streamer [*McIntosh*, 1972]. These are the so-called crown filaments in the high latitudes of the Sun. Crucial to this process is the condition, to be elaborated further below, that the active region magnetic fields have a certain net magnetic twist upon their emergence.

Whether the emerging field is a coherent magnetic flux rope or a chaotic field carrying a net twist, the essence of this process is that a main part of the magnetic field passes bodily clear through the photosphere. The rise is due to magnetic buoyancy [*Parker*, 1979]. The high electrical conductivity of the plasma does not allow fluids to move across the magnetic field. Combined with the effect of magnetic pressure, the magnetic field behaves like a weightless, pressurized fluid which readily, buoyantly wiggles its way upward through the stratified medium. This process proceeds with siphon flows and draining of plasma along the lines of force to low points in the magnetic field. This is the classical Parker instability which allows the plasma trapped in the magnetic field to sink while the evacuated parts of the field rise buoyantly in a runaway fashion. The two parts may eventually separate by forced magnetic reconnection so that a main part of the flux system escapes upward with significantly less mass than it originally trapped within itself.

The observational signature of this process is that initially when the top of the flux system penetrates the photosphere, we see new bipolar, inverted-U, fields anchored to fresh growing patches of equal and opposite magnetic polarities on the photosphere [*Zirin*, 1988]. As the main part of the magnetic flux system lifts bodily above the photosphere, the patches of opposite polarities will shrink, each with decreasing magnetic flux. At this stage, the bottoms of the flux system, locally in the form of U-shaped lines of force, are rising through and above the photosphere [*Lites et al.*, 1995; *Spruit, Title, and Van Ballegooijen*, 1987]. This last stage removes magnetic fluxes of the opposite signs in equal amounts from the photosphere. The same process takes place at successively higher levels above the photosphere. As the process progresses, each level is reduced in fluxes of the two signs while the rising flux system takes on the form of the flux rope trapped in the larger-scale bipolar fields of a coronal helmet. This explanation of the observed convergence and annihilation of opposite fluxes at a fixed atmospheric level is physically simple. The phenomenon is just a consequence of buoyantly rising fields and not the result of complex photospheric motions and reconnection processes. Ultimately observations need to determine whether one or the other mechanism, or neither, is the origin of this phenomenon. One obvious way of testing whether a rising U-loop or a submerging inverted U-loop is the cause of a mutual annihilation of two patches of opposite magnetic polarities observed at the photosphere, is to observe simultaneously for the similar process in the chromosphere above.

If we hypothesize that the buoyant body transfer of magnetic flux systems into the upper atmosphere is significant, the implication is that a substantial portion of the new flux generated by the dynamo escapes from the interior into the corona over the eleven-year magnetic cycle. This implication is important and merits quantitative examination, theoretically and observationally. As an order of magnitude estimate, the interesting question can be addressed here whether this process is capable of rejuvenating the entire flux of a given sign threading the solar surface at any one time. Modifying such an estimate presented by *Hundhausen* [1996], start with the fact that a typical helmet can fill up an angular size of 30° in width and 60° in length so that its coronal base has an approximate area of $\frac{1}{(8\pi)}$ in units of the total solar surface. The solar magnetic field averages $B_0 = 10$ gauss at the solar surface. Take a field of this strength to thread through the surface everywhere. It will be seen that the result we are deriving is actually independent of this field strength. Suppose the cavity flux rope contains about 1% of the flux threading through the helmet, which is a conservative estimate. This flux has been removed from the photosphere by the buoyant rise of the flux rope forming the helmet cavity. This takes place over evolutionary time scales. As the bottom of a local U-loop rises above the solar surface, pairs of opposite magnetic polarities on the surface mutually annihilate. When the cavity flux rope is in place, the CME blows off with this flux rope. This takes place dynamically in the corona. Each CME may carry away the fraction $\frac{1}{(800\pi)}$ of the total flux threading across the solar surface. With a modest rate of one CME a day, it will take about 7 years to remove the total solar flux, well within the eleven-year cycle.

The premise central to the above synthesis is that magnetic fields emerge into the solar atmosphere in a significantly twisted state. The observations of magnetic structures, either through direct field measurements or by inference, say, from H_α; have amassed data over more than four magnetic cycles. These data, suitably interpreted in terms of twisted magnetic fields, show that the magnetic fields in the solar atmosphere are predominantly of negative twist in the northern hemisphere and of positive twist in the southern hemisphere, with the predominant sense of twist in each hemisphere fixed and Independent of solar cycle [*Martin, Bilimoria, and Tracadas*, 1993].

This robust observational result is significant in the context of the property that magnetic twist in a highly conducting plasma cannot be readily destroyed *in situ* but is transferred from one set of lines of force to another set during magnetic reconnection. In this context, a local creation of a twist of one sense must imply the creation of the twist of the opposite sense. The fact that a majority of magnetic structures in a given hemisphere is of a given sign in twist then implies that whatever the hydromagnetic process creating this distribution of the magnetic twist, it is not a local process. It has to be a global process which puts twists of the opposite signs out into the two hemispheres, negative twists in the north and positive twists in the south. The only possibility that comes to mind is some process associated with solar differential rotation [*van Ballegooijen and Martens*, 1990]. In order not to detract from our synthesis, let us simply accept that such a global process operates in the Sun. It then follows that not only must the emerged fields begin with twists predominantly of a fixed sign in a given hemisphere but, in due course when the emerged fields have coalesced in the corona, either by quiescent or explosive reconnections, the fields form larger ordered structures, with their twists of like signs adding up accumulatively. Flux ropes are simple states of magnetic fields with a considerable amount of twist. This would explain why prominences form throughout the solar cycle. They form naturally in flux ropes, and flux ropes are the coronal accumulations of magnetic twists (or magnetic helicity) continually brought up with predominantly the same sign on each hemisphere.

Since the predominant magnetic twist in each hemisphere does not change with solar cycle, the twists of each successive magnetic cycle need to be removed to avoid unbounded accumulation. Given the uncompromising constraint that twist cannot be removed *in situ* in the corona, it follows that the only mechanism of removal is body ejection which is clearly what the CME does when it takes a magnetic flux rope out into interplanetary space.

This, we suggest, is the role for CMEs in solar activity. Each CME ends an evolutionary process which takes a part of the magnetic flux threading through the photosphere and moves that flux bodily up to form a flux rope at the base of the helmet streamer. When a CME is ready to erupt, the photosphere has already lost some of its emerged flux and, in consequence, the helmet has developed a clearly formed prominence cavity, all charged with energy to erupt. The CME then sets off to remove not only the flux transported from the photosphere into the corona but also the considerable magnetic twist accumulated in the flux rope. Associated with this twist is the energy needed to fuel the enormous energy budget of the CME and its post-CME flare. The physical distinction between the magnetic topologies in Figures 1 and 2 now becomes clear. Helmet streamers of the type in Figure 1 are extremely stable, an example of which is the helmet streamer newly formed by a post-CME flare that has reclosed the fields previously forced open by a CME [*Hiei, Sime, and Hundhausen*, 1994]. Helmet streamers of the type in Figure 2 are the ones which are highly energized to produce a CME. Each time a CME blows off, the coronal magnetic field suffers a permanent change, the change being the body transport of magnetic flux through the photosphere and then out of the corona.

Finally, we offer an insight into the destruction of the huge polar coronal holes taking place when activity picks up from minimum. In response to the new flux generated by the solar dynamo, the unipolar fields in the coronal hole eventually lose their deep rooted feature. The previously deeply rooted parts of the old fields then move buoyantly to the solar surface to be transported into the corona under helmet-streamers, so as to be systematically taken out by the CMEs.

4. CONCLUSION

This review summarizes a synthesis of solar activity as a hydromagnetic process in which the CME plays the central role to relate a diversity of otherwise disjoint phenomena. This point of view took its most complete form in Low [1996]. Here, we emphasize the possibility that the CME is the basic mechanism of coronal magnetic reconfiguration, taking magnetic flux and helicity from the photosphere and ejecting them, together with the enormous CME mass, into interplanetary space. In this manner each CME does something permanent to the coronal magnetic field and the accumulative effect of all the CMEs over a magnetic cycle is the removal of the global magnetic field in the corona and heliosphere to make room for the new fields belonging to the new cycle. This is only a tentative proposal at the present because many of the hydromagnetic processes making up this global picture, centered around the CME, need to be made quantitative. Also, many of its implications for observation need to be pursued and tested. The attraction of this global picture is that it relates a diversity of disjoint observations and theoretical ideas, and that it poses testable hypotheses.

REFERENCES

Aly, J. J., How much energy can be stored in a three-dimensional force-free magnetic field?, in *Astrophys. J. Lett.*, 375, L61, 1991.

Berger, M. A., Rigorous new limits on magnetic helicity dissipation in the corona, in *Geophys. Astrophys. Fluid Dyn.*, 30, 79, 1984.

Burlaga, L., E. Sittler, F. Mariani, and R. Schwenn, Magnetic loop behind an interplanetary shock: Voyager, Helios and IMP 8 observations, in *J. Geophys. R.*, 86, 6673, 1981.

Choe, G. S. and L. C. Lee, Formation of solar prominence by photospheric shearing motions, in *Solar Phys.*, 138, 291, 1992.

Dryer, M., Coronal transient phenomena, in *Sp. Sci. Rev.*, 33, 233, 1982.

Fisher, R. R. and A. I. Poland, Coronal activity below 2 R$_\odot$: February 15 - 17, in *Astrophys. J.*, 246, 1004, 1981.

Harrison, R. A., Solar coronal mass ejections and flares, in *Astron. Astrophys.*, 162, 283, 1986.

Hiei, E., D. G. Sime, and A. J. Hundhausen, Reformation of a coronal helmet streamer by magnetic reconnection after a coronal mass ejection, in *Geophys. Res. Lett.*, 20, 2785, 1994.

Hirayama, T., Theoretical models of flares and prominences. I. Evaporating flare model, in *Sol. Phys.*, 34, 323, 1974.

Howard, R. A., N. R. Sheeley, Jr., M. J. Koomen, and D. J. Michels, Coronal mass ejections 1979 - 1981, in *J. Geophys. Res.*, 90, 8, 173, 1985.

Hundhausen, A. J., Coronal mass ejections: A summary of SMM observations from 1980 and 1984 - 1989, in *The many faces of the sun*, edited by K. Strong, J. Saba and B. Haisch, in press, 1995.

Hundhausen, A. J., J. T. Burkepile, and O. C. St.Cyr, The speeds of coronal mass ejections: SMM observations from 1980 and 1984 - 1989, in *J. Geophys. Res.*, 99, 6543, 1994.

Hundhausen, A. J., T. E. Holzer, and B. C. Low, Do slow shocks precede some coronal mass ejections?, in *J. Geophys. Res.*, 92, 11173, 1987.

Illing, R. M. E. and A. J. Hundhausen, Disruption of a coronal streamer by an eruptive prominence and a coronal mass ejection, in *J. Geophys. Res.*, 91, 10951, 1986.

Inhester, B., J. Birn, and M. Hesse, The evolution of line-tied coronal arcades including a converging footpoint motion, in *Solar Phys.*, 138, 257, 1992.

Kahler, S., Solar flares and coronal mass ejections, in *Ann. Rev. Astron. Astrophys.*, 30, 113, 1992.

Kopp, R. and G. W. Pneuman, Magnetic reconnection on the corona and the loop prominence phenomenon, in *Sol. Phys.*, 50, 85, 1976.

Lites, B. W., B. C. Low, V. Martínez Pillet, P. Seagraves, A. Skumanich, Z. A. Frank, R. A. Shine, and S. Tsuneta, The possible ascent of a closed magnetic system through the photosphere, in *Astrophys. J.*, 446, 877, 1995.

Low, B. C., Eruptive magnetic fields, in *Astrophys. J.*, 251, 352, 1981.

Low, B. C., Equilibrium and dynamics of coronal magnetic fields, in *Ann. Rev. Astron. Astrophys.*, 28, 491, 1990.

Low, B. C., Magnetohydrodynamic processes in the solar corona: Flares, coronal mass ejections, and magnetic helicity, in *Plasma Phys.*, 1, 1684, 1994.

Low, B. C., Solar activity and the corona, in *Solar Phys.*, 167, 217, 1996.

Low, B. C. and J. R. Hundhausen, Magnetostatic structures of the solar corona. II. The magnetic topology of quiescent prominences, in *Astrophys. J.*, 443, 818, 1995.

Low, B. C. and D. F. Smith, The free energies of partially-open coronal magnetic fields, in *Astrophys. J.*, 410, 413, 1993.

Low, B. C., R. H. Munro, and R. R. Fisher, in *Astrophys. J.*, 254, 335, 1982.

Martin, S. F., Lecture notes on prominences, in *Dynamics of Quiescent Prominences*, edited by V. Ruzdjak and E. Tandberg-Hanssen (Springer Verlag, Berlin), 1, 1989.

Martin, S. F., R. Bilimoria, and P. W. Tracadas, Magnetic field configurations basic to filament channels and filaments, in *Solar Surface Magnetism*, edited by R.J. Rutten and C.J. Schrijver, (Kluwer Acad. Publ., Dordrecht), 303, 1993.

McIntosh, P. S., Inference of solar magnetic polarities from H-alpha, in *Prog. Astronault. Aeronault.*, 30, 65, 1972.

Mikic, Z., J. A. Linker, Disruption of coronal magnetic field arcade, in *Astrophys. J.*, 430, 898, 1994.

Parker, E. N., *Interplanetary Dynamical Processes*, (Interscience: New York), 1963.

Parker, E. N., *Cosmical Magnetic Fields*, (Oxford University Press: Oxford), 1979.

Parker, E. N., *Spontaneous Current Sheets in Magnetic*

Fields, (Oxford University Press: Oxford), 1994.

Rust, D., Spawning and shedding helical fields in the solar atmosphere, in *Geophys. Res. Lett.*, *21*, 241, 1994.

Sime, D. G., Coronal mass ejection rate and the evolution of the large-scale K-coronal density distribution, in *J. Geophys. Res.*, *94*, 151, 1989.

Spruit, H. C., A. M. Title, and A. A. van Ballegooijen, Is there a weak mixed polarity background field? Theoretical arguments, in *Solar Phys.*, *110*, 115, 1987.

Taylor, J. B., Relaxation and magnetic reconnection in plasmas, in *Rev. Mod. Phys.*, *58*, 741, 1986.

van Ballegooijen, A. A. and P. C. H. Martens, Magnetic fields in quiescent prominences, in *Astrophys. J.*, *361*, 283, 1990.

Wolfson, R. L. T., Equilibria and stability of coronal magnetic arches, in *Astrophys. J.*, *255*, 774, 1982.

Wolfson, R., and B. C. Low, Energy buildup in sheared force free magnetic fields, in *Astrophys. J.*, *391*, 353, 1992.

Zirin, H., *Solar Astrophysics*, (Cambridge Univ. Press: Cambridge), 1988.

B. C. Low, High Altitude Observatory, National Center for Atmospheric Research, P. O. Box 3000, Boulder, CO 80307-3000

Evolving Magnetic Structures and Their Relation to Coronal Mass Ejections

Joan Feynman

Jet Propulsion Laboratory, California Institute of Technology, Pasadena, CA

Solar activity regions are frequently concentrated into clusters which persist for many solar rotations. These activity complexes are associated with regions of weaker background magnetic fields which are most apparent after the activity itself has ceased. We call this combination of persistent activity and weaker background fields Evolving Magnetic Structures (EMS). Here we show examples of EMSs and describe the evolution of an EMS associated with major Coronal Mass Ejections (CME) and other solar and magnetospheric disturbances. We show that CMEs occurred in association with this EMS both when there were sunspots and when only the weaker background fields were present. We find that the time scale for important evolutionary changes in the EMS is about 3 to 6 weeks, i.e. about a half solar rotation or more. Thus studies of individual EMSs and their CMEs require that each EMS be observed continuously for several weeks, necessitating observations of both sides of the Sun.

1. INTRODUCTION

Statistical studies of solar activity have demonstrated conclusively that solar activity and emerging solar flux are not distributed at random on the solar surface (*Bumba and Howard* 1965; *Svestka*, 1976, *Gaizauskas et al.*, 1983; *Brouwer and Zwaan*, 1990). Instead solar activity appears to cluster; new bipoles tend to emerge in regions where flux has previously emerged *(Harvey and Zwaan, 1993)*; sunspots occur in groups; sunspot groups occur within activity centers, and these activity centers in turn cluster among themselves (*Brouwer, and Zwaan*, 1990).

Harvey and Zwaan (1993) studied newly emerging bipolar regions and found that at least 55% of magnetic bipoles (active regions) larger than 3.5 square degrees emerge as part of activity centers. New magnetic flux emerges over and over again within the same area.

The distribution of active regions and of weaker magnetic fields on the Sun was studied by Bumba and Howard (1965) who concluded that active regions were concentrated into complexes of activity associated with the development of larger regions of weaker background magnetic fields. Brouwer and Zwaan (1990) carried out a statistical study of the positions of 1,236 sunspot groups. They defined a sunspot nest as a relatively small space on the solar surface within which a succession of spot groups appears. About 1/3 of the sunspot groups in their study belonged to nests and these nests themselves had a strong tendency to cluster. The typical lifetime of the individual nests was from one to seven rotations. Gaisauskas et al., (1983) gave an example of a pattern that lasted for 6 months and contained 29 major active regions. Brouwer and Zwaan discussed a nest that lasted 12 rotations.

Magnetograms also clearly show the associated distinct patterns in the photospheric magnetic field. When the emergence of new sunspots ceases and the sunspots disappear a bipolar pattern of weaker magnetic fields containing a neutral line may still be seen. Sometimes, after one or two rotations with only the weaker fields, new sunspots again appear and flaring resumes.

It has been proposed that these clustered sunspots may represent the submerged flux loop arches from which active regions repeatedly emerge (*Zwaan and Harvey*, 1994). Parker (1984) argued that the clustering of solar activity found by Gaisauskas et al. (1983) requires that the surface of the Sun approximates an impenetrable barrier to magnetic flux, constrains possible dynamo models and gives essential information concerning the dynamics of solar convection.

There is also some evidence for even longer term patterning which may result in active longitudes and solar "hot spots" (c.f. *Bai,* 1990), however neither of these concepts are made use of in this paper. The relation of sunspot nests to active longitudes is uncertain [*Petrovay and Abuzeid*, 1991).

Here we will be concerned with clusters of activity centers, and the related regions of weaker fields. To emphasize the close relation of the complexes of activity and the regions of weaker fields it has been suggested (*Feynman and Hundhausen*, 1994) that they be considered as a single entity called an Evolving Magnetic Structures (EMS). This name emphasizes the fundamentally magnetic and evolutionary nature of this phenomena. EMSs are probably closely related to the evolutionary structures described much earlier by Kiepenheuer (1953) from Hα observations of sunspots and filaments. He reported that in their early stages sunspots emerged and there were few if any filaments. Later, as more active regions emerged, flaring occurred and filaments began to form, lengthen and rotate toward an east-west direction. In the final stage, several rotations later, the spots had all disappeared, flaring had ceased and only large scale quiescent filaments remained.

2. A QUIET SUN EMS

Figure 1 shows an example of an EMS at low solar activity during the approach to solar cycle minimum. Each of the six panels is a synoptic map of the photospheric magnetic field intensity as measured by the National Solar Observatory at Kitt Peak. A synoptic map is produced from daily measurements by showing the magnetic flux at central meridian passage collected over the 27 days of the solar rotation. The figure is a gray scale presentation in which the flux is represented in a dark (field into the Sun) and light (field out of the Sun) scale. Low field intensities appear as gray. The six panels of the figure show fields observed between the equator and a latitude of 40 degrees south during six successive rotations of the Sun. Note that between 70 and 120 degrees solar longitude there is structure in the magnetic field that returns with each rotation but evolves from one rotation to the next. Sunspots and newly emerging magnetic fields are observed in the first 3 or 4 rotations shown. This EMS contained several spot groups on some rotations. For example during the January 10 - February 6 rotation four south hemisphere spot groups (7645A, 7646, 7647, and 7649) passed the central meridian during approximately the same 24 hour period. In the final rotation shown in figure 1 only dispersed fields are seen. The magnetic field pattern has simplified and a magnetic neutral line is clearly seen. A filament more than 60 degrees in length formed in this neutral line.

3. CMEs AND EMSs

CMEs typically occur in association with flares and/or with filament eruptions (*Gosling et al.*, 1974; *Munro et al.*, 1979; *Webb and Hundhausen,* 1987, see however *McAllister et al*, 1996, for a discussion of a counter example). Likewise CMEs are associated with sheared magnetic fields and with the emergence of new magnetic flux (*Feynman and Martin*, 1995). In all of these cases, apparently the same type of large-scale coronal structure is destabilized and leaves the Sun (*Hundhausen,* 1994). Typical latitudinal widths of these structures are about 45 degrees (*Burkepile and St.Cyr*, 1993), much larger than the scale of the underlying flare or filament. It has been argued (*Feynman and Martin*, 1995) that, with the possible exception of the initial acceleration profile (*MacQueen and Fisher*, 1983), no clear characteristic has been identified that distinguishes between CMEs associated with flares, and those associated with quiescent filament eruption, or magnetic shear or newly erupting flux. This suggests that most CMEs, whatever their association, are initiated within the same type of large-scale, well organized, magnetic field regions (*Hundhausen*, 1988). Feynman and Hundhausen (1994) pointed out that an Evolving Magnetic Structure (EMS) is a single entity that involves flaring, active regions and quiescent filaments, magnetic flux eruption and magnetic shearing. Feynman and Martin (1995) suggested that CMEs take place during all stages of the evolution of the EMS and the CMEs are associated with flares or rising prominences (disappearing filaments), and erupting flux or shear depending on the stage of evolution of the EMS. An example of an EMS is given below and the hypothesis that CMEs take place throughout the evolution of EMS is tested.

4. AN ACTIVE SUN EMS

In March 1989 an activity center produced a series of a major CMEs, a series of large solar energetic proton enhancements and a very intense geomagnetic storm (c. f. *Feynman and Hundhausen*, 1994 and references therein). This activity center was a return of activity that appeared on at least one previous solar rotation and reappeared on the following rotation (*Joshi*, 1993), thus qualifying it as an EMS. Here we describe the evolution of that EMS and its relation to CMEs throughout its evolution.

Figure 2 shows stacked synoptic maps from the north solar hemisphere from November, 1988 (Carrington rotation 1809) through May, 1989 (rotation 1815). This

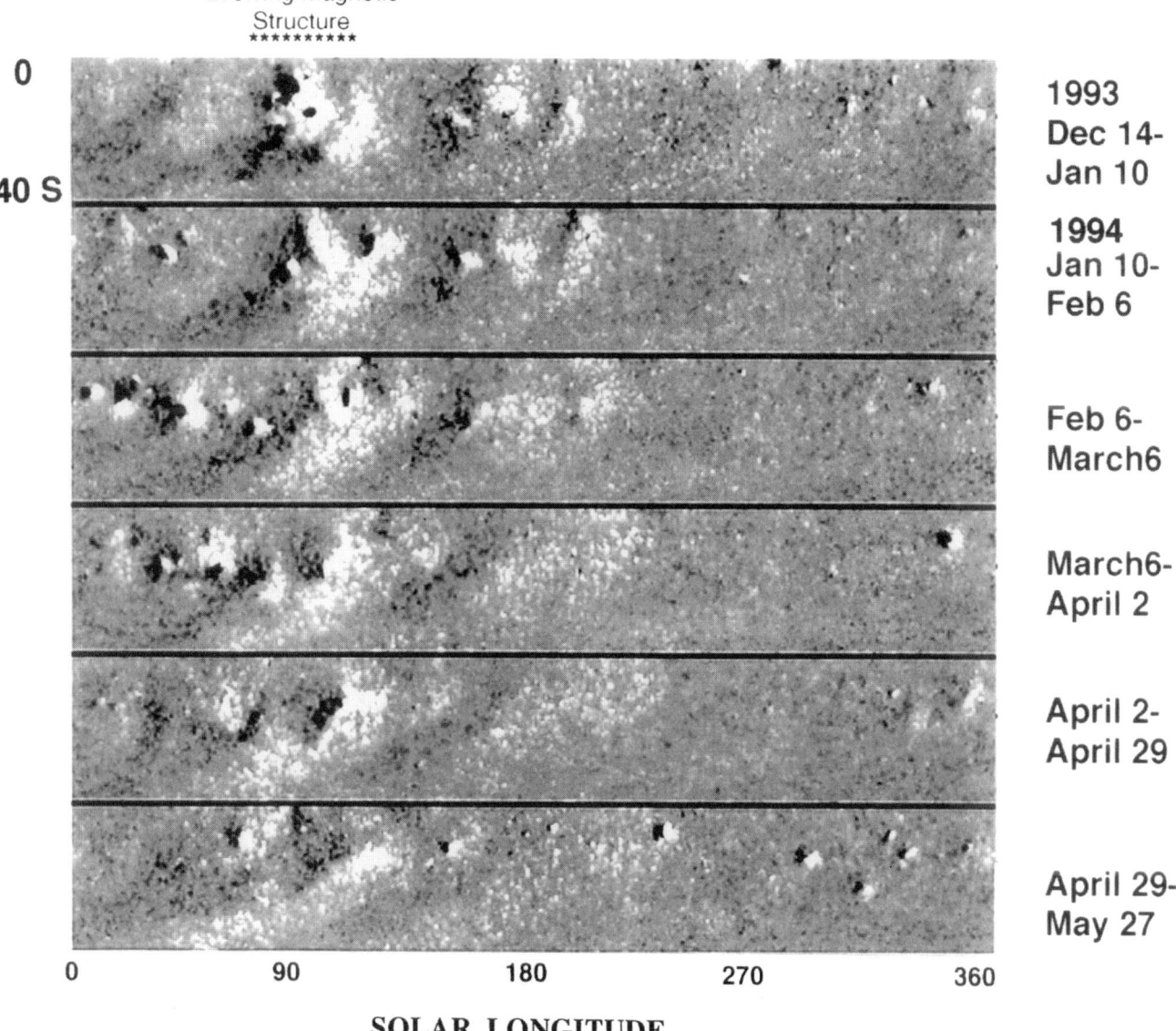

Figure 1. Stacked synoptic maps of the photospheric magnetic field during a period of low solar activity. The earliest rotation appears at the top of the figure. Latitudes between zero and forty degrees south are shown. An EMS is seen at about 90 degrees solar longitude. The data are from NSO/Kitt Peak.

was a very active period of the Sun and there were many activity centers and EMS candidates. As in the case of the EMS shown in Figure 1, changes in each of the activity centers took place during a single rotation. Near solar minimum it was simple to unambiguously identify the returns of a particular EMS because there was very little other activity. However, in Figure 2, although it is clear that activity returns, it is sometimes difficult to choose among the possible recurrences of activity complexes. The region that produced the March events appears on the strip labeled E and at a solar longitude of about 250 degrees. The returns on the previous and the following rotation identified by Joshi (1993) are shown between the white lines in the figure. The lines have been extended to earlier and later rotations. The data show evidence that this EMS existed during three rotations before those identified by Joshi. The activity during the rotation after the March events (second strip from the bottom) was much diminished. On the next rotation only weaker fields were seen. The rotation rate of this EMS was about 29 days. This is in agreement with the findings of Gaizauskas et al. (1983) who reported that each EMS has its own rotation rate.

52 EVOLVING MAGNETIC STRUCTURES AND CMES

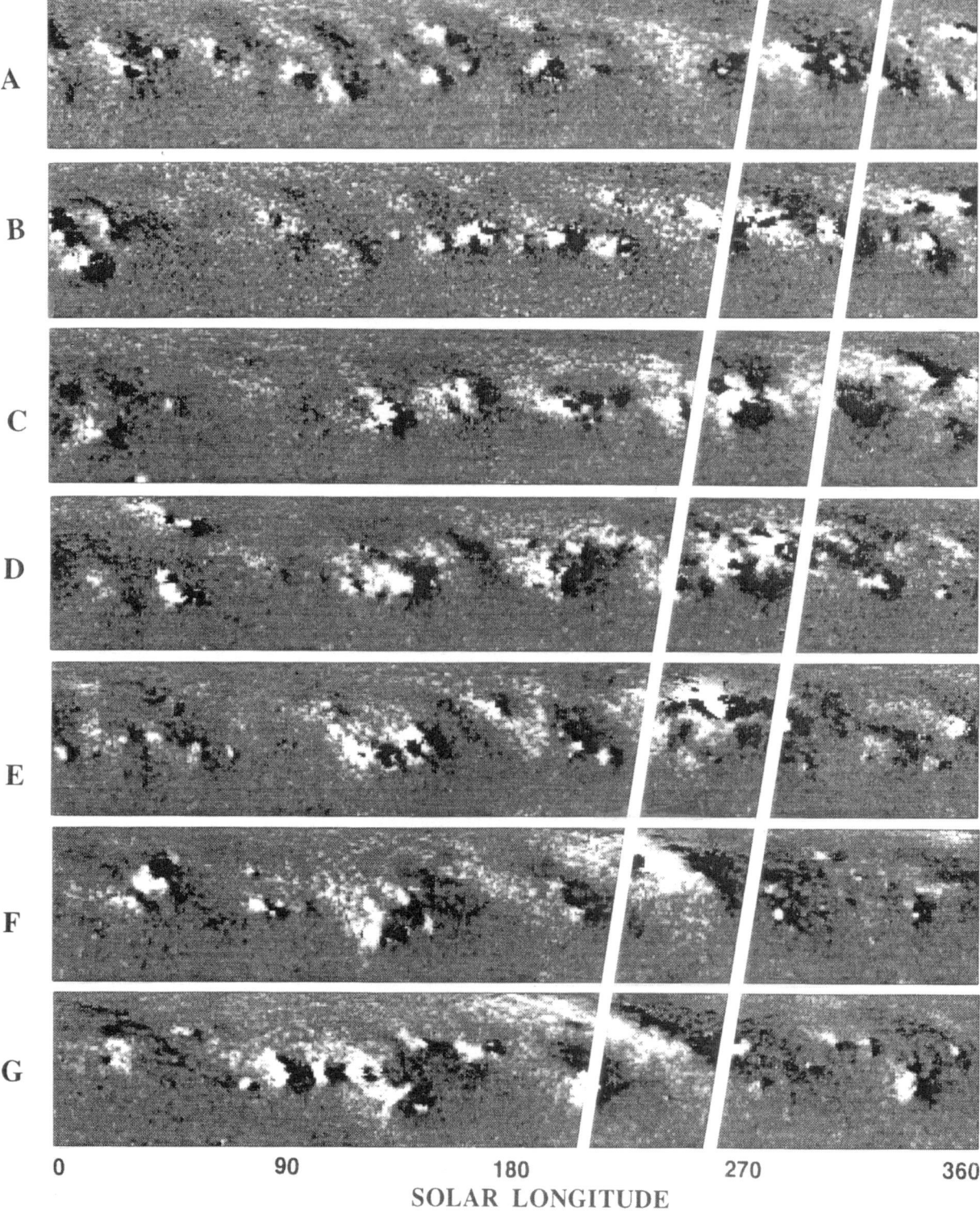

Figure 2. The same as Fig. 1 for an active period. The EMS is between the lines. Panel A, rotation -4, Nov. 15 -Dec. 12, 1988; B, rot. -3, Dec. 12, 1988-Jan 9, 1989; C, rot. -2, Jan 9-Feb. 5, 1989; D, rot. -1, Feb. 5-March 4, 1989; E, rot. 0, March 4-April 1, 1989; F, rot. +1, April 1- 28, 1989; G, rot. +2, April 28-May 25, 1989. See text.

4.1. Magnetic flux and flaring

EMS magnetic flux changes between returns and during disk passage. A rough estimate of the flux can be obtained from the sums of the areas of all the associated sunspots. Sunspot areas are routinely determined each day from full disk observations. Figure 3 gives an indication of the flux history for this EMS. The figure gives an estimate of the total area of the sunspots when the active region was at 60 degrees East and 60 degrees West. The East and West longitudes are chosen to be equal so that projection effects will not affect the relative flux determination. The sunspot areas are given in units of 10^{-6} of the visible solar hemisphere. The sunspot areas for the March rotation, the preceding two rotations, and the following rotation are shown. A solid line connecting points indicates that the region was between 60 degrees East and West on the face of the Sun. If there is a dashed line the region was closer than 60 degrees to the limb or on the far side of the Sun. In the discussion that follows, the rotation in which the March events took place will be referred to as rotation 0, the preceding rotations will be called -1, -2, etc. and the following rotations +1, +2 etc.

During rotation -2 the sunspot area increased from about 150 to about 1400 and the spot count increased from about 8 to 27. Remarkably, in spite of this large quantity of emerging flux, there was only one class M X-ray flare during that rotation. (X-ray events are classified as C, M, and X. C indicates 10^{-6} w/m^{-2} and M and X are each one factor of 10 higher.)

By the time the EMS reappeared (rotation -1) the area of the sunspots had further increased to about 1800 and there were about 32 individual spots. During rotation -1 nine X-ray flares of class M or above took place and both the area and spot count decreased by about a factor of two. The EMS appeared to be declining.

Although rotation -1 showed waning activity, by rotation 0 the area had increased to 2000. Even before the spot groups had come over the Eastern limb a series of high speed CMEs had begun (*Feynman and Hundhausen*, 1994).

During the 0 disk passage the area reached 3000 and the number of spots in the activity center reached 50. The sunspot magnetic configuration was highly complex and Joshi (1993) and Bai (1990) classified the region as "superactive". This activity center was the most prolific X-ray flare producer in the preceding 15 years. There were several spectacular events as the center of activity was carried across the face of the Sun (*Abbott et al.*, 1993). These include a March 6 event during which the GOES X-ray monitors went off scale. This was estimated to be a class X15 event, the largest ever seen to that date. On March 9 an X4 event occurred. March 10 had an X4.6 event associated with the CME that caused the geomagnetic storm of March 13. This storm included the most disturbed 24 hour interval in 120 years (*Allen et al.*, 1989). On March 16 a 3B flare occurred accompanied by bright surges over 34

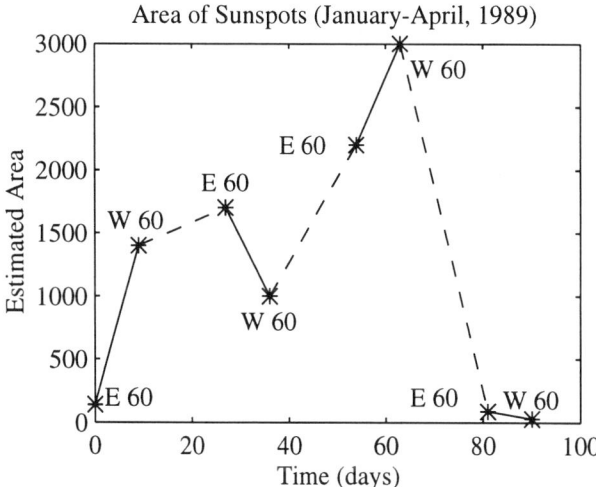

Figure 3. An estimate of sunspot area in the EMS during four returns. The time scale is in days, and the zero day is chosen to be the time when the region was at 60 degrees East on rotation -2. Two data points are given for each rotation, when the active region was 60 degrees East and West of central meridian passage.

degrees in extent. Altogether during disk passage 0 there were 35 class M flares and an additional 10 class X events. At least 195 individual optical flares were observed. Three major CMEs were observed (March 6, March 9, and March 17). The associated solar energetic particle events were the largest since 1972. None of this activity could have been predicted because the re-energization of the EMS took place while it was on the far side of the Sun.

When this EMS was at 60 degrees West and about to go over the limb its area was almost the same as it had been at central meridian passage. However, while it was unobserved, on the far side of the Sun, the area and number of sunspots declined precipitously. Only a small area (90) and a few spots (~3) where present on the + 1 return. It would have been extremely interesting to observe the phenomena associated with this precipitous decline of a "superactive" region.

Not unexpectedly, there were no sunspots during the +2 return. However a huge quiescent filament had developed in the neutral line of the dispersed fields. The filament was over 160 degrees in longitude and over 60 degrees in latitude. During this disk passage a section of the filament erupted and a section also disappeared during rotation +3 (not shown).

4.2. CME History

The association of CMEs with various indicators of solar activity has been studied in detail for the March (0) passage of this EMS (*Feynman and Hundhausen*, 1994). It was found that neither disappearing filaments, nor two ribbon flares, nor active dark filaments, nor major filament activity were statistically associated with CMEs at this time. Type II and Type IV radio bursts, long duration x-ray events and post flare loop systems showed strong associations with CMEs. The radio events and long duration X-ray events could not be used for CME identification in the present study because these phenomena relied on full disk observations which did not allow identification of the specific source of the event. There were too many other active regions on the Sun at this time. Although post flare loops showed a strong association with CMEs, they are rare and not commonly used as proxies for CMEs. Thus we have relied on coronal observations supplemented by observations of disappearances of the quiescent filaments that form in the late spotless stages of the EMS. For a discussion of the use of quiescent filament disappearances as proxies for CME see Feynman and Martin (1995).

Studies of the CME history of this EMS were hampered by the usual difficulty that CMEs are most clearly visible when they occur within 2.5 days of limb passage. If they are closer to center disk they may not be projected against the plane of the sky or, if seen, it may be difficult to ascribe them to one active region rather than another. Furthermore, CMEs are often not centered over the associated active region (*Harrison and Sime*, 1989). Table 1 gives the number of CMEs that appear to be associated with the activity complex near the limb on the basis of the CME position and time of eruption. The data are from the handbook compiled by Burkepile and St.Cyr (1993). During this period coronal observations were taken at approximately 90 minute intervals. In the seven rotations studied a total of 33 CMEs were observed at about the correct place and time. The East and West limb CMEs during rotation 0 have been described at length in the literature (*Feynman and Hundhausen*, 1994). (Note that, in addition to the CMEs counted in the table, the region is known to have produced a high speed CME on March 10. This CME caused a major magnetic storm. It is not included in the table because it occurred when the EMS was at 22 degrees East and the line of sight projection against the sky was far to the north of the 31 degrees latitude of the EMS. We wish to treat all the rotations the same in this study and so we omitted this CME from the table.). During the returns after rotation 0 there were 7 CMEs listed when the center of activity would have been near the limb if it had still been active. In addition, continuing CME activity can also be inferred from the two quiescent filament eruptions that occurred during rotations +2 and +3. Thus CME activity occurred during seven rotations of this EMS including rotations in which it contained spots and flaring and rotations when it contained only quiescent filaments.

The rate of CME production of this region can be compared with the average CME production rate. During 1988 and 1989 the average observed CME rate was a little

TABLE 1. Coronal Mass Ejections

Appearance Number	East Limb	West Limb
-3	4	4
-2	7	3
-1	2	1
0	3	2
+1	1	0
+2	0	1
+3	0	5

less than 1 per day (*Burkepile and St.Cyr*, 1993). The median width of the CMEs was about 42 degrees. In a given 40 degree interval we would expect to see a CME at a rate of about 1 per 10 days. The activity complex is near one limb or the other for 10 days per rotation. Thus the expected rate is 1 CME/rotation. The observed rate is between 4 and 5 CMEs / rotation.

5. FUTURE EMS STUDIES - OBSERVATIONAL DIFFICULTIES AND THEIR SOLUTIONS

EMSs have typical lifetimes of 6 months or more during which they may repeatedly give rise to coronal mass ejections and have important solar-terrestrial effects. The extraordinary tendency for repeated flux emergence over such long time periods and the well ordered background fields suggest that understanding EMSs is necessary for understanding the solar cycle and the solar dynamo (*Parker*, 1984). In the example studied here daily observations showed that the EMS flux built to a maximum while it was on the far side of the Sun, major CMEs took place when the region was on this side of the Sun and the EMS exhausted itself precipitously while on the far side of the Sun. These three important steps in the evolution, build up, CMEs and decline, took place during a period of about one and a half solar rotations. Thus there is a serious observational difficulty. Assuming one and a half solar rotations is a typical time scale, not all of the crucial evolutionary steps for a single EMS can be observed. In the case discussed here the buildup to maximum and the rapid decline where not actually observed.

This observational problem has plagued studies of EMSs and contributed to the curious situation that, although EMSs appear to be fundamental to the operation of the solar dynamo, very little is known about them except that they exist and are common. This is because only statistical studies of the general properties of EMS can be made using data from a single solar hemisphere., Individual EMSs have lifetimes of many months and important evolutionary changes within them take place on time scales of weeks. Thus, if they are observed hourly or daily from only one side of the Sun, the evolutionary changes within any one individual evolving structure can not be satisfactorily followed. In fact, without observing the other side of the Sun, it is often difficult to determine whether or not an individual sunspot group appearing on the east limb is the return of an EMS last seen 13 days earlier disappearing over the west limb. Since these phenomena include some of the most important entities for understanding the operation of the solar activity cycle, the initiation and acceleration of CMEs and forecasting and monitoring solar weather, it is vital that the handicap of single-solar-hemisphere observations be overcome.

The most straightforward way to accomplish this would be to station a spacecraft so that it views the other side of the Sun continuously. An orbit that accomplishes this has been identified. Data from the other side of the Sun (combined with whatever observations are taking place on this side of the Sun) would allow the EMS to be observed 20 of the 27 days of a solar rotation. (During the other 7 days the EMS would be too close to the limb.) Observations would then be continuous enough to study the evolution of individual EMSs. Such observations would very much facilitate studies of the role EMSs play in the solar cycle and in the operation of the dynamo [*Parker, 1984*; *Zeldovich et al., 1983*].

Acknowledgments: I thank the referees for their comments and suggestions. I also thank Karen Harvey and Jack Harvey for help in obtaining the magnetic field measurements from the National Solar Observatory at Kitt Peak. NSO/Kitt Peak data used here are produced cooperatively by NSF/NOAO, NASA/GSFC, and NOAA/SEC. The research reported in this paper was carried out by the Jet Propulsion Laboratory, California Institute of Technology, under a contract with the National Aeronautics and Space Administration.

REFERENCES

Abbott, M. S., et al., Region 5395: A harbinger of solar activity cycle 22, G. R. Heckman, M. S. Abbott, L. L. Orita and R. L. Bass, NOAA TM ERL SEL, 1989.

Allen, J., H. Sauer, L. Frank, and P. Reiff, Effects of the March 1989 Solar Activity, EOS, Trans. of the *AGU, 70,* 1479, 1989.

Bai, T., Solar "hot Spots" are still hot, *Ap. J. Letters, 364,* L17-L20, 1990,

Brouwer, M. P. and C. Zwaan, Sunspot nests as traced by a cluster analysis, *Solar Phys., 129,* 221, 1990.

Bumba, V. and R. Howard, Large-scale distribution of solar magnetic fields, *Astrophys. J., 141,* 1502, 1965.

Burkepile, J. T. and O. C. St.Cyr, A revised and expanded catalogue of mass ejections observed by the Solar Maximum Mission Coronagraph, NCAR/TN-369+STR, NCAR, Boulder Colo., 1993.

Feynman J. and A. J. Hundhausen, Coronal mass ejections and major solar flares: the great active center of March 1989, *J. Geophys. Res., 99,* 8451, 1994

Feynman, J. and S. F. Martin, The initiation of coronal mass ejections by newly emerging magnetic flux, J. *Geophys. Res., 100,* 3355, 1995.

Gaizauskas, V., K. L. Harvey, J. W. Harvey and C. Zwaan, Large-scale patterns formed by solar active regions during the ascending phase of cycle 21, *Astrophys. J. 265,* 1056, 1983.

Gosling, J. T., E. Hildner, R. M. MacQueen, R. H. Munro, A. I. Poland and C. L. Ross, Mass ejections from the sun: A view from Skylab, *J. Geophys. Res. 79,* 4581, 1974.

Harrison, R. A. and D. G. Sime, The launch of coronal mass ejections : White light and X-ray observations in the low corona, *J. Geophys. Res. 94,* 2333, 1989.

Harvey, K. L. and Zwaan, C. Properties and emergence patterns of bipolar active regions: Size distributions and emergence frequencies, *Solar Phys., 148,* 85, 1993.

Hundhausen, A. J., *The origin and propagation of coronal mass ejections, p181,* Proceedings of the Sixth International Solar Wind Conference, V. J. Pizzo, T. E. Holzer, and D. G. Sime, eds. NCAR/TN 306+Proc , 1988.

Joshi, A., Superactive region AR 5395 of solar cycle 22, *Solar Physics 147,* 269, 1993

Kiepenheuer, K. O., *Solar Activity, p322,* The Sun, G. P. Kuiper, ed., University of Chicago Press, Chicago Ill. 1953.

MacQueen, R. M. and R. R. Fisher, The kinematics of solar inner coronal transients, *Solar Phys. 89,* 89, 1983.

McAllister, A. H., M. Dryer, P. McIntosh, H Singer, and L. Weiss, A large polar crown mass ejection and a "problem " geomagnetic storm: April 14-23, 1994, *J. Geophys. Res., 101,* 13,497, 1996.

Munro, R. H., J. T. Gosling, E. Hildner, R. M. MacQueen, A. I. Poland, C. L. Ross, The association of coronal mass ejection transients with other forms of solar activity, *Solar Physics, 61,* 201, 1979.

Parker, E. N., Magnetic buoyancy and the escape of magnetic fields from stars, *Astrophys. J. 281,* 839, 1984.

Petrovay K. and B. K. Abuzeid, Cluster analysis of the space time distribution of sunspot groups during solar cycle no. 20. *Solar Physics 131,* 231, 1991.

Svestka, Z., *Solar Flares,* D. Reidel Publ. Co. Dordrecht, Holland, 1976.

Webb, D. F. and A. J. Hundhausen, Activity associated with the solar origin of coronal mass ejections, *Solar Phys., 108,* 353, 1987.

Zeldovich, Ya, B., A. A. Ruzmaikin, and D. D. Sokoloff, *Magnetic Fields in Astrophysics,* Gordon and Breach, New-York, Paris, London, 1983.

Zwaan, C. and K. L. Harvey, Patterns in the solar magnetic field, p27, in *Solar Magnetic Fields,* M. Schussler and W. Schmidt, eds., Cambridge University Press, 1994.

───────────

Joan Feynman, Jet Propulsion Laboratory, California Institute of Technology, Pasadena, CA 91109.

The Initiation of Coronal Mass Ejections by Magnetic Shear

Zoran Mikić and Jon A. Linker

Science Applications International Corporation, San Diego, California

Theoretical MHD models have shown that magnetic shear, which can be induced in coronal magnetic fields by photospheric flows (including differential solar rotation), can lead to the destabilization of large-scale coronal magnetic arcades and coronal streamers. When the shear exceeds a critical threshold, helmet streamers erupt, with characteristics that are generally similar to observations of coronal mass ejections (CMEs). These results indicate that magnetic shear can be considered to be a candidate for the initiation of CMEs. We discuss the concept of magnetic shear in the corona, and we describe its role in the energization of the coronal magnetic field. We review some theoretical results on the shearing of axisymmetric coronal arcades and streamers, and we present preliminary results on the evolution of a three-dimensional model of the solar corona at solar minimum, including the eruption of magnetic fields that resemble a CME.

1. INTRODUCTION

Although CMEs have now been observed for over two decades, we still do not known how they are initiated. The principal reason for this lack of understanding is that CMEs have been observed on the solar limb, where it is difficult to correlate them with changes in the photospheric magnetic field. Since the coronal magnetic field cannot be measured in general, the magnetic structure of coronal streamers prior to, and during eruption, is not known. Two promising developments in observations of CME initiation have occurred recently. First, a significant correlation has been shown to exist between erupting filaments observed on the solar disk and emergence of new flux [*Feynman and Martin*, 1995], which implies that emerging flux may initiate CMEs, since CMEs frequently (but not always) contain an erupting filament. Second, interpretations of Yohkoh SXT observations have shown that CMEs may be associated with "dimming" of the X-ray corona [e.g., *Hudson et al.*, 1996]. These observational techniques can relate CMEs to conditions in the photosphere and corona, and may help to identify the mechanism of CME initiation.

In order to develop a complete understanding of CMEs, a corresponding *theoretical* study of CME initiation is required. Since a CME expels a large amount of mass (up to 10^{16} g) into the solar wind and liberates a substantial amount of thermal energy in the corona, a successful theoretical model must demonstrate how this energy can be stored in the corona prior to eruption, in addition to showing how this energy can be released impulsively.

Over the last several years, theoretical models of the large-scale corona have evolved considerably. Simple idealized models have given way to sophisticated models which, in the near future, will allow us to follow the three-dimensional evolution and eruption of coronal streamers that match measured photospheric magnetic fields. Such models are useful for developing intuition about the properties of coronal streamers, testing hypotheses about CME initiation mechanisms, understanding the signatures of interplanetary CME observations, and eventually, being able to forecast the trajectory of CMEs, a principal goal of the National Space Weather Program.

In this paper we describe the use of theoretical models to study the role of magnetic shear in the initiation of CMEs.

Our results indicate that, from a theoretical point of view, an increase of magnetic shear may be the mechanism by which coronal mass ejections are initiated. The role of magnetic shear will undoubtedly be clarified in the future by more realistic theoretical calculations and with better observations of CMEs.

2. MAGNETIC SHEAR

The magnetic field is the principal mechanism by which energy can be stored in the lower solar corona. The term "magnetic shear" has been used loosely to refer to the energized state of the magnetic field in the solar corona. To illustrate the definition of magnetic shear, consider the idealized situation of a force-free field [e.g., *Priest*, 1982, p. 119], a good approximation to the state of strong magnetic fields in active regions, in which magnetic forces dominate other forces. In this approximation, the magnetic field is the only source of energy. The potential magnetic field is the lowest energy state for a given flux distribution in the photosphere. Therefore, in order to release energy, the magnetic field needs to be "energized" above the potential state. The magnetic field can be energized by being twisted; the electric currents associated with this twist provide a source of free energy that can, in principle, be released during an eruption. (Within the force-free model, the electric currents associated with this twist would flow along the magnetic field.) One way in which the coronal magnetic field can be twisted is by (non-uniform) photospheric flows. Convective flows in the dense photosphere tend to move the footpoints of the magnetic field lines that penetrate it through the effect of line tying [e.g., *Raadu*, 1972; *Einaudi and Van Hoven*, 1981; *Priest*, 1982]. We refer to the twist introduced by this effect as *photospheric shearing*. Magnetic twist can also arise in the corona from the eruption of twisted (current-carrying) fields from below the photosphere [e.g., *Leka*, 1995; *Lites et al.*, 1995]. The twist that energizes the magnetic field above the potential state has been termed *magnetic shear*. Note that magnetic shear includes twisting by photospheric shear flows as well as the twist introduced by eruption of twisted magnetic flux from below the photosphere.

In the context of coronal mass ejections, the situation is more complicated than in the idealized case of force-free magnetic fields, since additional sources of energy, including the kinetic energy in the flowing solar wind, gravitational potential energy, and thermal plasma energy, become important. Nevertheless, the concept of magnetic shear can still be used to express the energization of the coronal magnetic field.

In this paper we discuss magnetic shear introduced by the effect of photospheric flows only. The important effect of increasing magnetic shear through emerging flux is not addressed here. We describe how photospheric shear can store energy in the coronal magnetic field, and how it can lead to eruption. Such photospheric flows can arise from differential rotation and other large-scale flows in the photosphere. We therefore refer to the "shearing" of the large-scale coronal field by photospheric flows.

Magnetic field observations that are readily available today (e.g., synoptic maps of the normal component of the magnetic field in the photosphere deduced from line-of-sight magnetograms, from Wilcox Solar Observatory and the National Solar Observatory at Kitt Peak) cannot specify the magnetic shear in the corona. In order to fully specify the state of the corona, and in particular, its level of energization, it is necessary to specify the transverse component of the magnetic field. Thus, a fundamental aspect of the state of solar magnetic field is not provided by these magnetograms, as they do not allow configurations with different levels of magnetic shear to be distinguished. As is well known from studies of magnetic fields in active regions, a *vector* magnetogram is required to uniquely specify the magnetic field [*Mikić and McClymont*, 1994; *Mikić et al.*, 1996]. In principle, full-disk vector magnetograms can provide information about the transverse component of the magnetic field. Whether this can be done with sufficient accuracy to determine the relatively weak large-scale field is not known at present. In the future, the development of this capability will improve our ability to model and assess the state of magnetic shear in the solar corona.

Let us illustrate this situation by means of an idealized example. Consider the case of an axisymmetric equilibrium. The equilibrium magnetic field in the corona for a given normal magnetic field distribution in the photosphere (e.g., that corresponding to a dipole) can be found by solving the steady-state MHD equations with finite resistivity. We normally find the steady-state solution using a relaxation scheme, by integrating the time-dependent MHD equations to steady state with a fixed normal component of the magnetic field in the photosphere. The resulting solution is a "minimum magnetic shear" solution (loosely speaking) for the given boundary conditions, since it is found by a relaxation procedure in the presence of plasma resistivity. It will be the solution with the smallest twist in the magnetic field compatible with the specified normal magnetic field. The solution consists of the canonical coronal streamer configuration described by Pneuman and Kopp [1971], containing a dense closed-field region, in which the solar wind flow is arrested, surrounded by open magnetic field lines along which the solar wind flow streams to supersonic velocities. The magnetic field is potential (current-free) nearly everywhere, except in a narrow layer surrounding the open-closed field boundary, and along a sheet that extends from the tip of the coronal streamer to infinity, at which the current is concentrated into a sheet. Figure 1 shows an example of a configuration we computed with a dipole magnetic flux distribution [*Linker and Mikić*, 1995]. The current sheet that borders the open/closed field

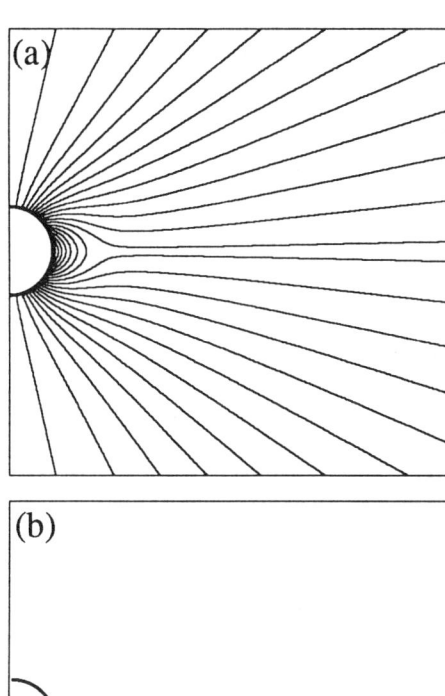

Figure 1. (a) Magnetic field lines (contours of constant flux ψ), and (b) contours of the azimuthal current density (J_ϕ) in an axisymmetric helmet streamer. A current sheet separates fields of opposite polarity and bounds the closed field region; away form the current sheet the field is nearly potential. The solar wind plasma flows outward along field lines in the open-field region, but is essentially stationary inside the closed-field region.

boundary is caused by the discontinuity of the magnetic field pressure, which in turn is induced by the discontinuity in plasma pressure between the open and closed field regions.

In this idealized axisymmetric case, the field lines have an arcade-like geometry in the closed-field region, where the field is entirely poloidal, with zero azimuthal (B_ϕ) magnetic field. In addition, the parallel component of the electric current density is everywhere zero. It is this component that is associated with the twist in the magnetic field. Therefore, this state is not likely to represent the state of the corona at any time, since it corresponds to a magnetic field that has no twist in it. This field is not likely to erupt without additional energization.

This equilibrium can be energized by twisting the magnetic field. In general, photospheric flows will twist the magnetic field, introducing magnetic shear. As noted above, without detailed observations of the transverse magnetic field it is not possible to uniquely determine the magnetic shear in the coronal field, so that at present we are limited to using phenomenological photospheric shear flow profiles to increase the twist of the magnetic field to a level that will cause eruption.

Therefore, the magnetic shear in our "initial" equilibrium state is not likely to correspond to that of the corona at any time, but is merely a starting point for our computations. In particular, this is the reason that a significant amount of shear is needed to cause the first eruption in our simulations (described below), since we start with a state which has minimum magnetic shear. Subsequent eruptions do not require as much shear. In principle, once vector magnetic field measurements become available, it may be possible to construct a coronal magnetic field equilibrium with a level of magnetic shear that corresponds to the state of the corona corresponding to the particular time of the magnetic field observation. At present, lacking more complete observational information about the state of the magnetic field, our best alternative is to energize the field (by applying phenomenological photospheric flows) to a level of magnetic shear that causes eruption to occur.

3. AN MHD MODEL OF THE SOLAR CORONA

A self-consistent description of the large-scale solar corona requires the coupled interaction of magnetic, plasma, and solar gravity forces, including the effect of the solar wind. In the magnetohydrodynamic (MHD) model, the coronal plasma is described by the following equations:

$$\nabla \times \mathbf{B} = \frac{4\pi}{c} \mathbf{J} , \qquad (1)$$

$$\nabla \times \mathbf{E} = -\frac{1}{c} \frac{\partial \mathbf{B}}{\partial t} , \qquad (2)$$

$$\mathbf{E} + \frac{1}{c} \mathbf{v} \times \mathbf{B} = \eta \mathbf{J} , \qquad (3)$$

$$\frac{\partial \rho}{\partial t} + \nabla \cdot (\rho \mathbf{v}) = 0 , \qquad (4)$$

$$\rho \left(\frac{\partial \mathbf{v}}{\partial t} + \mathbf{v} \cdot \nabla \mathbf{v} \right) = \frac{1}{c} \mathbf{J} \times \mathbf{B} - \nabla p - \nabla p_w + \rho \mathbf{g} + \nabla \cdot (\nu \rho \nabla \mathbf{v}) , \qquad (5)$$

$$\frac{\partial p}{\partial t} + \nabla \cdot (p \mathbf{v}) = (\gamma - 1)(-p \nabla \cdot \mathbf{v} + S) , \qquad (6)$$

where $\mathbf{B}$ is the magnetic field intensity, $\mathbf{J}$ is the electric current density, $\mathbf{E}$ is the electric field, $\mathbf{v}$, p, and ρ are the plasma velocity, pressure, and mass density, $\mathbf{g}$ is the gravitational acceleration, η is the plasma resistivity, ν is the kinematic plasma viscosity, S represents energy source terms, and the wave pressure p_w represents the acceleration due to Alfvén waves.

The application of this model to the structure and dynamics of the solar corona and inner heliosphere is discussed by Linker and Mikić [1997, in this volume]. Our thermodynamic model is presently being improved by including the effects of coronal heating, parallel thermal conduction, radiation loss, and acceleration due to Alfvén waves. We do not consider these improvements here. In the present application, we use the polytropic model [*Parker*, 1963] with an adiabatic energy equation ($S = 0$) with $\gamma = 1.05$, and no Alfvén waves ($p_w = 0$).

We have developed a three-dimensional code to solve equations (1)–(6) in spherical coordinates, as described by Mikić and Linker [1994]. This time-dependent model has been used to study the evolution of axisymmetric magnetic arcades [*Mikić and Linker*, 1994] and helmet streamers [*Linker and Mikić*, 1995], as well as the three-dimensional structure of the corona [*Mikić and Linker*, 1996; *Linker et al.*, 1996; *Linker and Mikić*, 1997].

4. DISRUPTION OF AXISYMMETRIC CORONAL ARCADES AND STREAMERS

Our efforts to understand CMEs have focused on first distilling the essential physics from the simplest model possible (disruption of axisymmetric coronal magnetic arcades), and then incorporating these effects into more realistic models of the solar corona (including the effect of the solar wind, differential rotation, and three-dimensional geometry).

The properties and stability of coronal magnetic fields have been studied extensively [see the references in *Mikić and Linker*, 1994]. In order to study the theoretical aspects of CME initiation, we started with the simplest model possible: we assumed zero beta [i.e., magnetic forces dominate plasma forces, so that we can neglect ∇p in Eq. (5)], a fixed density, we neglected gravity, and we modeled two-dimensional (axisymmetric) variation. We investigated the dynamical evolution of an initially dipolar magnetic field arcade subjected to idealized photospheric shearing motions [*Mikić and Linker*, 1994]. The calculations were performed using both the ideal and resistive MHD equations. When an arcade is subjected to a photospheric shear flow profile, the arcade evolves quasi-statically for small amounts of shear. During this phase, poloidal magnetic field (B_r, B_θ) is converted into azimuthal magnetic field (B_ϕ), and the magnetic energy increases with increasing magnetic shear as the field becomes more and more twisted; energy is therefore stored in the magnetic field. However, when a critical shear is exceeded, the field expands rapidly and produces a concentration of the electric current density. An ideal MHD (i.e., zero resistivity) calculation shows that a transition to a partially open configuration occurs at the critical shear value. In this state a small fraction of the magnetic field lines are closed but the majority of field lines (97% of the flux) are open. The open field lines of opposite polarity are separated by a tangential discontinuity. The magnetic energy of this partially open configuration is close to but less than the energy in a fully open field [*Aly*, 1984, 1991; *Sturrock*, 1991].

The transition to a partially open field requires an initially smooth magnetic field to evolve into one with discontinuities; this process has been described as magnetic nonequilibrium [*Parker*, 1972, 1979; *Priest*, 1981; *Vainshtein and Parker*, 1986]. The appearance of a discontinuity implies that even a small amount of plasma resistivity is important. When we included finite resistivity, the discontinuity was resolved into a current sheet which was subsequently the site of rapid magnetic reconnection, leading to fast flows and the ejection of a plasmoid [*Mikić and Linker*, 1994]. These results suggest that CMEs may be initiated by the destabilization of magnetic arcades by photospheric shear. The rapid inflation of the field at the critical shear value was also found by Roumeliotis et al. [1994].

Having found the underlying cause of the disruption of magnetic configurations, we applied our model to a more realistic equilibrium. A comparison with CME observations requires the important effect of the solar wind to be included. We started with an axisymmetric helmet streamer equilibrium corresponding to a dipole magnetic field distribution at the solar surface [*Pneuman and Kopp*, 1971], which was developed by integrating the time-dependent resistive MHD equations to steady state [*Steinolfson et al.*, 1982; *Linker et al.*, 1990; *Wang et al.*, 1993]. The evolution of this field in response to photospheric shear flow is qualitatively similar to that of the arcade. The closed-field region initially expands slowly as the field evolves quasi-statically; when a critical shear is reached, the magnetic field lines erupt outward, driving plasma into the outer corona. While the underlying reason for the disruption is the same in both cases (ideal MHD magnetic nonequilibrium), once the streamer begins to rise, the plasma within it accelerates into the solar wind, stretching and opening the magnetic field lines, and creating a current sheet at which the low-lying loops subsequently reconnect [*Linker and Mikić*, 1995]. As the reconnection proceeds, the closed-field region grows in size as successively higher loops reconnect [*Kopp and Pneuman*, 1976], a phenomenon that has been observed in Yohkoh soft X-ray images [*Hiei et al.*, 1993; *Tsuneta*, 1996].

These studies of arcades and helmet streamers used idealized photospheric flow profiles to induce the magnetic shear. One component of photospheric motions that may contribute to the energization of large-scale coronal fields is differential rotation. We have investigated how differential rotation [*Snodgrass*, 1983] affects axisymmetric coronal streamers over many rotations. As in the case of the idealized shear profile, the streamer disrupts when a critical

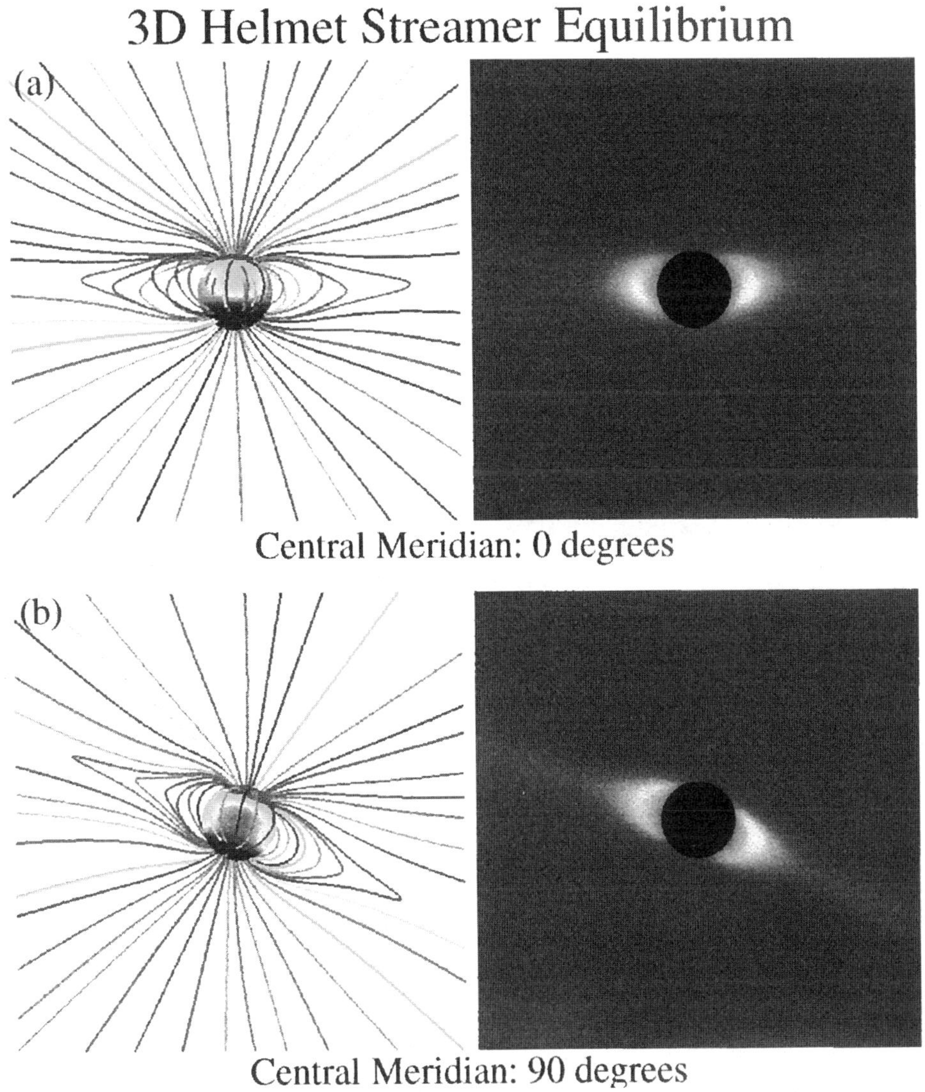

3D Helmet Streamer Equilibrium

(a)

Central Meridian: 0 degrees

(b)

Central Meridian: 90 degrees

Figure 2. An idealized three-dimensional model of the solar minimum corona. Shown on the left are traces of magnetic field lines; on the right is the corresponding computed plane-of-sky polarization brightness. (a) The view at a central meridian of 0, and (b) the view at a central meridian of 90°. It is apparent that the helmet streamer belt is tilted and warped.

shear is exceeded. With continued differential rotation, the streamer disrupts recurrently [*Linker et al.*, 1994]. A more complete understanding of the role of differential rotation in initiating CMEs requires three-dimensional calculations with more realistic fields, as described below.

5. DISRUPTION OF THREE-DIMENSIONAL CORONAL STREAMERS

The axisymmetric results provide, at best, a qualitative argument that magnetic shear may be a plausible mechanism for the initiation of CMEs. To proceed beyond this qualitative agreement with the properties of CMEs, it will be necessary to compare the details of the eruption with observations (e.g., frequency of eruption, requirements on the photospheric shear profile, the signature of the eruption in coronagraphs, the properties of the plasmoid in interplanetary space). We have already begun this task by studying the propagation of an erupted three-dimensional plasmoid through interplanetary space to 1 A.U. [*Linker and Mikić*, 1997]. Here we present preliminary results on the extension of our studies to the shearing and eruption of three-dimensional helmet streamers.

Our first studies of the evolution and stability of three-dimensional magnetic arcades were made using a zero-beta

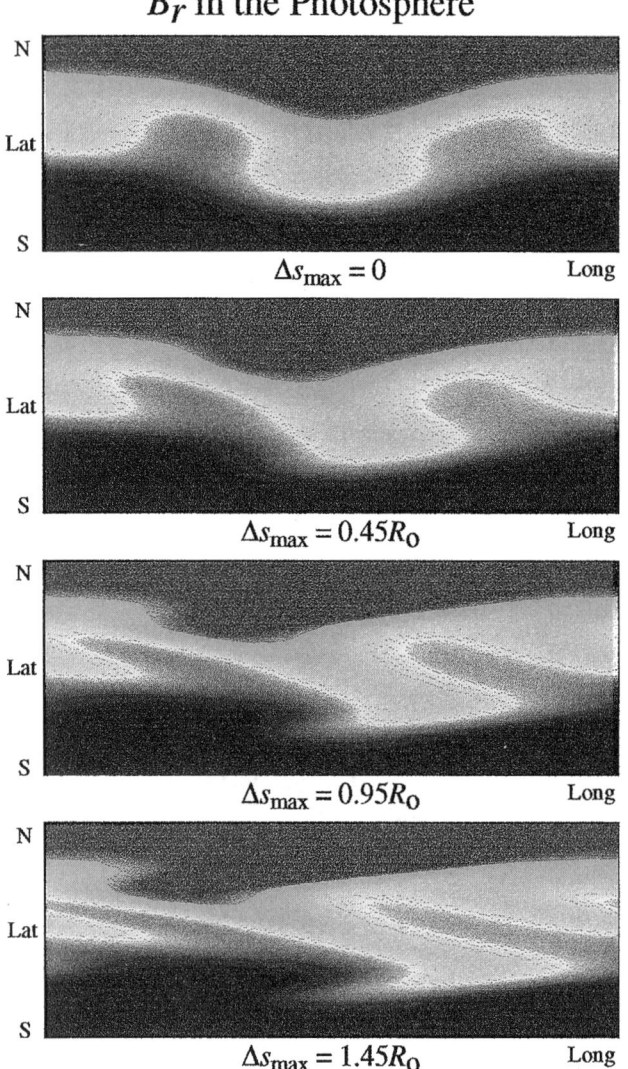

Figure 3. The evolution of the radial component of the magnetic field in the photosphere, B_r, as a function of the maximum footpoint displacement, Δs_{max}. Note that the photospheric flow changes the magnetic flux in the solar surface.

model without the solar wind. The results indicated that arcades erupt beyond a critical shear, as in the axisymmetric case. Here we describe the extension of this model to include the effect of the solar wind. To generate a 3D helmet streamer equilibrium, we first find the potential field in the corona that corresponds to the following radial magnetic field in the photosphere:

$$B_r = B_0(A_0 \cos^3\theta + A_2 \sin^2\theta \cos 2\phi + A_3 \sin^3\theta \sin 3\phi),$$

with $B_0 = 13.3$ Gauss, $A_0 = 1$, $A_2 = 0.1$, and $A_3 = 0.025$, rotated by $-20°$ about the y-axis. Here (r,θ,ϕ) are spherical coordinates: θ is the co-latitude ($\theta = 0$ is the North pole; $\theta = 180°$ is the South pole), and ϕ is the longitude. (The y-axis is the line $\theta = 90°$, $\phi = 90°$.) This is an idealization of the magnetic field at solar minimum, with a warped, asymmetric neutral line and relatively smooth fields, that resembles the large-scale magnetic field for the observational data we have studied, producing a tilted, warped heliospheric current sheet.

The potential field corresponding to this normal field at the photosphere was computed as an initial state, and the resistive MHD equations were integrated to steady state to compute a 3D coronal streamer equilibrium. Figure 2 shows the magnetic field lines in the equilibrium, along with the computed polarization brightness, at two choices of central meridian longitude. It is readily apparent that this is a three-dimensional configuration.

The evolution of this equilibrium was then followed in response to an applied photospheric shear flow. We used an idealized shear profile similar to that used previously for axisymmetric arcades [*Mikić and Linker*, 1994], with $\mathbf{v} = v_\phi(\theta)\hat{\phi}$, and a width $\Delta\theta_m = 30°$. (Note that this profile does not correspond to differential rotation; the evolution of this field for a differential rotation profile is discussed by Linker and Mikić [1997].) In this three-dimensional equilibrium, the normal component of the magnetic field in the photosphere changes as a result of the advection of magnetic flux. Figure 3 shows the evolution of B_r in the photosphere. We parameterize the amount of shear introduced at any particular time in terms of the maximum displacement of a field line footpoint from its initial position, Δs_{max}, as in [*Mikić and Linker*, 1994].

In order to minimize the computational time required to perform this 3D numerical simulation, we used a shear flow velocity that is ~ 10 times larger than flows that are typically observed in the photosphere. The maximum shear flow velocity was 4.8 km/s, compared to typical photospheric flow velocities of 0.5–1 km/s; the photospheric flow velocity associated with solar rotation is 2 km/s. Our previous results [*Mikić and Linker*, 1994] imply that magnetic fields evolve quasi-statically prior to eruption. Our enhanced shear flow velocity is therefore expected to shorten the time required to reach a level of shear that leads to an eruption, and is not expected to affect the nature of the eruption significantly (as long as the shear flow velocity is small compared to the Alfvén speed). It is possible to use a physical value of the shear flow velocity in our simulations (at greater computational expense), and this will be done in future refinements of this calculation.

Figure 4 shows the evolution of selected magnetic field lines as a function of the footpoint displacement. Note that the field lines rise slowly as they become twisted. When the shear reaches a critical value ($\Delta s_{max} \sim 1.6 R_0$, where R_0 is the solar radius), the field lines begin to expand outward rapidly. This behavior is qualitatively similar to that observed in the axisymmetric case, although the field

Field Line Evolution

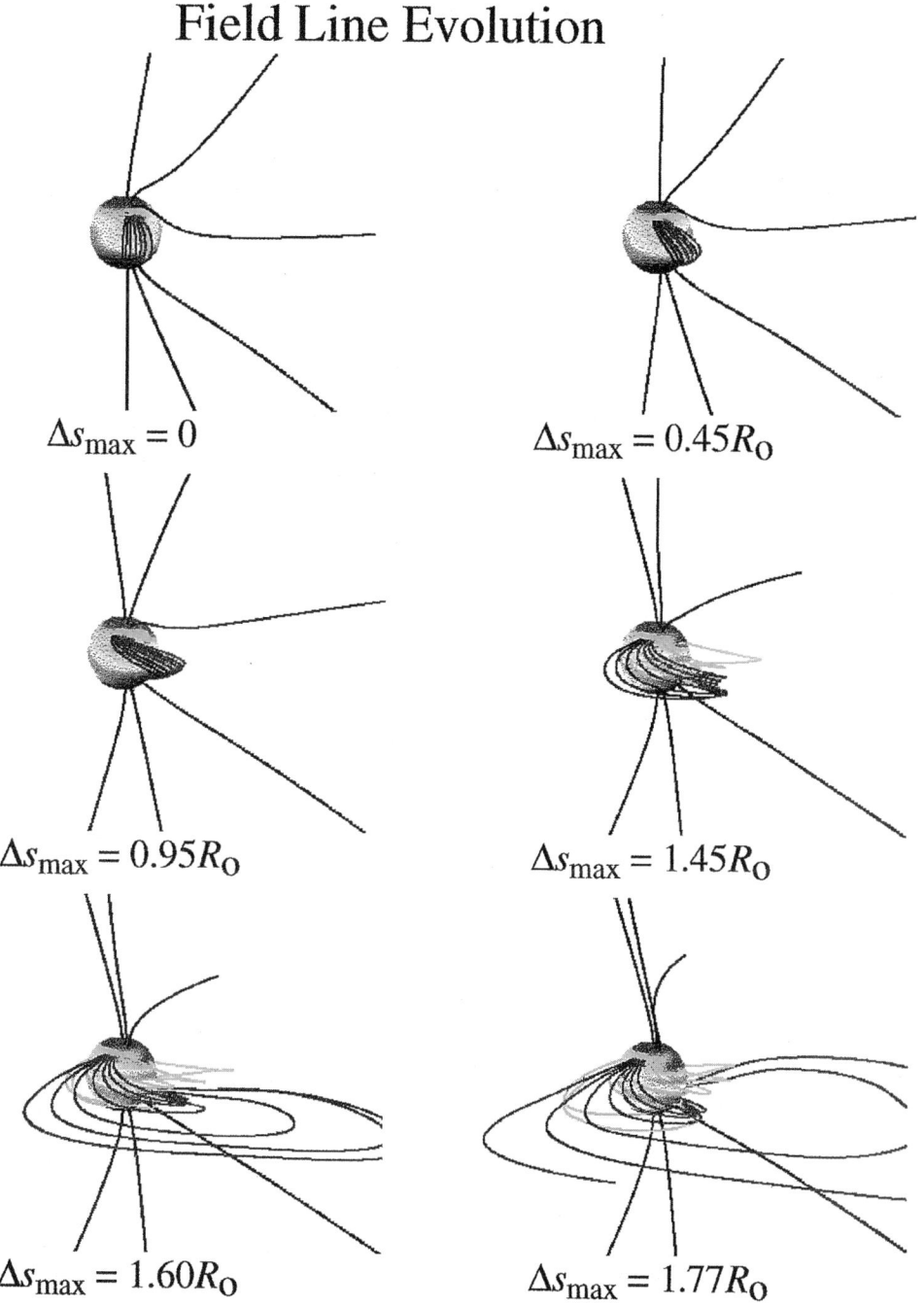

Figure 4. The evolution of selected magnetic field lines in a three-dimensional configuration as a function of the maximum footpoint displacement, Δs_{max}. The field lines initially evolve quasi-statically as they become twisted by the applied photospheric shear flow. However, when the twist approaches $\Delta s_{max} \sim 1.6 R_o$, there is a rapid upward motion of the field lines. This eruption is characteristic of a coronal mass ejection.

line geometry is considerably more complicated. The erupting magnetic field lines have a long wavelength structure, indicating that the eruption is global, with a large longitudinal extent. This feature of the eruption agrees with recent LASCO coronagraph images of CMEs, in which CME ejecta appear to leave the East and West solar limbs simultaneously.

The magnetic energy of the configuration decreases following the eruption, as magnetic reconnection takes place. The nature of the reconnection is complicated, and is not easily diagnosed. (We deduce that reconnection is taking place because the magnetic footpoint connectivity in the photosphere changes rapidly after the eruption.) The magnetic flux that is ejected has been followed as it propagates into interplanetary space in a related simulation [*Linker and Mikić*, 1997].

6. DISCUSSION

We have studied the evolution of 2D and 3D helmet streamer configurations in the presence of photospheric shearing flows. The streamers initially evolve quasi-statically, causing the closed field region to grow in size. When a critical shear is reached, the configurations erupt, sending plasma and magnetic fields into the outer corona. This eruption appears to be an ideal process, related to the tendency of the field to open when the energy of the field approaches the open field energy. Subsequent to this outward eruption of the field lines, reconnection of the magnetic field occurs. The nature of this reconnection process in three dimensions is quite complicated and is presently under investigation.

Our results indicate that magnetic shear may initiate coronal mass ejections. At present, this evidence must be regarded as qualitative, since it is based on the study of idealized magnetic field configurations. In future work it will be necessary to compare the predictions from our theoretical model with available observational data, especially the properties of plasmoids and "flux ropes" observed in the interplanetary medium, to obtain a more quantitative evaluation of the role of magnetic shear in CME initiation.

Acknowledgments. The research described here was supported by NASA, NSF, and AFPL. The computations were performed at the National Energy Research Supercomputer Center and the San Diego Supercomputer Center.

REFERENCES

Aly, J. J., *Ap. J.*, *283*, 349, 1984.
Aly, J. J., *Ap. J.*, *375*, L61, 1991.
Einaudi, G., and G. Van Hoven, *Phys. Fluids*, 24, 1092, 1981.
Feynman, J., and S. F. Martin, *J. Geophys. Res.*, *100*, 3355, 1995.
Hiei, E., A. J. Hundhausen, and D. G. Sime, *Geophys. Res. Lett.*, *20*, 2785, 1993.
Hudson, H. S., L. W. Acton, D. Alexander, S. L. Freeland, J. R. Lemen, and K. L. Harvey, Yohkoh/SXT Soft X-ray Observations of Sudden Mass Loss from the Solar Corona, in *Solar Wind Eight: Proceedings of the Eight International Solar Wind Conference*, edited by D. Winterhalter, J. T. Gosling, S. R. Habbal, W. S. Kurth, and M. Neugebauer, pp. 88–91, AIP Conf. Proceedings, *382*, AIP Press, Woodbury, N. Y., 1996.
Kopp, R. A., and G. W. Pneuman, *Sol. Phys.*, *50*, 85, 1976.
Leka, K. D., Ph. D. Thesis, University of Hawaii, 1995.
Linker, J. A., and Z. Mikić, *Ap. J.*, *438*, L45, 1995.
Linker, J. A., and Z. Mikić, Extending Coronal Models to Earth Orbit, in *Coronal Mass Ejections: Causes and Consequences* (this volume), 1997.
Linker, J. A., Z. Mikić, and D. D. Schnack, in *Proc. Third SOHO Workshop — Solar Dynamic Phenomena and Solar Wind Consequences*, Estes Park, Colorado, USA (ESA SP-373), 249, 1994.
Linker, J. A., Z. Mikić, and D. D. Schnack, in *Solar Drivers of Interplanetary and Terrestrial Disturbances*, edited by K. S. Balasubramaniam, S. L. Keil, and R. N. Smartt, Astronomical Society of the Pacific, Conference Series, *95*, 208, 1996.
Linker, J. A., G. Van Hoven, and D. D. Schnack, *Geophys. Res. Lett.*, *17*, 2281, 1990.
Lites, B. W., B. C. Low, V. Martínez Pillet, P. Seagraves, A. Skumanich, Z. A. Frank, and R. A. Shine, *Ap. J.*, *446*, 877, 1995.
Mikić, Z., and J. A. Linker, *Ap. J.*, *430*, 898, 1994.
Mikić, Z., and J. A. Linker, The Large-Scale Structure of the Solar Corona and Inner Heliosphere, in *Solar Wind Eight: Proceedings of the Eight International Solar Wind Conference*, edited by D. Winterhalter, J. T. Gosling, S. R. Habbal, W. S. Kurth, and M. Neugebauer, pp. 104–107, AIP Conf. Proceedings, *382*, AIP Press, Woodbury, N. Y., 1996.
Mikić, Z., J. A. Linker, and D. D. Schnack, in *Solar Drivers of Interplanetary and Terrestrial Disturbances*, edited by K. S. Balasubramaniam, S. L. Keil, and R. N. Smartt, Astronomical Society of the Pacific, Conference Series, *95*, 108, 1996.
Mikić, Z., and A. N. McClymont, in *Solar Active Region Evolution: Comparing Models with Observations*, edited by K. S. Balasubramaniam and G. W. Simon, Astronomical Society of the Pacific, Conference Series, *68*, 225, 1994.
Parker, E. N., *Interplanetary Dynamical Processes*, Wiley-Interscience, New York, 1963.
Parker, E. N., *Ap. J.*, *174*, 499, 1972.
Parker, E. N., *Cosmical Magnetic Fields*, 841 pp., Clarendon, Oxford, 1979.
Pneuman, G. W., and R. A. Kopp, *Sol. Phys.*, *18*, 258, 1971.
Priest, E. R., in *Solar Flare Magnetohydrodynamics*, edited by E. R. Priest, Gordon and Breach, London, 1981.
Priest, E. R., *Solar Magnetohydrodynamics*, 469 pp., Reidel, Dordrecht, 1982.
Raadu, M. A., *Sol. Phys.*, *22*, 425, 1972.
Roumeliotis, G., P. A. Sturrock, and S. K. Antiochos, *Ap. J.*, *432*, 847, 1994.
Snodgrass, H. B., *Ap. J.*, *270*, 288, 1983.
Steinolfson, R. S., S. T. Suess, and S. T. Wu, *Ap. J.*, *255*, 730, 1982.
Sturrock, P.A., *Ap. J.*, *380*, 655, 1991.
Tsuneta, S., *Ap. J.*, *464*, 1055, 1996.
Vainshtein, S. I., and E. N. Parker, *Ap. J.*, *304*, 821, 1986.
Wang, A. H., S. T. Wu, S. T. Suess, and G. Poletto, *Sol. Phys.*, *147*, 55, 1993.

Zoran Mikić and Jon A. Linker, Science Applications International Corporation, 10260 Campus Point Drive, San Diego, CA 92121

Coronal Mass Ejections: Causes and Consequences
A Theoretical View

James Chen

Plasma Physics Division, Naval Research Laboratory, Washington, D.C.

A physical model of coronal mass ejections (CMEs) and eruptive prominences is discussed in the framework of ideal magnetohydrodynamics (MHD). The initial structure is assumed to be an equilibrium magnetic flux rope embedded in the corona. The flux rope is identified with the prominence cavity and is assumed to contain a hot, low-density plasma component of average density $\bar{n}_c$ and a cold component of average density $\bar{n}_p$ corresponding to the prominence. The eruption of the model flux rope is caused by a form of emerging flux, i.e., an increase in the poloidal flux on the time scale of eruption. The initial eruption and subsequent propagation of the flux rope are investigated. It is assumed that the prominence material eventually falls back to the Sun. It is found that the initial cavity magnetic field and hot plasma component ($\bar{n}_c$) can evolve into a structure closely resembling observed interplanetary magnetic clouds at 1 AU and beyond. The flux ropes are stable to the ideal MHD kink mode throughout the eruption and subsequent propagation. In this picture, the classic bright rim of a CME is coronal gas plowed ahead of the expanding flux rope, and the prominence material is dragged along by the flux rope until much of it returns to the Sun. Thus, the CME-cavity-prominence structure represents three parts of one magnetically organized structure. The model results consistently reproduce the observed speeds of CMEs and eruptive prominences near the Sun, magnetic cloud expansion as a function of heliocentric distance, and the macroscopic properties of observed magnetic clouds.

1. INTRODUCTION

The Sun has fascinated scientists for centuries. To the casual observer, the Sun is constant and placid. To the more persistent (and fortunate) observer, it reveals a wealth of intriguing phenomena. Drawings of sunspots date back to at least the 17th century. (The existence of sunspots has been known for over 2000 years.) Solar flares were first discovered during routine observations [*Carrington*, 1860; *Hodgson*, 1860], and occasional prominences have been observed during eclipses for hundreds of years. The dramatic nature of eruptive prominences was finally elucidated by modern observational techniques.

In the early 1970's, a new form of eruptive phenomenon was discovered, now referred to as coronal mass ejections (CMEs). They were observed by coronagraphs on Orbiting Solar Observatory 7 (OSO-7) [*Tousey*, 1973], Skylab [*MacQueen et al.*, 1974], the P78-1 satellite (SOLWIND, 1979–1981) [*Michels et al.*, 1980; *Howard et al.*, 1985], and Solar Maximum Mission (SMM) (1980, 1984–1989) [*MacQueen*, 1980]. In these observations, CMEs appear as bright (line-of-sight density-enhanced) features expanding outward at speeds ranging from several tens of kilometers per second up to 2000 km/s. They represent discrete events in which large-scale magnetic field is disrupted, and significant amounts of plasma ($\sim 10^{15}$–10^{16} g) and magnetic energy ($\sim 10^{31}$–10^{33} erg) are injected into the solar wind [e.g., *Gosling et al.*, 1974].

Each successive experiment has given rise to new questions and new understanding of the phenomenon of CMEs and solar activity. The recent Large Angle Spectrometric Coronagraph (LASCO) experiment on Solar and Heliospheric Observatory (SOHO) continues this trend. With its improved sensitivity and spatial resolution and a field of view covering a Sun-centered region from 1.1 $R_\odot$ to 32 $R_\odot$, LASCO has yielded observations of solar eruptions affecting the corona on global scales [*Howard et al.*, this volume].

Following their discovery, CMEs were soon found to be correlated with the occurrence of geomagnetic storms [*Gosling et al.*, 1974]. It has long been realized that there are two distinct categories of magnetic storms, those that recur with the solar rotation period of 27 days [*Maunder*, 1905] and those that are nonrecurrent. Recurrent storms tend to be moderate, and their frequency is anticorrelated with sunspot numbers [*Greaves and Newton*, 1929]. In contrast, large storms tend to be nonrecurrent and occur near solar maximum. For nonrecurrent storms, association with solar flares was found [e.g., *Hale*, 1931; *Newton*, 1943], leading to the view that solar flares produced the geoeffective solar wind disturbances (e.g., shocks) responsible for nonrecurrent geomagnetic storms. (Shortly after the 1859 solar flare observed by *Carrington* [1860], a "great" geomagnetic storm occurred, but Carrington cautioned against making a causal association.) However, it gradually became clear that the disturbed solar wind streams associated with the occurrence of CMEs provide a better correlation with large nonrecurrent magnetic storms [e.g., *Gosling et al.*, 1991; *Kahler*, 1992; *Gosling*, 1993]. Earlier, *Joselyn and McIntosh* [1981] found that nonrecurrent storms tend to be correlated with eruptive prominences. Indeed, CMEs are more closely associated with prominence eruptions than flares [*Gosling et al.*, 1974; *Munro et al.*, 1979; *Sheeley et al.*, 1983; *Webb and Hundhausen*, 1987]. Even in the recurrent class of storms, CMEs propagating in the streamer belt may be responsible for large storms near solar minimum [*Crooker and Cliver*, 1994]. It is now argued that the earlier notion that such storms are caused by solar flares should be replaced by the view that the CMEs provide the key plasma link between solar eruptions and nonrecurrent geomagnetic storms [e.g., *Kahler*, 1992; *Gosling*, 1993].

A key question then is what solar wind structures result from CMEs and eruptive prominences. Empirically, interplanetary magnetic clouds [*Burlaga et al.*, 1981] have been associated with CMEs [*Klein and Burlaga*, 1982; *Wilson and Hildner*, 1984] and with eruptive prominences (in the disk view) [*Wilson and Hildner*, 1986; *Rust*, 1994; *Bothmer and Schwenn*, 1994; *Bothmer and Rust*, this volume; *Marubashi*, this volume]. A close association between CMEs and eruptive prominences has been established, from both a statistical point of view [*Gosling et al.*, 1974; *Munro et al.*, 1979; *Sheeley et al.*, 1983; *Webb and Hundhausen*, 1987] and a morphological point of view [*Illing and Hundhausen*, 1985, 1986]. Theoretically, it has been shown that initial equilibrium flux ropes compatible with quiescent prominence/cavity structures can evolve to resemble magnetic clouds observed near 1 AU and beyond [*Chen and Garren*, 1993; *Chen*, 1996]. A similar model was used to study the scaling behavior of expanding magnetic clouds in terms of heliocentric distance [*Kumar and Rust*, 1996]. Magnetic clouds constitute perhaps 1/3 of all interplanetary CMEs [*Gosling*, 1990] and are associated with perhaps 1/2 or more of large nonrecurrent geomagnetic storms [*Burlaga et al.*, 1987; *Tsurutani et al.*, 1988]. A large fraction of the magnetic clouds impinging on the Earth cause storms [*Wilson*, 1987; *Zhang and Burlaga*, 1987]. This is because magnetic clouds can impose long periods of southward IMF on the magnetosphere.

Much has been learned about CMEs and eruptive prominences, and several reviews have been written on CMEs [e.g., *Gosling et al.*, 1974; *Howard et al.*, 1985; *Hundhausen*, 1987, 1996; *Kahler*, 1987] and prominences [e.g., *Tandberg-Hanssen*, 1995]. However, the fundamental question of what causes the eruption of CMEs and prominences remains unanswered. In spite of the tantalizing associations among eruptive prominences, CMEs, and magnetic clouds, the physical relationships among these structures have only begun to be elucidated. Observationally, these solar eruptions have been studied by remote imaging techniques, and accurate measurements of the local magnetic field and plasma density are not available. The propagation of ejecta to 1 AU and beyond is inferred from *in situ* plasma data during fortuitous encounters with satellites. Thus, the dynamics of individual eruptive structures or ejecta are not well known observationally. Clearly this is an area where quantitative theoretical understanding of possible magnetic field geometries, driving forces, and propagation properties can be invaluable. Yet the complexity in studying the dynamics of three-dimensional (3-D) magnetic structures and lack of definitive observational constraints on the initial solar fields have made it difficult to construct realistic theoretical models. In this paper, we describe a recent theoretical model in which the mechanism of eruption and subsequent propagation of the ejecta can be quantified in a unified manner.

2. CORONAL MASS EJECTIONS AND PROMINENCE ERUPTIONS

2.1. Magnetic Geometry

The classic defining feature of CMEs, as seen in coronagraphs, is the discrete density enhancement moving outward. These structures have large spatial scales of the order of the solar radius ($R_\odot$) near the Sun and apparently can maintain their structural integrity over at least several solar radii after eruption, suggesting that they are well organized by magnetic field. They often appear to be loop-like (the "bright rim"), and some early theoretical models have treated such loops as discrete magnetic flux ropes [*Mouschovias and Poland*, 1978; *Anzer*, 1978; *van Tend*, 1979]. Later it was realized that the loops may be a two-dimensional projection of shell-like structures in the plane of the sky [e.g., *Howard et al.*, 1982; *Webb*, 1988; *Hundhausen*, 1996]. A significant recent understanding is that the classic loop-like CMEs often occur in a three-part configuration consisting of the bright rim followed by a prominence embedded in a dark (low-density) "cavity." This morphological property was established based on the SMM data [*Illing and Hundhausen*, 1985, 1986; *Hundhausen*, 1996] and was also noted in different observations [e.g., *Fisher et al.*, 1981].

An attractive model geometry for the initial configuration is the "magnetic arcade," usually described in two dimensions (2-D). The response of an arcade to the quasi-static shearing of the photospheric footpoints has received much attention [e.g., *Low*, 1977; *Birn and Schindler*, 1981; *Mikić et al.*, 1988; *Priest and Forbes*, 1990; *Forbes*, 1990; *Finn and Chen*, 1990; *Forbes and Isenberg*, 1991; *Wu et al.*, 1991; *Finn et al.*, 1992; *Mikić and Linker*, 1994]. The underlying question is whether or not loss of equilibrium or instability can result. This line of investigation is ongoing, primarily using numerical simulation techniques. In arcade models, flux ropes may result as part of the eruption process [e.g., *Anzer and Pneuman*, 1982; *Mikić et al.*, 1988; *Priest and Forbes*, 1990; *Forbes*, 1990; *Wu et al.*, 1991; *Finn et al.*, 1992; *Mikić and Linker*, 1994; *Gosling et al.*, 1995]. In some models of prominence formation, the sheared arcades produce equilibrium flux ropes [e.g., *van Ballegooijen and Martens*, 1989]. The formation of flux ropes from arcades requires magnetic reconnection.

Another scenario provides that the magnetic topology underlying a cavity-prominence structure is that of a magnetic flux rope [e.g., *Kuperus and Raadu*, 1974; *Démoulin and Forbes*, 1992; *Chen and Garren*, 1993; *Low and Hundhausen*, 1995; *Kumar and Rust*, 1996; *Chen*, 1996]. In this conception, the CME would be a shell-like structure spatially corresponding to the upper boundary or the leading edge of the flux rope. This is an attractive configuration because eruptive prominences often exhibit helical features that are presumably indicative of the underlying "twisted" magnetic field lines [e.g., *Tandberg-Hanssen*, 1995]. In a somewhat different approach, *Wu and Guo* [this volume] have modeled a helmet streamer using a magnetic arcade, which "erupts" in response to the introduction of a magnetic flux rope that is forced up through the photosphere.

In this paper, we adopt the point of view that a flux rope exists as an initial structure. Perhaps a flux rope can emerge pretwisted [*Tanaka*, 1991; *Lites et al.*, 1995; *Leka et al.*, 1996] or can be formed [*van Ballegooijen and Martens*, 1989]. Flux rope field may thread the chromosphere, giving rise to characteristic fibril patterns [*Foukal*, 1971; *Martin et al.*, 1992]. One interpretation is that such fibril patterns are consistent with the prominences (filaments) being magnetic flux ropes with the prominence material located inside [*Démoulin and Forbes*, 1992; *Chen and Garren*, 1993; *Rust*, 1994; *Low and Hundhausen*, 1995; *Chen*, 1996]. However, the same chromospheric fibril patterns have been interpreted as a signature of arcades or the upper half of a flux rope with the lower half remaining below the photosphere [*Martin and Echols*, 1994]. In the latter conception, the prominence material is assumed to be trapped along bundles of field lines, which are nearly vertical at the footpoints and nearly horizontal at the top [*Martin and McAllister*, this volume]. This debate is ongoing and will not be discussed further in this paper.

2.2. Magnetohydrodynamics Forces

One of the most fundamental and intriguing questions is what causes a quiescent prominence to suddenly erupt (a process often referred to as "disparition brusque"). In this section, we will describe the forces using standard magnetohydrodynamics (MHD). In the next section, we will discuss a proposed mechanism of eruption. The discussion is based on the paper of *Chen* [1996].

Figure 1 shows a schematic of a flux rope, viewed end on at an oblique angle. At the core of the flux rope is a current loop with minor radius a. The top (or the leading edge) of the current loop apex is denoted by point A, the center of mass of the apex is denoted by point B, and the bottom (or the trailing edge) of the apex is denoted by point C. Outside the current loop ($r \geq a$, where r is the minor radial coordinate measured from point B), the magnetic field is poloidal (i.e., locally azimuthal as indicated by B_p), and is produced by the toroidal current I_t (locally parallel to the flux rope axis, the dash-dot curve through point B). Inside the current loop, the toroidal and poloidal magnetic field components are both nonzero, and the field lines are helical or "twisted." The short vertical lines indicate where the cold ($T \lesssim 10^4$ K) prominence

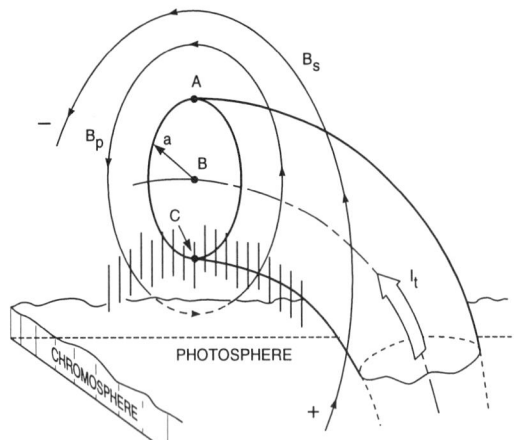

Figure 1. A model prominence-cavity structure. The current loop is embedded in its own poloidal field (B_p). The top (A), center of mass (B), and trailing edge (C) are indicated. The overlying ambient field B_s demarcates the boundary of the flux rope. (From *Chen* [1996].) The vertical short lines illustrate the prominence mass. The model, however, only requires that the mass be inside the flux rope, and no specific spatial distribution is assumed.

material may be located if the underlying magnetic configuration of prominences is that of flux ropes [e.g., *Kuperus and Raadu*, 1974]. This is the so-called "inverse polarity" configuration. Thus the flux rope consists of the current loop and its magnetic field, with the poloidal component (B_p) extending beyond the current loop itself. The entire system is embedded in the corona. An ambient magnetic field B_s is included. We assume that B_s is perpendicular to the plane of the flux rope, but this assumption is not essential. The current loop is assumed to be part of a large current system extending below the photosphere (perhaps connected to the dynamo). The photospheric footpoints of the flux rope are assumed to be stationary and separated by distance $2s_0$. This is a simple but fully 3-D configuration.

Given a current loop, we can define the poloidal flux enclosed by the loop and the photosphere by $\Phi \simeq cL_p I_t$, with the self-inductance approximated by [e.g., *Miyamoto*, 1976]

$$L_p = \frac{4\pi \Theta R}{c^2}\left[\ln\left(\frac{8R}{a}\right) - 2 + \frac{\xi_i}{2}\right], \quad (1)$$

where $2\pi\Theta R$ is the arc length of the loop and Θ is the angular extent of the flux rope above the photosphere. The quantity ξ_i is the internal inductance and will be defined below. For the initial loop, we take a to be constant along the loop. As the loop expands, this simplification becomes invalid because the minor radius a_a at the apex becomes significantly greater than the footpoint minor radius a_f, which is assumed to remain at the initial value a_0. We will use the effective loop inductance

$$L \equiv (L_a + L_f)/2, \quad (2)$$

where L_a and L_f are the values of L_p using, respectively, a_a and a_f. At $t = 0$, we have $a_a = a_f$ so that $L = L_a = L_f$, but for $t > 0$, we have $L_f > L_a$. For a more exact expression for the inductance of an arbitrary slender ($R/a \gg 1$) loop, see *Landau et al.* [1984]. Keeping these caveats in mind, the total poloidal flux of the loop above the photosphere is

$$\Phi_p = cLI_t, \quad (3)$$

and the total poloidal magnetic field energy is

$$U_p \equiv \frac{1}{2}LI_t^2. \quad (4)$$

The toroidal flux is $\Phi_t = \pi\overline{B_t}a^2$, and the toroidal energy is $U_t = (1/4)\pi\Theta a^2 R\overline{B_t}^2$. The three dimensionality of the flux rope is embodied in the finite values of Φ_p, U_p, and U_t.

The MHD force density acting on a plasma element is

$$\mathbf{f} = (1/c)\mathbf{J} \times \mathbf{B} - \nabla p + \varrho\nabla\varphi, \quad (5)$$

where ϱ is the mass density and φ is the gravitational potential. The current $\mathbf{J}$ has toroidal component $J_t(r)$ and poloidal component $J_p(r)$ inside the current loop ($r \leq a$). The loop above the photosphere may be thought of as a section of a torus characterized by major radius R and minor radius a. The integrated force acting in the major radial direction on a section of unit length can be written as

$$F_R = \frac{I_t^2}{c^2 R}\left[\ln\left(\frac{8R}{a}\right) + \frac{1}{2}\beta_p - \frac{1}{2}\frac{\overline{B_t}^2}{B_{pa}^2}\right.$$
$$\left. + 2\left(\frac{R}{a}\right)\frac{B_s}{B_{pa}} - 1 + \frac{\xi_i}{2}\right] + F_g + F_d, \quad (6)$$

where $\beta_p \equiv 8\pi(\overline{p} - p_a)/B_{pa}^2$, $\overline{p}$ is the average pressure inside the loop, p_a the ambient pressure, $\overline{B_t}$ the average toroidal field inside the loop, $B_{pa} \equiv B_p(a) = 2I_t/ca$, $\xi_i \equiv 2\int rB_p^2(r)dr/(a^2 B_{pa}^2)$ is the internal inductance, and I_t is the total toroidal current $I_t \equiv 2\pi\int J_t(r)rdr$. The actual form of $J_t(r)$ enters the analysis only through ξ_i and does not explicitly affect the other terms.

The term F_g is the gravitational force per unit length acting along the major radius. At the apex,

$$F_g = \pi a^2 mg(Z)(n_a - \overline{n}_T). \quad (7)$$

Here, $\overline{n}_T$ is the average density inside the current loop, and $g(Z)$ is the gravitational acceleration at height Z from the photosphere. The surface gravity is $g_s = 2.74 \times 10^4$ cm/s^2. The interaction of the flux rope with the ambient gas is modeled by the drag term F_d

$$F_d = c_d n_a m_i a (V_a - V)|V_a - V|, \quad (8)$$

where $V \equiv dZ/dt$ is the apex speed, $V_a(Z)$ is the speed of the ambient solar wind, and c_d is the drag coefficient. This term provides the momentum coupling.

In applying equation (6) to solar loops, we make a geometrical simplification that the loop above the photosphere is a section of a torus with uniform major radius R. The minor radius a will not be assumed to be uniform. The most mathematically difficult contribution to evaluate is the nonlocal $J_t B_p$ term, which has been rigorously treated with nonuniform R and a elsewhere [*Garren and Chen*, 1994]. The approximation in (6) was found to be good for slender flux ropes ($a/R < 1$) with smooth variations in the toroidal direction. However, unless the loop dynamics is calculated self-consistently in 3-D, one needs to prescribe the geometry. The simplification of uniform R can be written as

$$R = (Z^2 + s_0^2)/2Z, \quad (9)$$

where $2s_0 < R_\odot$ is assumed to remain fixed. This condition leads to the result that $R = 0.5$ AU when the apex reaches 1 AU, consistent with observationally inferred major radius of 0.3 AU [*Burlaga et al.*, 1990].

The apex motion is governed by

$$M \frac{d^2 Z}{dt^2} = F_R = \left(\frac{\Phi_p^2}{c^4 L^2 R} \right) f_R + F_g + F_d, \quad (10)$$

where f_R is simply the quantities in the square brackets in (6). The factor $(I_t^2/c^2 R)$ in (6) has been rewritten in terms of Φ_p and L using equation (3), where Φ_p, L, and R are functions of t. In (10), $M \equiv \pi a^2 \overline{n}_T m_i$ is the total mass per unit length of the loop, and Z is the height of the center of mass of the apex (point B in Figure 1). The expansion of the rest of the loop is determined by (9) according to Z and fixed s_0. The minor radius $a(t)$ of the apex satisfies

$$M \frac{dw}{dt} = \frac{I_t^2}{c^2 a} \left(\frac{\overline{B}_t^2}{B_{pa}^2} - 1 + \beta_p \right), \quad (11)$$

where $w \equiv da/dt$ is the minor radial expansion speed. This equation simply states that $\overline{B}_t$ and $\overline{p}$ tend to increase a while B_{pa} and p_a tend to decrease a. Previously, the expansion of clouds has been attributed to the pressure inside the clouds [*Klein and Burlaga*, 1988] or to spherical expansion of the solar wind [*Suess*, 1988]. The former is given by $\overline{B}_t^2/8\pi + \overline{p}$ and the latter by a decrease in p_a in the β_p term. The (-1) term shows that B_{pa}^2 provides a minor radially inward force, i.e., the pinch force, which can provide equilibrium even if $\overline{B}_t^2/8\pi + \overline{p}$ is greater than the pressure of the ambient solar wind [*Suess*, 1988]. The minor radial expansion occurs in near equilibrium [*Chen*, 1989].

The initial loop contains a cool component (i.e., prominence material) so that the average total density is

$$\overline{n}_T(t) = \overline{n}_c(t) + \overline{n}_p(t), \quad (12)$$

where the subscript c refers to "cavity" and p to "prominence." In Figure 1, the short vertical lines indicate the prominence material, resembling the equilibrium configurations modeled by *Démoulin and Forbes* [1992] and *Low and Hundhausen* [1995]. We point out that the present model uses only averaged densities, and the spatial distribution of cavity and prominence plasmas inside the flux rope does not enter the calculation. The two components are treated separately using (12) via the mass M in (10) and (11). Because quiescent prominences are at estimated temperatures of $\overline{T}_p \lesssim 10^4$ K $\ll \overline{T} \sim 1$–2×10^6 K [*Tandberg-Hanssen*, 1995], we neglect the pressure due to $\overline{n}_p$ so that $\overline{p} = 2\overline{n}_c k \overline{T}$.

To close the MHD equations, the equation of state inside the loop is taken to be

$$\frac{d}{dt}\left(\frac{\overline{p}}{\overline{\varrho}^\gamma} \right) = 0, \quad (13)$$

where $\overline{\varrho} = \overline{n}_c m_i$ is the average mass density of the hot plasma inside the loop, and γ is the ratio of specific heats at constant pressure and at constant volume ($1 \leq \gamma \leq 5/3$). In a previous calculation [*Chen and Garren*, 1993], $\gamma \simeq 1$ was used. Comparing the results with those obtained by using $\gamma = 5/3$, it was shown that the overall propagation properties are not significantly influenced by the γ values. However, it was also shown that if adiabatic expansion ($\gamma = 5/3$) is assumed, the temperature of the loop decreases to a few degrees K at 1 AU. It was argued that magnetic clouds and CMEs with the observed temperature of several times 10^4 K to 10^5 K must maintain magnetic and thermal connection with the Sun. This result is insensitive to the exact magnetic geometry. In the following discussion, $\gamma \simeq 1.2$ is used.

A quiescent prominence may persist with little net motion for days or even weeks. Demanding $d^2Z/dt^2 = 0$ and $d^2 a/dt^2 = 0$, we obtain

$$B_{pa} = (R_0/a_0)\Lambda_0^{-1} \left\{ -B_{s0} + \left[B_{s0}^2 - \left(\frac{a_0}{R_0} \right)^2 \Lambda_0 \right. \right.$$
$$\left. \left. \times \left[8\pi(\overline{p} - p_a) + 4\pi m_i R_0 g_0 (n_a - \overline{n}_c - \overline{n}_p) \right] \right]^{1/2} \right\}, \quad (14)$$

where $\Lambda_0 \equiv \ln(8R_0/a_0) - 1.5 + \xi_i/2$, $g_0 = g(Z_0)$, and $B_{s0} = B_s(Z_0)$. All other quantities also refer to the initial values. Here the equilibrium condition has been given for the apex. The sign convention is such that $B_{s0} < 0$ provides major radially inward $I_t B_{s0}$ (downward at the apex).

The basic physics of the terms in the square brackets in (6) is well understood [*Shafranov*, 1966] and was applied to the equilibrium and dynamics of solar flux ropes previously [*Xue and Chen*, 1983; *Chen*, 1989; *Garren and Chen*, 1994]. The same physics is also used in *Kumar and Rust* [1996] and *Wu and Guo* [this volume]. We only note here that $\mathbf{J} \times \mathbf{B}$ has two contributions: $R^{-1}[\ln(8R/a) - 1 + \xi_i/2]$ arising from $J_t B_p$ and $-R^{-1}(\overline{B_t}^2/2B_{pa}^2)$, the downward "tension" due to $J_p B_t$. Equation (6) neglects correction terms of order a/R.

Three points are worth mentioning. First, force balance is possible for flux ropes embedded in an ambient plasma with free boundaries. The constraint (14) admits a wide variety of quasi-stationary flux ropes under coronal conditions. Second, the motion of flux ropes determined by (10) and (11) has been analyzed in the linear and nonlinear regimes [*Chen*, 1989; *Cargill et al.*, 1994]. A flux rope can expand (rise) if Φ_p is increased but that for constant Φ_p, the flux rope can only oscillate about its initial equilibrium even if it is impulsively driven up from below by, for example, an upflow of plasma [*Chen*, 1990]. Third, equilibrium flux ropes satisfying (14) are stable to the sausage and kink modes under typical coronal conditions [*Xue and Chen*, 1983]. Furthermore, the flux rope remains stable during the flux injection process. This point will be elaborated on below. The stability of the flux rope with increasing poloidal flux (i.e., increasing winding numbers) is due to the fact that a 3-D flux rope rises as Φ_p is increased, increasing the length of the flux rope. In contrast, in a straight cylinder the magnetic field pitch increases as the azimuthal flux is increased, eventually becoming unstable to the sausage and kink modes when B_p becomes sufficiently greater than $\overline{B_t}$. Note also that the usual stability analyses of cylindrical pinches are carried out with no background plasma. An exception is the work of *Manheimer et al.* [1973] that considered the effects of ambient conducting and nonconducting gases on the kink stability of Z-pinches (in which $\overline{B_t} = 0$). They found that the presence of an ambient gas has a stabilizing influence even in this unstable regime ($\overline{B_z} = 0$) because of the added mass that has to be displaced by the pinch. Extrapolating this result to the solar flux rope case, we expect that the ambient corona and the dense prominence material suspended inside the flux rope should provide added stability to the already stable system ($\overline{B_t} \simeq B_{pa}$). In this respect, treating the prominence as a passive gas collisionally suspended inside the flux rope is different from treating the prominence as a current sheet. The former structure has no current-driven instability while the latter may exhibit instabilities of current sheets.

We define a measure of the characteristic pitch angle of the magnetic field at the apex by

$$\vartheta \equiv \tan^{-1}(\Gamma),$$

where

$$\Gamma \equiv \frac{B_{pa}}{\overline{B_t}}. \quad (15)$$

If we denote the pitch at the apex by Γ_a and that at the footpoint by Γ_f, then we have

$$\Gamma_f = \left(\frac{a_f}{a_a}\right)\Gamma_a, \quad (16)$$

where Γ_a and Γ_f are given by (15) using a_a and a_f, respectively. The relation (16) is general, depending only on the conservation of I_t and $\Phi_t = $ const regardless of the exact variation in a along the flux rope. A typical field line winds around the minor radius $\overline{N}$ times in traversing from one footpoint to the other, where

$$\overline{N} \simeq \frac{\Theta R}{a}\Gamma. \quad (17)$$

Initially, $a_a = a_f$ so that $\Gamma_a = \Gamma_f$, but (16) shows that Γ_f becomes much smaller than Γ_a after the flux rope erupts. We will estimate the winding number using (17) during the initial period.

3. MECHANISM OF ERUPTION

3.1. *Storage-Release of Magnetic Energy Versus Driven Mechanism*

In the popular theoretical view, an existing coronal magnetic field structure (e.g., an arcade or a flux tube with untwisted field) is stressed via the shearing of photospheric footpoints of field lines in a quasi-static manner: the magnetic energy is slowly deposited in the corona, and the stored energy is released abruptly due to loss of equilibrium or instability. We will refer to this scenario as the "storage-release" paradigm. Alternatively, magnetic energy may be "injected" into an existing structure, driving the structure driven out of equilibrium.

As a generic process, the shearing of the magnetic footpoints of arcades has been extensively investigated, as referenced in section 2.1. A different scenario is to hypothesize pressure pulses from flares as a trigger and driver of CMEs [*Dryer*, 1982; *Wu et al.*, 1983; *Steinolfson*, 1985]. However, the notion that solar flares cause or drive CMEs has since been criticized [e.g., *Harrison*, 1986; *Harrison and Sime*, 1989; *Hundhausen*, 1996]. Another scenario is that prominence eruption drives the CME [*Hu*, 1983]. In a similar scenario, MHD waves generated by flares have been used as a trigger of prominence eruption [*Sakai and Nishikawa*, 1983]. Observationally, these ideas have been criticized because when CMEs are observed with associated eruptive prominences, the prominences usually lag behind and are slower than the CMEs, making it difficult to think of prominences as the driver of CMEs [*Hundhausen*, 1987]. Instead, *Hundhausen* [1987] suggested that the occurrence of CMEs causes the eruption of prominences. Theoretically, recall that pressure pulses unaccompanied by an increase in Φ_p can only cause a flux rope to oscillate [*Chen*, 1990].

In the works of *Kuperus and Raadu* [1974], *Démoulin and Forbes* [1992], and *Low and Hundhausen* [1995], equilibrium flux rope structures containing current sheets were considered in 2D. In three dimensions, shearing the footpoints of flux ropes has been discussed by *Browning and Priest* [1986], who increased the twist of the field lines and found that such systems may encounter bifurcation in equilibrium during quasi-static evolution. In contrast, *Chen* [1990] specified increasing poloidal flux, which is equivalent to increasing the field line winding numbers, and found that an isolated flux rope does not exhibit bifurcation in the resulting quasi-static evolution. *Amari et al.* [1996] have recently investigated the dynamical evolution of a 3-D flux rope whose footpoints are anchored in the photosphere and subjected to horizontal twisting motion, with the condition that the fluid speed at the simulation boundaries be zero. They found that a flux rope can undergo rapid expansion. In this paper, we consider the response of flux ropes to injection of poloidal flux. For solar flux ropes, the difference between the driven process and the storage-release process is in the time scale on which the poloidal flux is increased. In the driven scenario, the flux increases on the dynamical time scale (tens of minutes to hours) instead of days or weeks.

3.2. Poloidal Flux Injection

Equation (10) provides a mathematical description of the response of a flux rope to increasing poloidal flux. In the conventional storage-release concept, Φ_p is quasi-statically increased. In response, the flux rope rises slowly, reaching another equilibrium [*Chen*, 1989, 1990]. On the other hand, if Φ_p is increased rapidly, the flux rope expands rapidly, exhibiting an eruptive behavior. The physical process will be described using a specific example, the "reference flux rope," for the purpose of quantifying the results in terms of solar parameters. The basic physics is generic.

The reference flux rope is initially in equilibrium and is specified as follows. Let the apex height be $Z_0 = 1.5 \times 10^5$ km at center of mass, the major radius $R_0 = 3.75 \times 10^5$ km, the minor radius $a_0 = 7.5 \times 10^4$ km, and the footpoint separation $2s_0 = 6 \times 10^5$ km. These dimensions are chosen to be consistent with quiescent prominences [e.g., *Tandberg-Hanssen*, 1995]. The ambient coronal temperature is taken to be $T_a = 2 \times 10^6$ K. We assume that the low-density cavity plasma is at $\overline{T}_0 = T_a$. The ambient density at $Z = Z_0$ is taken to be $n_a = 3 \times 10^8$ cm^{-3}. The plasma density inside the flux rope is averaged over the flux rope so that the spatial distribution of the prominence gas is not treated. We use $\overline{n}_c = (1/4)n_a = 7.5 \times 10^7$ cm^{-3} for the cavity plasma. These values correspond to $p_a = 0.16$ dyn/cm^2 and $\overline{p}_c = 0.04$ dyn/cm^2. The average density of the cold prominence component is $\overline{n}_p$. The parameter $\chi_0 \equiv \overline{n}_p/\overline{n}_c$ proves to be important in determining the structure of the flux rope at 1 AU and beyond [*Chen*, 1996], but we will not expand on this point. If we take $\chi_0 = 30$ for this example, the total mass of the flux rope is $M_T = 4.75 \times 10^{16}$ g. This corresponds to a rather massive prominence. Finally, we choose the ambient magnetic field to be $B_{s0} = -0.5$ G. Equation (14) then gives the poloidal magnetic $B_{pa} = 4.4$ G, so that $\beta_p \simeq -0.16$. By setting $d^2a/dt^2 = 0$ in (4), we find $\overline{B}_t = 4.8$ G. The magnetic field on the flux rope axis, which is entirely toroidal, is $\sim 3\overline{B}_t$ [*Chen*, 1996, equation (30)] so that $B_t(r=0) \simeq 15$ G. This value is consistent with those estimated in observed quiescent prominences [*Rust*, 1967; *Leroy et al.*, 1983]. Furthermore, the magnetic field strength increases with height inside the lower half of the flux rope, which is consistent with the findings of *Rust* [1967] and *Leroy et al.* [1983] that the magnetic field tends to increase with height inside prominences. It is satisfying that the flux rope dimensions chosen to be consistent with quiescent prominences indeed lead to magnetic field values via (14) in agreement with the observationally estimated values.

Suppose that we prescribe a flux injection profile $d\Phi_p/dt$. Figure 2 shows a profile of $d\Phi_p/dt$. The poloidal flux $\Phi_p(t)$ normalized to the initial value Φ_0 is also shown. The functional form is chosen merely to mimic a packet of poloidal magnetic flux $\Delta\Phi_p$ (i.e., magnetic energy) with different rise and fall off time scales. This flux injection profile corresponds to an increase of $\Delta\Phi_p = 2.7 \times 10^{22}$ Mx in the poloidal flux and $\Delta U_p = 2.1 \times 10^{32}$ erg in magnetic energy. Note that $d\Phi_p/dt \to 0$ for $t > 400$ min. Thus Φ_p is conserved thereafter. The toroidal flux Φ_t is taken to be conserved throughout the process. That is, the vertical magnetic field in the photospheric footpoints is held unchanged. We want to examine the dynamical consequences of this process.

Figure 3 describes the response of the flux rope during the first 300 min, showing the height-time curve (panel a) as

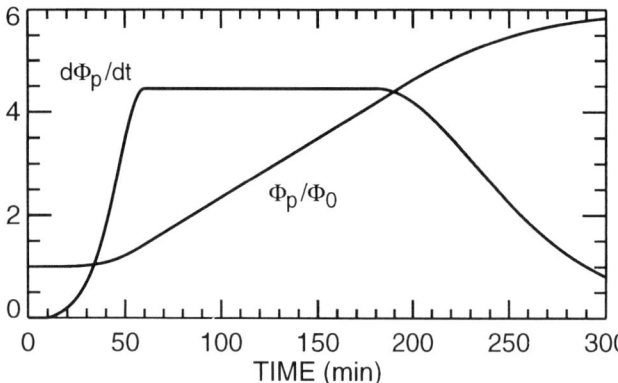

Figure 2. Flux injection profile $d\Phi_p/dt$ in 1.5×10^{17} Mx/s and $\Phi_p(t)/\Phi_0$. $\Phi_0 = 5.4 \times 10^{21}$ Mx.

well as the upward velocity (panel b) of the apex. In each panel, the solid (dashed) curve describes the dynamics of the leading (trailing) edge of the current loop (points A and C in Figure 1). First, we note that, while $d\Phi_p/dt = 0$ ($t < 10$ min), the loop is stationary, as required by the equilibrium condition. Although the flux injection rate reaches maximum at $t \simeq 50$ min, the loop exhibits little motion until $t \simeq 60$ min due to inertia. For the reference flux rope, the speed of the leading (trailing) edge of the apex reaches ~ 470 km/s (~ 390 km/s) at $t = 120$ min, with the center of mass speed at $V \simeq 430$ km/s. At $t = 300$ min, we find $Z \simeq 10 R_\odot$, $a \simeq 6 \times 10^5$ km, $V \simeq 540$ km/s for the center of mass, and a minor radial expansion speed of $w \simeq 51$ km/s.

The height-time plot shows that the leading edge outpaces the trailing edge. This figure bears considerable resemblance to Figure 6 of *Illing and Hundhausen* [1986] in which the trajectories of a white-light bright rim and an associated eruptive prominence are plotted. This figure is reproduced in Figure 4. The leading edge (Figure 3a) is identified with the SMM ring, and the trailing edge with the prominence. The calculated height-time profile is also consistent with those of several eruptive prominences analyzed by *Schmahl and Hildner* [1977] and *Kahler et al.* [1988]. Mathematically, the initial height and expansion speed exhibit exponential dependence on time with the inertial time scale proportional to $M^{1/2}/B_{pa}$ [*Chen*, 1989]. In Figure 3a, we have indicated the height $Z = 2R_\odot$ by a line. We see that the leading edge of the current loop crosses this height ahead of the trailing edge. With the denser material weighted toward the trailing edge, the cross-section would likely be elongated, and the trailing edge would be slower than indicated. In addition, the actual cavity size is larger than $2a$, with a greater velocity differential. These effects are not included, so that this expression significantly underestimates the time lag in a real-

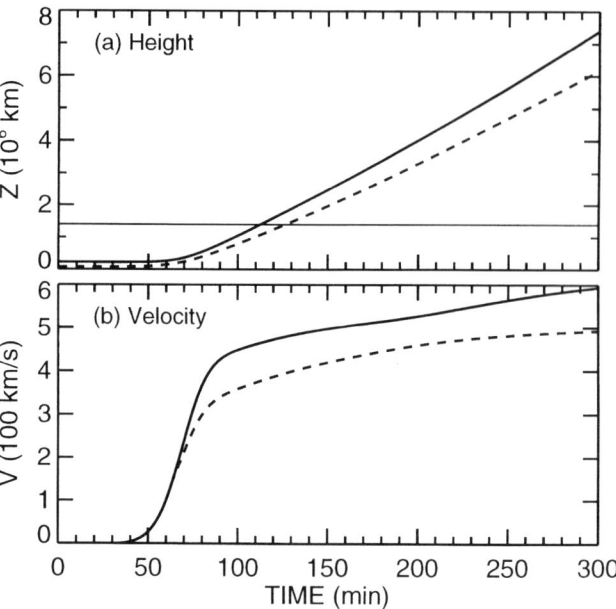

Figure 3. Dynamics of the flux rope in response to the flux injection shown in Figure 4. Solid curve: the leading edge of the current loop (point A in Figure 1). Dashed curve: the trailing edge of the current loop (point C in Figure 1). (a) Apex height versus time. (b) Apex speed versus time.

istic eruption of cavity-prominence structure. Nevertheless, if we identify the leading edge with loop-like bright rims (i.e., CMEs), the present result shows the physical reason for the lag time between a CME loop and prominence material.

3.3. Flux Rope Stability

We briefly explore further consequences of the injection of poloidal flux. It is interesting to evaluate the "typical" magnetic field pitch Γ defined by (15) during the process of poloidal flux injection. Figure 5 shows the pitch parameter for the apex Γ_a (dashed curve) and the footpoint Γ_f (solid curve). We see that Γ_f increases by $\sim 30\%$ for tens of minutes (30–40 min for this example) while Γ_a is nearly constant (although not shown here, Γ_a remains remarkably constant, decreasing to only $\Gamma_a \simeq 0.8$ at 1 AU). These properties have been noted and explained previously [*Chen*, 1996]. We have also plotted the typical winding number $\overline{N}$ using Γ_a in (17). For the initial flux rope, $\Theta \simeq 0.3$ so that $\overline{N} \simeq 1.4$. That is, the typical field lines wrap around the current loop approximately 1.4 times in going from one footpoint to the other. (Individual field lines can have zero twist at $r = 0$ to infinite twist for $r > a$.) Previously, *Hood and Priest* [1981] found that a straight cylindrical current loop

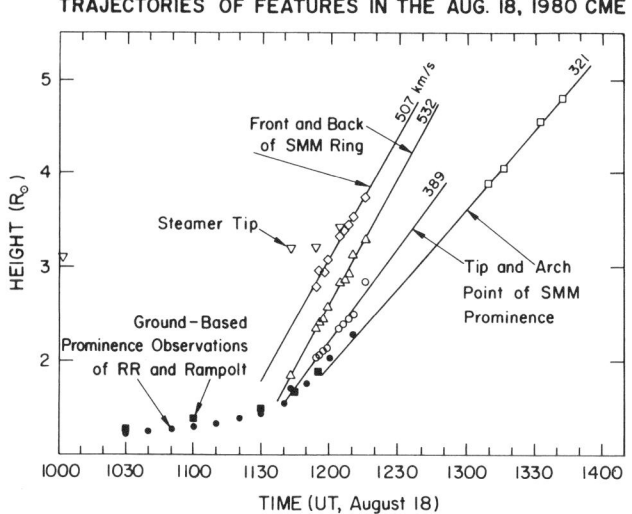

Figure 4. Height-time profiles of several features observed in a CME event observed by SMM. Reproduced from Figure 6 of *Illing and Hundhausen* [1986].

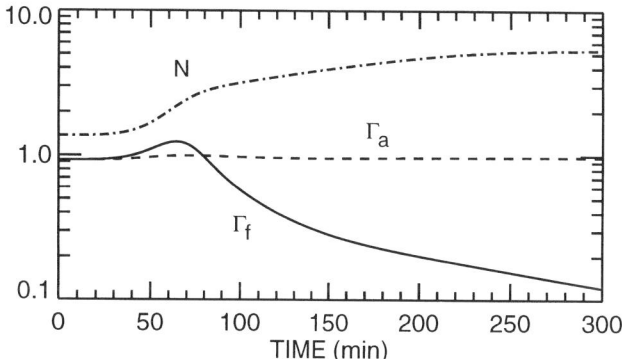

Figure 5. Typical magnetic field pitch of the footpoints (Γ_f) and the apex (Γ_a), given by equations (15) and (16). The winding number $\overline{N}$ for typical field lines. Injection of Φ_p corresponds to increasing the field line twist, which reaches maximum and remains constant when $d\Phi_p/dt$ vanishes (see Figure 4).

with uniform magnetic field twist is unstable to kinking if the twist exceeds a critical value even if ideal MHD line-tying is included. This critical value corresponds to $\overline{N} = 1.25$ (2.5π in their notation). The present result resembles this critical value but is an equilibrium condition, rather than a stability requirement. No comparable equilibrium constraint is present in the straight-cylinder geometry.

The full stability analysis of toroidal flux rope in the coronal environment has not been carried out, but we can illustrate the basic stability properties using a simple consideration. Consider the quantity [*Kadomtsev*, 1966],

$$\omega^2 = \frac{B_{pa}^2 k^2}{4\pi\rho} \Omega, \quad (18a)$$

where

$$\Omega \equiv \left[1 - \Gamma^2 \ln\left(\frac{1}{ka}\right)\right]. \quad (18b)$$

Here, k is the axial wave number and Γ is defined by (15). If $\omega^2 > 0$ ($\omega^2 < 0$), then the flux rope is stable (unstable) in ideal MHD according to the energy principle. If the footpoints are immobile, the longest wave length for the flux rope is $k = 1/\Theta R$ so that $1/ka = \Theta R/a$. Note that the minor radial profile affects the stability through the quantity Γ. In order to estimate the stability properties during the initial flux injection and subsequent expansion, we have evaluated the quantity Ω as a function of time. This is plotted in Figure 6, which shows that Ω is positive throughout the flux injection process. It does dip somewhat but thereafter becomes increasingly positive, i.e., more stable, as the flux rope expands. This is because the expansion causes an increase in ΘR and therefore in L as can be seen from (1). If the major radial expansion is fast enough, I_t and B_p decrease according to (3) even as poloidal flux is injected. This is due to the 3-D toroidicity of the flux rope manifested in the finiteness of Φ_p defined in (3). Although this argument is heuristic and approximate, it does indicate that the flux rope can remain stable during flux injection.

Observationally, *Vršnak et al.* [1991] found that among prominences exhibiting apparent helical structures, those at the onset of eruption and in the acceleration phase tend to have the greatest twist with quiescent and post-acceleration prominences exhibiting less twist per unit length. Their overall conclusion is that the main characteristic of the eruption is a decrease in the apparent twist. The theoretical results shown in Figure 5 are consistent with these observed properties. Note that $\overline{N}$ increases to ~ 5 as a result of poloidal flux injection. This exceeds the critical twist of ~ 1.5 turns established for straight cylinders [e.g., *Hood and Priest*, 1981], but the pitch per unit length of the flux rope decreases, leading to increasing $\Omega > 0$ and stability.

Based on analyses of straight cylindrical flux ropes of fixed length, MHD instabilities such as kink have been proposed as a candidate mechanism to trigger the initial eruption [e.g., *Hood and Priest*, 1981; *Vršnak*, 1990; *Rust and Kumar*, 1996]. Our results suggest an alternative interpretation that the increase in the pitch Γ_f and the winding number $\overline{N}$ results from the injection of poloidal flux, which is the actual cause of the onset of expansion. Thus, we reinterpret the observation of *Vršnak et al.* [1991] as a signature of the flux injection process. No kink instability is indicated for the onset of eruption. We point out that the new conclusion arises

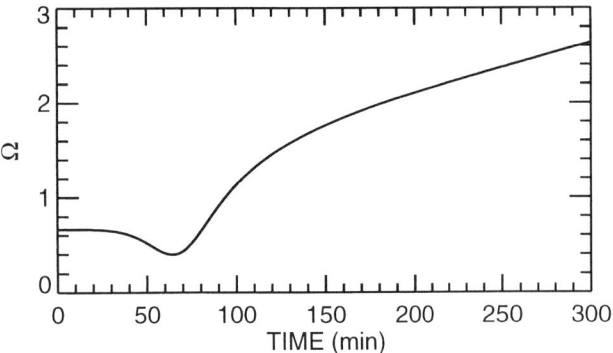

Figure 6. Stability of the flux rope against the "kink" ($m = 1$) mode. The flux rope is stable if $\Omega > 0$. The flux rope is also stable to the "sausage" model.

from consideration of the stability of the 3-D flux rope in toroidal geometry: increasing $\overline{N}$ is accompanied by the increasing length of the flux rope which prevents the magnetic field pitch from increasing. Equation (18b) shows that the stability condition is determined by the pitch rather than the winding number $\overline{N}$, and the key quantity Γ is insensitive to the specific profile of the magnetic field inside the flux rope.

As the flux rope expands, a becomes nonuniform along the flux rope, and the application of (18) is only approximate. We also note that Γ_f/Γ_a becomes much smaller than unity as the flux rope expands. Smaller values of Γ implies increased stability according to (18), implying that the legs of the flux rope are more stable. We therefore expect that the longest unstable wavelength to decrease relative to the total length of the flux rope, so that $1/ka$ decreases, further increasing the tendency toward stability. Empirically, *Bothmer and Schwenn* [1994] and *Marubashi* [this volume] found that most observed magnetic clouds have the same magnetic structure (i.e., orientation and polarity) as that inferred for the associated solar filaments. This is consistent with the apparent stability of solar flux ropes against the kink mode.

4. PHYSICAL OBSERVABLES

We have shown that it is possible to consistently explain the dynamics of the three-part structure of CMEs [*Illing and Hundhausen*, 1986] by postulating that the classic CME loop, prominence cavity, and the prominence correspond to different parts of one magnetically organized structure, viz., a flux rope. In this section, we examine a number of additional properties of the model flux rope that can be compared with observed properties of CME-prominence structures and interplanetary flux ropes.

Eruption Speeds Near the Sun. In a statistical study of CME speeds, *MacQueen and Fisher* [1982] found that CMEs (i.e., the bright rims) generally show only moderate or no acceleration beyond 2–3 $R_\odot$ from the solar surface, with most of the acceleration occurring much closer to the Sun. This is consistent with the height-time profiles shown in Figure 3. Figure 7 shows the driving force acting on the flux rope. We see that the force increases rapidly with the injection of Φ_p but that after several tens of minutes, it decreases rapidly even while the flux injection continues at the maximum rate. The duration of the peak acceleration is determined primarily by the inertia and the initial length of the flux rope. This can be understood as follows. Returning to (10), we see that the multiplicative factor in front of f_R approximately scales as $\Phi_p{}^2/\Theta^2 R^3$. If the expansion rate of R^3 exceeds the poloidal flux injection rate, then the driving force can decrease even if Φ_p is increasing. This is what happens for $t \gtrsim 70$ min shown in Figure 7. Figure 3 shows that during this period, the apex expansion speed has increased to $\sim$ 200–300 km/s and the apex height is beginning to increase significantly. It is easy to see that when the apex reaches 2–3 $R_\odot$, a quiescent prominence-like flux rope has increased its length (ΘR) and the inductance L roughly by a factor of 5–10. This decreases the driving force by approximately a factor of 25–100, accounting for the apparent cessation of rapid acceleration. The inertia of the flux rope determines the initial acceleration. If the initial flux rope length is significantly longer (shorter) than the reference flux rope used here ($2s_0 = 6 \times 10^5$ km), then the rapid reduction in the driving force occurs farther out (closer in). Thus the typical range in distance 2–3 $R_\odot$ is determined by the initial flux rope length and is consistent with flux rope lengths of hundreds of thousands of kilometers. Note that this analysis pertains to those events with rapid acceleration in the inner corona where the original CMEs tabulated in the statistical ensemble were observed. If a flux rope is accelerated slowly by relatively slow poloidal flux injection for, say, a few tens of hours, the acceleraton may persist beyond 2–3 $R_\odot$.

Poloidal Flux Injection. The modest increase in Γ_f ($T \simeq$ 40–80 min in Figure 5), which is equivalent to an increase in the horizontal component of the photospheric and chromospheric magnetic field, is a consequence of the injection of poloidal flux and is directly related to the peak in the force (Figure 7). This effect is potentially observable in vector magnetograms with sufficient accuracy and time resolution that may become available in the future. However, in one observation of eruptive prominence, *Engvold et al.* [1976] inferred an increase in the toroidal current near the footpoints lasting for 30–40 min during the early phase of eruption. This may be an indication of the effect discussed here. After such a period, the calculation predicts that Γ_f decreases monoton-

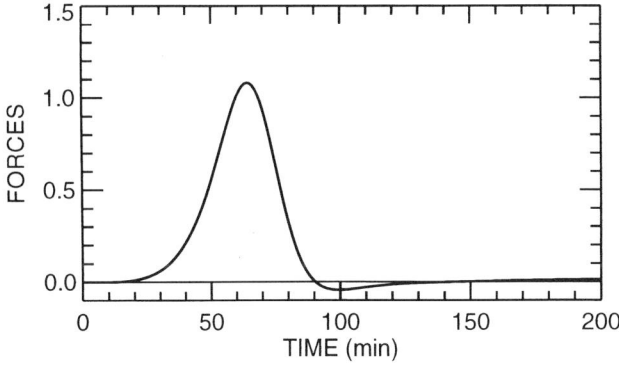

Figure 7. The total major radial force F_R given by (6) resulting from the injection of poloidal flux (Figure 4). It is significant only for ~60 min. As $F_R \to 0$, the acceleration decreases, and the flux rope coasts with a nearly constant speed V.

ically, so that the magnetic field near the footpoints become increasingly radial.

Although some possible scenarios have been mentioned [*Chen*, 1989, 1990], the mechanisms for poloidal flux injection have not been established. Nevertheless, the injection of poloidal magnetic flux as a trigger of eruption is compatible with the observation that CMEs appears to be preceded by flux emergence [*Feynman and Martin*, 1995], although the present model provides a different interpretation.

Minor Radial Expansion. In a statistical study of the size of interplanetary magnetic clouds observed between 0.3 AU and 4.2 AU, *Bothmer and Schwenn* [1994] found that magnetic cloud size exhibits a scaling relation given by

$$a \propto Z^\kappa, \qquad (19)$$

with the range of values $\kappa = 0.78 \pm 0.10$ obtained using linear regression. An examination of (11) reveals that $\overline{B_t}/B_{pa}$ increases as the flux rope propagates but that $\beta_p < 0$ increases in magnitude faster, giving $d^2a/dt^2 < 0$ for $t \gtrsim 30$ hr. For the reference flux rope, we find approximately $\kappa \simeq 1$ for 0.3–2 AU but $\kappa \simeq 0.71$ for 2–5 AU. Over the entire region 0.3–5 AU, we find $\kappa \simeq 0.88$. We suggest that the pressure effect ($\beta_p < 0$) explains the empirical result $\kappa = 0.78 \pm 0.10$ and that the decreasing exponent signifies the increasing importance of the pressure term $\beta_p = 8\pi(\overline{p} - p_a)/B_{pa}^2$. Thus, the minor radial structure is not force-free, although the minor radial expansion remains in near-equilibrium. The model results quoted here are based on the reference flux rope, whereas the result of *Bothmer and Schwenn* [1994] is obtained from an ensemble of observed clouds. It would be interesting to evaluate κ for a range of initial flux ropes.

Macroscopic Interplanetary Properties. The apex of the reference flux rope reaches 1 AU (1.5×10^8 km) at $t \simeq 74$ hr (3 days), where the speed of the center of mass is $V \simeq 570$ km/s and the minor radius is $a = 1.7 \times 10^7$ km. Thus, the current loop has a thickness $2a \simeq 0.23$ AU. The minor radial expansion speed is $w \simeq 60$ km/s. The magnetic field is $B_{pa} = 8.3$ nT and $\overline{B_t} = 9.7$ nT. The flux rope density is $\overline{n} \simeq 2.5$ cm^{-3} ($n_a = 5$ cm^{-3}), having evolved from the initial cavity plasma. The temperature is $\overline{T} \simeq 9 \times 10^4$ K, and the solar wind temperature is comparable at $T_a \simeq 9.5 \times 10^4$ K. These results are entirely consistent with observed magnetic clouds [e.g., *Burlaga et al.*, 1981; *Klein and Burlaga*, 1982; *Lepping et al.*, 1990]. The reference flux rope represents a fast cloud with a relatively strong magnetic field embedded in a slow solar wind stream ($V_{sw} = 400$ km/s), resembling the one detected on 14–15 January 1988 [e.g., *Farrugia et al.*, 1993]. By continuing the calculation, it can be shown that model flux ropes that have dynamical properties consistent with both prominence eruptions at the Sun and magnetic clouds at 1 AU remain in agreement with observed magnetic clouds even at ~5 AU. The calculations show that beyond this point, cloud fields resulting from various initial conditions become comparable to the typical ambient field in magnitude, and additional effects not included in equations (6) and (11) may be needed to analyze the dynamical evolution.

Magnetic Field Profile. The most basic field property of interplanetary magnetic clouds is the characteristic magnetic field profile, which has been fitted by force-free and nearly force-free pinches [*Marubashi*, 1986; *Burlaga*, 1988; *Suess*, 1988]. These studies assume that magnetic clouds can be modeled as locally straight cylinders. For $a/R \ll 1$, this is a good approximation. Here, we discuss the magnetic field structure constrained by the macroscopic dynamical quantities such as a, B_{pa}, and $\overline{B_t}$. The field need not be force-free. The constraints we impose are

(i) the spatial scale be of the order of the minor radius a,

(ii) the current distribution vanish smoothly at $r = a$.

Physically, constraint (i) is an assertion that a plasma structure tends to maximize the spatial scales in all directions consistent with applicable boundary conditions. For constraint (ii) to be valid, shocks and compression of fields due to stream interactions must be excluded. Simple non-force-free as well as force-free magnetic field profiles can be constructed based on the above constraints. Figure 8 shows an example, based on the reference flux rope as would be observed by a stationary observer at 1 AU. The slight asymmetry arises the minor radial expansion given by (11). Here, we have assumed that the magnetic cloud axis lies in the ecliptic plane and is perpendicular to the Sun-Earth line. The z-direction is normal to the ecliptic and points toward north in the Earth frame with the cylinder axis in the y (East-

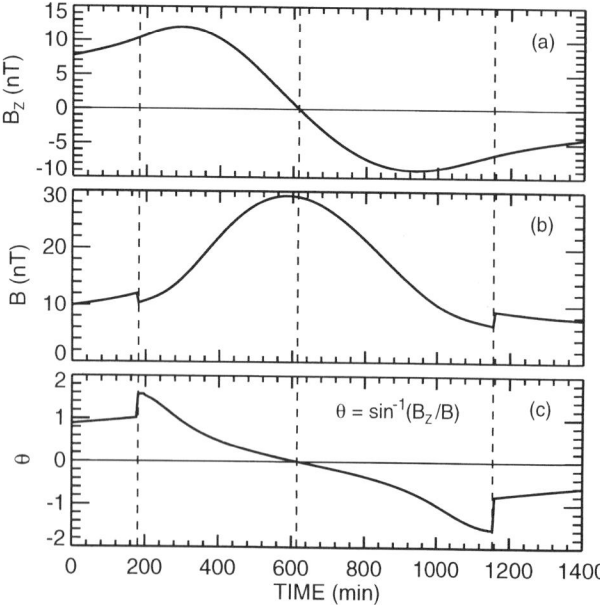

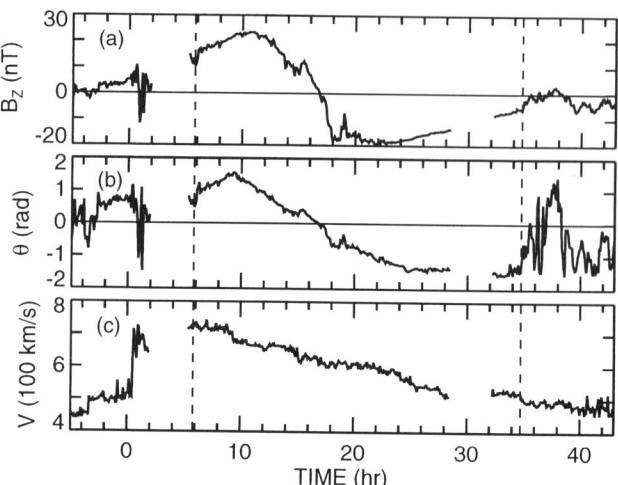

Figure 8. A possible magnetic field profile inside the flux rope. The definition of magnetic clouds given by *Burlaga* [1988] corresponds to the current loop inside the flux rope, indicated by the vertical dashed lines at $t \simeq 200$ min and 1200 min. The dashed line in the middle refers to the time when $B_z = 0$. From *Chen* [1996].

West) direction. The magnitude of the field has maximum value ($\sim 3\overline{B_t}$) of nearly 30 nT for this example. The toroidal component (B_y) peaks near where $B_z = 0$. Figure 8c shows the magnetic field rotation angle $\theta \equiv \sin^{-1}(B_z/B)$. In displaying the field profile, we have included $B_y \simeq 6$ nT, with no attempt to match the fields smoothly. Figure 9 shows B_z and B of a magnetic cloud observed near 1 AU on January 14–15, 1988. The model profile closely resembles that of the observed cloud in all essential aspects. The above non-force-free profile (21) has no remarkable differences from the Bessel function profile truncated at the first zero of J_0 [*Burlaga*, 1988], which was obtained based on the assumption that magnetic clouds evolve to the relaxed "Taylor" state [*Taylor*, 1974] constrained by the conservation of the magnetic helicity. Some pros and cons of this postulate have been given [*Chen*, 1996], but the force-free configurations constitute a subset of solutions given by constraints (i) and (ii). It should be noted, however, that neither the force-free Lundquist solution nor this non-force-free class of solutions is unique or necessarily in equilibrium for given $\overline{p}$, p_a, $\overline{B_t}$, and B_{pa}.

The vertical dashed line in the middle of Figure 8 denotes the center of the flux loop as defined by $B_z = 0$. The dashed lines on the left and right represent the leading and trailing edges of the current loop where the extrema of θ occur and J_t,

Figure 9. An interplanetary magnetic cloud observed by IMP-8 near 1 AU on January 14–15, 1988.

$J_p = 0$. These points correspond to the edges of a magnetic cloud as defined by *Burlaga* [1988]. This definition identifies the cloud as the heliospheric counterpart of the current loop that is embedded inside the initial cavity. The poloidal field (B_p) extends beyond the current loop. The magnetic field immediately outside magnetic clouds [*Zhang and Burlaga*, 1988] is often suggestive of the r^{-1} dependence of B_p. Thus, the initial prominence cavity extends beyond the magnetic cloud as defined by Burlaga. The magnetic field shown in Figure 9 exhibits foreshortening or skewing typical of observed magnetic clouds [*Burlaga*, 1988]. One issue that arises is what happens to the initial prominence/filament. We have modeled the observable prominence gas as a cold component that is passively contained inside the flux rope and that falls back to the Sun. The result based on this model construction yields a flux rope density of 2–5 cm^{-3} depending on the assumed density of the hot cavity plasma component. This density is quite consistent with the typical densities of magnetic clouds, but a given magnetic cloud may have a few tens of particles per cubic centimeter. In addition, the density may vary considerably inside a cloud. If a fraction of the initial prominence material is assumed to be carried by the flux rope, then the calculated range of density will be greater than 2–5 cm^{-3}. Quantitatively, however, this effect depends on the ionization processes and detailed coupling between the prominence material and the cavity. These processes cannot be studied within the framework of ideal MHD used here.

5. DISCUSSION

We have discussed a new theoretical approach that treats the initial equilibrium, eruption, and propagation in a unified manner. The results have been discussed with the emphasis on elucidating the physical process. Several important issues have been discussed in the literature in connection with the dynamics of magnetic flux ropes. These include magnetic helicity and irreversible release of magnetic energy to heat the interplanetary plasmas. Here, we summarize a few points. The reader is referred to *Chen* [1996] for a treatment of these issues and how different models address them.

Magnetic Helicity. A number of ideas have been suggested regarding the use of magnetic helicity conservation as a means to constrain models of coronal and solar wind magnetic structures. One has been mentioned earlier in connection with the Bessel function (Lundquist solution) magnetic field [*Burlaga*, 1988]. More recently, *Kumar and Rust* [1996] and *Rust* [this volume] advanced a model in which both the dynamics and energetics of an expanding flux rope are governed by the conservation of the global helicity $K = \oint \mathbf{A} \cdot \mathbf{B}\, dx^3$ based on the conjecture of *Taylor* [1974]. They argue that in addition to the conservation of toroidal and poloidal fluxes, helicity conservation can be imposed as a separate constraint to calculate the amount and the timescale of magnetic energy released during the expansion of flux ropes. They further argue that the magnetic energy is released inside the flux rope in the form of plasma heating. However, no dissipative mechanisms are specified or included in the equations. This approach has been critiqued by *Chen* [1996] based on physical and mathematical considerations. The basic points are that (1) in the solar wind with free boundaries, conservation of the helicity K is equivalent to the conservation of Φ_p and Φ_t so that it does not constitute an independent constraint and (2) helicity conservation does not provide a mechanism to determine the amount and rate of magnetic energy released in the form of irreversible heating of the plasma inside the flux rope. It is not known whether the nearly collisionless solar wind provides the necessary "anomalous" dissipation.

Thermal Energy of Expanding Flux Ropes. It is interesting to contrast the consequences of the different assumptions in the model of *Kumar and Rust* [1996] and that of *Chen and Garren* [1993] and *Chen* [1996]. In the former, the plasma inside the flux rope, which is identified with the initially cold prominence gas, is heated by magnetic energy released during the expansion. The flux rope is assumed to be adiabatic with $\gamma = 5/3$ in (13) and with the legs of the flux rope anchored in a cold photosphere. The magnetic energy required to obtain the typical temperatures of magnetic clouds is 60–80% of the magnetic energy of the initial flux rope. This is of the order to 10^{32}–10^{33} erg. In the latter, a significant fraction of the initial magnetic energy is dissipated by drag (using $c_d \simeq 1$–3 based on MHD simulations of flux ropes under heliospheric conditions). This dissipation occurs outside the flux rope. Inside the flux rope, the equation of state (13) uses $\gamma \simeq 1.2$, and the legs of the flux rope are assumed to be embedded in the hot (2×10^6 K) corona as the thermal reservoir. The thermal energy required to explain the observed temperatures of magnetic clouds is of the order of 10^{30} erg, which is a small fraction ($\lesssim 10^{-2}$) of the magnetic energy of the initial flux rope. The basic magnetic energy release mechanism of the expanding flux rope, i.e., the Lorentz force doing work, is the same for both models and has been previously discussed [*Chen*, 1989].

There has been a recent suggestion that interplanetary magnetic clouds require an adiabatic index of $\gamma < 1$ to expand [*Osherovich et al.*, 1995; *Farrugia*, this volume]. This choice implies that as the solar wind expands, its temperature rises. This is unphysical. The requirement of $\gamma < 1$ is removed if 3-D flux ropes are used in an ambient medium in which the pressure decreases with the distance from the Sun (see section 2.6.3 of *Chen* [1996]).

It should be noted that the thermal input implied by $1 \leq \gamma < 5/3$ is reversible whereas plasma heating by the dissipation of magnetic energy is irreversible.

6. SUMMARY

The phenomenon of coronal mass ejections is of fundamental importance for understanding eruptive solar activities as well as their geomagnetic consequences. Much new knowledge has been gained regarding the empirical relationships among CMEs, eruptive prominences, solar flares, and interplanetary magnetic clouds. However, the physics of the initial magnetic field topology, eruption of the initial quiescent structure, and subsequent evolution represents a major gap in the understanding of this phenomenon. This gap is a significant theoretical challenge because of the intrinsic complexity and large spatial scales. The present paper has described an ongoing effort to construct a physics-based theoretical model that can quantify the entire process of eruption of the initial flux rope and the subsequent propagation through the interplanetary medium in a unified manner. The emphasis is to elucidate the physics at each step.

Using an integrated MHD approach, the eruption and propagation of flux ropes have been reduced to two coupled second-order differential equations and a number of ancillary equations providing physical constraints. The mathematical solutions are expressed in terms of macroscopic quantities such as the speed of expansion, average magnetic field, and other spatially averaged quantities. The solutions describe the evolution of flux ropes given the initial structure and a specified poloidal flux injection profile. The results are consistent with the eruption of CMEs/prominences near the Sun and with interplanetary magnetic clouds observed near 1 AU and beyond. It is remarkable that the solutions exhibit no qualitative deviations in macroscopic quantities from those of observed magnetic clouds.

In the proposed model described here, the underlying magnetic topology is that of a flux rope. The cause of eruption is the injection of the poloidal flux Φ_p with a nonzero $d\Phi_p/dt$ on the time scale of eruption. In the driven scenario described here, the profile of rapidly increasing flux is prescribed. No specific subphotospheric mechanisms for this flux increase have been modeled, but it has been suggested that flux ropes can act as conduits of magnetic energy in any process that redistributes magnetic flux [*Chen*, 1989, 1990]. Subsequent to the eruption, the flux rope remains stable, and can expand through the interplanetary medium maintaining its integrity. The resulting structure closely resembles observed interplanetary magnetic clouds. This dynamical model suggests that (1) magnetic clouds are the interplanetary counterpart of the magnetic field and hot plasma of the initial prominence cavity, (2) the plasma density inside the clouds observed near 1 AU and beyond is consistent with much of the cold prominence returning to the Sun (or small amounts being ejected), and (3) the bright CME loop is plowed ahead of the flux rope, not necessarily producing an identifiable interplanetary counterpart. This picture is consistent with those CME events that have the CME-cavity-prominence morphology [*Illing and Hundhausen*, 1986].

The injection of poloidal flux on the dynamical time scale (hours), as opposed to the quasi-static storage-release of magnetic energy through footpoint shearing, is regarded as controversial. In both paradigms, however, it is the poloidal flux that is increased. The difference is in the time scales. One possibly observable model discriminator has been identified, i.e., the modest increase in Γ_f during the initial eruption (Figure 5). Further observational consequences should be identified and quantified. In this regard, we suggest that the 1 AU consequences of any proposed model are directly observable model output that should be compared with available observation. In a recent parameter study, the initial conditions, especially the value of χ_0 (section 3.2), were found to influence the properties of interplanetary flux ropes. This is potentially a stringent constraint on theoretical models.

Acknowledgments. This work was supported by the Office of Naval Research. The IMP-8 data used in Figure 9 is courtesy of NASA/GSFC IMP-8 team and UCLA Space Physics Data Center.

REFERENCES

Amari, T., J. F. Luciani, J. J. Aly, and M. Tagger, Very fast opening of a three-dimensional twisted magnetic flux tube, *Astrophys. J.*, 466, L39, 1996.

Anzer, U., Can coronal loop transient be driven magnetically?, *Solar Phys.*, 57, 111, 1978.

Anzer, U., and G. W. Pneuman, Magnetic reconnection and coronal transients, *Solar Phys.*, 79, 129, 1982.

Birn, J., and K. Schindler, Two-ribbon flares: magnetostatic equilibria, in *Solar Flare Magnetohydrodynamics*, edited by E. R. Priest, p. 337, Gordon and Breach, New York, NY, 1981.

Bothmer, V., and R. Schwenn, Eruptive prominences as sources of magnetic clouds in the solar wind, *Space Sci. Rev.*, 70, 215, 1994.

Bothmer, V., and D. M. Rust, The field configuration of magnetic clouds and the solar cycle, this volume.

Browning, P. K., and E. R. Priest, The magnetic nonequilibrium of buoyant flux tubes in the solar corona, *Solar Phys.*, 92, 173, 1984.

Burlaga, L. F., E. Sittler, F. Mariani, and R. Schwenn, Magnetic loop behind an interplanetary shock: Voyager, Helios, and IMP 8 observations, *J. Geophys. Res.*, 86, 6673, 1981.

Burlaga, L. F., Magnetic clouds and force-free fields with constant alpha, *J. Geophys. Res.*, 93, 7217, 1988.

Burlaga, L. F., W. Behannon, and L. W. Klein, Compound streams, magnetic clouds, and major geomagnetic storms, *J. Geophys. Res.*, 92, 5725, 1987.

Burlaga, L. F., R. P. Lepping, and J. A. Jones, Global configuration of a magnetic cloud, in *Physics of Magnetic Flux Ropes*, Geophys. Monogr. Ser., vol. 58, edited by C. T. Russell, E. R. Priest, and L. C. Lee, p. 373, AGU, Washington, DC, 1990.

Cargill, P. J., J. Chen, and D. A. Garren, Oscillations and evolution of current-carrying loops in the solar corona, *Astrophys. J.*, 423, 854, 1994.

Carrington, R. C., Description of a singular appearance seen on the Sun on September 1, 1859, *Mon. Not. R. Astron. Soc.*, 20, 13, 1860.

Chen, J., Effects of toroidal forces in current loops embedded in a background plasma, *Astrophys. J.*, 338, 453, 1989.

Chen, J., Dynamics, catastrophe and magnetic energy release of toroidal solar current loops, in *Physics of Magnetic Flux Ropes*, Geophys. Monogr. Ser., vol. 58, edited by C. T. Russell, E. R. Priest, and L. C. Lee, p. 269, AGU, Washington, DC, 1990.

Chen, J., Theory of prominence eruption and propagation: Interplanetary consequences, *J. Geophys. Res.*, 101, 27,499, 1996.

Chen, J., and D. A. Garren, Interplanetary magnetic clouds: Topology and driving mechanism, *Geophys. Res. Lett.*, 20, 2319, 1993.

Crooker, N. U., and E. W. Cliver, Postmodern view of M-regions, *J. Geophys. Res.*, 99, 23383, 1994.

Démoulin, P., and T. G. Forbes, Weighted current sheets supported in normal and inverse configurations: A model for prominence observations, *Astrophys. J.*, 387, 394, 1992.

Dryer, M., Coronal transient phenomena, *Space Sci. Rev.*, 33, 233, 1982.

Engvold, O., J. Malville, and B. M. Rustad, The eruptive prominence of June 8, 1974, *Solar Phys.*, 48, 137, 1976.

Farrugia, C. J., M. P. Freeman, L. F. Burlaga, R. P. Lepping, and K. Takahashi, The Earth's magnetosphere under continued forcing: Substorm activity during the passage of an interplanetary magnetic cloud, *J. Geophys. Res.*, 98, 7657, 1993.

Farrugia, C. J., Recent work on modeling the global field line topology of interplanetary magnetic clouds, this volume.

Feynman, J., and S. F. Martin, The initiation of coronal mass ejection by newly emerging magnetic flux, *J. Geophys. Res.*, *100*, 3355, 1995.

Finn, J. M., and J. Chen, Equilibrium of solar coronal arcades, *Astrophys. J.*, *349*, 345, 1990.

Finn, J. M., P. N. Guzdar, and J. Chen, Fast plasmoid formation in double arcades, *Astrophys. J.*, *393*, 800, 1992.

Fisher, R., C. J. Garcia, and P. Seagraves, On the coronal transient-eruptive prominence of 1980 August 5, *Astrophys. J.*, *246*, L 161, 1981.

Forbes, T. G., Numerical simulation of a catastrophe model for coronal mass ejections, *J. Geophys. Res.*, *95*, 11,919, 1990.

Forbes, T. G., and P. A. Isenberg, A catastrophe mechanism for coronal mass ejections, *Astrophys. J.*, *373*, 294, 1991.

Foukal, P., Morphological relationships in the chromospheric H_α fine structure, *Solar Phys.*, *19*, 59, 1971.

Garren, D. A., and J. Chen, Lorentz self-forces on curved current loops, *Phys. Plasmas*, *1*, 3425, 1994.

Gosling, J. T., Coronal mass ejections and magnetic flux ropes in interplanetary space, in *Physics of Magnetic Flux Ropes*, Geophys. Monogr. Ser., vol. 58, ed. by C.T. Russell, E.R. Priest, and L.C. Lee, p. 343, AGU, Washington, DC, 1990.

Gosling, J. T., The solar flare myth, *J. Geophys. Res.*, *98*, 18,937, 1993.

Gosling, J. T., J. Birn, and M. Hesse, Three-dimensional magnetic reconnection and the magnetic topology of coronal mass ejection events, *Geophys. Res. Lett.*, *22*, 869, 1995.

Gosling, J. T., E. Hildner, R. M. MacQueen, R. H. Munro, A. I. Poland, and C. L. Ross, Mass ejections from the sun: A view from Skylab, *J. Geophys. Res.*, *79*, 4581, 1974.

Gosling, J. T., J. McComas, J. L. Phillips, and S. J. Bame, Geomagnetic activity associated with Earth passage of interplanetary shock disturbances and coronal mass ejections, *J. Geophys. Res.*, *96*, 7831, 1991.

Greaves, W. M. H., and H. W. Newton, On the recurrence of magnetic storms, *Mon. Not. R. Astron. Soc.*, *89*, 641, 1929.

Hale, G. E., The spectrohelioscope and its work, Part III. Solar eruptions and their apparent terrestrial effects, *Astrophys. J.*, *73*, 379, 1931.

Harrison, R. A., Solar coronal mass ejections and flares, *Astron. Astrophys.*, *162*, 283, 1986.

Harrison, R. A., and D. G. Sime, Comments on coronal mass ejection onset studies, *Astron. Astrophys.*, *208*, 274, 1989.

Hodgson, R., On a curious appearance seen in the Sun, *Mon. Not. R. Astron. Soc.*, *20*, 16, 1860.

Hood, A. W., and E. R. Priest, *Solar Phys.*, *73*, 289, 1981.

Howard, R. A., D. J. Michels, N. R. Sheeley, and M. J. Koomen, The observation of a coronal transient directed at earth, *Astrophys. J.*, *263*, L101, 1982.

Howard, R. A., N. R. Sheeley, M. J. Koomen, and D. J. Michels, Coronal mass ejections: 1979–1981, *J. Geophys. Res.*, *90*, 8173, 1985.

Howard, R. A., et al., Observations of CMEs from SOHO/LASCO, this volume.

Hu, W.-R., The dynamic process of a coronal transient associated with an eruptive prominence. I. Basic mechanism, *Astrophys. Space Sci.*, *92*, 373, 1983.

Hundhausen, A. J., The origin and propagation of coronal mass ejections, in *Proceedings of the Sixth International Solar Wind Conference*, edited by V. J. Pizzo, T. Holzer, and D. G. Sime, p. 181, 1987.

Hundhausen, A. J., Coronal mass ejections: A summary of SMM observations from 1980 and 1984–1989, in *The Many Faces of the Sun*, edited by K. Strong, J. Saba, and B. Haisch, Springer-Verlag, 1996.

Illing, R. M. E., and A. J. Hundhausen, Observation of a coronal transient from 1.2 to 6 solar radii, *J. Geophys. Res.*, *90*, 275, 1985.

Illing, R. M. E., and A. J. Hundhausen, Disruption of a coronal streamer by an eruptive prominence and coronal mass ejection, *J. Geophys. Res.*, *91*, 10,951, 1986.

Joselyn, J. A., and P. S. McIntosh, Disappearing solar filaments: A useful predictor of geomagnetic activity, *J. Geophys. Res.*, *86*, 4555, 1981.

Kadomtsev, B. B., Hydromagnetic stability of a plasma, in *Reviews of Plasma Physics*, vol. 2, edited by M. A. Leontovich, p. 153, Consultants Bureau, New York, 1966.

Kahler, S., Coronal mass ejections, *Rev. Geophys.*, *25*, 663, 1987.

Kahler, S. W., Solar flares and coronal mass ejections, *Annu. Rev. Astron. Astrophys.*, *30*, 113, 1992.

Kahler, S. W., R. L. Moore, S. R. Kane, and H. Zirin, Filament eruptions and the impulsive phase of solar flares, *Astrophys. J.*, *328*, 824, 1988.

Klein, L. W., L. F. Burlaga, Interplanetary magnetic clouds at 1 AU, *J. Geophys. Res.*, *87*, 613, 1982.

Kumar, A., and D. M. Rust, Interplanetary magnetic clouds, helicity conservation, and current-core flux ropes, *J. Geophys. Res.*, *101*, 15,667, 1996.

Kuperus, M., and M. A. Raadu, The support of prominences formed in neutral sheets, *Astro. Astrophys.*, *31*, 189, 1974.

Landau, L. D., E. M. Lifshitz, and L. P. Pitaevskii, *Electrodynamics of Continuous Media*, 2nd edition, p. 124, Pergamon Press, Oxford, UK, 1984.

Leka, K. D., R. C. Canfield, A. N. McClymont, and L. van Driel-Geszteleyi, Evidence for current-carrying emerging flux, *Astrophys. J.*, accepted, 1996.

Lepping, R. P., J. A. Jones, and L. F. Burlaga, Magnetic field structure of interplanetary magnetic clouds at 1 AU, *J. Geophys. Res.*, *95*, 11,957, 1990.

Leroy, J. L., V. Bommier, and S. Sahal-Brechot, The magnetic field in the prominences of the polar crown, *Solar Physics*, *83*, 135, 1983.

Lites, B. W., B. C. Low, V. Martínez Pillet, P. Seagraves, A. Skumanich, Z. A. Frank, R. A. Shine, and S. Tsuneta, The possible ascent of a closed magnetic system through the photosphere, *Astrophys. J.*, *446*, 877, 1995.

Low, B. C., Evolving force-free magnetic fields. I. The development of the preflare stage, *Astrophys. J.*, *212*, 234, 1977.

Low, B. C., and J. R. Hundhausen, Magnetostatic structures of the solar corona. II. The magnetic topology of quiescent prominences, *Astrophys. J.*, *443*, 818, 1995.

MacQueen, R. M., Coronal transients: A summary, *Philos. Trans. R. Soc. London, Ser. A*, *297*, 605, 1980.

MacQueen, R. M., and R. R. Fisher, The kinematics of solar inner coronal transients, *Astrophys. J.*, *254*, 335, 1982.

MacQueen, R. M., J. A. Eddy, J. T. Gosling, E. Hildner, R. H. Runro, G. A. Newkirk, Jr., A. I. Poland, and C. L. Ross, The outer corona as observed from Skylab: Preliminary results, *Astrophys. J. Lett.*, *187*, L85, 1974.

Manheimer, W. M., M. Lampe, and J. P. Boris, Effect of a surrounding gas on magnetohydrodynamic instabilities in Z pinch, *Phys. Fluids*, *16*, 1126, 1973.

Martin, S. F., W. H. Marquette, and R. Bilimoria, The solar cycle pattern in the direction of the magnetic field along the long axes of polar filaments, in *The Solar Cycle, ASP Conference Series*, vol. 27, edited by K. L. Harvey, p. 53, 1992.

Martin, S. F., and C. R. Echols, in *Solar Surface Magnetism*, edited by R. J. Rutten and C. J. Schrijver, pp. 339–346, NATO Advanced Science Institutes Series, Kluwer Academic, Boston, Mass., 1994.

Martin, S. F., and A. McAllister, Predicting the sign of helicity in erupting filaments and coronal mass ejection, this volume.

Marubashi, K., Structure of the interplanetary magnetic clouds and their solar origins, *Adv. Space Res.*, *6*, 335, 1986.

Marubashi, K., Interplanetary flux ropes and solar filaments, this volume.

Maunder, E. W., Magnetic disturbances, 1882 to 1903, as recorded at the Royal Observatory, Greenwich, and their association with sunspots, *Mon. Not. R. Astron. Soc.*, *65*, 2, 1905.

Michels, D. J., R. A. Howard, M. J. Koomen, and N. R. Sheeley, Jr., The solar mass ejection of 8 May 1979, in *Solar and Interplanetary Dynamics*, edited by M. Dryer and E. Tandberg-Hanssen, pp. 387–391, 1980.

Mikić, Z., D. C. Barnes, and D. D. Schnack, Dynamical evolution of a solar coronal magnetic field arcade, *Astrophys. J.*, *328*, 830, 1988.

Mikić, Z., and J. A. Linker, Disruption of coronal magnetic field arcades, *Astrophys. J.*, *430*, 898, 1994.

Miyamoto, K., *Plasma Physics for Nuclear Fusion*, MIT Press, Cambridge, MA, p. 186, 1976.

Mouschovias, T. C., and A. Poland, Expansion and broadening of coronal loop transients: a theoretical explanation, *Astrophys. J.*, *220*, 675, 1978.

Munro. R. J., J. T. Gosling, E. Hildner, R. M. MacQueen, A. I. Poland, and C. L. Ross, The association of coronal mass ejection transients with other forms of solar activity, *Solar Phys.*, *61*, 201, 1979.

Newton, H. W., Solar flares and magnetic storms, *Mon. Not. R. Astron. Soc.*, *103*, 244, 1943.

Osherovich, V. A., C. J. Farrugia, and L. F. Burlaga, Nonlinear evolution of magnetic flux ropes, 2. Finite beta plasma, *J. Geophys. Res.*, *100*, 12,307, 1995.

Priest, E. R., and T. G. Forbes, Magnetic field evolution during prominence eruptions and two-ribbon flares, *Solar Phys.*, *126*, 319, 1990.

Rust, D. M., Magnetic fields in quiescent solar prominences. I. Observations, *Astrophys. J.*, *150*, 313, 1967.

Rust, D. M., Spawning and shedding helical magnetic fields in the solar atmosphere, *Geophys. Res. Lett.*, *21*, 241, 1994.

Rust, D. M., and A. Kumar, Evidence for helically kinked magnetic flux ropes in solar eruptions, *Astrophys. J.*, *464*, L199, 1996.

Rust, D. M., Helicity conservation, this volume.

Sakai, J., and K.-I. Nishikawa, A model of 'disparitions brusques' as an instability driven by MHD waves, *Solar Phys.*, *88*, 241, 1983.

Schmahl, E., and E. Hildner, Coronal mass-ejections-kinematics of the 19 December 1973 event, *Solar Phys.*, *55*, 473, 1977.

Shafranov, V. D., Plasma equilibrium in a magnetic field, in *Reviews of Plasma Physics*, vol. 2, edited by M. A. Leontovich, p. 103, Consultants Bureau, New York, 1966.

Sheeley, N. R., Jr., R. A. Howard, M. J. Koomen, and D. J. Michels, Associations between coronal mass ejections and soft X-ray events, *Astrophys. J.*, *272*, 349, 1983.

Steinolfson, R. S., Theories of shock formation in the solar atmosphere, in *Collisionless shocks in the heliosphere: Reviews of current research*, *Geophys. Res. Monogr. Ser.*, vol. 35, edited by B. T. Tsurutani and R. G. Stone, p. 1, AGU, Washington, 1985.

Suess, S. T. Magnetic clouds and the pinch effect, *J. Geophys. Res.*, *93*, 5437, 1988.

Tanaka, K., Studies of a very flare-active delta group – peculiar delta spot evolution and inferred subsurface magnetic rope structure, *Solar Phys.*, *136*, 133, 1991.

Tandberg-Hanssen, E., *Solar Prominences*, D. Reidel, Dordrecht, Holland, 1995.

Taylor, J. B., Relaxation of toroidal plasma and generation of reverse magnetic fields, *Phys. Rev. Lett.*, *33*, 1139, 1974.

Tousey, R., The solar corona, *Space Res.*, *13*, 713, 1973.

Tsurutani, B. T., W. D. Gonzalez, F. Tang, S. I. Akasofu, and E. Smith, Origin of Interplanetary southward magnetic fields responsible for major magnetic storms near solar maximum (1978–1979), *J. Geophys. Res.*, *93*, 8519, 1988.

van Ballegooijen, A. A., and P. C. H. Martens, Formation and eruption of solar prominences, *Astrophys. J.*, *343*, 971, 1989.

Van Tend, W., The onset of coronal transients, *Solar Phys.*, *61*, 89, 1979.

Vršnak, B., Eruptive instability of cylindrical prominences, *Solar Phys.*, *129*, 295, 1990.

Vršnak, B., V. Ruždjak, and B. Rompolt, Stability of prominences exposing helical-like patterns, *Solar Phys.*, *136*, 151, 1991.

Webb, D. F., Erupting prominences and the geometry of coronal mass ejections, *J. Geophys. Res.*, *93*, 1749, 1988.

Webb, D. F., and A. J. Hundhausen, Activity associated with the solar origin of coronal mass ejections, *Solar Phys.*, *108*, 383, 1987.

Wilson, R. M., Geomagnetic response to magnetic clouds, *Planet. Space Sci.*, *35*, 329, 1987.

Wilson, R. M. and E. Hildner, Are interplanetary magnetic clouds manifestations of coronal transients at 1 AU?, *Solar Phys.*, *91*, 169, 1984.

Wilson, R. M., and E. Hildner, On the association of magnetic clouds with disappearing filaments, *J. Geophys. Res.*, *91*, 5867, 1986.

Wu, S. T., et al., Magnetohydrodynamic simulation of the coronal transient associated with the solar limb flare of 1980, June 29, 18:21 UT, *Solar Phys.*, *85*, 351, 1983.

Wu, S. T., M. T. Song, P. C., H. Martens, and M. Dryer, Shear-induced instability and arch filament eruption: A magnetohydrodynamic (MHD) numerical simulation, *Solar Phys.*, *134*, 353, 1991.

Wu, S. T., and W. P. Guo, A self-consistent numerical magnetohydrodynamic model of helmet streamer and flux-rope interactions: Initiation and propagation of coronal mass ejections, this volume.

Xue, M. L., and J. Chen, MHD equilibrium and stability properties of a bipolar current loop, *Solar Phys.*, *84*, 119, 1983.

Zhang, G. and L. F. Burlaga, Magnetic clouds, geomagnetic disturbances, and cosmic ray decreases, *J. Geophys. Res., 93*, 2511, 1988.

J. Chen, Code 6790, Naval Research Laboratory, Washington, DC 20375. (e-mail: chen@ppdu.nrl.navy.mil)

A Self-Consistent Numerical Magnetohydrodynamic (MHD) Model of Helmet Streamer and Flux-Rope Interactions: Initiation and Propagation of Coronal Mass Ejections (CMEs)

S. T. Wu and W. P. Guo

Center for Space Plasma and Aeronomic Research and Department of Mechanical and Aerospace Engineering, University of Alabama, Huntsville, Alabama

We present results for an investigation of the interaction of a helmet-streamer arcade and a helical flux rope under the helmet dome. These results are obtained by using a three-dimensional axisymmetric, time-dependent ideal magnetohydrodynamic (MHD) model. Because of the physical nature of the flux-rope, we investigate two types of flux ropes; (1) high-density flux rope (i.e. flux rope without cavity, [*Wu et al.*, 1996]), and (2) low-density flux rope (i.e. flux rope with cavity [*Guo* and *Wu* 1996]). When the streamer is disrupted by the flux rope, it will evolve into a configuration resembling the typical observed loop-like Coronal Mass Ejection (CME) in both cases. The streamer-flux-rope system with cavity is easier to disrupt, and the propagation speed of the CME is faster than in the streamer-flux-rope system without cavity. Our results demonstrate that magnetic buoyancy plays an important role in disrupting the streamer.

1. INTRODUCTION

Coronal Mass Ejections (CMEs) were first observed by the OSO-7 white light coronagraph in the 1960's. Space and ground-based observations established CMEs as an important component of solar coronal and interplanetary physics. These fascinating features have speeds ranging from less than 100 km s^{-1} to more than 1,000 km s^{-1}, and up to 10^{16} g of coronal plasma with accompanying magnetic field is ejected away from the sun. They are believed to be the cause of interplanetary shocks and geomagnetic storms [*Kahler*, 1992; *Gosling, et al.*, 1991]. A number of studies [*Kahler*, 1987; *Hundhausen*, 1993; *Dryer*, 1994] give some insight into the physical mechanisms which cause the CME initiation and propagation, although there is still much to be understood.

In the late 70's and early 80's, theoretical efforts to study the dynamics of CMEs were treated as an initial boundary-value problem in the context of magnetohydrodynamic (MHD) simulations [*Nakagawa et al.*, 1978; 1981; *Steinolfson et al.*, 1978; *Wu et al.*, 1978, 1982]. The works in this period focused on the dynamical response of the corona to a thermal pulse introduced at the coronal base. The initial states were static coronae with open or closed potential magnetic fields. The thermal pulse added to this idealized background state was believed to be released by magnetic-to-thermal energy conversion during a flare. *Dryer et al.* [1979] made a comparison directly with a particular CME event using Skylab-observed flare parameters as input. This approach was questioned by *Sime et al.* [1984] because of the absence of several important observed characteristics that were seen in four Skylab CMEs.

Observations during the mid-80's found that many CMEs appear to leave the solar surface earlier than the onset of associated flares [*Harrison*, 1986], and CMEs seemed to be more closely associated with erupting prominences than with flares [*Kahler et al.*, 1989]. Recently, it is widely held that it is the destabilization of large-scale coronal magnetic fields that initiates CMEs [*Hundhausen*, 1993].

The evolutionary progress in modeling CMEs can be summarized as follows. In the "first-generation" modeling work [*Wu, et al.*, 1978; 1982; *Steinolfson, et al.*, 1978, *Nakagawa, et al.*, 1978, 1981], the initial coronae were usually assumed to be static with potential or force-free fields. Observations, however, showed that many CMEs originate from disruption of large-scale quasi-static structures in coronal helmet streamers [*Illing and Hundhausen*, 1986]. Hence, in the "second-generation" modeling work, coronal helmet-streamers were, and are presently, considered to be suitable as an initial state to study CME initiation. *Steinolfson, Suess* and *Wu* [1982] first constructed a self-consistent numerical helmet streamer solution including the solar wind using a relaxation method. The importance of the initial corona in CME simulations was pointed out by *Steinolfson* and *Hundhausen* [1988]. They constructed three initial coronal models and showed that only the heated helmet streamer can reproduce the major observed characteristics of loop-like CMEs. However, they still used a thermal driver as in the "first generation" studies. Using a magnetic driver, *Guo et al.* [1991] also reproduced the major observed characteristics of loop-like CMEs in an ordinary helmet streamer. Recently, *Wang et al.* [1995] have shown again that the pre-event model atmosphere plays a key role in the simulation of CMEs.

An additional solar driver mechanism was also recognized because of the fact that photospheric shear can store magnetic free energy in coronal magnetic fields. Accordingly, *Wu et al.* [1983] performed the first numerical 2D MHD simulation of the coronal response to photospheric line-tied footpoint motion. Further numerical studies [e.g. *Mikic et al.*, 1988; *Biskamp* and *Welter*, 1989] demonstrated that shearing may cause the coronal magnetic field to erupt.

Wu et al. [1991] demonstrated a scenario of arch-filament eruption due to photospheric shearing which may lead to the initiation of CMEs. In a more recent simulation, *Linker* and *Mikic* [1995] studied the dynamics of a helmet streamer when photospheric shearing is imposed. They found that the streamer erupts when a critical shear is exceeded. However, it takes an unrealistically long time for the shear to exceed the critical value. Thus, two important points emerged during this "second generation" of numerical studies: (1) an appropriate steady-state helmet streamer had to be constructed; and (2) a variety of solar "drivers" demonstrated potential mechanisms for causal CME generation.

A fundamental theoretical issue concerning the energy source of CMEs has been discussed in the recent work of *Aly* [1984, 1991], *Sturrock* [1991], *Low* [1994] and *Low* and *Hundhausen* [1995]. *Aly* [1984, 1991] and *Sturrock* [1991] showed that if a force-free magnetic field is anchored to the surface of the sun, it cannot have an energy in excess of that in the corresponding fully open configuration. *Low* [1994] and *Low* and *Hundhausen* [1995] proposed that magnetic energy in the form of detached magnetic fields with cross-field currents may be the source of the total mass-ejection energy. As discussed in *Low* [1994] and *Low* and *Hundhausen* [1995], there is much indirect observational evidence indicating the existence of detached magnetic structures in the closed region of helmet streamers.

Recently, we have extended our two-dimensional planar MHD model [*Wu, Guo* and *Wang*, 1995] to investigate the dynamical evolution of a coronal streamer. This work, containing a detached magnetic structure (bubble) in its closed field region, has been extended to a three-dimensional axisymmetric geometry (two-dimensional geometry with three-component vector fields) [*Wu, Guo* and *Dryer*, 1996]. This extension enables us to study the dynamical response of a helmet streamer to the emergence of a helical magnetic flux-rope as proposed by *Low* [1994]. In the present study, we shall use this model to investigate the dynamical interactions of a helmet streamer and a flux rope with different properties. The models for the streamer and flux-rope system are described in Section 2, numerical results are given in Section 3, and concluding remarks are given in Section 4.

2. MODELS FOR THE STREAMER AND FLUX-ROPE SYSTEM

According to observations, the helmet streamer reflects a global-scale coronal magnetic field topology which consists of three parts: the high density dome, the low-density cavity and the prominence within the cavity. *Low* [1994] suggested that this global-scale coronal magnetic field topology could be represented by a two-part magnetic system; (1) the cavity containing a detached magnetic flux rope running above the polarity-inversion line and anchored at its two ends in the photosphere, and (2) the streamer arcade in the other direction linking bipolar regions, as shown in Figure 1. *Low* and *Hundhausen* [1995] have constructed an analytical solution without the solar wind and with emphasis on the magnetic topology of quiescent prominences.

Recently, *Guo* and *Wu* [1997] have constructed a numerical MHD solution to represent the above scenario, viz. a quasi-static helmet streamer containing a flux rope with cavity (i.e. low-density flux rope) in its closed field region. This solution is based on our previous solution of

Figure 1. Schematic representation of a streamer arcade and flux-rope system.

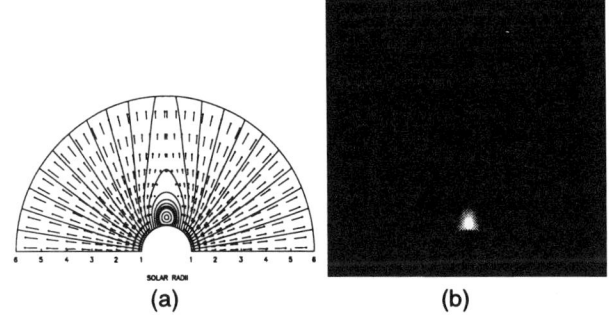

Figure 2. (a) Computed magnetic field lines and velocity vectors; and (b) the corresponding polarization brightness for a streamer-flux-rope system without cavity.

the streamer-high-density flux rope described by *Wu* et al. [1996]. Since the methods used to construct these solutions are presented elsewhere [*Wu et al.*, 1996; *Guo and Wu*, 1997], we only describe the physical models for these two cases in the following sections.

2.1 Streamer-Flux-Rope System Without Cavity

This case considers a helmet streamer containing a high-density flux rope in its closed field region. It has been shown, observationally, that there is a bright core in the streamer. Figure 2 shows the numerical solution for (a) magnetic field lines and velocity vectors, and (b) polarization brightness, which is obtained by the line-of-sight integration of the Thomson scattering of photospheric light using computed coronal density from the model output. This numerical solution is obtained by solving a set of standard ideal MHD equations in a three-dimensional axisymmetric geometry using the relaxation method [*Steinolfson, et al.*, 1982, *Wu, et al.*, 1995, 1996]. The physical parameters at the solar surface are $n_o = 3.2 \times 10^8$ cm^{-3}, $T_o = 1.8 \times 10^6$ K, $B_o = 2.0$ G. The center of the flux rope has $B_\varphi = 0.67$ G, $\beta = 1.9$.

2.2 Streamer-Flux-Rope System With Cavity

This case, for a low-density flux rope, is obtained by simultaneously decreasing density and increasing the strength of the azimuthal component of the magnetic field (B_φ) of the quasi-static solution for the streamer-flux rope without the cavity. The final solution is a quasi-static helmet streamer containing a flux rope with cavity. The physical parameters at the solar surface are the same as given in section 2.1, but the flux rope is different. At the center of the flux rope, $B_\varphi = 0.97$ G and $\beta = 0.12$. The magnetic field lines, velocity vectors, and polarization brightness for this case are shown in Figure 3. By comparing Figures 2 and 3, we immediately recognize that the core of the high-density flux rope is much brighter than in the case of the low-density flux rope which shows void regions at the core of the streamer. The comparison also shows that the plasma beta (β) is much smaller for the low-density flux rope in comparison to the high-density flux rope.

3. NUMERICAL RESULTS

In order to understand the dynamical interactions of streamers and these two types of flux-rope systems, we have performed self-consistent numerical MHD computations using three-dimensional axisymmetric ideal MHD equations [*Wu et al.*, 1996] with those cases mentioned in the previous section as the initial state. The computational domain extends from 1 to 6 solar radii, and from the pole to the equator, which gives a grid of 106 (radial direction) × 92 (meridional direction). To initiate the evolutionary computation, we increase the strength of the azimuthal component of the magnetic field B_ϕ of the flux rope as follows:

$$B_\phi^{n+1} = B_\phi^n \left(1 + \delta\left(1 - \frac{r^*}{0.85 r_f}\right)\right), \qquad (1)$$

where r_f is the radius of the flux rope, δ is an arbitrary constant related to the magnitude of the increasing field strength, r^* is the distance between the center of the flux rope and the point where the strength of B_ϕ is raised, with $r^* < 0.85\, r_f$ for this study, and the superscript "n" indicates the time step. Once the flux rope starts to move upward, we stop increasing B_ϕ in this unphysical manner and allow B_ϕ to be determined by the MHD equations. During this initiation process, we keep the solution smooth to make sure the state is at quasi-equilibrium. The temporal evolution for these two cases is summarized in the following.

For the purpose of making direct comparison between these two cases, we have set up the magnetic energy

86 HELMET STREAMER AND FLUX ROPE INTERACTIONS

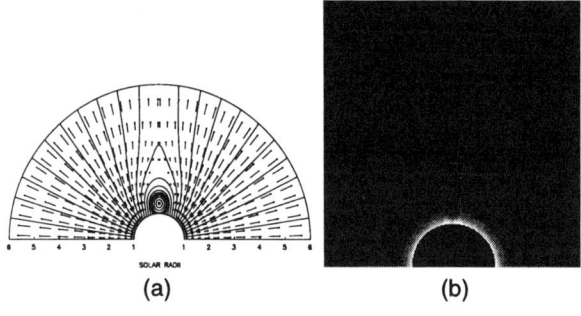

Figure 3. (a) Computed magnetic field lines and velocity vectors; and (b) the corresponding polarization brightness for a streamer-flux-rope system with cavity.

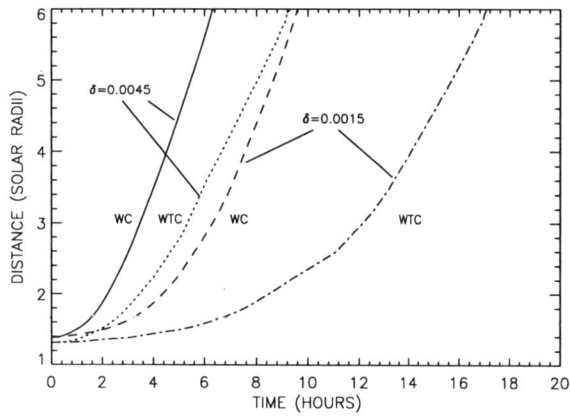

Figure 4. The location of the flux-rope center versus time for the streamer-flux-rope system with and without cavity of fast ($\delta = 0.0045$) and slow (($\delta = 0.0015$) events. (wc = with cavity, wtc = without cavity).

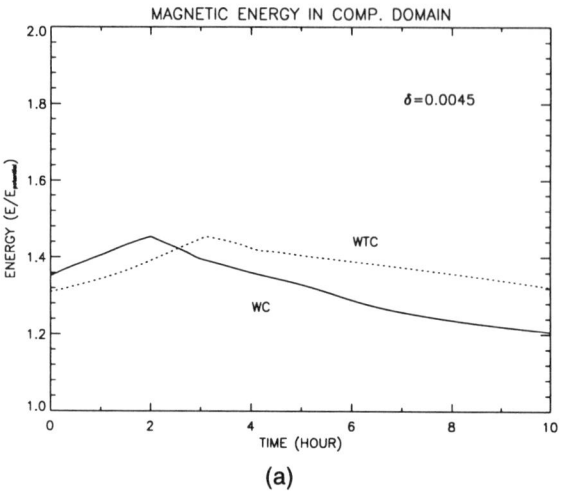

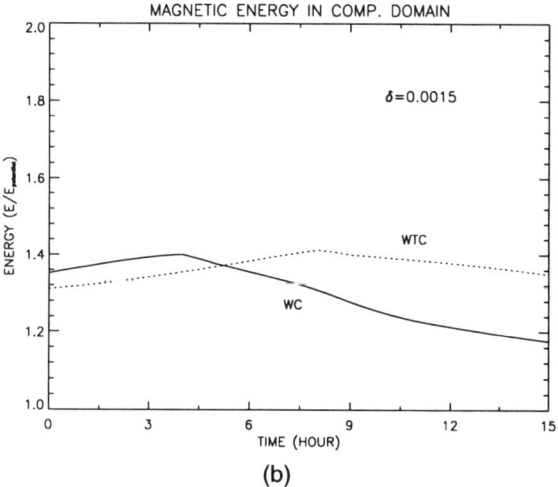

Figure 5. The evolution of the magnetic energy in the computational domain normalized by the magnetic energy of the corresponding potential field for the streamer-flux-rope system with and without cavity of fast ($\delta = 0.0045$) and slow ($\delta = 0.0015$) events.

contents in each of these two types of streamer-flux-rope systems for two different values of the prescribed B_ϕ. To implement this situation, we simply give two different values of δ in Eq. (1). For illustration of this process and understanding of the physical consequences, we have tailored our choice of δ into two categories: (1) fast propagation and (2) slow propagation events. It was determined that $\delta = 0.0045$, leads to a fast propagation event, and the slow event corresponds to $\delta = 0.0015$. Figure 4 shows the helioradial position of the flux-rope center versus time for these two values of δ and two types of streamer-flux-rope systems. The magnetic energy, in units of the magnetic energy of the corresponding potential field for these two values of δ and two types of streamer-flux-rope system, is shown in Figure 5. For $\delta = 0.0045$, we increase B_ϕ during a period of 2 hours for the streamer-flux

rope with cavity as compared to 3.1 hours for the streamer-flux rope without cavity. This made the magnetic energy for the two cases almost the same, as shown in Figure 5a. For $\delta = 0.0015$, the corresponding times are 4 hours and 8 hours. The magnetic energy for these two cases are shown in Figure 5b. We note the following:

(1) the streamer-flux-rope system with cavity responds to the perturbation as given in Eq (1) much faster than the streamer-flux-rope system without cavity, as shown in Figure 5. Figure 4 also shows that the propagation speed for the streamer-flux-rope system with cavity is higher than for the structure without cavity (i.e. 232 km s^{-1} versus 155 km s^{-}

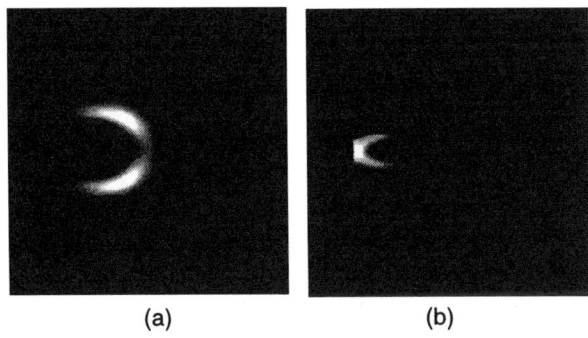

Figure 6. The polarization brightness for the streamer-flux rope system (a) with and (b) without cavity of fast event ($\delta = 0.0045$) at (a) 2 hours and (b) 3 hours.

¹ for $\delta = 0.0045$). This is understandable, because structure with cavity has less mass and a stronger magnetic field, thereby triggering the magnetic buoyancy force into play, as we can observe by comparing Figures 6 and 7. In an earlier work, *Wu, Guo* and *Wang*, [1995] demonstrated quantitatively that the non-equilibrium state of the streamer-bubble system is due to the nonlinear interactions of the Lorentz, pressure, and gravitational forces. The mass is less and the field is stronger in the present streamer-flux-rope system with cavity. These factors will cause a more readily available activation of the magnetic buoyancy force and less gravitational down pull, which together enable the streamer-flux-rope system with cavity to easily reach non-equilibrium.

(2) The results shown in Figure 4 and 5 demonstrate the effects of emerging flux on the streamer-flux rope system with and without cavity. That is, the streamer structure with cavity is more fundamental to the occurrence of coronal mass ejections (Figure 6), as suggested by *Low* [1994]. It is worth noting that the CME is propagating in front of the flux rope with a speed that is much faster than the speed measured at the center of the flux rope, because there is local expansion of the flux rope (Figure 7) due to the buoyant force. In the present calculation, the CME loop-front speeds are ~ 305 km/s and ~ 280 km/s, respectively, for the case with and without cavity for a fast event and ~ 250 km/s and ~ 210 km/s for the case with and without cavity for a slow event.

It is understood that if the simulated models are meaningful, they must exhibit simulated features which resemble observed characteristics. In order to examine the present models on this issue, we have constructed the polarization brightness for these two types of streamer-flux-rope systems for $\delta = 0.0045$, as shown in Figure 6. We note that the observed loop-like CMEs [*Burkepile* and *St. Cyr*, 1993] are simulated by the model of streamer-flux rope with cavity (Figure 6a). In the case of the streamer-flux rope

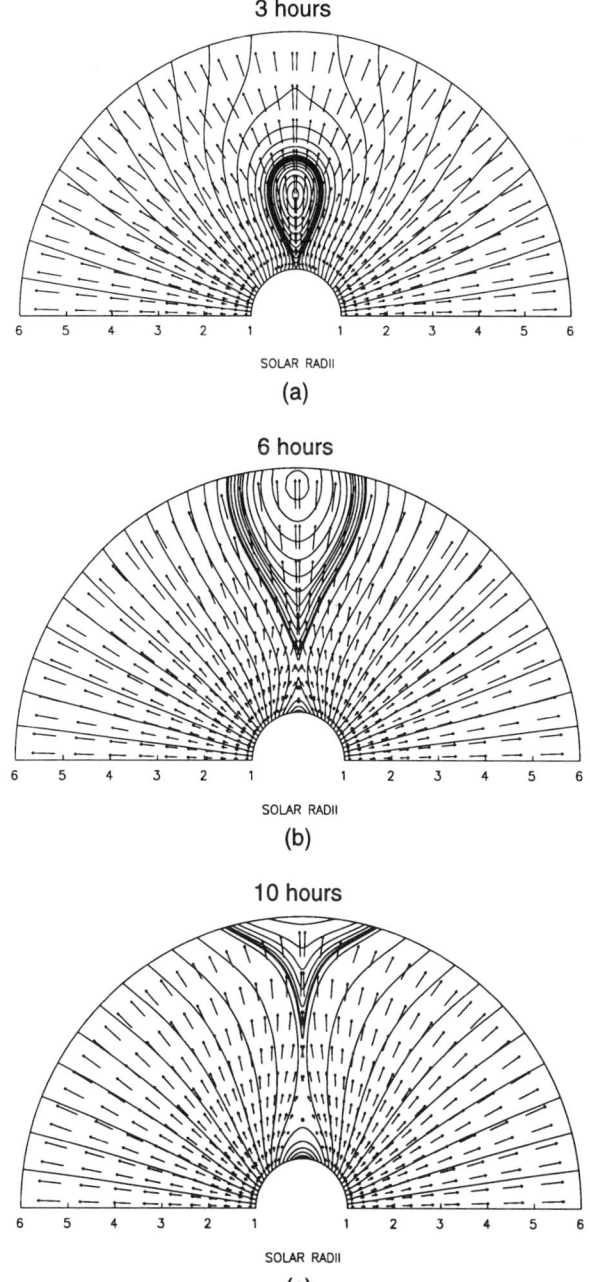

Figure 7. The evolution of magnetic field topology and velocity vector for the streamer-flux-rope system with cavity of fast event, at 3, 6, and 10 hours.

without cavity (Figure 6b), the inner part is a little different, although the frontal loop is similar to observation,. Because this is a high-density flux rope, the trailing edge of the flux rope shows much brighter features than the loop-like CME itself. A void region is shown in the core of the high-density flux rope because of the increase of B_ϕ, as discussed above. Assuming the third dimension thickness to be 0.1 solar radius, the mass involved in the two fast cases are 7.9 × 10^{15} g for the streamer-flux-rope without cavity and 4.6 × 10^{15} g for the streamer-flux-rope with cavity.

For completeness, the evolution of the magnetic field topology and velocity vectors for the streamer-flux- rope system with and without cavity for the fast event (δ = 0.0045) are shown in Figure 7 and Figure 8, respectively. By looking at the magnetic field topology, we clearly see the buoyancy effects in the case of the streamer-flux-rope system with cavity. It is also worth noting that there is no magnetic reconnection occurring in this model. The topology change shown in Figure 7 and 8 is due to the emergence of an "X" type neutral point from the lower boundary. *Wu et al.* [1995] examined and discussed this topology change in detail. The drops in magnetic energy shown in Figure 5 are due to the expansion and upward motion of the flux rope. As the flux rope expands and moves upward, the magnetic force does work on the plasma. Part of the magnetic energy is converted into other forms of energy. As a final remark, this model could give a reasonable approximation to a streamer arcade system, where the length of the arcade is much larger than the diameter of the flux rope.

4. CONCLUDING REMARKS

In this study we have employed two models [*Wu et al*, 1996; *Guo* and *Wu*, 1997] to investigate the dynamical interactions of streamers and two types of flux ropes. On the basis of this simulation study, we conclude:

1. The numerical results for the streamer-flux-rope with cavity reproduce most of the three parts of the global coronal streamer features, as suggested theoretically by *Low* [1994].

2. The helmet streamer-flux-rope system has more magnetic energy than the conventional streamer without flux-rope. Because of its high magnetic energy content, the streamer-flux-rope system is easily disrupted by disturbances, as demonstrated in this paper.

3. With the same magnetic energy contents, the model of streamer-flux-rope system with cavity responds faster in comparison to the model without cavity. Also, the propagation speed is higher with cavity than without cavity, which demonstrates that the magnetic buoyancy force plays an important role in disrupting the streamer. For both cases, the erupted streamer-flux-rope system reproduces the major observed characteristics of loop-like CMEs.

In summary, using these two models, we are capable of performing quantitative analyses of the complex CME event.

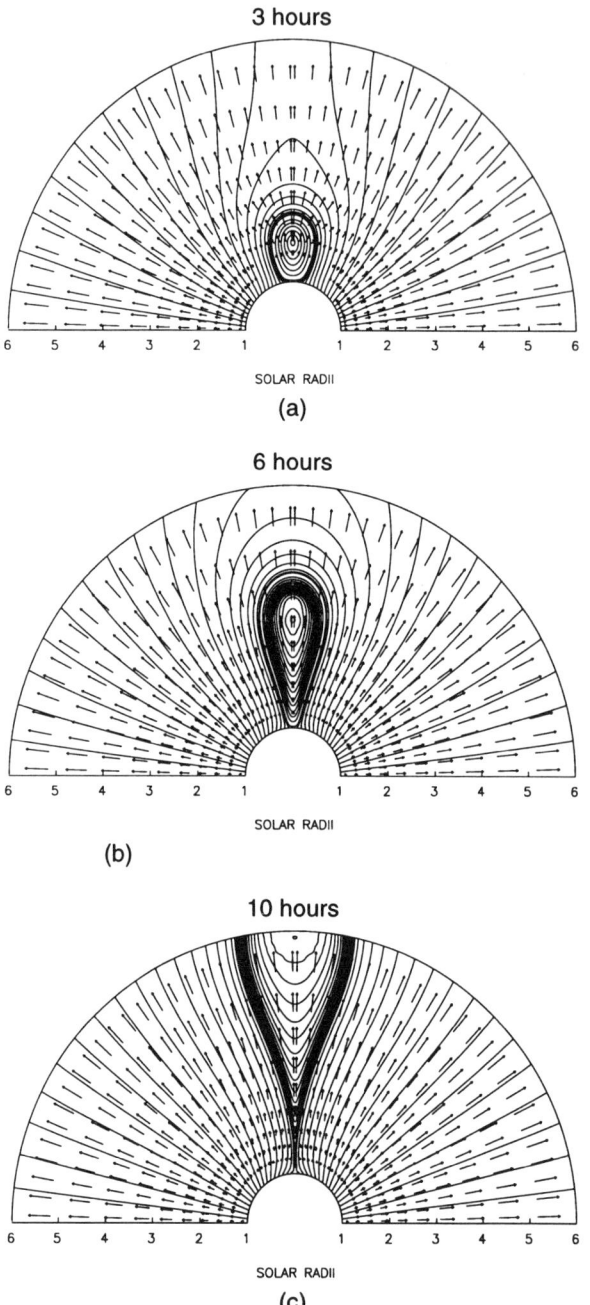

Figure 8. The evolution of magnetic field topology and velocity vector for the streamer-flux-rope system without cavity of fast event, at 3, 6, and 10 hours.

In addition we may now predict the north-south change of the B_z-component of the interplanetary magnetic field (IMF) associated with CME events.

Acknowledgments. The authors wish to thank Dr. E. Tandberg-Hanssen for reading the manuscript and giving critical suggestions. The work performed by STW and WPG is supported by NASA Grant NAG8W-4665 and NRL through USRA N00014-C-95-2058.

REFERENCES

Aly, J. J., Some properties of force-free magnetic fields in infinite regions of space, *Astrophys. J., 283*, 349, 1984.

Aly, J. J., How much energy can be stored in a three-dimensional force-free magnetic field?, *Astrophys. J., 374*, L61, 1991.

Biskamp, D. and H. Wolther, Magnetic arcade evolution and instability, *Solar Phys., 120*, 49, 1989.

Burkepile, J. T. and C. D. St. Cyr, A revised and expanded catalogue of mass ejections observed by the Solar Maximum Mission coronagraph, NCAR/TN-369-STR, 1993.

Dryer, M., S. T. Wu, R. S. Steinolfson and R. M. Wilson, Magnetohydrodynamic models of coronal transients in the meridional plane. II. Simulation of the coronal transient of 1973 August 21, *Solar Phys., 227*, 1059, 1979.

Dryer, M., Interplanetary studies: Propagation of disturbances between the sun and the magnetosphere, *Space Sci. Rev., 67*, 363, 1994.

Gosling, J. T., D. J. McComas, J. L. Philips, and S. J. Bome, Geomagnetic activity associated with earth passage of interplanetary shock disturbances and coronal mass ejections, *J. Geophys. Res., 96*, 7831, 1991.

Guo, W. P. J. F. Wang, B. X. Liang and S. T. Wu, A numerical simulation of magnetically driven coronal mass ejections, in Z. Svestka, B. V. Jackson, and M. E. Machado (Eds.) "Eruptive Solar Flares", IAU Colloquium 133, 381, 1991.

Guo, W. P. and S. T. Wu, A magnetohydrodynamic description of coronal helmet streamers containing cavity, *Astrophys. J.* (to be submitted), 1997.

Harrison, R. A., Solar Coronal Mass Ejections and Flares, Astron. Astrophys., 162, 283, 1986.

Hundhausen, A. J., Sizes and locations of coronal mass ejections: SMM Observations from 1980 and 1984 - 1989, *J. Geophys. Res., 98*, 13177, 1993.

Illing, R. M. E. and A. J. Hundhausen, Disruption of coronal streamer by an eruptive prominence and coronal mass ejection, *J. Geophys. Res., 91*, 10951, 1986.

Kahler, S. W., Solar Mass Ejections, *Rev. Geophys., 25*, 63, 1987.

Kahler, S. W., N. R. Sheeley, Jr. and M. Liggett, Coronal mass ejections and associated X-ray flare durations, *Astrophys. J., 344*, 1026, 1989.

Kahler, S. W., Solar flares and coronal mass ejections, Ann. Rev. *Astron. Astrophys., 30*, 113, 1992.

Linker, J. A., and Z Mikic', Disruption of a helmet streamer by photospheric shear, *Astrophys. J., 438*, L45, 1995.

Low, B. C., Magnetohydrodynamic processes in the solar corona: flares, coronal mass ejections, and magnetic helicity, *Phys. Plasmas, 1*, 1684, 1994.

Low, B. C., and J. R. Hundhausen, Magnetostatic structures of the solar corona. II. The magnetic topology of quiescent prominences, *Astrophys. J., 443*, 818, 1995.

Mikic', Z., D. C. Barnes, and D. D. Schnack, Dynamic evolution of a solar coronal magnetic field arcade, *Astrophys. J., 328*, 830, 1988.

Nakagawa, Y., S. T. Wu, and S. M. Han, Magnetohydrodynamics of atmospheric transients. I. Basic results of two-dimensional plane analyses, *Astrophys. J., 219*, 314, 1978.

Nakagawa, Y., S. T. Wu, and S. M. Han, Magnetohydrodynamics of atmospheric transients. III. non-plane, two-dimensional analysis, *Astrophys. J., 244*, 331, 1981.

Sime, D. G., R. M. MacQueen, and A. J. Hundhausen, Density disruption in looplike coronal transients: a comparison of observations and theoretical model, *J. Geophys. Res., 89*, 2113, 1984.

Steinolfson, R. S., S. T. Wu, M. Dryer, and E. Tandberg-Hanssen, Magnetohydrodynamic models of coronal transients in the meridional plane. I. The effect of the magnetic field, *Astrophys. J., 225*, 259, 1978.

Steinolfson, R. S., S. T. Suess, and S. T. Wu, The steady global corona, *Astrophys. J., 255*, 730, 1982.

Steinolfson, R. S., A. J. Hundhausen, Density and white lite brightness in looplike coronal mass ejections: Temporal evolution, *J. Geophys. Res., 93*, 14269, 1988.

Sturrock, P. A., Maximum energy of semi-infinite magnetic field configurations, *Astrophys. J., 380*, 655, 1991.

Wang, A. H., S. T. Wu, and G. Poletto, Numerical modeling of coronal mass ejections based on various pre-event model atmospheres, *Solar Phys., 161*, 365, 1995.

Wu, S. T., M. Dryer, Y. Nakagawa and S. M. Han, Magnetohydrodynamics of atmospheric transients. II. two-dimensional numerical results for a model solar corona, *Astrophys. J., 219*, 324, 1978.

Wu, S. T., Y. Nakagawa, S. M. Han, and M. Dryer, Magnetohydrodynamics of atmospheric transients. IV. nonplane two-dimensional analyses of energy conversion and magnetic field evolution, *Astrophys. J., 262*, 369, 1982.

Wu, S. T., Y. Q. Hu, Y. Nakagawa, and E. Tandberg-Hanssen, Induced mass and wave motions in the lower solar atmosphere. I. Effects of shear motion on flux tubes, *Astrophys. J., 266*, 866, 1983.

Wu, S. T., M. T. Song, P. C. H. Martens, and M. Dryer, Shear induced instability and arch filament eruption: A magnetohydrodynamic (MHD) numerical simulation, *Solar Phys., 134*, 353, 1991.

Wu, S. T., W. P. Guo and J. F. Wang, Dynamical evolution of a coronal streamer-bubble system, I. A self-consistent, planar magnetohydrodynamic simulation, *Solar Phys., 157*, 325, 1995.

Wu, S. T., W. P. Guo, and M. Dryer, Dynamical evolution of a coronal streamer-flux rope system. II. A self-consistent non-planar magnetohydrodynamic simulation, *Solar Phys.*, 1996. (in press).

S. T. Wu and W. P. Guo, Center for Space Plasma and Aeronomic Research and Department of Mechanical and Aerospace Engineering, The University of Alabama in Huntsville, Huntsville, Alabama 35899 USA

Solar Magnetic Topologies and Reconnection

P. Démoulin

Paris Observatory, DASOP, URA 2080 (CNRS)

A review is presented on the topology and eruption of twisted magnetic configurations in connection with eruptive flares, dynamical "disparition brusque" of prominences, and coronal mass ejections. A classical view of magnetic reconnection is mainly based on the MHD physics of 2-D configurations with an X-type neutral point, or on the extension of it to 3-D, and it is thought to be accompanied by flux transport across separatrices (places where the field-line mapping is discontinuous). This view is too restrictive because energy release can also occur in configurations without a magnetic null point both in the corona and terrestrial magnetosphere. These configurations have continuous field-line linkage; it implies that the classical picture of reconnecting field lines at the intersection of separatrices should be revised. We present here such an attempt applied to twisted configurations. We define a function N that gives a quantitative estimate of the field line linkage. While the field line linkage is continous, there are extremely thin layers, called quasi-separatrix layers (QSLs), where the gradient of the field-line linkage from one part of a boundary to another exeeds its typical value by many orders of magnitude. Even for highly conductive media these extremely thin layers behave like separatrices. Thus reconnection without null points can occur in QSLs with a breakdown of ideal MHD and a change in connectivity of plasma elements.

1. HOW CAN WE DEFINE 3-D MAGNETIC RECONNECTION?

Magnetic reconnection is currently thought to be the main way magnetic energy is released in highly conductive plasmas. The full set of partial differential equations governing the mechanism is, however, so complex that most of the theoretical studies have been limited so far to two-dimensional (2-D) systems. A 2-D magnetic configuration may be split up into topologically distinct regions by separatrices, which are special field lines that generally intersect at X-type null points. A key feature is that the mapping of field lines from one footpoint to another on a boundary becomes discontinuous at the separatrices. During reconnection, field lines approach an X-point from opposite domains and eventually reach the separatrices: then they break, reconnect and move away from the X-point into the remaining two domains that border the separatrices.

When the three components of the magnetic field vanish at some points it is possible to generalise the 2-D picture in a straightforward way. The separatrix curves become separatrix surfaces which divide up the volume into topologically distinct regions. *Lau and Finn* [1990] follow *Greene* [1988] in restricting the term reconnection to phenomena arising in resistive MHD associated with isolated null points. They consider kinematic reconnection near a single null point and a pair of nulls, in which the magnetic field is prescribed and the velocity normal to the magnetic field is deduced from the electric potential imposed at one footpoint of the field lines. Later *Lau and Finn* [1991] applied their approach to a long arcade containing a flux tube and found that local separatrices show up in plots of the delay function (or flux tube volume) as a function of footpoint position. Recently, *Priest and Titov* [1996] have discovered two basic types of reconnection modes around 3-D nulls; they have singular flows along the spine axis and the fan surface associated with the null, respectively. (The local topology around a null point is defined by the eigenvectors of the matrix describing the local linearized magnetic field around the null; the fan surface is defined by the two eigenvectors associated with the eigenvalues with the same sign, and the spine axis is defined by the third eigenvector.)

May reconnection occur in more general configurations where the magnetic field norm does not vanish? If yes, how can we define this finite-B reconnection? What are its properties? A step to finite-B reconnection has been made by analysing magnetic configurations invariant in one direction (i.e., the magnetic field has three components but depends on only two spatial coordinates). This is usually called 2.5-D reconnection. The previous null point of the 2-D configuration is here transformed into a magnetic field line, with a constant field directed along it. In 2.5-D configurations new phenomena appear, such as the development of current sheets all along the separatrices [*Low and Wolfson*, 1988; *Vekstein et al.* 1991] and the presence of strong jetting along them [*Priest and Forbes*, 1992]. These 2.5-D configurations can clearly be only an approximation of observed solar configurations; so we must analyze configurations with no symmetry and show that the phenomena described in 2.5-D configurations still exist in 3-D configurations. *Lau and Finn* [1991] showed that this is true for "long plasmoid" configurations, where the width of the plasmoid is no longer translationally invariant but still goes to an infinite distance. The separator (and separatrices) remain because there are infinitely long field lines. However, when the plasmoid has a finite extension ("short plasmoid" configurations), there no longer are separatrices, and we can pass continously from field lines inside to outside the plasmoid, as pointed out by *Schindler et al.* [1988]. This is mathematically justified by the so-called theorem about continous dependence on the initial conditions of ordinary differential equations [see, e.g., *Arnol'd* [1992]. The solution of the field line equations can have a localized discontinuity in the mapping only in the presence of a magnetic null point (or of field lines tangent to the boundary).

Faced with this (apparent) structural instability of separatrices when going from 2.5-D to 3-D configurations with finite-B, several groups have investigated ways to replace the topological approach that is familiar in 2-D. *Hesse and Schindler* [1988] used a local description of the field given by Euler potentials (U, V) such that $\mathbf{B} = \nabla U \times \nabla V$. Because, at a given time, Euler potentials are constant along each field line, they allow us to determine whether two plasma elements stay on the same field line (no reconnection) or not (reconnection). In their view, 3-D reconnection occurs when and where the resistivity is locally enhanced, but they do not explain self-consistently why and when it should occur at some precise locations. Another point of view was given by *Priest and Forbes* [1989]. They generalised the 2-D reconnection approach around an X point to 3-D by looking for a field line which has an X-type topology in a plane orthogonal to it. They called such a line a "potential singular field line." They showed that, in general, there is a continuum of such lines, and they proposed that the imposed boundary flows select the particular field-line on which reconnection occurs.

In both previous approaches, the spatial location of reconnection has no relationship to magnetic field-line linkage. However, analysis of X-ray bright points and solar flare observations show that the release of energy is linked to the magnetic field line linkage [*Démoulin et al.*, 1997b, *Mandrini et al.*, 1996,1997; *Parnell et al.*, 1994; and references therein], but the presence of null points in the magnetic configuration is not a necessity. At this point we feel that the field line linkage is important in flares to determine the localisation of the energy release, and that we need to characterise the connectivity of the coronal field in a general way, without any reference to null points or field lines tangent to the photosphere; however, this definition should include these as particular cases.

2. QUASI-SEPARATRIX LAYERS

We propose that magnetic reconnection may occur in 3-D in the absence of null points at quasi-separatrix layers (QSLs). These are regions where initially close field lines separate widely in the corona. QSLs are defined in terms of a dimensionless function that we call N. If we integrate, over a distance s in both directions, a field line passing at a point P(x, y, z) of the corona, the end points of coordinates (x', y', z') and (x'', y'', z''), define a vector $\vec{D}(x, y, z) = \{X_1, X_2, X_3\} = \{x'' - x', y'' - y', z'' - z'\}$. A drastic change in field-line linkage means that for a slight shift of point P(x, y, z), $\vec{D}(x, y, z)$ varies greatly. In the solar case, the distance s to be used is the distance to the photosphere ($z' = z'' = 0$, see *Démoulin et al.* [1996b] for a justification), and the expression for $N(x,y)$ is:

$$N(x,y) = \sqrt{\sum_{i=1,2}\left[\left(\frac{\partial X_i}{\partial x}\right)^2 + \left(\frac{\partial X_i}{\partial y}\right)^2\right]}. \quad (1)$$

This function is evaluated on the photosphere and represents the norm of the displacement gradient tensor defined when connecting, by field lines, the positive to the negative flux region. The displacement of the two footpoints is in general of the same order of magnitude, meaning that $N \approx 1$ in general; this is true, except in localized regions where N takes higher values. These regions characterize the field lines involved in the QSLs, and following these lines we can locate the coronal portion of these layers.

A feature common to all the flaring regions studied is the presence of QSLs where release of energy occurs [*Démoulin et al.*, 1997b]. This is illustrated in Figure 1 for one of the active regions studied. In order to determine the regions of rapid change in field-line linkage, the coronal magnetic field is extrapolated from the observed photospheric field assuming a linear force-free field configuration. The Hα brightenings are found along restricted regions of very thin QSLs; an upper bound of their thickness is 1 Mm, but it is several order of magnitude smaller in most of the cases. Figure 1 shows a simple bipolar region where the vertical component of the photospheric field is divided in only two polarities. Nevertheless the region has complex field line linkage: the connectivity between flare brightenings, as indicated by the four sets of field lines, is similar to that found in quadrupolar regions (where two positives and two negatives are clearly separated by inversion lines).

According to *Priest and Démoulin* [1995], when smooth boundary motions are imposed, 3-D magnetic reconnection occurs preferentialy along QSLs. The basic reason is that the field-line velocity at coronal heights is proportional by a factor N to the value of the velocity at the photosphere, thus it may become very large at QSLs. The consequence of this is that the field-line velocity greatly exceeds the possible plasma velocity; under these conditions, the field lines slip rapidly through the plasma, and a component of the electric field appears along the magnetic field in the layer. Subsequently, magnetic reconnection is forced at QSLs by boundary motions. A different thing happens when the evolution of the magnetic configuration can be considered as quasi-static. In this case, *Démoulin et al.* [1996b] have

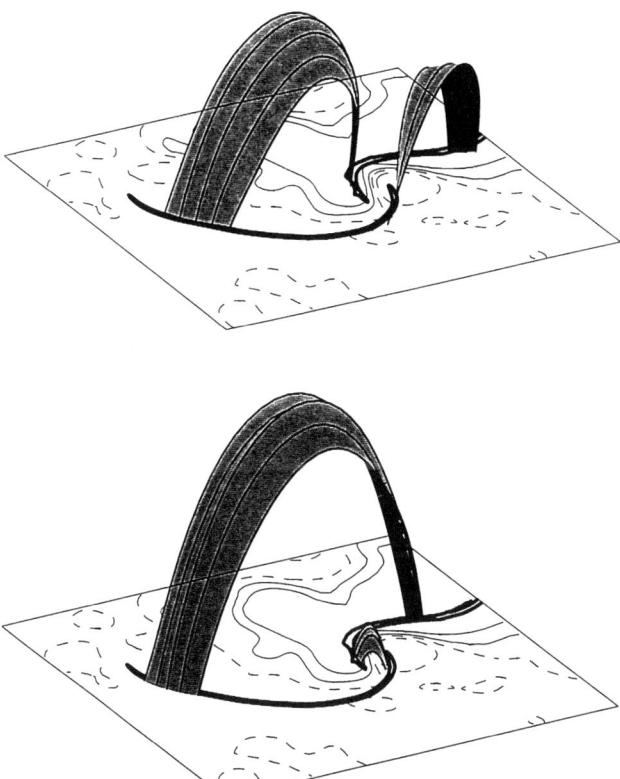

Figure 1. Example of the magnetic topology found in a flaring active region (AR 2776). The photospheric vertical field is shown at the base by the isocontours $\pm 100, 400$ G (continous and dashed thin lines). The intersection of the QSL with the photosphere is shown by an isocontour ($N = 10$) of the function N defined by Equation 1 (dark black features). The four kinds of magnetic coronal linkage, at the borders of QSL, are drawn as surfaces with field lines; they link the Hα ribbons (which are found on the QSL) [from Démoulin et al., 1997b].

shown that concentrated currents are naturally formed even by smooth photospheric motions at QSLs. This is so because two neighbouring field lines are subjected to different photospheric motions since their opposite footpoints are separated by a great distance, and, therefore, electric currents with strong density are created at QSLs. Under the quasi-static hypothesis, the current density is expected to be larger where the QSL is thinner. However, at the place where QSLs are too thin, say $\delta < \delta_{crit}$, a quasi-static evolution cannot be achieved, and reconnection is initiated. Where $\delta > \delta_{crit}$, electric currents accumulate until a threshold is reached. A brusque release of magnetic energy then follows because of either a current-driven instability or an ideal instability in the configuration.

3. TOPOLOGY OF TWISTED-FLUX TUBES

Now we illustrate the notion of QSL in twisted configurations because they are likely to be involved in ejections from the Sun of magnetized plasma, as in prominence "disparition brusques", eruptive flares, and coronal mass ejections. In 2-D the simplest configuration has the topology of Figure 2. When going to 3-D, we need not only to add a third component B_y to the magnetic field but also to make the twisted field region of finite extent (i.e., for $|y|$ large enough, field lines should connect the two photospheric polarities, as in an arcade). While the separatrices are kept in the first transformation (2.5-D configuration), they are not, in general, in the second one. Faced with this (apparent) structural instability when going from 2.5-D to 3-D configurations, we have analysed the field line linkage in 3-D analytical models of twisted configurations [Démoulin et al., 1996a]. We design the 3-D field so that we keep the 2-D and 2.5-D configurations for some parameter values. We do not constrain the configuration to be in equilibrium (there is presently no 3-D analytical model), although we do suppose the field is divergence-free.

Typically the intersection of the QSL with the photosphere takes place along two parallel elongated strips on both sides of the neutral line that end with a kind of J-shape (Figure 3). Field lines change rapidly from the lower arcade to the upper arcade when their feet cross one of the elongated parallel part of the strips. They are the remnants of field lines located below and above the twisted flux tube in 2-D configurations (Figure 2). The field lines of the twisted flux tube have their foot points inside the J-shaped ends of the strips. As B_y decreases, maintaining the finite extent of the twisted region in the y direction, the elongated parallel part of the strips stays basically unchanged, while the J-shaped ends wrap on themselves to become part of a folded-like spiral. From a global view, it seems that the J-shaped ends of the strips close on themselves (they are nearly closed in Figure 3). It is worth noting here that this closing is not complete in the sense that the field line linkage is still continuous when going from the inside of the twisted flux tube to the outside arcades, but as B_y decreases, the transition becomes sharper. When the maximum twist is greater than 4π, the QSL also separates field-lines having n turns from those having $n + 1$ turns (as shown in Figure 4). However, the drastic change in the field line linkage becomes exponentially less important as n increases, and the sharper part of the QSL corresponds to the transition from arcade-like field lines to twisted ones with one turn; that is the part

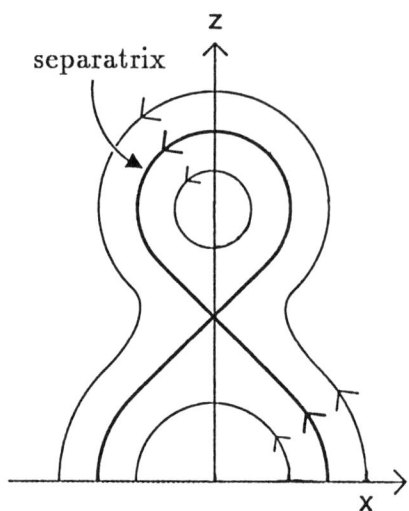

Figure 2. A typical twisted configuration in 2-D above a bipolar photospheric field. The presence of important electric currents permit the creation of a closed magnetic region in the centre of a classical arcade. One field line, called the separatrix, divides the configuration into three distinct magnetic regions: the central island, the lower and upper arcades.

of this QSL which is shown in Figure 3 and to which the text below refers. This pattern is much more complex than in 2.5-D configurations (and can only be partially described here)! In 3-D the QSL of twisted configurations is formed by a flat magnetic flux tube. (The tube's intersection with the lower boundary is shown in Figures 3 and 4.) The complexity of the field-line path prevents us from having a fine drawing of such a volume. The thinner part of the QSL volume is at the limit of both twisted, lower and upper-arcade field lines. In the limit of an infinite extension in the y direction with a translationnal symmetry (2.5D configuration), the QSL volume degenerates to the usual separatrices.

QSLs are characterized by localized high values of the function N. Their thickness is inversely proportionnal to the local maximum value of N. How thin is a QSL in twisted configurations? When the maximum twist in the configuration is lower than one turn, the QSL is relatively thick, but its thickness decreases exponentially with increasing twist. The range of QSL thickness is typically 10^{-5} to 10^{-15} (relative to the main plasmoid size) when a few field line turns are present in the twisted region (in the range 2 to 5 with the above values). For a typical scale length of, say, 10^8 m of the whole configuration, the QSL can be thinner than 1 km. This implies that the field line linkage changes so drastically

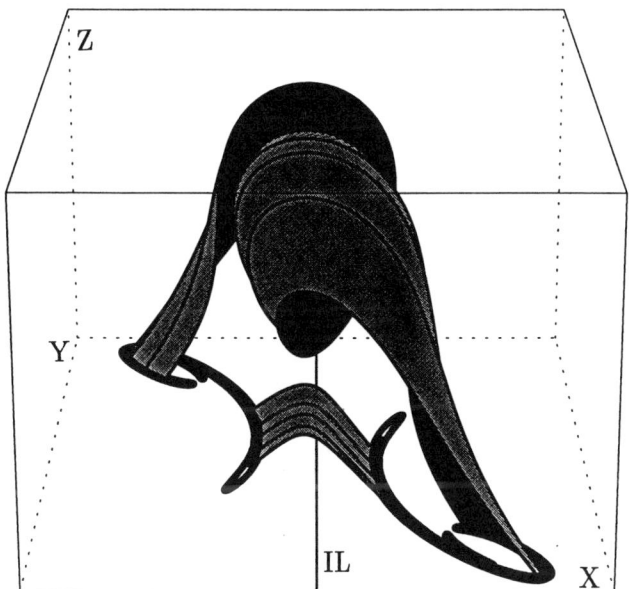

Figure 3. Perspectivic view of a 3-D twisted configuration with a finite spatial extension and without nulls or field line tangent to the photosphere. Separatrices are no longer present, but there is a very thin volume (QSL) where the field line linkage changes rapidly. The intersection of the QSL with the lower boundary (plane $z=0$) is shown by an isocontour of the function N defined by Equation 1. This intersection forms two elongated strips on both sides of the boundary inversion line (IL). From these strips the QSL extends above, following magnetic field lines (the complexity of this elongated volume precludes a clear drawing of it). Two representative sets of field lines have been included; they belong to the periphery of the twisted flux tube and to the lower arcade.

in the QSL that the physical system behaves as if it has separatrices, even for a very low resistivity. This is in agreement both with *Lau and Finn* [1991], who show that in their short-plasmoid model there are indeed well-localized peaks in the electric field potential, and with *Longcope and Strauss* [1994], who find thin current layers formation at QSLs in a coalescence instability of twisted flux tubes.

Very thin QSLs indeed appear with relatively large values of B_y; these are in the range of 5–10 % of the maximum field strength for the above reported thickness (and within the assumption of the simple analytical field expressions used so far). A much broader range of field strengths can be found in the field of simple magnetic configurations (such as a simple arcade) where QSLs are clearly not present. Therefore we cannot associate separatrices with low field strength regions. QSLs are a global property of given magnetic configurations, and cannot be determined by local criteria. Very thin QSLs can be present without the presence of any structure similar to a null point. This result supports and strengthens previous results from analysis of flares which show that flare occurrence does not require the presence of null points in the magnetic configuration [*Démoulin et al.*, 1994a].

The configurations analysed so far are simple analytical examples of 3-D twisted configurations. Emphasis has been given to simple fields in order to understand their basic properties. While the field expressions are simple, the field line linkage properties can nevertheless be quite complex! Investigating several field models, we found that, for a given number of turns in the twisted region, the magnitude of the QSL thickness dependents only slightly on the magnetic configuration chosen. Because these results come from the global field-line linkage, and the QSLs have structurally stable properties, we believe that these basic properties are characteristic of twisted configurations.

4. OBSERVATIONS OF TWISTED CONFIGURATIONS

Most of coronal twisted magnetic field configurations escape being observed because there is not enough plasma to trace the field lines nor magnetic field measurement in the corona. However, evidence for the existence of twisted magnetic configurations can be obtained when they loose their equilibrium.

The dynamics of eruptive prominences showing helical-like fine structures at the limb has been extensively studied by several authors [e.g., *Rompolt*, 1990; *Vršnak*, 1990; and *Vršnak et al.*, 1992]. The high conductivity of the solar atmosphere implies that the magnetic field is frozen in the plasma (except in some local places where reconnection is taking place). In the quiescent phase of prominences, the cold plasma is caught in the magnetic-configuration dips. It is only during the eruption of prominences that the cold plasma gains sufficient kinetic energy to move along field lines. During this short phase (compared to the prominence life-time), the cold plasma fills the magnetic structure. Many observations suggest a helical-like pattern well represented by a twisted flux tube with a global half-torus shape initially anchored in the photosphere at both ends.

Following the developments of Section 3, the breakdown of ideal MHD will mainly occur at the QSL of an erupting twisted configuration. As a matter of fact, Hα flare ribbons with a curved J-shape have been previously

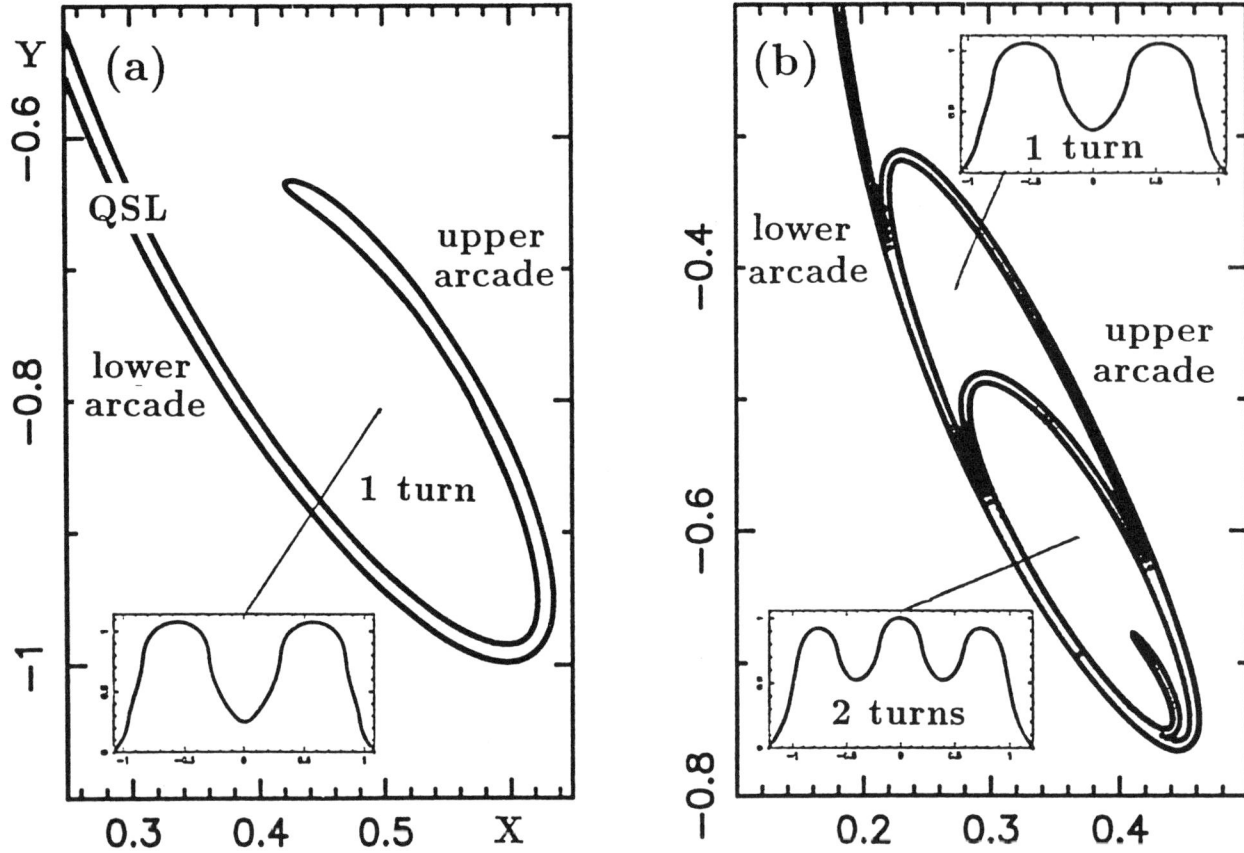

Figure 4. Local view from the J-shaped ends of the intersection of the QSL with the plane $z = 0$. As the twist increases, the QSL becomes more folded (here only the isocontour $N = 20$ is shown). The QSL delineates regions with characteristic field line shapes like separatrices do in Figure 2 (lower and upper arcade, twisted region), and also field lines with different numbers of turns (shown in separate boxes with horizontal and vertical distances along the axes) [adapted from Démoulin et al., 1996a].

reported in several prominence eruptions and flares (e.g., *Martin*, 1979; *Zirin*, 1988, p. 281, 345 and 365, *Moore et al.*, 1995, *Pevtsov et al.*, 1996). The end shape of ribbons may give a good indication of how much the eruptive configuration is twisted. However, it would be useful to observe the upper chromospheric plasma in spectral lines more sensitive to energy deposition than Hα because the eruption of quiescent prominences is usually less energetic than flares (and so does not always give well-marked ribbons in the Hα line).

At the coronal level, the observations of Yohkoh are now providing many examples of S (or inverse-S) shaped coronal loops [see, e.g., *Pevtsov et al.*, 1996; *Rust and Kumar*, 1996]. They are intrepreted as the signature of erupting twisted configurations in association with filament and coronal mass ejections. The event of 25 Oct. 1994 was particularly well observed, since there are magnetic, X-ray and radio data [*Manoharan et al.*, 1996]. Twisted X-rays loops together with small arcade-like loops underneath were produced during a long-duration C4.7 flare. These loops have a shape similar to the set of field lines drawn in Figure 3; they are interpreted as reconnected loops.

Another way to probe coronal configurations is to extrapolate the magnetic field from vector magnetograms. At present, however, even in the force-free field assumption, it is not yet clear what is the best way to deal with this nonlinear and non-local problem, in particular when the field line linkage is as complex as in twisted-field configurations [see, e.g., *Démoulin et al.*, 1997a and *Amari et al.*, 1997]. The extrapolation technique is now well known only if the computation is restricted to a linear force-free field. *Van Driel-Gesztelyi et al.* [1997] have attempted such computation for the long duration flare of 25 Oct. 1994. Within the range of observed nonpotentiality (from Mees observatory magnetograms), their extrapolated field lines approach the shape of the inverse-S shaped and lower arcade loops observed in soft X-rays. In particular, the field lines associated with the inverse-S shaped loops are mostly parallel to the photospheric inversion line when the average observed shear is used in the extrapolation. However, the linear force-free extrapolation does not permit the field to reach a twisted configuration, mostly because in the extrapolation the electric currents are not as concentrated as they are in reality. The present extrapolation techniques are too limited to use the photospheric vector magnetograms as a diagnostic of twisted configurations in the corona.

5. ERUPTIONS OF TWISTED CONFIGURATIONS

The eruption of a twisted stucture can be triggered by a loss of equilibrium or a rapid injection of new flux. In the last case, the eruption is driven directly from the emergence without need of pre-stored energy in the corona [see *Chen*, 1996,1997], while in the first case, a slow photospheric evolution permits accumulation of energy in the corona long before the loss of equilibrium occurs. A catastrophe may happen in the configuration when a cusp in the equilibrium curve (e.g., twist versus height) is present.

The loss of equilibrium for a line-current was first proposed by *Van Tend and Kuperus* [1978] as a mechanism for eruptive flares, and since, has been improved by several authors [e.g., *Martens and Kuin*, 1989; *Isenberg et al.*, 1993; *Forbes and Priest*, 1995, and references therein]. The present development considers the ideal MHD evolution of a twisted-flux rope in a magnetic field configuration invariant by translation. The internal solution of the flux rope is treated self-consistently with the external field evolution. A non-equilibrium point can be reached either by photospheric converging motions or by increasing the flux-rope twist. In ideal MHD, a new equilibrium at higher height is theoretically possible after the catastrophe. However, a long current sheet is formed underneath; consequently, with finite resistivity, reconnection is driven there, and the flux rope can go to infinite height. With a bipolar photospheric field, *Forbes et al.* [1994] showed that the maximum energy released by the loss of equilibrium is only 9% of the free energy available. The main effect of the non-equilibrium is rather the formation of a current sheet with its associated release of energy.

Can we generalise this approach in 3-D twisted configurations? It seems that there is no conceptual difficulty to generalise this non-equilibrium in 3-D configurations. However, there are great technical difficulties in building a time dependent 3-D configuration. At present we can only guess the evolution from the above knowledge of non-equilibrium in 2-D configurations and the structure of static magnetic 3-D configurations (Section 3). In 3-D twisted magnetic configurations, a QSL will replace the usual separatrices of 2-D and 2.5-D configurations. High current densities form naturally along a QSL in ideal MHD, in particular if the configuration is forced to evolve by a global instability. As the current density increases, the threshold for a current-driven instability is exceeded, giving turbulence and anomalous resistivity. Then rapid energy release via reconnection can start. The eruption of a 3-D twisted magnetic on the sun may have close analogy with flux rope formation occurring in the Earth's magnetotail [*Birn and Hesse* 1990, *Hesse et al.*, 1996].

At this point it is noteworthy that there is no necessity to open the magnetic field: reconnection transforms large arcade-like field lines to the ones at the border of the twisted flux-tube and the arcade-like ones underneath. In this scenario, the open field constraint is avoided (the open field has the highest energy of all force-free fields with the same photospheric vertical field, and so the open field cannot be reached from a force-free state by releasing energy [*Aly*, 1991; *Sturrock*, 1991]). However, there is still the possibility of a partial opening of the field if the large arcade-like field lines are not reconnected (if their field strength is weak enough, they can be carried upward by the underlying erupting twisted configurations). Indeed, that is what *Manoharan et al.* [1996] observed: the erupting twisted configuration interacts with an overlying large-scale magnetic arcade, partially reconnects with it, and partially opens it. The opening is traced by the formation of two coronal holes, beginning about 30 minutes after the eruption, at remote monopolar regions. There are other recent observations of such coronal holes formed in relationship with long duration events [e.g., *Hudson et al.*, 1996; *McAllister et al.*, 1997; *Watari et al.*, 1997; see *Hudson*, 1997 for a review].

In the above scenario of the eruption of a twist magnetic structure, the ideal evolution forms small-scale currents in the system which can be dissipated by a low resistivity. However, are the small scales thin enough in 3-D configurations to allow reconnection at a rate fast enough? Which magnitude of anomalous resistivity is required? The thickness of the current layers are unknown for all observed configurations, and the theoretical configurations analysed so far are too much idealized. Furthermore, observations do not provide much information on the real magnetic configurations: soft-X-ray observations bring only 2-D projection of a few loops which are at the periphery of the twisted region with no information on the more twisted core or the surrounding arcades, and photospheric magnetic data would require at least a nonlinear force-free extrapolation. However, QSLs in coronal twisted configurations are expected to be thinner than in the idealized models analysed so far, because in these models the photospheric field has no fragmentation in intense flux tubes, while it is known that this fragmentation gives sharp QSLs [Démoulin et al., 1996b]. What remains to be investigated is the exact importance of the total twist and of the nonlinear development of a global instability in the creation of very thin current layers, in order to understand where and when magnetic reconnection starts. This is a huge challenge, but it would be worth the effort involved. Some recent optical and X-ray observations, as well as theoretical developments, indicate that eruption of twisted magnetic configurations are likely to be the common origin of eruptive flares, dynamical "disparition brusque" of prominences, and coronal mass ejections.

Acknowledgments. I would like to thank L. Van Driel-Gesztelyi, J.C. Hénoux and two anonymous referees for their help to improve the manuscript. Financial support to attend the Chapman meeting was provided by U.S. National Science Foundation and Institut National des Sciences de l'Univers.

REFERENCES

Amari, T., J.J. Aly, J.F. Luciani, T. Boulmezaoud, and Z. Mikić, Reconstructing the solar coronal magnetic field as a force-free magnetic field, *Sol. Phys.*, in press, 1997.

Aly, J.J., How much energy can be stored in a three-dimensional force-free magnetic field?, *Astrophys. J.*, 375, L61-L64, 1991.

Arnol'd, V.I., *Ordinary Differential Equations*, 334 pp., Springer Verlag, 1992.

Birn, J. and M. Hesse, The magnetic topology of the plasmoid flux rope in a MHD-simulation of magnetotail reconnection, in Physics of Magnetic Flux Ropes, *Geophys. Monogr. Ser.*, vol. 58, edited by C.T. Russell, E.R. Priest and L.C. Lee, pp. 655-661, AGU, Washington, 1990.

Birn, J., M. Hesse, and K. Schindler, MHD simulations of magnetotail dynamics, *J. Geophys. Res.*, 101, 12936-12954, 1996.

Chen, J., Theory of prominence eruption and propagation: interplanetary consequences, *J. Geophys. Res.*, 101, 27499-27519, 1996.

Chen, J., Coronal mass ejections: causes and consequences. A theoretical View, this issue, 1997.

Démoulin, P., J.C. Hénoux, and C.H. Mandrini, Are magnetic null points important in solar flares?, *Astron. Astrophys.*, 285, 1023-1037, 1994.

Démoulin, P., E.R. Priest, and D.P. Lonie, 3D magnetic reconnection without null points - 2. Application to twisted flux tubes, *J. Geophys. Res.*, 101, 7631-7646, 1996a.

Démoulin P., J.C. Hénoux, E.R. Priest, and C.H. Mandrini, Quasi-separatrix layers in solar flares - I method, *Astron. Astrophys.*, 308, 643-655, 1996b.

Démoulin, P., J.C. Hénoux, C.H. Mandrini, and E.R. Priest, Can we extrapolate a magnetic field when its topology is complex?, *Sol. Phys.*, in press, 1997a.

Démoulin, P., L.G. Bagalá, C.H. Mandrini, J.C. Hénoux and M.G. Rovira, Quasi-separatrix layers in solar flares - II Observed magnetic configurations, *Astron. Astrophys.*, in press, 1997b.

Forbes, T.G., and E.R. Priest, Photospheric magnetic field evolution and Eruptive Flares, *Astrophys. J.*, 446, 377-389, 1995.

Forbes, T.G., E.R. Priest, and P.A. Isenberg, On the maximum energy release in flux-rope models of eruptive flares, *Sol. Phys.*, 150, 245-266, 1994.

Greene, J.M., Geometrical properties of 3D reconnecting magnetic fields with nulls, *J. Geophys. Res.*, 93, 8583-8590, 1988.

Hesse, M. and K. Schindler, A theoretical foundation of general magnetic reconnection, *J. Geophys. Res.*, 93, 5559-5567, 1988.

Hesse, M., J. Birn, M.M. Kuznetsova, and J. Dreher, A simple model of core field generation during plasmoid evolution, *J. Geophys. Res.*, in press, 1996.

Hudson, H.S., L.W. Acton, and S. L. Freeland, A long-duration solar flare with mass ejection and global consequences, *Astrophys. J.*, 470, 629-635, 1996.

Hudson, H.S., Solar antecedents of geomagnetic storms, in Magnetic Storms, edited by B.T. Tsurutani et al., American Geophysical Union, Washington, D.C., in press, 1997.

Isenberg, P.A., T.G. Forbes, and P. Démoulin, Catastrophic evolution of a force-free rope: a model for eruptive flares, *Astrophys. J.*, 417, 368-386, 1993.

Lau Y.T. and J.M. Finn, 3-D kinematic reconnection in the presence of field nulls and closed field lines, *Astrophys. J.*, 350, 672-691, 1990.

Lau Y.T. and J.M. Finn, 3-D kinematic reconnection of plasmoids, *Astrophys. J.*, 366, 577-591, 1991.

Longcope, D.W. and H.R. Strauss, The form of ideal current layers in line-tied magnetic fields, *Astrophys. J.*, 437, 851-859, 1994.

Low, B.C. and R. Wolfson, Spontaneous formation of electric current sheets and the origin of solar flares, *Astrophys. J.*, 324, 574-581, 1988.

Mandrini, C.H., P. Démoulin, L. Van Driel-Gesztelyi, B. Schmieder, G. Cauzzi, and A. Hofmann, 3D magnetic reconnection at an X-ray bright point, *Sol. Phys.*, 168, 115, 1996.

Mandrini, C.H., P. Démoulin, L.G. Bagalá, L. Van Driel-

Gesztelyi, J.C. Hénoux, B. Schmieder and M.G. Rovira, Evidence of magnetic reconnection from H_α, soft X-rays and photospheric magnetic field observations, *Sol. Phys.*, in press, 1997.

Manoharan, P.K., L. Van Driel-Gesztelyi, M. Pick, and P. Démoulin, Evidence for large-scale solar magnetic reconnection from radio and X-ray measurements, *Astrophys. J.*, *468*, L73-76, 1996.

Martens, P.C.H., and N.P.M. Kuin, A circuit model for filament eruptions and two-ribbon flares, *Sol. Phys.*, *122*, 263-302, 1989.

Martin, S.F., Study of the post-flare loops on July 1973 – II Dynamics of the Hα loops, *Sol. Phys.*, *64*, 165-176, 1979.

McAllister A., H. Kurokawa, K. Shibata, and N. Nitta, A filament eruption and accompanying coronal field changes on November 5, 1992, *Sol. Phys.*, *169*, 123-149, 1997.

Moore, R.L., T.N. La Rosa and L.E. Orwig, The wall of reconnection, driven magnetohydrodynamic turbulence in a large solar flare, *Astrophys. J.*, *438*, 985-996, 1995.

Parnell C.E., E.R. Priest and L. Golub, The 3-D structure of X-ray bright points, *Sol. Phys.*, *151*, 57-74, 1994.

Pevtsov, A.A., R.C. Canfield, and H. Zirin, Reconnection and helicity in a solar flare, *Astrophys. J.*, *473*, 533-538, 1996.

Priest, E.R. and P. Démoulin, Three dimensional magnetic reconnection without null points. I Basic theory of magnetic flipping, *J. Geophys. Res.* 100, 23.443-23.463, 1995.

Priest, E.R. and T.G. Forbes, Steady reconnection in three dimensions, *Sol. Phys.*, *119*, 211-214, 1989.

Priest, E.R. and T.G. Forbes, Magnetic flipping - reconnection in three dimensions without null points, *J. Geophys. Res.*, *97*, 1521-1531, 1992.

Priest, E.R. and V.S. Titov, Magnetic reconnection at 3D null points, *Phil. Trans. of Roy. Soc. Lond.*, *354*, 2951-2992, 1996.

Rompolt, B., Small scale structure and dynamics of prominences, *Hvar Obs. Bull.*, Vol. 14 No. 1, 37-102, 1990.

Rust, D.M., and A. Kumar, Evidence for helically kinked magnetic flux ropes in solar eruptions, *Astrophys. J.*, *464*, L119-L202, 1996.

Schindler, K., M. Hesse and J. Birn, General magnetic reconnection, parallel electric fields and helicity, *J. Geophys. Res.*, *93*, 5547-5557, 1988.

Sturrock, P., Magnetic energy of semi-infinite magnetic field configurations, *Astrophys. J.*, *380*, 655-659, 1991.

Van Driel-Gesztelyi, L., P.K. Manoharan, M. Pick, and P. Démoulin, Reorganisation of the Solar Corona by a C 4.7 Flare, *Ad. Space Res.*, in press, 1997.

Van Tend, W., and M. Kuperus, The development of coronal electric current systems in active regions and their relation to filaments and flares, *Sol. Phys.*, *59*, 155, 1978.

Vekstein, G.E., E.R. Priest and T. Amari, Formation of current sheets in force-free magnetic fields, *Astron. Astrophys.*, *243*, 492-500, 1991.

Vršnak, B. 1990, Dynamics and internal structure of an eruptive prominence *Sol. Phys.*, *127*, 129-137, 1990.

Vršnak, B., V. Ruždjak, and B. Rompolt, Stability of prominences exposing helical-like patterns, *Sol. Phys.*, *136*, 151-167, 1992.

Watari, S., Z. Smith, H.A. Garcia, T. Detman, and M. Dryer, Coronal change at the south-west limb observed by Yohkoh on 9 November 1991 and the subsequent interplanetary schock at Pionner-Venus-Orbiter, *Sol. Phys.*, in press, 1997.

Zirin, H., Astrophysics of the Sun, 433 pp., Cambridge University Press, 1988.

P. Démoulin, Paris Observatory, DASOP, F 92195 Meudon Cedex, France, DEMOULIN@OBSPM.FR

Opening Solar Magnetic Fields:
Some Analytical and Numerical MHD Aspects

T. Amari [1,3], J.F Luciani [2], J.J Aly [1], and Z. Mikic [4]

We present a review of results centered around analytical and numerical aspects in the theory of the evolution of force-free magnetic fields (meaning that the pressure forces and gravity are not taken into account) driven by photospheric boundary motions, with the issue of possible evolution towards the so called (partially or totally) open field. Some of these aspects are : (i) the nature and existence of sequences of equilibria, (ii) does opening occur ? (and in what sense) , (iii) if yes, at which rate (slow, fast, very fast or even in finite time), (iv) what type of motions (shear-twist) can lead to opening, as well as (v) energy build up and release.

1. INTRODUCTION

To make this presentation simple, from the theoretical point of view, let us divide the solar atmosphere in two parts. The corona, filled up with a low density magnetized plasma, in which the magnetic pressure dominates the plasma pressure ($\beta \leq 10^{-2}$). On the contrary, the photosphere, denser (about a million time the coronal value), is characterized by a gas pressure that dominates the magnetic pressure. Photospheric motions are governed by the underlying convection and turbulence, occurring on various scales. The electric conductivity is so high that in good approximation the magnetic field can be considered frozen into the plasma [*Priest*, 1982]. Whence, in the photosphere, the dense plasma drives the anchored magnetic foot points, while in the corona, the corresponding field lines move, to reduce the stresses imposed at the photospheric level, driving the plasma with them.

This defines an evolutionary problem with some constraints for the system of magnetic fields and plasma. The characteristic photospheric evolution time is large, compared to the time taken by an Alfven wave to travel a typical corona magnetic loop structure. Thus the corona can rapidly adjust to photospheric changes, and the resulting slow evolution can be considered as quasi-static to a good approximation. The mechanical energy injected at the photospheric level is converted into magnetic energy in the corona, at least in a first stage. Associated electric currents are generated into the corona. The system evolves until very short length scales (far smaller than the photospheric motions length scales) appear in the corona. At this stage even a tiny resistivity plays an important role in these layers. The system then liberates part of the previously stored magnetic energy by magnetic reconnection, and this energy, is then converted in particle acceleration, heating .... The modeling of this kind of evolution, slow in a first stage, and then dynamic later on, plays an important role in the understanding of coronal mass ejections (CME's) and flares. Determining whether reconnection is a driver of

---

[1] *CEA, DSM/DAPNIA, Service d'Astrophysique (URA 2052 associée au CNRS), Centre d'Etudes de Saclay, F-91191 Gif sur Yvette Cedex, France*

[2] *CNRS, Centre de Physique Théorique de l'Ecole Polytechnique, F-91128 Palaiseau Cedex, France*

[3] *CNRS, L.P.S.H. Observatoire de Paris, F-92195 Meudon Principal Cedex, France*

[4] *S.A.I.C., 10260 Campus Point Drive, San Diego, CA 92121, USA*

the phenomena or rather a consequence of a mainly ideal process (non equilibrium or instability), is one of the key problem addressed in the theory of large scale eruptive phenomena.

This modeling mostly take place in the background of the fluid approximation that represent Magnetohydrodynamics (MHD). One must solve in general a so called *continuous problem* that consists of the MHD equations

$$\rho \frac{\partial \mathbf{v}}{\partial t} = -\rho(\mathbf{v} \cdot \nabla \mathbf{v}) + \mathbf{j} \times \mathbf{B} - \nabla p$$
$$+ \nabla \cdot (\nu \rho \nabla \mathbf{v}) + \rho \mathbf{g}, \quad (1)$$
$$\frac{\partial \mathbf{B}}{\partial t} = \nabla \times (\mathbf{v} \times \mathbf{B}) - \nabla \times (\eta \mathbf{j}), \quad (2)$$
$$\frac{\partial p}{\partial t} = -(\mathbf{v} \cdot \nabla)p - \Gamma p(\nabla \cdot \mathbf{v}) + H, \quad (3)$$
$$\frac{\partial \rho}{\partial t} = -\nabla(\rho \mathbf{v}), \quad (4)$$
$$\nabla \cdot \mathbf{B} = 0, \quad (5)$$
$$\mathbf{j} = \nabla \times \mathbf{B}, \quad (6)$$

where $\mathbf{B}$ is the magnetic field, $\mathbf{g}$ the gravitational field, $\mathbf{v}$ the velocity, $\mathbf{j}$ the electric current density, ρ the mass density, p the plasma pressure, H the heating term (resulting from resistive and viscous dissipation), ν the kinematic viscosity, η the plasma resistivity, and Γ the adiabatic index; in a domain Ω, bounded or not, complemented by a set of boundary conditions on the border $\partial\Omega$, which mainly translates the fact that a slow a photospheric boundary velocity field ($\mathbf{v}\mid_{\partial\Omega}$) is imposed on $\{z = 0\}$ (see *Forbes and Priest* [1987] for the minimal number of independent boundary conditions for a viscous plasma).

To address in this paper some of the important issues associated with this evolution problem, we first consider, from the analytical and numerical point of view, the *model problem* that represents the axi-symmetric cartesian sheared arcade problem that has lead to a large amount of research. In Section III we present some results concerning fully three dimensional MHD models for simple and more complex magnetic topologies, while Section IV presents concluding remarks.

2. THE MODEL PROBLEM : SHEARING OF A 2D CARTESIAN ARCADE

2.1 *Analytical Approaches*

Most of the analytical studies have been concerned wuith the case of invariance under translation along the $\hat{x}$-axis. Shearing motions along the $\hat{x}$-axis are applied on $\{z = 0\}$, to the footpoints of an initially current free arcade. The system is assumed to evolve through a sequence of equilibria that are force-free. The imposed shear, X, and the toroidal component B_x of the magnetic field are related by :

$$X(a,t) = B_x(a,t) \int_{\mathcal{C}_p(a,t)} \frac{\mathrm{d}s}{|\nabla A(a,s,t)|}, \quad (7)$$

where A is the poloidal flux function. The toroidal component is null at $t = 0$, and increases with shear, at least in a first stage.

A first approach that has been extensively used is the so called *generating function* approach. The idea is to take B_x to depend in an a priori prescribed way on A and on a parameter $\lambda \geq 0$: $B_x = B_x(A, \lambda)$, with $B_x(A, 0) = 0$. A continuous sequence of regular equilibria (belonging to the space $C^2(\Omega) \cap C^0(\bar{\Omega})$) is constructed by solving the resulting well known Grad-Shafranov equation [*Grad* 1975] for each value of λ, starting from the current-free configuration obtained for $\lambda = 0$. The shear is thus computed a posteriori by using Eq. (??), and it is expected that, for a well chosen function $B_x(A, \lambda)$, its value increases uniformly (in some sense) along the sequence. This is indeed the case if one adopts the simplest possible parameterization: $B_x(A, \lambda) = \lambda f(A)$, with f a given function. The sequence stops existing once λ reaches some critical value λ_c [*Aly*, 1984, and *Heyvaerts et al.*, 1983, for a detailed mathematical analysis]. This "loss of equilibrium" was taken to be physically meaningful by many authors [see *Low* 1990, and references therein], as naturally associated with the onset of a dynamics phase, possibly leading to a violent energy releasing event.

Jockers [1978] first remarked that the meaning of λ_c was not so clear, as the problem which is actually solved is not the original one, in which it is the shear X rather than B_x which is imposed to the field. He pointed out the possibility that large values of B_x would never be never produced in the actual physical problem (a point that was proven to be the case indeed).The method was used recently by *Aly* [1994] to construct an explicit analytical sequence of equilibria corresponding to an evolution driven by both x- and y-motions on ∂D (opening of the field and formation of a current sheet). Of course, the sequence was obtained by using a parameterization $B_x = f(A, \lambda)$ more complicated than the one used previously.

Later on, following *Jockers* [1978], the relevant boundary value problem that imposes the shear at the photospheric level rather than the functional dependency as in the generating function approach, was reconsidered more mathematically [*Aly* 1985, *Aly and Amari* 1985]. The dependency between the toroidal component and the shearing profile then implies a highly non linear and non local problem for which only few exact analytical results exist. This shearing problem has a solution which extends to arbitrarily large shears. The regularity of the variational solution, which is a condition for it to be also a classical solution of the equation, is still an open and very difficult problem, but it can be reasonably conjectured that smoothness holds indeed for the simple arcade configuration. In a first phase the poloidal structure of the field does not change very much, while the strength $|B_x|$ of the toroidal component increases with time as t and the free-energy as t^2. Then, in a second phase, the poloidal field starts expanding at approximately the shearing velocity, $|B_x|$ decreases as t^{-1} and the energy keeps increasing as $\ln t$, i.e., at a smaller rate. Asymptotically, the field approaches an open configuration, totally or partially open, depending on the velocity profile on the boundary, that is potential everywhere, except on a singular surface on which the current has concentrated into a current sheet. Analytic computations of the asymptotic state are reported in *Amari and Aly* [1990].

On the other hand, if changes in the topology of the lines are made possible by introducing a slight resistivity, then for $t \leq t_c$, the arcade has the lowest possible energy under the imposed constraints. But for $t_c < t$, reconnection becomes energetically favorable [*Aly* 1990]. The arcade configuration becomes metastable, and may lead to the ejection of a plasmoid under the influence of a sufficiently large perturbation. However, this theory is not able to predict if reconnection is actually going to develop – it gives only a necessary condition for this phenomenon to occur. To address this problem, it is necessary to solve the full set of MHD equations, which can be done only numerically.

2.2 Numerical Approaches

Once the limits of the analytical approach have been identified, as well as the useful generality of the results, let us turn now towards numerical simulations that represent a complementary powerful tool to address some of the unsolved previous issues.

Now, one wants to solve the *discretized problem* associated to the previous *continuous problem*. In the domain Ω_h (necessarily bounded), one solves the MHD equations spatially discretized by a finite elements, spectral, collocation, or finite differences method (where the subscript h stands for the spatial discretization. The time discretization is performed by any scheme that allows one to compute at a given time $\mathbf{v}_h, \mathbf{b}_h, p_h, \rho_h$ as a function of the value of these variables at previous time steps (explicit methods) or as a function of their current values (implicit methods) or as a mixture of the two (semi-implicit methods) [*Amari et al.* 1996a]. Some discrete boundary conditions on the border $\partial\Omega_h$ are added.

This problem has been considered in both the quasistatic ideal case [*Zwingmann* 1987, *Finn and Chen* 1990, *Klimchuk and Sturrock* 1989, *Wolfson and Verma* 1991, *Platt and Neukirch* 1994] and the dynamics (ideal and resistive) case [*Mikic et al.* 1988, *Biskamp and Welter* 1989, *Inhester et al.* 1992, *Finn et al.* 1992]. In particular, the dynamics evolution of a single sheared arcade has been first studied by *Biskamp and Welter* [1989], who found that the configuration evolves slowly, with no formation of current concentrations and no occurrence of reconnection. But their simulation time was only of the order of $400\tau_A$ with a maximum shearing velocity field of 10^{-2} in units of Alfven velocity. It should be noted that they showed the formation of plasmoids in a system of three interacting arcades. *Inhester et al.* [1992] reconsidered the problem with a different numerical code based on an explicit scheme, and got similar results: For a simulation time up to a few tens of τ_A (corresponding to a moderate amount of shear too), the evolution of the sheared arcade is very quiet. The situation turns out to change, however, if the velocity field on the boundary is given a y-component which drives the footpoints towards the neutral line. In that case, the system exhibits a catastrophic behavior at some stage: A vertical current sheet develops, through which the lines subsequently reconnect, which leads to the formation of a plasmoid that is ejected at about the Alfven speed.

As remarked by *Schnack et al.* [1990] and *Kerner* [1992], a temporal stiffness property results from the large value of the Lundquist number $S = \tau_D/\tau_A$, which is a measure of the difference between the longest evolution time-scale and the shortest one. As the value of S also determines the choice of the grid size, we thus need to limit ourselves with the moderate value of S which has been already quoted. Taking a realistic coronal value ($S \approx 10^{10}$) would imply indeed several hundreds (and even thousands) of τ_A to reach the nonlinear evolution, and a number of discretized nodes that is impossible to achieve with current computer memory sizes. More specific to our problem here are the requirements on the numerical schemes which arise from the analytical results obtained for the truly half-space situation discussed above. The field has to undergo a slow longtime evolution before small scales appear. This implies the necessity for the numerical algorithm to use a large time-step Δt, while simultaneously keeping a good spatial resolution.

Because of the small size of the cells, a nonuniform mesh would also necessitate the introduction of small time-steps if classical explicit methods were used as in *Inhester et al.* [1992]. Semi-implicit schemes allow advance by larger steps, and a class of improved such schemes was proposed and extensively used by *Schnack et al.* [1990] and more recently by *Mikic and Linker* [1994]. We rather use a more general new class that has been first proposed in the framework of fusion research by *Lerbinger and Luciani* [1992] and by *Amari et al.* [1996a-b] in astrophysics . This scheme only affects the fast modes, that are not relevant to describe the slow longtime arcade evolution. A consequence is that we can use simulation time-steps that are larger than in the previous calculations, while the adaptation to the smallest expected spatial scales is done once and for all at the very beginning, which avoids annoying smoothing effects as in *Biskamp and Welter* [1989].

When a slow photospheric shear profile (corresponding to a velocity of 10^{-2} in units of Alfven velocity) is applied to the the footpoints of the initial arcade, with no resistivity, it has been shown by *Amari et al.* [1996b] that the configuration evolves slowly at the same speed as the photospheric driver, and that electric current builds up progressively in the central vertical region, in agreement with the analytical estimates of energy build up and toroidal magnetic field component presented above. In the limit of the spatial resolution compatible with the maximum electric current found, the obtained shear is about $85°$ (remember that $90°$ corresponds to an infinite shear). No non equilibrium behavior occurs, since a neighbouring force-free equilibria always exists. Then when a small amount of resistivity is introduced by setting $\eta = 10^{-4}$, the field lines reconnect rapidly in the central current layer in a few tens of Alfven times, ejecting a plasmoid.

One may argue that this splitting of a slow ideal stage (building electric current) followed by a fast resistive stage is rather artificial, and that with a constant resistivity and continual dissipation, a stable magnetic island with no fast reconnection would have resulted *Choe et al.* [1996a-b], considered the same kind two stage splitting approach, reducing progressively the first ideal stage until no plasmoid was formed, showing the existence of a minimum shear to create a plasmoid. However once again this approach can still be considered as not self consistent, justifying the use of full resistive MHD simulations with a finite resistivity from the very beginning of the simulation.

This has been recently done with two different numerical codes (based upon different numerical schemes) [*Amari et al.* 1997a]. The configuration still evolves until small length scales associated with high electric currents are created, for a shear angle of about $80°$, and while still shearing, a plasmoid is ejected. Further shearing leads to the expulsion of more energetic plasmoids. When looking at the magnetic energy build up and release, the former plasmoid is not associated with an important release of magnetic energy, whereas the following ones lead to a subsequent decrease of this quantity. Note that the fact that magnetic energy is still increasing during the ejection of the first plasmoid (while the slope is decreasing) must be due to the fact that mechanical energy is still injected while shearing. The understanding of how the ejection of the plasmoids of third and later generations may be due to boundary effect is currently under study.

3. THREE DIMENSIONNAL MHD OF TWISTED FLUX ROPES

One limitation of the previous class of x-invariant cartesian model problems is certainly the fact that one needs necessarily high magnetic shear (an ingredient that has been reported by many observers for flare occurrence [*Hagyard et al.* 1990]), but at the same time a correspondingly unrealistic high displacement of the footpoints of the field lines. 3D models therefore appear as a step further towards true modeling. It has not yet been defined a 3D model problem as in 2D, since more generally the fully 3D problem hast not been dealt with very much so far. *Klimchuk and Sturrock* [1992] presented examples of evolving equilibria, with their calculations showing no evidence for the development of nonequilibrium phenomena. But they were able to impose shears of only a moderate value. *Dahlburg et al.* [1991] also computed the evolution of some structures, which turned out to have a gentle behaviour. But they made an assumption of incompressibility of the plasma, which is not realistic and certainly prevents a large expansion of the system. Even in the case of a simple arcade like 3D magnetic configuration the situation is not straightforward for defining a simple model problem since photospheric twist and shear now become two possile candidates for injecting mechanical energy.

3.1 *Simple Magnetic Topology Configurations*

Simple bipolar configurations that consist of two spots have been observed as during the 1994, 24 October flare [*Van Driel-Gesztelyi* 1995], and has indeed lead

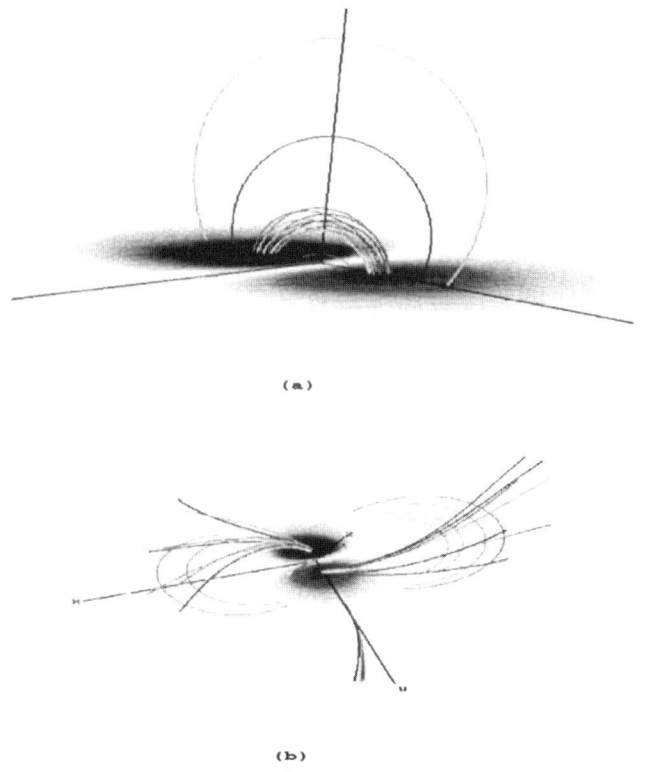

Figure 1. Very fast opening of a twisted magnetic flux tube. The initial magnetic configuration (a) is current free, and consists of two spots of opposite polarities. A slow tangential boundary velocity field, that is vortex like, is applied to the footpoints of this configuration. The two vortices, whose supports are localized to each spot, turns symmetrically. The evolution is obtained by solving numerically the MHD equations. (b) Beyond some critical time (twist), the evolution is no longer quasi-static. The initially stronger central flux ropes emerges very rapidly, pushing the outer weaker field, that lean sidewards, leading to partial opening of the configuration . Note the high shear in the flux tube close to the neutral line that was initially perpendicular to the neutral line, and that is now almost aligned with it, as well as its r S-shaped structure, both often observed in large scale energetic eruptive phenomena, . The role of resistivity in this phenomena is non existent since it is a purely ideal MHD result.

to several morphological studies for understanding the evolution of such a region. A model of this type of configuration may be obtained in a first approximation by taking an initial current-free magnetic field above a simulated photospheric magnetogram modeling these two spots [*Amari et al.* 1996c]. Applying slow photospheric twisting boundary motions corresponding to two coherent vortices, shows that the ideal MHD evolution can be split in two stages : i) first slow stage, during which the coronal configuration responds with a velocity that is of the order of the photospheric driving velocity one, and during which the magnetic energy increases monotonically, ii) second, a fast stage when the magnetic energy tends toward a limit, of about 1.8 in units of the energy of the potential field (while the fully open field is 2.3 in the same units), the evolution experiences a dynamic transition, inflating at a very fast rate, with no existing neighboring equilibria. Note that a new phenomenon that does not exist in the axi-symmetric case, now occurs in 3D. In the 2D axis-symmetric case, any field line above sheared inflating field lines, also necessarily inflates, but in the fully 3D case the third available degree of spatial freedom, allows these field lines to line up sideways and let the inner flux rope emerge from below [Figure 1]. At the same time the classical S shape of 3D X-ray loops appears in the fields that experienced the most shear.

3.2 *Fully-Partially Open Field*

In agreement with the analytical results, *Amari et al.* [1996c] found that the energy of the force-free state stays below the upper bound that represent the fully open field [*Aly* 1991 and *Sturrock* 1991].*Partial opening* occurs in the following sense : the central flux rope opens (experiences a very fast dynamics transition, in a sense that will be discussed later) while the weaker surrounding field lines remain still closed. These outer field lines should also, if one could wait long enough, experience opening, as for any field line whose footpoints are twisted (even if the boundary motions are small there). The support of the photospheric velocity field, is much more concentrated, almost as if it was compact, (while it is not actually) and therefore opening occurs initially for the central flux rope, whose trace on $\{z = 0\}$ coincides with the support of the applied boundary velocity field. Unfortunately it is not possible, with these numerical simulations, to analyze the opening of the outer region, because the flux rope has already reached the outer boundary. It is worth noting as a theoretical argument, that it is even not possible to construct a semi open field, in the same way as the fully open field can be constructed. This configuration would not be in equilibrium [*Amari et al.* 1997b]. This is just what the simulations seem to show. Let us just add that this opening occurs when the photospheric transverse magnetic field is almost aligned with the neutral line as expected from observational grounds.

3.3 Rate of Opening

Let us now discuss at which rate opening may occur, and to start, let us consider the axi-symmetric case again. The situation is however different, depending on the geometry of the system. In cartesian coordinates, *the energy of the open field is infinite*, whereas in spherical coordinates it is *finite* [Aly 1984-1990-1991]. Therefore, injecting a finite amount of mechanical energy in finite time, will never lead to opening since opening occurs by shearing motions, asymptotically [Aly 1985, Aly and Amari 1985, Amari and Aly 1990] in the cartesian case. However the energy of the open field could be approached from below in finite time in principle in spherical geometry.

To get an estimate of the the rate of opening in spherical coordinates system, one may start with the model introduced in Aly [1995], and then compute the " distance " to the open field [Amari et al. 1997b]. It is found that opening should occur in finite time. Note that some analytical estimates [Aly 1995] as well as axi-symmetric numerical equilibrium calculations of *Roumeliotis et al.* [1994], *Wolfson* [1995] and MHD simulations [*Mikic and Linker* 1994] show that opening could occur at a rate that is faster than exponential.

For three dimensional magnetic configurations, no equivalent analytical proof exists. A relevant quantity, namely the flux tube volume :

$$F_{vol} = \int_C \frac{ds}{|\mathbf{B}|}, \quad (8)$$

can however be used to measure the rate opening. It is possible to prove that *at least asymptotically* the magnetic configuration opens [Aly 1990], while numerically, for the previous bipolar configuration, F_{vol} fits like :

$$F_{vol} = \frac{\alpha}{(t - tc)^\beta}. \quad (9)$$

where α is a constant, and $\beta \approx 0.9$, which however shows that opening would rather occur in finite time [Amari et al. 1997a-b]. The transition is a break down of the quasi-static approximation. This *Global Singular Non Equilibrium* is different from the *Global Non Equilibrium* that occurred with the generating function approach [Aly 1995].

3.4 Twist or Shear ?

One concern that one may now have is about the nature of the photospheric boundary motions that are responsible for the opening property. As shown in Figure 1, the kind of magnetic configuration chosen corresponds to two spots that are very close to each other. The type of twisting motions used are such that the field lines that are very close to the neutral line, are mainly getting shear (in a way similar to the axi-symmetric cartesian case), whereas the other field lines are essentially being twisted. Moreover, as remarked above, this effect is also visible on the transverse photospheric component of the magnetic field which becomes nearly aligned with the neutral line. One way to address the issue whether twist or shear is more important, consists in splitting the two spots, so that the same kind of vortex like twisting motion do not create too much shear near the neutral line. After a maximum total twist between $(1.8 - 2)\pi$ (so that each foot point has turned less than half a turn) the same kind of transition as in the previous case with opening occurs [Figure 2] with a twist that mostly tends to stay in the weaker field line regions. That is to say, it stays at greater heights rather than close to the spots where the tendency of the field is to resist the imposed twist (note that this behavior was also observed in the previous case). Our conclusion would be that rather than distinguishing between shearing and twisting motions, it is more meaningful to discuss the amount and localization of coronal electric current that are created by the boundary motions.

5. CONCLUSION

In this paper we have presented some of the analytical and numerical results concerning recent and less recent progress in the theory of the evolution of a coronal magnetic configuration that is stressed at the photospheric level by a boundary velocity field. This problem has been used for modeling solar eruptive phenomena.

The axi-symmetric case has lead to a large amount of work and is now rather well known. A well posed model problem in cartesian or spherical geometry has been set and rigorous results have been obtained. Note that the cylindrical case is very close to the spherical one and is therefore ommitted in this discussion. In ideal MHD opening occurs asymptotically (infinite shear) in the cartesian case while at a rate faster than exponential (and even at finite shear in some numerical simulations) in the spherical case; the two cases differing essentially from the corresponding infinite (finite) value of the energy of the open field in the cartesian (spherical) case.

There are much less results in 3D because of the mathematical and numerical complexity. There is not a unique model problem, but several, in particular because of the possible complexity of the boundary velocity field that may at least contain shear, twist or both. A

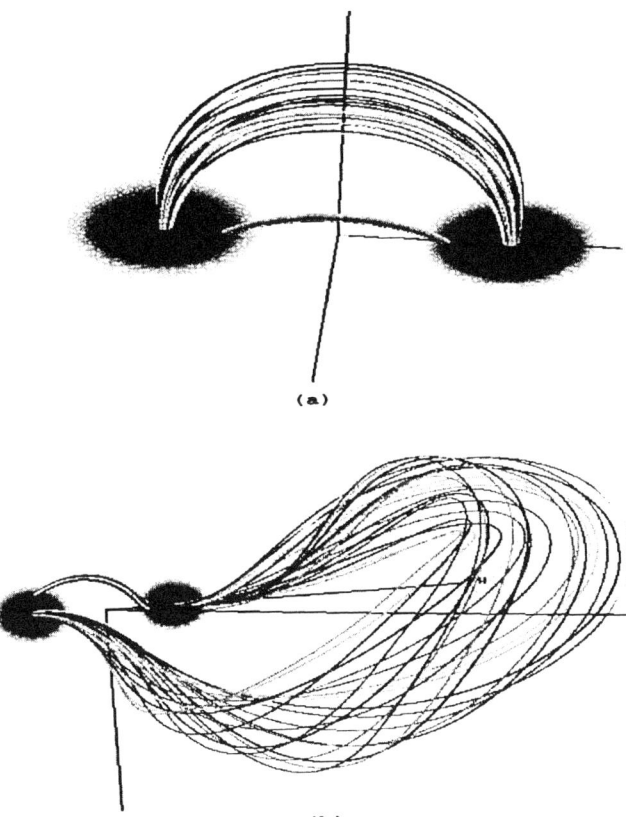

Figure 2. Effect of Twist/Shear in a three dimensional ideal MHD numerical simulation of a twisted initially current free bipolar magnetic configuration (a). The two spots are sufficiently far from each other, to minimize the shear near the neutral line when twisting as on Figure 1. (b) The same very fast opening behavior as on Figure 1 occurs, , beyond some critical time, and separates still slowly evolving closed field lines, from stronger kinked flux rope that expands at a fast rate. It seems that the amount and localization of electric currents is more important than distinguishing between shear or twist.

first model problem that has been considered is the slow photospheric twisting of a simple bipolar magnetic configuration. From the analytical point of view it is known that the configuration should at least open asymptotically while the numerical simulations exhibit very fast opening in almost finite time, this opening affecting only some of the flux ropes (depending on the photospheric velocity field support) because of a new 3D effect which gives more freedom to some of the field lines that may line sideways, remaining closed. This opening occurs always below the upper bound of the totally open magnetic field.

To built an adequate model for solar eruptive phenomena such as flares or CME's, several theoretical progress remain to be done, and for that first define relevant 3D model problems, as well as solving some yet unsolved problems even in the axi-symmetric case. Let us now discuss some of these problems.

One of the first theoretical issue concerns the amount of magnetic energy that can be released in a large eruptive phenomena. In most of the force-free models that have been considered so far, one encounter the open field energy upper bound. Since the existence of this bound rests on a mathematical argument that concerns any force-free configuration whose field lines are all connected to the photosphere (simply connected configurations), as well as a compactness assumption, would it be possible to pass beyond this limit if one would consider the slow shearing of a axi-symmetric configuration having a magnetic island? Whether this island could be created by reconnection of a previously sheared arcade-like magnetic configuration or by the rising of a sub-photospheric flux rope [see *Low* 1996], leads to consider another relevant model problem that would consist (in axi-symmetry) of the slow rising of a subphotsheric flux rope in a pre-existing overlying arcade. This difficult problem as not been yet been considered rigorously.

This last model problem is part of a more general class of problems that would need to be addressed in 3D (and rather well known in 2D [*Amari* 1991]) that is the MHD evolution of the a complex magnetic topology configuration. Although for most of the CME's and big two ribbons flares the bipolar configuration model problem discussed above seems to be relevant it is not the case for more common flares, that seem to occur in more "complex" magnetic topology configuration [*Demoulin* 1997]. A first generic case could be the shearing or twisting of a quadrupolar magnetic configuration that would have a complex or almost complex topology. Recent results [*Amari et al.* 1997b] show that the very fast opening phenomena described in previous sections still occured. At the dynamic transition the sharply varying regions of connectivity that were present in the initial configuration still persist. However opening reveals new regions associated with the separation between the closed and open field lines. More analysis should however be done to determine how these different regions depend on the applied boundary velocity field.

The resistive effects that occurs in such evolution are also not yet well understood although clearly responsable of several dissipative signature (such as X rays). Let us first compare the role of finite resistivity in 2D and 3D in simple topology configurations such as the bipolar one discussed above.

The physical mechanisms seems to be totally different in the 2D and 3D cartesian cases. In 2D, in absence of resistivity, the system evolves through a sequence of configurations that are always in equilibrium. In the presence of a small amount of resistivity (that might be considered as still high when compared to coronal values) these configurations lead to the ejection of a plasmoid. In 3D an ideal non equilibrium phenomenon, called a global singular non equilibrium, occurs! Recent results [*Amari et al.* 1997b] seems to show that in presence of a small value of the resistivity, the amount of magnetic energy that is released is small. Beyond the critical transition that we have discussed above (and whose nature is essentially ideal) *there is a competition between the very fast inflation of the structure and the generation of small length scale electric currents* corresponding to the current sheet that would be present in the open field. The global singular non equilibrium depends on the profile of the applied boundary velocity, and also on the the new spatial freedom in 3D. Unlike the 2D spherical case, where the axi-symmetry imposes a distribution of electric currents on the whole central plane of symmetry, the fast inflation of the central flux rope (while neighboring arcades lean sideways) shows that the configuration is still far from the open field when the dynamic transition occurs. Furthermore, if numerical simulations could go on longer, one might be able to address the issue whether the acceleration of the flux rope will make the magnetic energy decrease ? In any case we think that one of the main conclusions would be that resistivity may not be so important as a *trigger* of large eruptive phenomena [see *Klimchuck* 1997, for the role of reconnection in CME], while it may be dominant in in most flares that are mainly manifestation of reconnection (a model problem relevant to those flares is the MHD modeling of emergence of new magnetic flux, involving reconnection due to a change of topology). It is also finally worth noting that the magnetic configuration may look open at coronal length scales, but still be closed at the interplanetary medium length scales.

Acknowledgments. The Numerical computations presented in this paper have been performed on the CRAY supercomputers of the Commissariat à l'Energie Atomique and the Institut I.D.R.I.S of the Centre National de la Recherche Scientifique, as well as of the San Diego Supercomputer Center. This work has been supported by a NATO collaborative research.

REFERENCES

Aly, J.J., *Astrophys. J., 283*, 349, 1984.
Aly, J.J., *Astron. Astrophys., 143*, 19, 1985.
Aly, J.J., *Comp. Phys. Comm., 59*, 13, 1990.
Aly, J.J., *Astrophys. J., 375*, L61 , 1991.
Aly, J.J., *Astron. Astrophys., 288*, 1012, 1994.
Aly, J.J., *Astrophys. J., 283*, L63 , 1995.
Aly, J.J., Amari, T., in *Theoretical Problems in High Resolution Solar Physics* , edited by H.U. Schmidt,p. 319, MPA 212, 1985.
Aly, J.J., Amari, T., *Astron. Astrophys.,221* , 287, 1989.
Aly, J.J., Amari, T., *Astron. Astrophys.,in press*, 1996.
Amari, T., Aly, J.J., *Astron. Astrophys.,227*, 628, 287, 1990.
Amari, T.,in *Advances in Solar System Magnetohydrodynamics* , edited by E.R. Priest, and A. W. Hood,p. 173, Cambridge University Press, 1991.
Amari, T., Luciani, J.F. Joly P., *S.I.A.M. Journal of Scientific Computing* in Press, 1996a.
Amari, T., Luciani, J.F. Aly J.J, Tagger, M.,*Astron. Astrophys., 306* , 913, 1996b.
Amari, T., Luciani, J.F. Aly J.J, Tagger, M., *Astrophys. J., 466*, L39, 1996c.
Amari, T., Luciani, Mikic Z., Plasmoid Ejection in the MHD of Resistively Sheared Arcade, *preprint, Saclay*, 1997a.
Amari, T., Luciani, Aly., Three Dimensional MHD evolution of Twisted Magnetic Configurations, *Preprint, Saclay* , 1997b.
Antiochos, S. K., Magnetic Topology and Current Sheet Formation, in *Solar and Stellar Flares (IAU Cool.No 104)* edited by B.M. Haish and M. Rodono, p.271, Kluwer Academic Publishers, Dordrecht, 1989.
Berger, M. A., On the use of perturbation expansions in magnetic non equilibrium theory, *Internal Report*, University of Saint-Andrews, U.K, 1990.
Biskamp, D., Welter, H., *Sol. Phys., 120*,49, 1989.
Choe, G.S., Lee, L.C.,*Astrophys. J., 472*, 360, 1996a.
Choe, G.S., Lee, L.C.,*Astrophys. J., 472*, 372, 1996b.
Dahlburg, R.B., Antiochos, S. K., and Zang, T. K. *Astrophys. J., 383*, 420, 1991.
Demoulin, P. Solar Topologies, *this issue*, 1997.
Finn, J.M., Chen, J.,*Astrophys. J., 349*, 345, 1990.
Finn, J.M., Guzdar, P.N., Chen, J.,*Astrophys. J., 393*, 800, 1992.
Forbes, T., Priest, E.R., *Rev. Geophys. 25*, 1583, 1987.
Grad, H., Hu, P.N., Stevens, D.C.,*Proc. Nat. Acad. Sci. 72*,3789, 1975.
Heyvaerts, J., Lasry, J.M., Schatzman, M., Witomsky, P., *Astron. Astrophys., 111*, 104, 1982.
Inhester, B., Birn, J., Hesse, M., *Sol. Phys., 138*, 257, 1992.
Jockers, K.,*Sol. Phys., 56*, 37, 1978.
Klimchuk, J.A., *this issue*, 1997.
Klimchuk, J.A., Sturrock, P., *Astrophys. J., 345*, 1034, 1989.
Klimchuk, J.A., Sturrock, P., *Astrophys. J., 353*, 385, 1992.
Lerbinger, K., Luciani, J. F.,*J. Comput. Phys. 97*, 444 1992.

Low, B.C., *Astrophys. J., 323*, 358, 1987.
Low, B.C., and Wolfson, R. *Astrophys. J., 324*, 574, 1988.
Low, B.C., it Ann. Rev. Astron. Astrophys., 28, 205, 1990.
Low, B.C., *Phys. Plasmas 1*, 1684, 1994.
Low, B.C., *Solar Physics 167*, 217, 1996.
Mikic, Z., Barnes, D. C., Schnack, D. D., *Astrophys. J., 328*, 830, 1988.
Mikic, Z., and Linker, J. A., *Astrophys. J., 430*, 898, 1994.
Mikic, Z., and Linker, J. A., This issue, 1997.
Parker, E. N., Cosmical Magnetic Fields, Oxford Universty Press, 1979.
Platt, U., Neukirch, T., *Sol. Phys., 153*, 287, 1994.
Priest E.R., *Solar Magnetohydrodynamics*, D. Reidel, Norwell, Mass., 1982.
Roumeliotis, G., Sturrock, P.A., Antiochos, S.K., *Astrophys. J., 423*, 847, 1994.
Schnack, D. D., Mikic, Z., Barnes, D. C., Van Hoven, G., it Comput. Phys. Comm. 59, 21, 1990.
Sturrock, P. A. and Woodbury, E. T., in *Plasma Astrophysics*, edited by P.A. Sturrock, Academic Press, 1969.
Sturrock, P. A. *Astrophys. J., 380*, 655, 1991.
Van Ballegooijen, A. A., *Geophys. Astrophys. Fluid Dyn., 41*, 18, 1988.
Van Driel-Gesztelyi, L., Private communication, 1995.
Vekstein, G.E., E.R. Priest and T. Amari, Formation of current sheets in force-free magnetic fields, *Astron. Astrophys., 243*, 492–500, 1991.
Wolfson, R., Verma, R., *Astrophys. J., 375*, 254, 1991.
Wolfson, R., *Astrophys. J., 443*, 810, 1994.
Zwingmann, W., *Sol. Phys., 111* 309, 1987.

T. Amari CNRS: CE de Saclay, and Observatoire de Paris, FRANCE amari@gemini.saclay.cea.fr

The Topology and Instability of Complex Magnetic Fields

Aaron William Longbottom

Mathematics Department, University of St. Andrews, St. Andrews, KY16 9SS, Scotland

The magnetic structures of active-region and quiescent filaments and their overlying magnetic arcades which erupt to give a CME are extremely complex. New models are presented for such structures and for the complex field produced by many solar magnetic sources. Their properties are described, including the nature of their skeletons, which consist of the 3D null points and a web of spine curves and separatrix surfaces. In particular, the nature of the magnetic field near a switch-back in the global polarity inversion line is described where CME eruptions have been found with Yohkoh to be common. The nature of the instabilities that may lead to prominence eruption is examined, indicating the conditions for the onset of such a process. Finally, the evolution of such fields through a series of equilibria is analysed.

1. INTRODUCTION

Recent observations of chromospheric and coronal structures are giving a wealth of new detailed information regarding the configuration of the magnetic field and confined plasma above the solar surface. Not only do we have magnetograms at the photospheric level, but also a myriad of other observations at greater heights indicating the complex morphology that the solar magnetic field exhibits. Such observations show the form of the magnetic field overlying active regions to be fully three-dimensional and, in many cases, to contain wildly varying length-scales. Such states provide a challenging problem for the theorist, but by their very nature such structures are hard to model realistically. It is only with an understanding of the simpler one- and two-dimensional cases, and the rapid development of computational tools and techniques, that at last some initial attempts can be made to unlock the topology and mechanisms of realistic three-dimensional fields.

In section 2 we examine how complex coronal fields can be modelled theoretically in a three-dimensional configuration, while section 3 examines a method of determining their stability. The evolution of such models, through a series of equilibria, is discussed in section 4. We close with our conclusions in section 5.

2. CORONAL EQUILIBRIA

In general, static MHD equilibria satisfy the force balance equation

$$\mathbf{j} \times \mathbf{B} = \nabla p + \mathbf{F} , \quad (1)$$

where $\mathbf{B}$ is the magnetic field, $\mathbf{j} = \nabla \times \mathbf{B}$ is the current density, p is the gas pressure, and $\mathbf{F}$ represents any external forces (such as gravity). For the solar corona, magnetic forces dominate the other forces. Thus, to a first approximation, we can obtain coronal equilibria by solving solely for the magnetic field via

$$\mathbf{j} \times \mathbf{B} = 0 ,\qquad(2)$$

or

$$\nabla \times \mathbf{B} - \alpha \mathbf{B} = 0 ,\qquad(3)$$

where α is some function of space. If, α is zero then $\mathbf{j} = 0$ and the resulting equilibrium is said to be current-free or potential. If, α is a constant then the field is said to be linear force-free; if, α is a true function of position then it is nonlinear force-free. Here we describe two methods of calculating such equilibria.

As a first approximation potential equilibria can be constructed. This may be done analytically. Thus, models relating to a vast region of parameter space may be investigated with relative ease, while the essential nature of the problem is retained.

For example, (D. Mackay, personal communication, 1997) a potential model has been developed to explain the switchback of a coronal arcade observed with Yohkoh (A. McAllister, Boulder). The general layout of flux sources and sinks is shown in Figure 1. There is a local imbalance of flux sinks near the corner of the switchback which is balanced by a series of equally spaced sources at a large distance. The resulting potential field is plotted in Figure 2. The local imbalance in flux leads to a shearing of field lines on each arm (horizontal arm left skew, inclined arm right skew). This shear as a function of length along the polarity inversion line, from the corner of the switchback, for a number of heights is plotted in Figure 3. The general increase in the angle of the magnetic shear with height, as well as the global arcade structure, agrees well with orientations of the magnetic field deduced from the coronal arcade observations.

Whilst potential field models are simple to calculate they cannot fully represent the nature of coronal fields. In reality we would like to construct three dimensional force-free equilibria. However, apart from a number of overly simplistic cases, this must be done using numerical methods. Computational constraints in terms of cpu time and machine memory now become important, and these restrict the number of fields that may be calculated. With this in mind A. Longbottom et al., [submitted paper, *Astron. Astrophys. Supp.*, 1997] have used multigrid methods to enable the fast generation of linear and nonlinear force-free equilibria.

Figure 4 shows a line of sight magnetogram (K. Harvey, Kitt Peak) of an emerging active region together with the field lines calculated from the linear force-free field. Here the length scale is $0.01 units = 1$". It was found that a potential model was unable to give the correct foot point connectivity. A large value of α was needed to rectify this, indicating that in this develop-

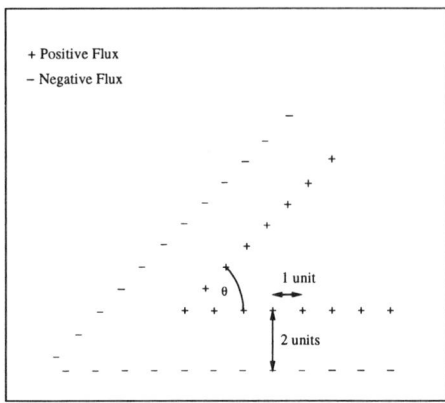

Figure 1. The positions of the flux sources and sinks used to model the switchback. Note the local imbalance in flux at the apex.

ing active region there are significant currents, with the overall topology in a force-free configuration.

It could be argued that the field in an active region is far from being force-free (but may possibly be described by a nonlinear force-free field). However, such models have some drawbacks. Unless a great deal of data is known about the initial field and changes to the boundary conditions then there are problems of nonuniqueness of the computed solution. In addition, in order to model a nonlinear force-free field, values of all three components of the field are needed at the photospheric boundary. See [*McClymont et al.*, 1997] for a critical discussion of methods for calculating force-free fields.

Here we only have a line of sight magnetogram. Thus unless some assumptions about the structure of the current are made, in this case the choice of α, an equilibrium field cannot be calculated. It should be noted, however, that we are able to calculate a linear force-free field that approximates the actual observed coronal structures to a high degree. For more details of such modelling the reader is referred to D. Mackay et al., [submitted paper, *Astrophys. J.*, 1997].

3. LINEAR STABILITY

The role of MHD instabilities in coronal magnetic fields is twofold. Firstly, for equilibria to model long-lived coronal structures they must be stable to those modes which act on time scales shorter than their lifetimes. Secondly, if, the initiation mechanism for the ejection of material is to be an instability, then once some critical parameter has been exceeded a sufficiently fast mode (of the order of minutes) must be triggered. In the corona these instabilities are ideal MHD modes.

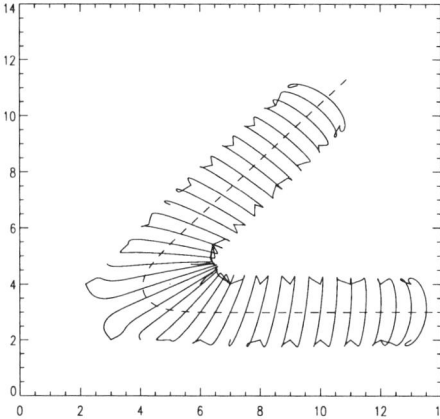

Figure 2. A plot of magnetic field lines at a given height above the switchback. The shear of the field varies with height.

In general, finding the parameters for the onset of an ideal mode is a non-trivial problem involving the solution of a system of three, coupled, second-order partial differential equations in three independent variables. If, a significant proportion of parameter space is to be examined then this is a very time consuming process. However, [*Longbottom et al.*, 1993], using an energy method, have developed a technique that can tightly bound these critical values. This method requires the solution of three coupled second-order ordinary differential equations, a much more tractable problem, and thus allows a large amount of parameter space to be investigated within a short time.

The change in potential energy due to a small perturbation ξ may be written as the sum of two integrals

$$\delta W = \frac{1}{2\mu}\{\mathcal{I} + \mathcal{J}\}, \qquad (4)$$

where $\mathcal{I}$ contains only derivatives of the perturbation in the direction of the magnetic field and $\mathcal{J}$ is strictly positive. The parameters defining marginal stability may be bounded by two conditions, one necessary and one sufficient for stability/instability. These conditions are determined by the solution of Euler-Lagrange equations which take the form of two sets of second order ordinary differential equations, that may be integrated along individual field lines, with the appropriate boundary conditions. Here line-tied boundary conditions are implemented, this acts as a stabilizing effect on the ideal modes.

A necessary condition for stability (sufficient for instability) is derived by choosing a particular trial function, that varies rapidly *across* field lines

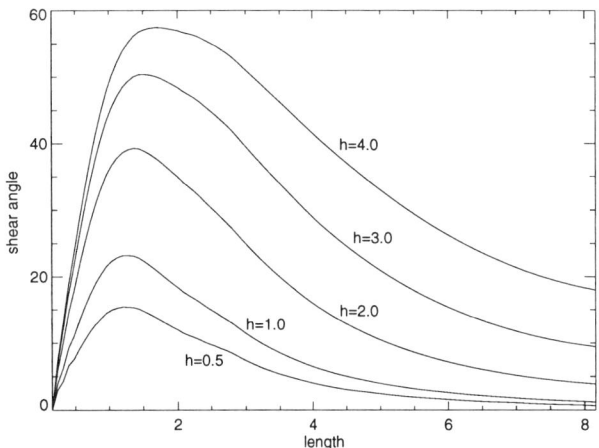

Figure 3. The shear of the magnetic field lines, as a function of length along an arm of the switchback, for a number of different heights.

$$\xi(x, y, z) = \xi(x, z)\cos(mS), \qquad (5)$$

where y is the egnorable direction, the wavenumber (m) is large and (for a $2\frac{1}{2}$D equilibria)

$$S = y - \int_0^s \left.\frac{B_y}{B}\right|_A ds. \qquad (6)$$

Here the field is defined by

$$\mathbf{B} = \nabla A(x, z)\mathbf{e_y} + B_y\mathbf{e_y}. \qquad (7)$$

Then, the parameter regime for which $\delta W < 0$ for this trial function may be determined, giving a condition for definite instability. A sufficient condition for stability (necessary for instability) is derived by a partial minimization of the energy integral through ignoring $\mathcal{J}$, which is strictly positive. Then a parameter regime for which $\delta W > 0$ may be determined, thus giving a condition for definite stability.

The method can be used to examine the stability of prominence equilibria [see *Longbottom and Hood*, 1994 and references there in]. Here we present the analysis of two prominence equilibria due to [*Cartledge and Hood*, 1993] and [*Hood and Anzer*, 1990]. The first is a finite width loop model, the second a finite width arcade model. Figure 5 shows necessary and sufficient conditions for the stability of the Cartledge-Hood model as the angle of the magnetic field shear at the bottom of the loop is varied. The upper curve gives the length of loop beyond which the equilibrium is definitely unstable, while the lower curve gives the length of loop up to

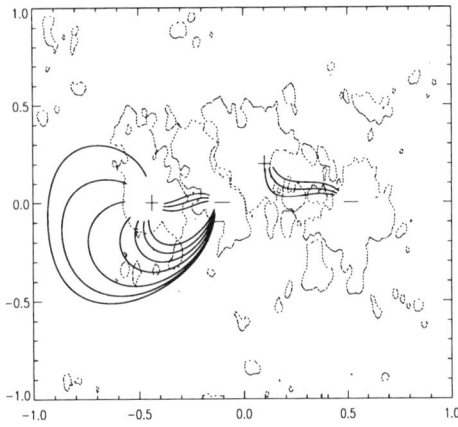

Figure 4. A plot of the magnetogram, overlaid with projections of the low lying field lines, giving the correct foot point connectivity.

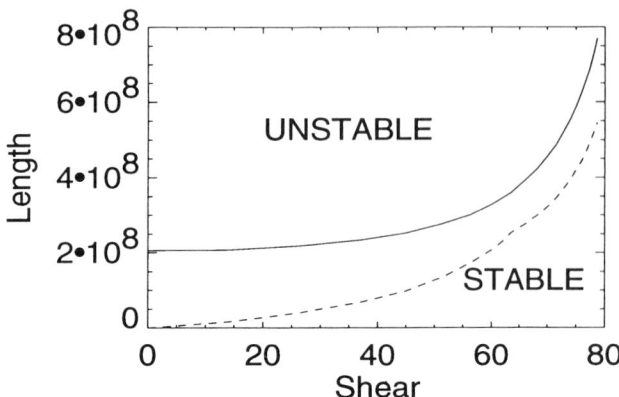

Figure 5. The necessary and sufficient conditions for stability for the Cartledge-Hood prominence model. This shows the maximum length (in metres) of the loop for a given shear. Above the solid line the equilibrium is definitely unstable, below the dashed line it is definitely stable.

which the equilibrium is definitely stable. Between the two curves the stability properties are undetermined by this method. It can be seen that increasing the shear angle at the base of the loop (and thus decreasing the twist and current in the equilibrium) has a significant stabilizing effect.

Figure 6 shows the necessary and the sufficient conditions for stability in the case of the Hood-Anzer model for a three prominence densities as the parameter l is varied. l is a measure of how much the magnetic shear angle varies with height. For a given density the lower curve defines a sufficient condition for stability, below which the prominence would be definitely stable, while the upper curve defines a necessary condition for stability, above which the prominence would be definitely unstable. It can be seen that, for moderate values of l, increasing the density, and thus the mass, of the prominence is destabilising. For larger values of l the effect of density becomes uncertain as the region of undetermined stability, between the necessary and the sufficient conditions, becomes greater. Increasing l, i.e. having a magnetic shear angle that changes rapidly with height, is stabilising. This can be explained by noting that the instability triggered in this case has a short wavelength across the field, raising and lowering neighbouring field lines, but minimizing their bending. Thus, if the magnetic field direction varies with height, this mode will be a much harder to excite.

4. EVOLUTION THROUGH EQUILIBRIA

If, the response time of the coronal field is much faster than any changes at the photospheric boundary, such as plasma motion or flux emergence, then the field

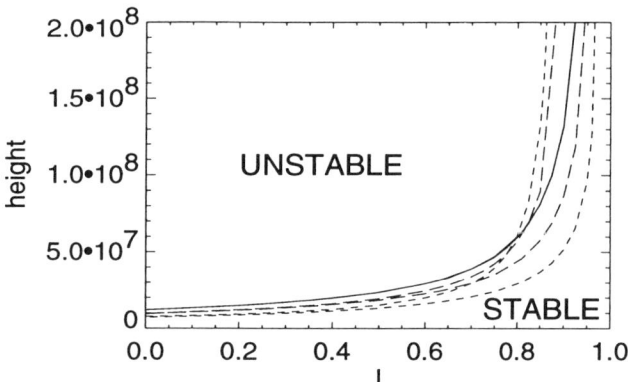

Figure 6. The necessary (upper curves) and sufficient (lower curves) conditions for stability for the Hood-Anzer prominence model. For a given shear l the necessary condition gives the height (in metres) above which the prominence is definitely unstable. The sufficient condition gives the height (in metres) below which the prominence is definitely stable. The curves correspond to prominence number densities of 10^{15} (solid), 10^{17} (long dashed), and 2.5×10^{17} (short dashed).

will effectively evolve though a series of equilibria. These equilibria are determined by the boundary conditions at a given time. If, as a consequence of a gradually changing boundary condition, the topology of the equilibrium undergoes a dramatic change then this may be indicative of an eruptive event. Care must however be taken when interpreting the true nature of such evolutionary events modelled by a changing equilibrium parameter. Examples of such an process leading to emerging flux break-out, are described below.

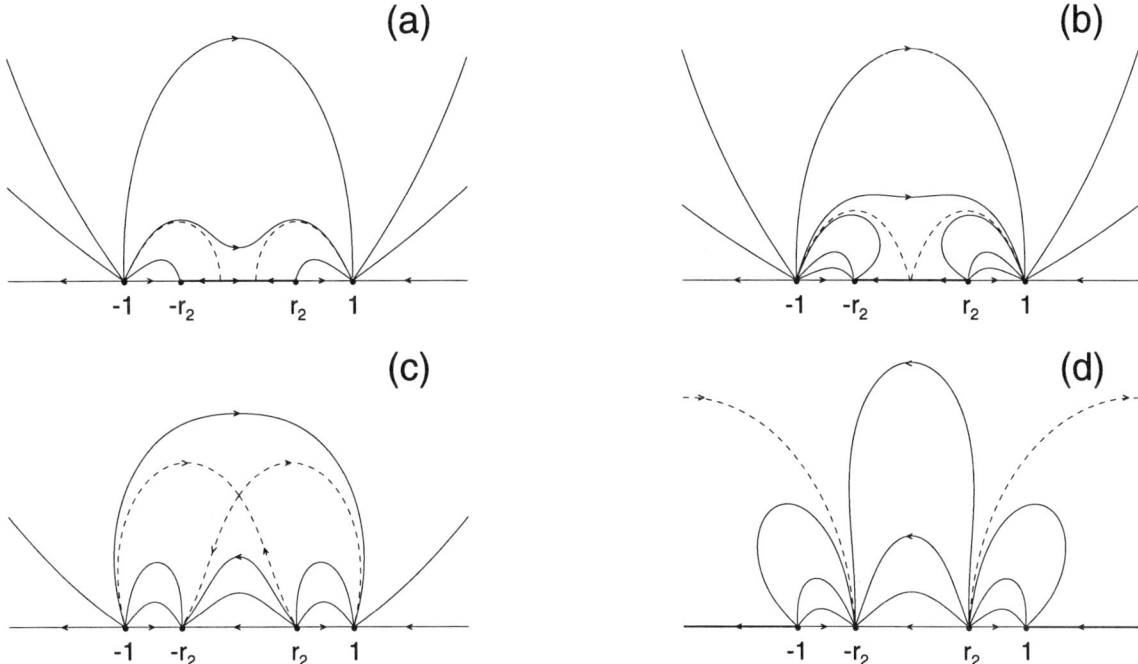

Figure 7. 2D cuts through the field in the $x - z$ plane, as ϵ is increased, for the case of colinear sources. Field lines are shown as solid curves, the separatrix as dashed curves. Lines along the x-axis joining null points to sources are photospheric spine curves. (a) Two distinct nulls. (b) Coalescence at origin. (c) Formation and rise of separator (the intersection of the two separatrix curves). (d) Break out of internal flux.

In Figure 7 the simple case involving the calculation of the three dimensional potential field associated with four varying point sources is considered (E. Priest and T. Bungey, personal communication, 1997). Initially, the sources are colinear. Sources of strengths -1 and 1 are placed at $x = 1$ and $x = -1$, respectively. The second pair of sources of strengths $\pm\epsilon$ are placed at $x = \pm r_2$. With $\epsilon = 0$ there exists just a bipole field. However, as ϵ increases, two nulls appear between $x = \pm r_2$ with two separatrix surfaces dividing space into three distinct regions of field line connectivity, Figure 7a. As ϵ increases, the two nulls coalesce at the origin, Figure 7b. Any further increase in ϵ results in the formation of a separator, at the intersection of the two separatrix surfaces, dividing space into four distinct regions of field line connectivity, Figure 7c. The height of the separator defines the maximum height of the internally confined flux. As ϵ increases the separator rises, until at some critical value ($\epsilon = 1/r_2$) the internal flux has broken through the external flux, Figure 7d.

The case for non-colinear sources is shown in Figure 8. The general behaviour is similar to that of the colinear case. The initial configuration consists of two pairs of equal and opposite sources. One pair of strength ∓ 1 is aligned along the x axis at $x = \pm 1$ while the other pair of strength $\pm\epsilon$ at an angle (θ) to the x axis, in the $x - y$ plane, at radii of r_2. Figure 8a shows the initial position of the flux sources, photospheric spine curves and four separatrix curves. For this case two separatrix surfaces exist independently, dividing the volume into three distinct regions of field line connectivity. As ϵ increases, the two separatrix surfaces grow and meet in the $x - y$ plane (Figure 8b). Further increase of the strength of the internal bipole leads to the formation of a separator at the intersection of the two separatrix surfaces. This rises (Figure 8c) until, eventually, the interior field breaks through the exterior field (Figure 8d).

Figure 9 shows the height of the separator, as the strength of the internal bipole is increased, for a number of angles (θ) between the two pairs of sources. Two observations may be made. For a gradually increasing ϵ, the separator initially rises slowly until some critical value is reached, at which point a much more rapid rise ensues. Also, for a given ϵ the maximum separator height is obtained by having the two sources inclined to one another at 45 deg in the $x - y$ plane.

116 COMPLEX MAGNETIC FIELDS

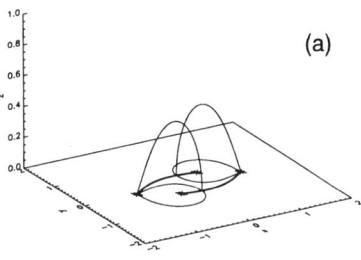
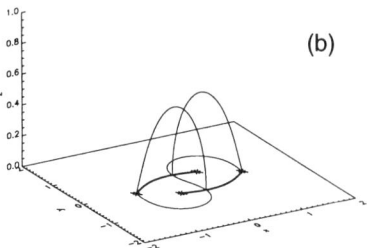
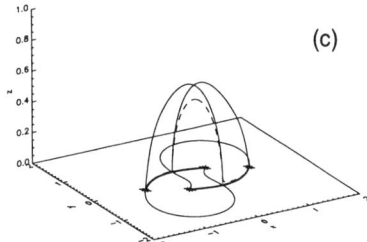
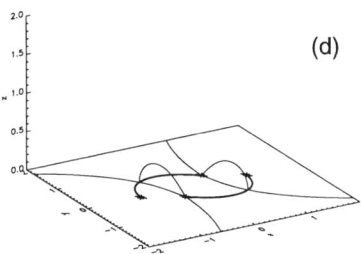

Figure 8. The 3D topology of non-colinear sources as ϵ is increased. The two pairs of sources (*) are shown together with the photospheric spine curves (dark solid lines), joining the null points to the sources in the $x - y$ plane, and four separatrix curves (solid lines), defining the separatrix surface. (a) Two distinct separatrix surfaces. (b) Meet at origin. (c) Formation and rise of separator (dashed line at the intersection of the two separatrix surfaces). (d) Break out of internal flux.

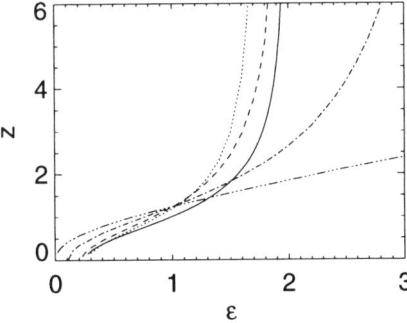

Figure 9. The height of the separator as ϵ is increased. Curves represent (solid, dotted, dashed, dot-dashed, dot-dot-dot-dashed) $\theta = 0, \pi/4, \pi/2, 3\pi/4, \pi$ respectively.

5. CONCLUSIONS

In this paper we have examined how the complex three-dimensional magnetic field observed in the solar corona may be modelled. In section 2 examples of how potential and force-free fields can be computed were outlined and their resulting structure discussed. Employing a potential model for the field allows the essential aspect of the structure to be examined for large portions of parameter space. The example given here enables the general overlying sheared arcade of a switchback to be neatly explained by a local flux imbalance. The more realistic force-free case is much harder and time-consuming to calculate. However, with an efficient numerical technique the nature and role of the current in magnetic fields above active regions may be studied. A method of determining the linear stability of equilibria was outlined in section 3 and applied to two prominence equilibria. The effect of a varying angle of magnetic shear was shown to be critical to equilibrium stability properties for both arcade and loop models of prominences. An example of the evolution of the coronal field through a series of equilibria, in response to changes at the photospheric boundary, was studied in section 4. It was found that gradual emergence of flux from beneath an existing dipole can lead to a break out of the internal flux through the existing field.

Acknowledgments The author would like to thank the two referees, who made a number of of useful comments for the improvement of this text, and also to acknowledge the following people for their useful discussions and help in producing this paper: T. N. Bungey, R. A. S. Fiedler, A. W. Hood, D. H. Mackay, E. R. Priest, G. J. Rickard.

REFERENCES

Cartledge, N., and A. W. Hood. External and internal solutions for the twisted, flux-tube, prominence model. *Sol. Phys. 148*, 253-275, 1993.

Hood, A.W., and U. Anzer. A model for quiescent solar prominences with normal polarity, *Sol. Phys. 126,* 117-133, 1990.

Longbottom, A.W., and A.W. Hood. The MHD stability of a twisted, flux-tube prominence model, *Sol. Phys. 154,* 51-68, 1994.

Longbottom, A.W., J.P. Melville, and A.W. Hood. Bounds on the stability of 3D magnetic equilibria in the solar corona, *Sol. Phys. 146,* 93-118, 1993.

McClymont, A.N., Jiao, L., and Mikic, Z., Problems and progress in coronal three-dimensional coronal active region magnetic fields from boundary data, *Sol. Phys. (Huntsville Workshop special issue),* in press, 1997.

A.W. Longbottom, School of Mathematical and Computational Sciences, University of St. Andrews, KY16 9SS, Scotland.

Helicity Conservation

D. M. Rust

Applied Physics Laboratory, The Johns Hopkins University, Johns Hopkins Road, Laurel, MD

Observations of solar and interplanetary magnetized features are examined and compared with the behavior expected of flux ropes under the constraint of helicity conservation. A model for interplanetary magnetic clouds is reviewed, in which conservation of magnetic helicity allows explanation of the clouds' thermodynamic and magnetic properties. Elevated temperatures in the clouds are attributed to magnetic energy dissipation during expansion under the constraint of helicity conservation. The initial phase of cloud evolution is identified with filament eruptions, which often coincide with transient, high-temperature, sigmoid brightenings in coronal images. These sigmoid features obey the global rule of field chirality segregation according to hemisphere: right-handed fields (positive helicity) in the south and left-handed fields (negative helicity) in the north. Detailed measurements of 50 sigmoid brightenings have shown that filament eruptions and probably coronal mass ejections are due to a helical kink instability. The ejection of twisted magnetic fields from the Sun in filament eruption/CME events represents a loss of magnetic flux and helicity that is consistent with measurements at Earth and with the generative power of the solar dynamo.

1. INTRODUCTION

The concepts of magnetic helicity and conservation of magnetic helicity are becoming important tools for interpreting coronal mass ejections (CMEs) and their interplanetary effects. In this paper, I provide an elementary discussion of magnetic helicity with references to papers where the subject is treated in a more rigorous manner. My principal aim is to make the reader comfortable with magnetic helicity and then to show how helpful it is in understanding the physics of coronal mass ejections.

Magnetic helicity is a strictly conserved quantity in ideal MHD, and laboratory plasma experiments [*Ji et al.*, 1995] have shown that even in resistive plasmas, magnetic helicity decays much more slowly than magnetic energy. An observer of space plasmas can reasonably assume that helicity is conserved in his laboratory as well.

In the interpretations of observations presented in this paper, I will assume that magnetic helicity is effectively conserved in space plasmas from the base of the solar convection zone to Earth orbit and beyond. I know of no proof that this is so, but *Berger* [1984] showed that helicity decays in the corona on a diffusion time scale. Therefore, helicity decay cannot be important in CMEs or normal coronal evolution.

I will show that the helicity measured at 1 AU in the course of an 11-year cycle is approximately equal to the helicity generated by the solar dynamo. I will bring other calculations of helicity to the reader's attention, but it is very important to remember that all such calculations may be quite unreliable at this point. The interpretation of vector magnetograms is still a developing and controversial undertaking, and the solar dynamo is not well understood.

Nevertheless, I think it is helpful to begin to get a sense of helicity quantified. Even without quantification, helicity conservation suggests natural explanations for prominence eruptions and CMEs, assuming that twisted flux ropes are their fundamental structural elements. Some models quite different from the twisted flux rope model can also account for these phenomena, but it is useful to ask whether these models are consistent with helicity conservation. We will find that the familiar sheared arcade models, for example, do not pass this test.

2. MAGNETIC HELICITY AND TOPOLOGY

Magnetic helicity density h_m is defined as $\mathbf{A}\cdot\mathbf{B}$, where $\mathbf{A}$ is the vector potential and $\mathbf{B}$ is the vector field. Within a volume V, the magnetic helicity is

$$H_m = \int_V h_m \, dV. \qquad (1)$$

Magnetic helicity is usually introduced this way, and since practically no one knows how to calculate it uniquely, only the theoretically inclined press on, but for help, see *Finn and Antonsen* [1985] I will fall back on the topological interpretation of H_m developed by *Berger and Field* [1984]. They show that H_m is a measure of magnetic fieldline twisting, linking and kinking (also called writhing). For example, in a closed, twisted flux rope, $H_m = \pm T\Phi^2$, where T is the number of twists and Φ is the magnetic flux. The sign of H_m, i.e., the *chirality*, depends on whether the twist is in the sense of a right-hand or left-hand screw. Chirality can be a very convenient tracer of fields, since even when H_m is not strongly conserved, chirality is.

Figure 1(a) is a sketch of a very complex magnetic field which has many links and kinks. In ideal MHD, each kind of helicity is conserved [*Woltjer*, 1958], and it would not be possible to change the number of fieldline linkages nor the number of kinks. This is equivalent to saying there are no reconnections. But kinks and twists can be interchanged, and this is likely to be an important feature of solar magnetic fields.

Figure 1(b) illustrates the topological equivalence of an untwisted, writhing flux rope and a twisted toroidal flux rope. Similarly, Figure 1(c) shows that an untwisted toroid is equivalent to a kinked toroid with a writhing helicity of $\pm\Phi^2$ and a twist helicity of $\mp\Phi^2$. It will be useful to keep this picture in mind when considering how fields might become twisted as they rise to the photosphere from the base of the convection zone.

3. RECONNECTION

Berger [1984] showed that H_m in the corona is approximately conserved even during rapid reconnections. Consider a simple illustration: the familiar two-loop flare

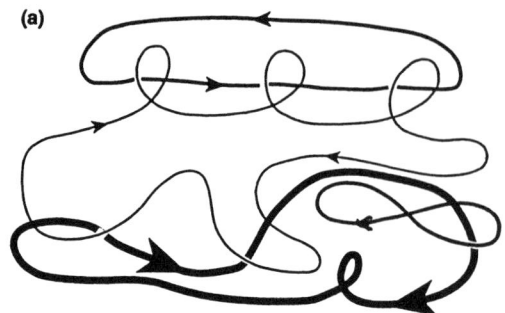

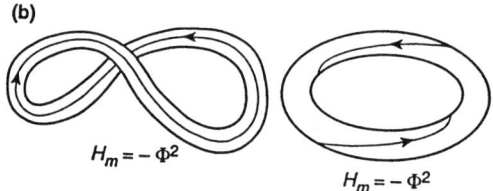

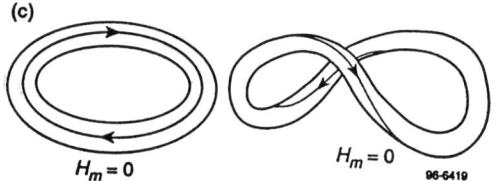

Figure 1. Topology and helicity conservation: (a) Tangled magnetic field with writhing (kink) and linking helicities. (b) A flux rope with kinks can be flattened out, but the resultant toroid will have a twist. (c) A toroid with no helicity can be distorted to exhibit twist helicity equal and opposite to writhing helicity.

model [*Gold and Hoyle*, 1960] (Figure 2). It focuses on the azimuthal fields of adjacent loops, which would form an X-type neutral point, in two dimensions, if the twists of the loops have the same chirality. Each loop shown in the figure resembles half a toroid from the laboratory plasmas known as "spheromaks" [*Rosenbluth and Bussac*, 1979]. Two such toroids can indeed merge by reconnection [*Ono et al.*, 1993], and H_m is conserved in the process [*Taylor*, 1986]. But from the laboratory experiments, we now know that the Gold and Hoyle model is misleading, because the field is not annihilated in such co-helical mergers. And *Canfield et al.* [1996] showed that the fields of active regions on opposite sides of the solar equator but of like chirality will reconnect to form a complex of equator-crossing coronal loops. The process seems to be a gentle one on the Sun, as it is in the laboratory, so it seems an unlikely explanation for flares. On the other hand, when two flux ropes with opposite H_m merge, the net helicity goes to zero, leaving

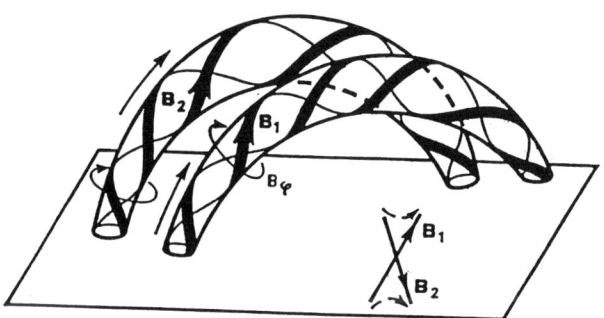

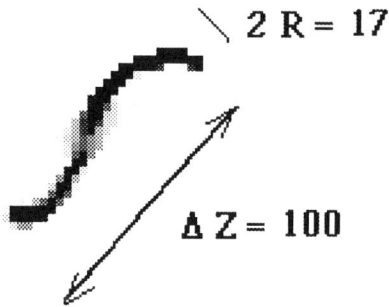

Figure 2. *Gold and Hoyle* [1960] flare model with reconnecting, co-helical flux ropes. Observations in the corona and in the laboratory show that such reconnections are not explosive.

only a current-free field and a heated gas. This process has not yet been positively observed on the Sun.

When two similarly twisted flux ropes merge, their helicities add, and the length of the resultant flux rope will increase [*Rust and Kumar*, 1994]. If the end-to-end twist of the flux rope exceeds ~ 2.5π radians, the flux rope will be unstable, and it will kink [*Hood*, 1991]. If it is approximately force-free and rooted in the photosphere, its kink assumes a characteristic sigmoid (S) shape [*Rust and Kumar*, 1996]. The predicted height-to-width ratio of the S is 5.4. It resembles an integration symbol ∫, in agreement with observations (Figure 3). Such coronal brightenings are usually associated with CMEs [*Webb et al.*, 1976].

Nishikawa et al. [1994] modeled the kinetic processes associated with the co-helical merging just described. They showed that there is strong heating of the entrained plasma as the two flux ropes lose their separate identities. In laboratory experiments, *Wysocki et al.* [1990] showed that when H_m is increased, a flux rope can expand, kink and heat at the same time. *Pevtsov et al.* [1996] describe what appears to be a similar physical process in a solar flare.

4. EVIDENCE OF HELICAL FLUX ROPES ON THE SUN

We have already seen good evidence for helical flux ropes, namely, those sigmoid brightenings whose average aspect ratio is what we expect from a field that is twisted until it kinks. One of the most direct and important consequences of helicity conservation is that flux ropes should retain their chirality throughout their lifetime and during their passage from the convection zone to beyond Earth's orbit. Thus, we expect, for example, that right-handed fields in sunspots, filaments and coronal arcades will remain right-handed. If flux ropes from one such feature eventually appear in another, we should expect right-handedness in both features. In particular, if we can

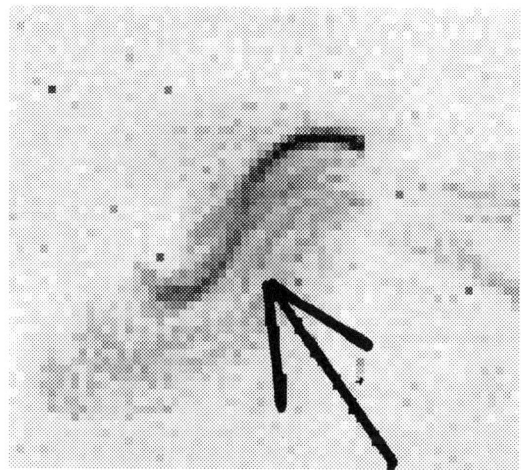

Figure 3. (bottom) Yohkoh image of a southern-hemisphere sigmoid coronal brightening on October 14, 1992. (top) Contrast-enhanced version suitable for determining the aspect ratio $\Delta Z/2R$ (in arbitrary units).

determine the chirality of the fields in CMEs at the Sun, we expect that these same fields will have the same chirality when they pass Earth.

While it is hard to actually measure H_m in a sunspot or filament or in coronal loops, it is often quite easy to determine the chirality. In sunspots, for example, the fibrils betray the chirality by their spiral shape. If we approximate a sunspot's magnetic field as a force-free twisted flux rope standing vertically in the photosphere, then, following *Nakagawa et al.* [1971], we can find a solution in cylindrical coordinates to the force-free field equation:

$$\nabla \times \mathbf{B} = \alpha \mathbf{B}. \qquad (2)$$

Assuming the field decreases exponentially with height z, the solution is

$$B_r = \beta J_1(kr) e^{-\beta z}, \quad (3)$$
$$B_\phi = \alpha J_0(kr) e^{-\beta z}, \quad (4)$$
$$B_z = k J_0(kr) e^{-\beta z}, \quad (5)$$

where B_r, B_ϕ and B_z denote the radial, azimuthal and vertical field components, respectively, J_0 and J_1 are Bessel functions, and

$$k = \sqrt{\alpha^2 + \beta^2}. \quad (6)$$

To find the sign of α, we need only determine the sense of the spiral, as shown in Figure 4, where

$$\tan\theta = \frac{B_\phi}{B_r} = \frac{\alpha}{\beta}. \quad (7)$$

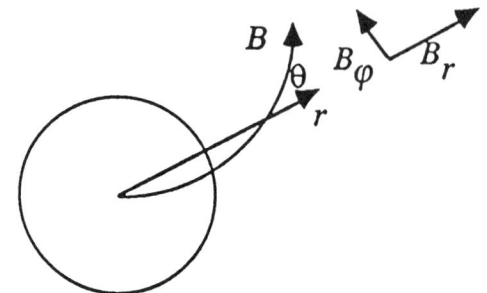

Figure 4. When fibrils around a sunspot (indicated by a circle) bend counterclockwise (as shown by the curved arrow), the field's chirality is positive (see eq. 7).

Since β is positive, α must be positive in the case shown in Figure 4. Sunspots that look like they are rotating clockwise (CW) have positive chirality. *Richardson* [1941] and *Ding et al.* [1987] found that, in the southern hemisphere, sunspots resembling clockwise whorls outnumber counter-clockwise (CCW) ones by 3:1; the situation is reversed in the north. The direction of spot rotation does not depend on magnetic field polarity or solar cycle number or phase.

Now let us see whether the hemispherical segregation of helicity implicit in the sunspot results can be found in other features. *Martin et al.* [1992] discovered that all Hα filaments are either "dextral" or "sinistral." Most dextral filaments lie in the northern solar hemisphere; sinistral filament predominate in the south. A dextral filament looks like a right-hand screw. But theoretical considerations suggest that dextral filaments are threaded by *left-handed* fields [*Rust and Kumar*, 1994]. The Hα-absorbing material that makes filaments visible should ride on the *underside* of the helix where the upward-curving fields can support it. Then a filament that looks like a right-hand screw would be threaded by a left-handed helical field, and vice versa. Although *Vrsnak et al.* [1991] performed a beautiful analysis of prominences that clearly establishes that their fields are helical, they did not determine chirality. On this point, there is only the brief report [*Gigolashvili*, 1978] that, of 19 prominences having helical structure, the 16 observed in the northern hemisphere all rotated like left-hand screws. Three southern hemisphere prominences showed right-handed twisting.

Although *Martin et al.* did not explicitly argue that dextral filaments are threaded by right-handed fields, their terminology suggests a different interpretation than *Rust and Kumar's*. To try to resolve the issue, *Rust and Martin* [1994] studied sunspots that showed conspicuous spiral structure and also were at one end of a filament or filament channel. They found a 1:1 relationship linking sunspot spirals with the two types of filaments. Those filaments with an end curving into CCW sunspots were invariably "dextral" while those curving into CW sunspots were invariably "sinistral." These observations are easy to understand if spots composed of left-handed flux ropes are connected by filaments with left-handed fields. In a general way, this result is consistent with *Seehafer* [1990] and *Pevtsov et al.* [1995] who found that most AR fields are predominately of either positive or negative chirality.

5. FROM THE CORONA TO MAGNETIC CLOUDS

Helical fields can be better understood in the context of helicity conservation. If left-handed fields emerge in the north and form sunspots, then shouldn't the filaments that spring from them be left-handed? Shouldn't the coronal arcades be left-handed? Finally, shouldn't the fields ejected from the northern hemisphere be left-handed? On this final point, there is supporting evidence. More than 80% of magnetic clouds (CMEs whose fields are measured *in situ* at 1 AU [*Lepping et al.*, 1990]) have the chirality predicted from the hemispherical segregation rule [*Rust and Kumar*, 1994, *Bothmer and Schwenn*, 1994]. Hence, the picture that emerges is one of H_m (and energy) buildup in a filament and its coronal surroundings, either through flux rope mergers [*Martin and Livi*, 1992, *Pevtsov et al.*, 1996], a series of minor relaxation events [*Low*, 1996], or helicity transfer from the submerged, compressed parts of fieldlines to the emerged, expanded parts [*Parker*, 1979]. Whichever, when H_m becomes too large for the confines of the AR or coronal streamer, we get a CME.

Shearing coronal loops by moving the footpoints [*Mikic and Linker*, 1996] does not seem a viable option for energizing CMEs for the simple reason that the chirality is wrong. Most filaments are more-or-less parallel to the lines of latitude. Differential rotation in the northern hemisphere will shear loops crossing any such filament channel into loops with positive helicity. Nearly all magnetic clouds from the northern hemisphere have left-handed fields.

The principle of H_m conservation can help us understand how an erupted filament/coronal arcade complex evolves between the Sun and Earth When a filament, modeled as a twisted flux rope, erupts and expands, H_m conservation leads to rapid heating within 0.1 solar radius and enough gradual heating thereafter to largely offset the effects of adiabatic expansion [*Kumar and Rust*, 1996]. The rapid heating causes the high-temperature sigmoid brightenings seen in X-ray images at CME onset, and the gradual heating explains how magnetic clouds can still have a temperature of 10^5 °K at 1 AU, even after immense expansion.

Let us assume that during a magnetic cloud's evolution, H_m, Φ and mass are conserved. Let us follow the conservation laws wherever they might lead with minimum recourse to dynamics. Then, $H_m \sim \lambda E_m$, where λ is a scale factor (e.g., cloud diameter) and E_m is the total magnetic energy [*Kumar and Rust*, 1996]. With H_m constant, E_m must decrease as the cloud expands. About one-third of the energy is converted into kinetic and potential energy as the CME lifts out of the corona. If the rest of the magnetic energy is converted to heat, the axial field strength B_o, pressure nT, density n, cloud radius r_o and proton temperature T_p agree with a least-squares fit to data from about 50 clouds observed from 0.3 AU to 4 AU (Table 1).

Table 1. Interplanetary Magnetic Clouds: Flux Rope Model vs. Observations

Observables	Power Law Fit	Fit for Flux Rope Model	Number of Clouds
B_0, nT	$18.4/d^{1.8}$	$18.8/d^2$	52
n, cm^{-3}	$7.2/d^{2.8}$	$5.8/d^3$	25
r_0, AU	$0.148 d^{0.97}$	$0.155 d$	34
$T_p, 10^4$ °K	$3.4/d^{0.7}$	$3.6/d$	18

The heating is probably due to turbulence in the cloud. The fields are always trying to relax to a minimum-energy force-free state, such as that first described by Lundquist [1950]. The H_m spectrum in the cloud gradually shifts to emphasize longer wavelengths, through reconnections, but $H_m = \int h_m dV$ remains the same.

Although the flux rope model of magnetic clouds agrees well with observations, it is not entirely clear how the complex mix of hot coronal loops and a cool filament develops into a remarkably seamless magnetic flux rope at 1 AU. One possibility is that the magnetic fields in a filament/corona loop arcade already form a giant flux rope and that the apparent complexity is simply due to relatively unimportant local temperature and density variations. This is the view behind the model discussed above, but the physics of transformation from complex to elementary appearance has not yet been worked through.

6. ORIGIN OF HELICITY IN THE SUN

Why are helical fields segregated by hemisphere? From the classical [*Babcock*, 1961] model (Figure 5), it is obvious that the toroidal field winding is left-handed in the north and right-handed in the south. This naturally accounts for fields of opposite chirality in the two hemispheres. More sophisticated mean-field dynamo models also produce equal and opposite magnetic helicity north and south of the equator [*Glatzmaier*, 1985]. H_m generated in each hemisphere in a solar cycle can be calculated from the difference in rotation rate from pole to equator and the polar flux of $\sim 10^{22}$ Mx: it is of order 10^{45} Mx^2.

Another way that subsurface fields may gain helicity is through the action of the Coriolis force on flux tubes as they rise from the base of the convection zone [*Fan et al.*, 1994]. In the north, the sense of rotation due to the Coriolis force is right-handed, and consequently, spot groups have a tilt, i.e., the leader spots are closer to the equator than the follower spots. If we now invoke helicity conservation, this writhing helicity must be compensated by an opposite twist helicity. Hence, the twist of fields in the north will be left-handed. The situation is essentially that shown in Figure 1(c), where a planar toroid is made to writhe by some outside force, but because H_m is conserved, twist appears in the flux rope.

We can calculate how much helicity is generated by the Coriolis forces. Consider the 3000 ARs that appear in each cycle. They carry a total of 10^{25} Mx of flux, or about 3.3 x 10^{21} Mx per AR. The average tilt T, or writhing angle, of the fields, due to the Coriolis force is 6° or about 0.017 turns. From the formula $H_m = T \Phi^2$, and $\Phi = 3.3$ x 10^{21} Mx, one finds $H_m = 2$ x 10^{41} Mx^2 per AR. Counting all 3000 ARs, $H_m \approx 0.5$ x 10^{45} Mx^2 released into interplanetary space per solar cycle because of the Coriolis force.

7. FATE OF HELICAL FIELDS

The Sun systematically sheds twisted fields. The amount of twisted magnetic flux ejected in CMEs in the course of a solar cycle is roughly equal to the independently measured [*Schrijver and Harvey*, 1994] flux lost from the photosphere [*Bieber and Rust*, 1995]. H_m in interplanetary fields is approximately 2 x 10^{45} Mx^2 per cycle. The chirality of the IP fields is negative in the north and positive in the south, as on the Sun [*Bieber et al.*, 1987]. Thus, we find that H_m

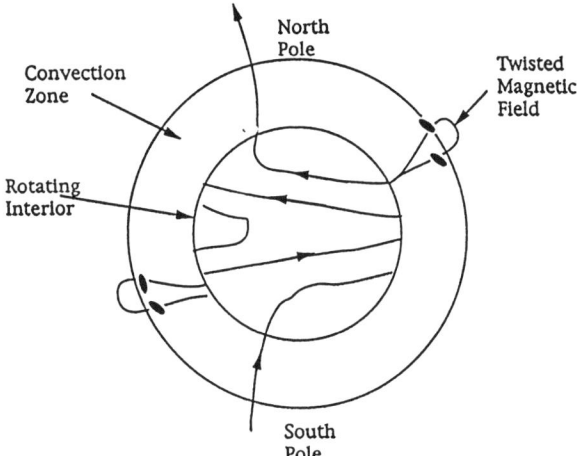

Figure 5. Toroidal fields of the solar dynamo and an emerged flux rope writhing because of the Coriolis force experienced during its rise from the convection zone.

from all eruptions in a solar cycle, estimated from the action of the Coriolis force, is of the correct sign and of the same order as H_m measured in interplanetary space. About as much helicity is generated by flux rope winding in the Babcock picture of the solar dynamo. It remains to be seen which process is more important.

8. MEASURING AND USING HELICITY

Magnetic helicity generated in the solar interior emerges in ARs, so one may say that ARs undergo helicity charging, at least during their growth phase. Using vector magnetic field observations and the force-free field approximation, *Wang* [1996] calculated dH_m/dt in a rapidly growing AR to be 1.9×10^{39} Mx^2 s^{-1}. He attributed the helicity increase to cross-helicity, i.e., field twisting by vortical mass motions. He estimated that $H_m \approx 4 \times 10^{43}$ Mx^2 for the whole AR. This value may be too high, considering that ~ 2×10^{45} Mx^2 is generated per solar cycle. The value depends on a very uncertain estimate for the height of the helicity-charged fields. But it is especially interesting to note that Wang found almost no helicity dissipation due to currents. These first results have to be regarded with a great deal of caution, but they are consistent with the suggestion above that helicity is conserved except as it is carried off in eruptions.

H_m is difficult to measure, but it is probably the most important index of energy buildup and imminent instability. H_m can be estimated from a time series of vector magnetograms or from observations of twisted or sigmoid features [*Pevtsov et al.*, 1996, *Rust and Kumar*, 1996, *Vrsnak et al.*, 1991].

In summary, helicity charging in coronal flux ropes and helicity conservation in ejected flux ropes is a useful paradigm for exploring the physics of solar activity and of CMEs in particular. After a half century of mostly naïve fascination with coiled solar features, we are finally beginning to quantify what we see. The past 20 years of laboratory plasma physics, especially the work of J. B. Taylor, has opened our eyes.

Acknowledgments. I am pleased to acknowledge very helpful discussions with Sara Martin. This research was sponsored by the National Science Foundation Division of Polar Programs, grant DPP-9119870, and by the Air Force Office of Scientific Research, grant AFOSR-90-0102.

REFERENCES

Babcock, H. W., The topology of the Sun's magnetic field and the 22-year cycle, *Astrophys. J.*, *133*, 572, 1961.

Berger, M. A., Rigorous new limits on magnetic helicity dissipation in the solar corona, *Geophys. Appl. Fluid Dyn.*, *30*, 79, 1984.

Berger, M. A. and G. B. Field, The topological properties of magnetic helicity, *J. Fluid Mech.*, *147*, 133, 1984.

Bieber, J. W., P. Evenson and W. H. Matthaeus, Magnetic helicity of the Parker field, *Astrophys. J.*, *315*, 700, 1987.

Bieber, J. W. and D. M. Rust, The Escape of Magnetic Flux from the Sun, *Astrophysical Journal*, *453*, 911, 1995.

Bothmer, V. and R. Schwenn, Eruptive prominences as sources of magnetic clouds in the solar wind, *Space Sci. Rev.*, *70*, 215, 1994.

Canfield, R. C., A. A. Pevtsov and A. N. McClymont, Magnetic chirality and coronal reconnection, in *Magnetic Reconnection in the Solar Atmosphere*, edited by R. D. Bentley and J. T. Mariska, pp 341 - 346, Astron. Soc. Pacific, Bath, U.K., 1996.

Ding, Y. J., Q. F. Hong and H. Z. Wang, A statistical study of the spiral spots on the solar disc, *Solar Phys.*, *107*, 221, 1987.

Fan, Y., G. H. Fisher and A. N. McClymont, Dynamics of emerging active region flux loops, *Astrophys. J.*, *436*, 907, 1994.

Finn, J. M. and J. Antonsen, T. M., Magnetic helicity: what is it and what is it good for?, *Comments Plasma Phys. Controlled Fusion*, *9*, 111-126, 1985.

Gigolashvili, M. S., An investigation of macroscopic motions using the Ca$^+$ lines in the prominence of 15 October 1969, *Solar Phys.*, *60*, 293, 1978.

Glatzmaier, G. A., Numerical simulations of stellar convective dynamos. II Field propagation in the convection zone, *Astrophys. J.*, *291*, 300, 1985.

Gold, T. and F. Hoyle, *Monthly Notices Roy. Astron. Soc.*, *120*, 89, 1960.

Hood, A. W., MHD of solar flares, in *Advances in Solar System Magnetohydrodynamics*, edited by E. R. Priest and A. W. Hood, pp 307, Cambridge Univ. Press, 1991.

Ji, H., S. C. Prager and J. S. Sarff, Conservation of magnetic helicity during plasma relaxation, *Phys. Rev. Lett.*, *74*, 2945, 1995.

Kumar, A. and D. M. Rust, Interplanetary magnetic clouds, helicity conservation and intrinsic-scale flux ropes, *J. Geophys. Res., 101*, 15667, 1996.

Lepping, R. P., J. J. A. and L. F. Burlaga, Magnetic field structure of interplanetary magnetic clouds at 1 AU, *J. Geophys. Res., 95*, 11957, 1990.

Low, B. C., Solar activity and the corona, *Solar Phys., 167*, 217, 1996.

Lundquist, S., Magnetohydrostatic fields, *Ark Fys, 2*, 361, 1950.

Martin, S. F. and S. H. B. Livi, The role of cancelling magnetic fields in the buildup to erupting filaments and flares, in *Eruptive Solar Flares,* edited by Z. Svestka, B. V. Jackson and M. E. Machado, pp 33, Springer-Verlag, Berlin, 1992.

Martin, S. F., W. H. Marquette and R. Bilimoria, The solar cycle pattern in the direction of the magnetic field along the long axes of polar filaments, in *The Solar Cycle,* edited by K. L. Harvey, pp 53, 1992.

Mikic, Z. and J. A. Linker, The role of magnetic shear in initiating CMEs, *Coronal Mass Ejections: Causes and Consequences (these proceedings),* 1996.

Nakagawa, Y. and e. al., On the topology of filaments and chromospheric fibrils near sunspots, *Solar Phys, 19*, 72, 1971.

Nishikawa, K.-I., J.-I. Sakai, J. Zhao, T. Neubert and O. Buneman, Coalescence of two current loops with a kink instability simulated by a three-dimensional electromagnetic particle code, *Astrophys. J., 434*, 363, 1994.

Ono, Y., A. Morita and M. Katsurai, Experimental investigation of three-dimensional magnetic reconnection by use of two colliding spheromaks, *Phys. Fluids B, 5*, 3691, 1993.

Parker, E. N., *Cosmical Magnetic Fields,* Oxford Univ. Press, 1979.

Pevtsov, A. A., R. C. Canfield and T. R. Metcalf, Latitudinal variation of helicity of photospheric magnetic fields, *Astrophys. J., 440*, L109-L112, 1995.

Pevtsov, A. A., R. C. Canfield and H. Zirin, Reconnection and Helicity in a Solar Flare, *Astrophysical Journal, 473*, 533, 1996.

Richardson, R. S., The nature of solar hydrogen vortices, *Astrophys. J., 93*, 24, 1941.

Rosenbluth, M. N. and M. N. Bussac, *Nucl. Fusion, 19*, 489, 1979.

Rust, D. M. and A. Kumar, Helical magnetic fields in filaments, *Solar Physics, 155*, 69-97, 1994.

Rust, D. M. and A. Kumar, Helicity charging and eruption of magnetic flux from the sun, in *Proc. 3rd SOHO Workshop - Solar Dynamic Phenomena and Solar Wind Consequences,* pp 39, European Space Agency, 1994.

Rust, D. M. and A. Kumar, Evidence for Helically Kinked Magnetic Flux Ropes in Solar Eruptions, *Astrophysical Journal Letters, 464*, L199, 1996.

Rust, D. M. and S. F. Martin, A correlation between sunspot whirls and filament type, in *Proc. 14th Int. Summer Workshop,* pp 337, National Solar Observatory, Sunspot, NM, 1994.

Schrijver, C. J. and K. L. Harvey, The photospheric magnetic flux budget, *Solar Phys., 150*, 1, 1994.

Seehafer, N., Electric current helicity in the solar atmosphere, *Solar Phys., 125*, 219, 1990.

Taylor, J. B., Relaxation and magnetic reconnection in plasmas, *Rev. Mod. Phys., 58*, 741, 1986.

Vrsnak, B., V. Ruzdjak and B. Rompolt, Stability of prominences exposing helical-like patterns, *Solar Phys., 136*, 151, 1991.

Wang, J., A note on the evolution of magnetic helicity in active regions, *Solar Phys., 163*, 319, 1996.

Webb, D. F., A. S. Krieger and D. M. Rust, Coronal X-ray enhancements associated with Hα filament disappearances, *Sol. Phys., 48*, 159, 1976.

Woltjer, J., A theorem on force-free magnetic fields, *Proc. Nat. Acad. Sci. USA, 44*, 489, 1958.

Wysocki, F. J., J. C. Fernández, I. Henins, T. R. Jarboe and G. J. Marklin, Improved energy confinement in spheromaks with reduced field errors, *Phys Rev. Lett., 65*, 40, 1990.

D. M. Rust, Applied Physics Laboratory, Johns Hopkins Road, Laurel, MD 20723

Predicting the Sign of Magnetic Helicity in Erupting Filaments and Coronal Mass Ejections

S. F. Martin

Helio Research, 5212 Maryland Ave., La Crescenta, California

A. H. McAllister[1]

High Altitude Observatory, PO Box 3000, Boulder, Colorado

The one-to-one relationships previously found between dextral fila ments and left-skewed coronal arcades and corresponding relationships between sinistral filaments and right-helical coronal arcades are shown to have utility in predicting some magnetic field changes that will occur when filaments and their overlying coronal fields erupt. The predictions are model dependent. We use empirical models for both coronal arcades and filaments which are based on observational evidence that the fine structure and mass motions in the coronal structures and in filaments are always parallel to the local magnetic field. Under these conditions and assuming that magnetic reconnection will occur in the corona beneath ris ing stretched magnetic structures which have oppositely-directed and ad jacent magnetic fields, we make the following predictions: (1) Right- skewed coronal arcades will develop into CMEs with right-helical struc ture which will subsequently be detectable as interplanetary clouds with right-helical magnetic fields; left-skewed coronal arcades will develop in to CMEs with left-helical structure later detectable as interplanetary clouds with left-helical magnetic fields. (2) During eruption, sinistral fil aments will acquire a low degree of left-helical structure (of the order of 180 degrees or less) and dextral filaments will develop right helical struc ture. Without further knowledge of the evolution of erupting filament magnetic fields within CMEs as both features partially detach from the Sun, it is not yet possible to know whether the helical structure in fila ments, if developed in this predicted way, will be preserved or identifi able in interplanetary clouds. Hence, the test of prediction (2) currently rests on thorough Doppler or magnetic field observations of erupting fila ments near the Sun.

---

[1]Now at Helio Research, La Cresenta, California

Coronal Mass Ejections
Geophysical Monograph 99
Copyright 1997 by the American Geophysical Union

1. INTRODUCTION

A close geometric relationship has recently been established between two mutually exclusive types of filaments, known as "dextral" or "sinistral", and the relative direction of the coronal arcades overlying these two types [*Martin and McAllister*, 1996; *Martin, S. F. and*

A. H. McAllister, The Skew of X-ray Coronal Loops Overlying Hα Filaments, unpublished, 1997]. By definition, when a filament is viewed from the positive network side, it is "dextral" if the magnetic field component along its long axis points to the right. Similarly, if the magnetic field component along its axis points to the left, while still viewed from the positive network side, it is "sinistral" [*Martin, Bilimoria, and Tracadas*, 1994]. Dextral and sinistral filaments also have a structural identity. When viewed from the positive network side, the barbs (also known as legs) of dextral filaments extend from the filament axis to the chromosphere and are angled to the right of vertical. Similarly, from the same view, the barbs of sinistral filaments are angled to the left of vertical. Examples are shown in Figure 1.

The identity of dextral and sinistral filaments often can be made from other viewing perspectives as well. When seen from above, looking away and along the axis, the barbs of dextral filaments veer to the right away from the axis as if they were right-hand exits on an automobile speedway. Similarly, the barbs of sinistral filaments veer to the left away from the axis as if they were left-hand exits on a high-speed roadway. Examples are schematically shown in Figure 13 in Martin, Bilimoria and Tracadas [1994].

Martin and McAllister [1966; *Martin, S. F., and A. H. McAllister*, The Skew of X-ray Coronal Loops Overlying Hα Filaments, unpublished 1997] have similarly identified two categories of coronal arcades overlying filaments and named them "right-skewed" and "left-skewed". When viewed from above and in relation to the filament axis, skewed coronal loops within a coronal arcade appear to lie at an angle with respect to the filament below. Looking away and along the filament axis, if the acute angle (made by the loops with respect to the filament axis) is forward and to the right, the coronal loops and the arcade as a whole are "right-skewed". If the acute angle is forward and to the left, the coronal loops and arcade are "left-skewed". An example each of a right-skewed and a left-skewed X-ray arcade is shown respectively in the left and right panels in Figure 2.

The relationship between filaments and their overlying coronal arcades is such that dextral filaments lie beneath left-skewed arcades of coronal loops and sinistral filaments lie beneath right-skewed coronal arcades [*Martin and McAllister*, 1996]. The respective chirality (handedness) of filaments and their overlying arcades is thus an inverse relationship as these chiralities have been defined. A corollary to these one-to-one relationships is: skewed coronal loops always have a component in common with the magnetic field along the axis of a filament. Although this result was found for quiescent filaments, the authors believe it is applicable to all fila-

Figure 1. A dextral filament (upper image) is recognized when viewed from the side by the barbs (pronounced legs or fine structure) which are angled to the right with respect to any vertical. Similarly, the sinistral filament (lower image) has fine structure which is angled to the left of vertical. Among quiescent filaments, the dextral category is dominant in the northern solar hemisphere and the sinistral category is dominant in the southern hemisphere.

ments; among numerous individual examples in active regions that we have checked, no exceptions have been found.

The invariability of these observed relationships provides a new tool for inferring physical factors about the 3-dimensional nature of the magnetic fields of quiescent filaments and their coronal implications. These relationships are especially useful when combined with the observational deduction that the mass flows in chromo-

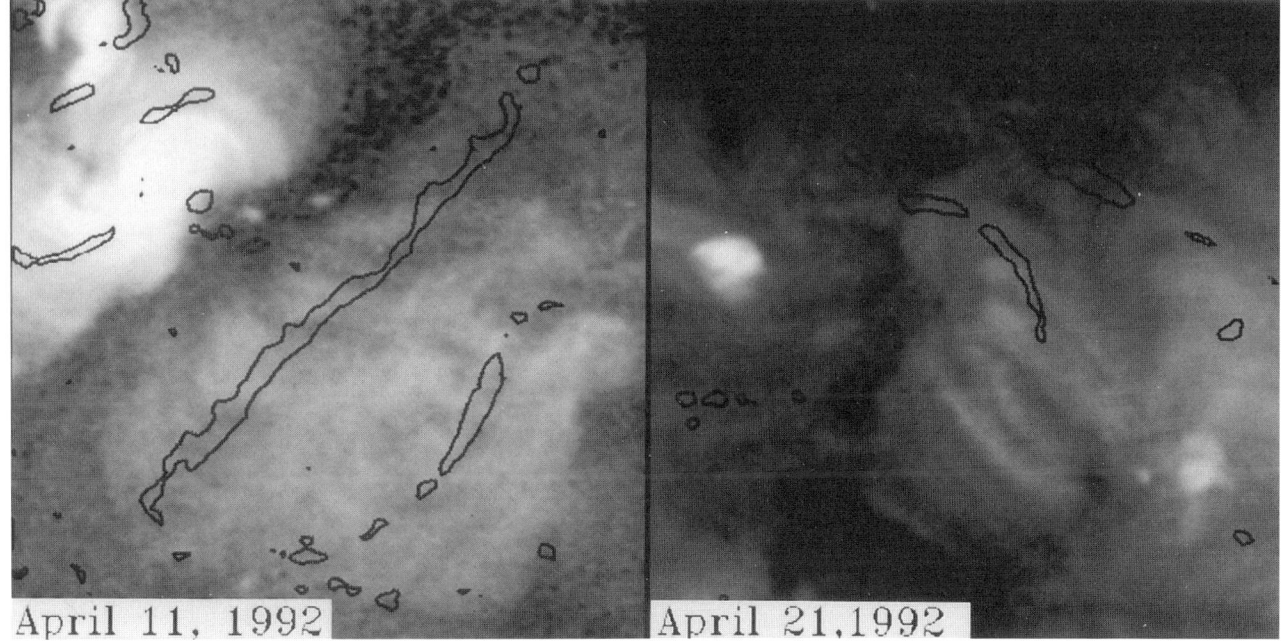

Figure 2. X-ray arcades often straddle filaments at an angle which is skewed, as shown here, rather than perpendicular to the filament axis. The sense of skew is defined relative to the filament axis when looking away and along the axis. The two panels shown here have H-alpha contours plotted over the coronal arcade images. The images are from Big Bear Solar Observatory and Yohkoh SXT respectively. The left-hand panel shows two right-skewed arcades over sinistral filaments (southern hemisphere), while the right-hand panel shows left-skewed arcades over dextral filaments (northern hemisphere).

spheric structures (and in filaments) are parallel to the local magnetic field [Foukal, 1971].

The association, between filament type and coronal arcade skew, can be described as an inverse relationship in the sense that right-hand (dextral) filaments are associated with left-hand (left-skewed) arcades and vice-versa. This inverse relationship, together with the findings that both the overlying coronal arcade and most filament barbs have an axial field component in common with the filament axis, is sufficient to show that the same barbs and overlying arcade also have magnetic field components that oppose each other in directions orthogonal to the filament axis. From these relationships, and knowledge that barbs are anchored to the chromosphere, we can deduce the following: the chromospheric (and photospheric) footpoints of filament barbs should be anchored in magnetic fields opposite in polarity to the footpoints of the overlying coronal arcades on both sides of filaments. This is in agreement with the findings of Martin and Echols [1994] and Martin [1994]; the major barbs of a filament were rooted in the minority polarity on each side of a specific filament.

If we know the polarity of the photospheric magnetic fields to which filament barbs and ends are anchored, and if we have observations revealing the dextral or sinistral filament structure, we can deduce the direction of the magnetic field component along the axis of a filament and the skew of its overlying coronal arcade. Vice-versa, if we know the direction of the magnetic field component along the axis of a filament, we can deduce the polarity of the fields to which the barbs are anchored. We can also deduce the sense of skew and approximate direction of the magnetic field in the corona overlying the filament.

Similarly, knowledge of the orientation of a coronal arcade together with knowledge of the polarity of the photospheric magnetic fields at the arcade base, allows the prediction of some features of the magnetic field geometry of any associated filament. In many cases, this is enough information to construct a simple, empirical, 3-dimensional model of the magnetic field of the filament channel, the overlying arcade, and the filament. Such models are shown and discussed in Sections 2 and 3 for coronal arcades and filaments respectively. These simple models can then be used to predict certain geometric field changes that should occur when a coronal mass ejection (CME) occurs and a filament concurrently erupts. This includes predicting the sign of helicity that will develop in the erupting coronal arcade, and similarly in the filament, as both structures become parts of a coronal mass ejection (and later an interplanetary cloud). Following from the above relationships, this paper is an elucidation of the additional predictions that can be made about coronal and filament magnetic fields when they erupt, and in particular about their signs of helicity observable near the Sun and later in interplanetary space. Such predictions will have at least two practical consequences: (1) improved forecasts of the magnitude of geomagnetic storms at Earth from knowing in advance whether the leading edge of an interplanetary cloud has a northward or southward component relative to Earth's magnetic field, and (2) increased ability to ascertain the correctness of models of filaments and CMEs; verifiable or unverifiable predictions of the sign of helicity in erupting filaments and CMEs would be evidence, respectively, for or against the model employed.

2. GEOMETRIC PROPERTIES OF ERUPTING CORONAL ARCADES

Predicting the magnetic topology of a CME, or any part of it such as an embedded filament, requires some knowledge, or a model, of the pre-eruptive coronal structure and some knowledge, or a model, of the eruptive process. Observations of CMEs have shown that they generally consist of three parts, a bright front, a dark 'cavity', and often a bright core [*Hundhausen*, 1996]. The bright front corresponds to coronal plasma at the leading edge of erupting coronal magnetic fields. The cavity is believed by some to be a region of stronger magnetic fields (and thus lower density), which could be the main driver of the eruption [*Low*, 1994; *Hundhausen*, 1996]. The core, when present, is often strongly associated with an erupting prominence and with a signature of an erupting event in the Yohkoh Soft X-Ray Telescope (SXT) [*Hiei, Hundhausen and Sime*, 1993]. The filament, or core, often appears to be embedded within the lower part of the much larger cavity.

The large-scale pre-eruptive structure, when seen clearly at the limb in white light, generally has the form of a helmet streamer [*Hundhausen*, 1996]. The central portions of this structure are apparently closed loops which are rooted on either side of a polarity inversion line that lies under the streamer. The SXT images from the Yohkoh satellite have shown that the base of a high-latitude, helmet streamer often maps onto the base of a large polar-crown arcade (see Figure 3). In other cases, at lower latitudes or over active regions, there may be no large-scale structures visible in soft X-rays. However, there is also no evidence to contradict the assumption that the smaller-scale soft X-ray structures lie under the base of the large-scale white light structures.

In interpreting the soft X-ray images we rely on the

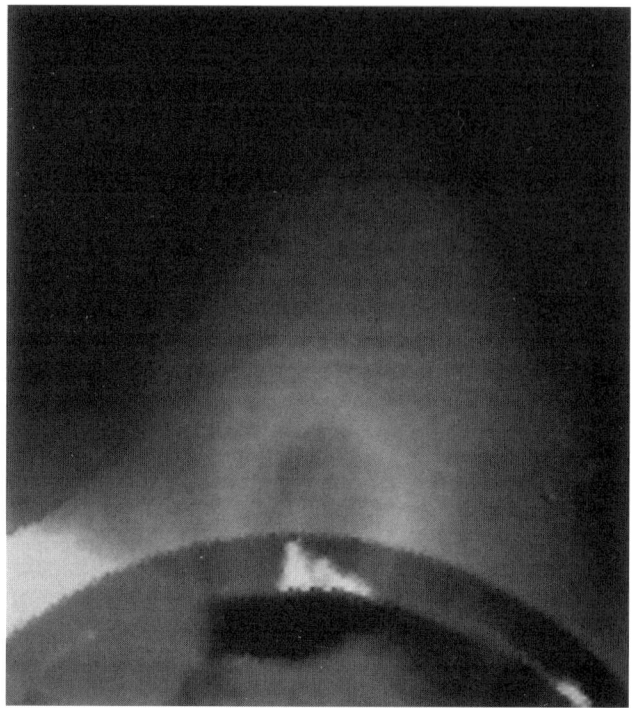

Figure 3. This composite of images shows a coronal streamer above a coronal arcade, which in turn is above a prominence (filament observed above the limb). The data was obtained from HAO's Mauna Loa Solar Observatory, the Yohkoh Soft X-ray Telescope, and Mauna Loa, respectively. The images shown are from January 22, 1992, two days prior to the CME of January 24, 1992 [*Hiei, 1993; Hundhausen, and Sime*, 1994]. North is to the lower left.

fact that the > 2 M K^o plasma flows readily along the field lines, but negligibly across them [*Poletto et. al.*, 1975]. Therefore it is accepted that the long tube-like structures seen in the SXT images, are closely aligned with the magnetic fields and can safely be taken as indicating the actual magnetic field topologies. In particular we can use the images of coronal arcades to indicate the orientation of the base of the overlying helmet streamers with respect to the polarity inversion lines. The profile of helmet streamers, and coronal arcades, seen projected in two dimensions over the limb, are those of arches or loops that can appear to be approximately perpendicular to the inversion lines. However, Martin and McAllister [1996] have found that many arcades are skewed with respect to the polarity inversion line, ie. they are often at an angle that is not perpendicular to it.

This skew can be accounted for in some cases by the non-uniform distribution of flux along the two sides of the magnetic inversion, such that even a potential field has skew [*Antiochos et al.*, 1994]. It may also be due to the creation of non-potential, current carrying, structures through various processes that form and cause the evolution of the filament channels and coronal streamers [*van Ballegooijen and Martens*, 1989; *Antiochos et al.*, 1994]. In either case the skew visible in the soft X-ray arcades will reflect the skew that exists in the larger coronal streamers. As suggested by Marubashi [1986; 1996], Gosling et. al. [1990], and Bothmer and Schwenn [1994], when a skewed set of erupting coronal loops reconnects near its base, it will begin to form a flux rope with a definite sense of helicity. When there is right-hand skew in the arcade, there will be right-hand helicity in the newly forming flux rope. Similarly left-hand skew in the pre-event structure will yield left-hand helicity in the resultant flux rope. If, in the latter stages of its development, this flux rope (CME) becomes an interplanetary cloud it will have the same sense of helicity.

Therefore, with knowledge of the skew of the pre-eruptive coronal structure based on the skew of the underlying soft X-ray arcade, and a model such as those proposed by Marubashi [1986, 1996], Gosling [1990], Bothmer and Schwenn [1996], we can predict the helicity of the interplanetary flux ropes formed as a result of a given CME. In Plate 1 the wire models depict how a right-skewed coronal arcade becomes a right-helical CME which would eventually be measured as a right-helical interplanetary cloud.

It has been previously demonstrated that dextral filaments and left-skewed arcades are dominant in the northern solar hemisphere [*Martin, Bilimoria and Tracadas*, 1994; *Martin and McAllister*, 1996; *Martin, S. F. and A. H. McAllister*, The Skew of X-ray Coronal Loops Overlying Hα Filaments, unpublished, 1997]. Similarly, sinistral filaments and right-skewed arcades are dominant in the southern solar hemisphere. From the relationships discussed above, it is expected that these hemispheric patterns will correspond to interplanetary clouds having the same dominant sign and distribution north and south of the ecliptic. This anticipated hemispheric distribution has been shown already by Rust and Kumar [1994] and Bothmer and Rust [this volume]. However, as explained in Section 3.3, Rust and Kumar interpret erupting filaments as having the opposite sign of helicity to that predicted herein.

3. GEOMETRIC PROPERTIES OF ERUPTING FILAMENTS

3.1 The Filament Model in its Pre-eruptive State

The model, that we use for predicting some key magnetic changes that will occur when filaments erupt, was initially illustrated by Martin and Echols [1994]. The most recent version of this model updated by Martin is shown in the upper part of Plate 2. Added to the initial wire model by Martin and Echols are representative bundles of field lines running the full length of the filament magnetic field (yellow wires) and representative bundles of skewed coronal fields (overlying white loops) typical of the full overlying arcade of coronal field. This model was conceived and further developed by combining observed features of filaments and their overlying coronal arcades and observed or inferred relationships between those features given in Section 1. As a concept derived solely from observations, this filament model is like the empirical models of erupting CMEs proposed by Marubashi [1986; this volume], Gosling et al. [1990], and Bothmer and Schwenn [1994] and adopted in Section 2 of this paper.

The heart of the model in Plate 2 is derived from observations of a filament in Hα and associated magnetic fields recorded by Martin at the Big Bear Solar Observatory from 11-17 May 1992. That filament is shown in the lower part of Plate 2. The model is scaled to the measured length of the filament and the calculated height of the filament at discrete points along its axis as observed from 15-17 May 1992. The height was calculated from the measured offset of the upper edge of the filament relative to a polarity inversion defined both by Hα fibril structure beneath the filament and by the adjacent opposite polarity fragments of network magnetic field seen in the videomagnetograms from Big Bear Solar Observatory. The videomagnetograms were recorded concurrently with the Hα filtergrams. The only assumption in the calculated height is that the axis of the filament lies in a plane vertically above the observed polarity inversion. The majority positive polarity in the magnetogram on the right side of the filament is represented by white, blue and green areas and the majority negative polarity on the left side is represented by black, red, and orange areas. The blue, orange and yellow wires are intended to represent bundles or concentrations of filament field. The model depicts the magnetic field of the filament rather than the mass. In concept, mass can occupy any part of the model and does not have to be dense enough or low enough in temperature to be visible in Hα everywhere along all representative bundles of field.

The key observational features contributing to the filament model are:

1. coincidence of the upper (southern) end of the filament with network fields of positive polarity and the lower (northern) end with network fields of negative polarity;
2. mass motions along the filament axis; these provide the concept of an axial magnetic field along the

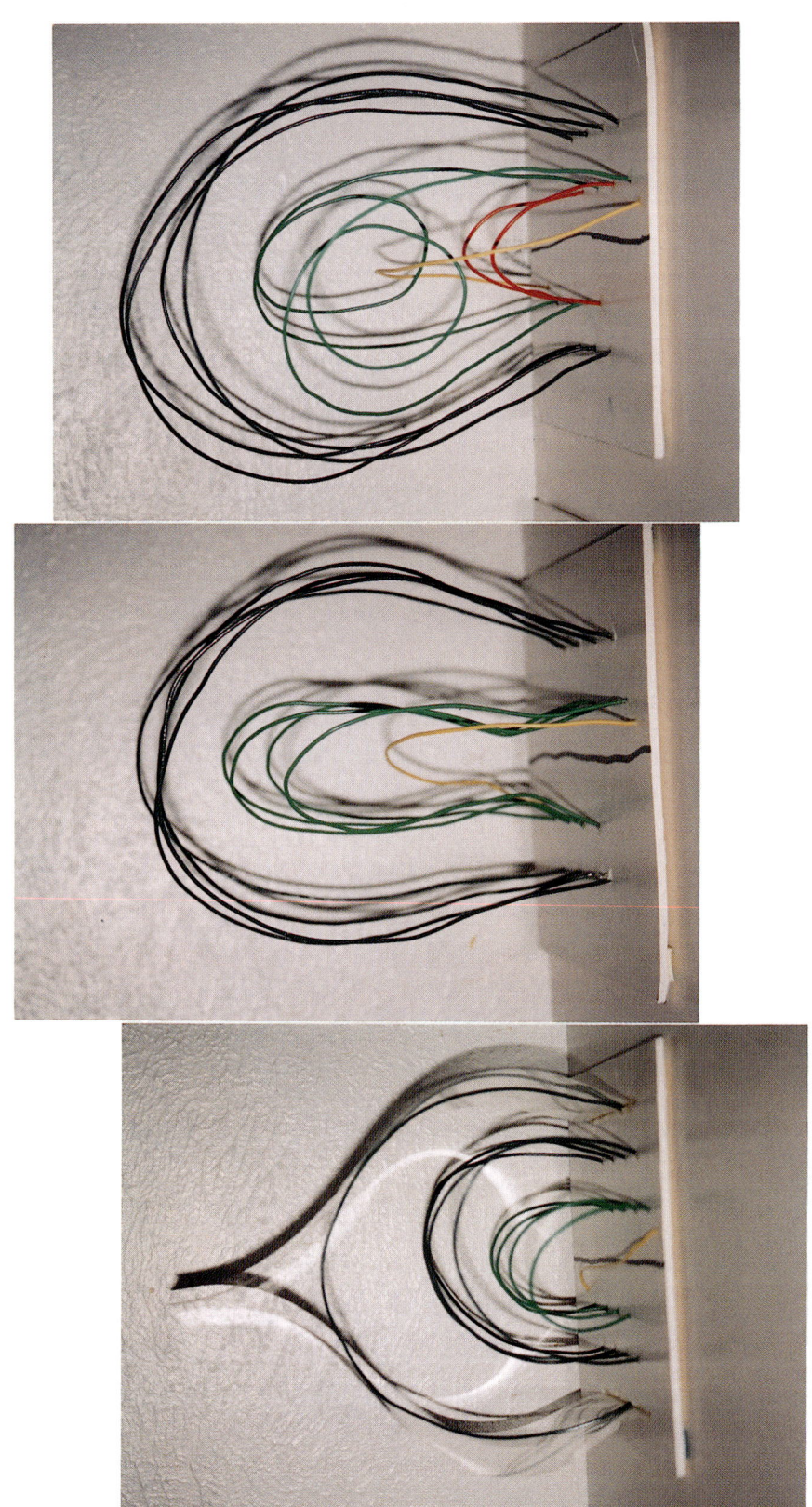

Plate 1. Wire illustration representing the magnetic field changes during a coronal mass ejection (CME). Magnetic reconnection between the legs of ascending and adjacent coronal loops in a left-skewed arcade result in a left-helical magnetic field configuration. Similarly right-skewed arcades would become right-helical CMEs.

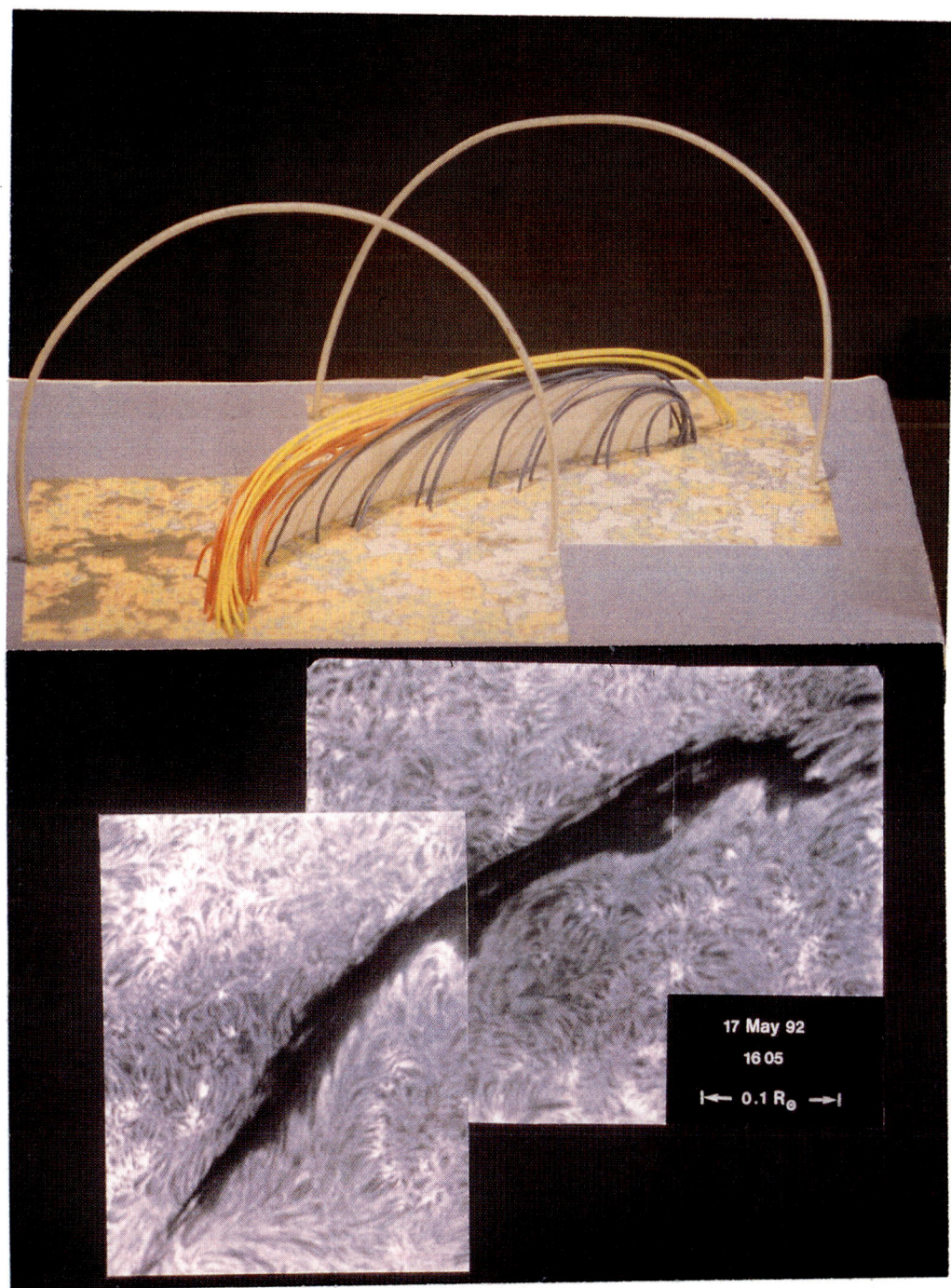

Plate 2. The three-dimensional wire model (lower image) is a schematic representation of the magnetic field of the filament (upper image). The filament axial magnetic field is denoted by yellow wires which run the full length of the filament. The blue wires represent bundles of magnetic field which emanate from positive polarity photospheric fields at the upper (southern end) of the filament and terminate in small photospheric magnetic fields of minority polarity (negative field) on the near side of the filament; the orange wires represent bundles of magnetic field which extend from the negative polarity end of the filament to pockets of minority polarity (positive field) along the far side of the filament.

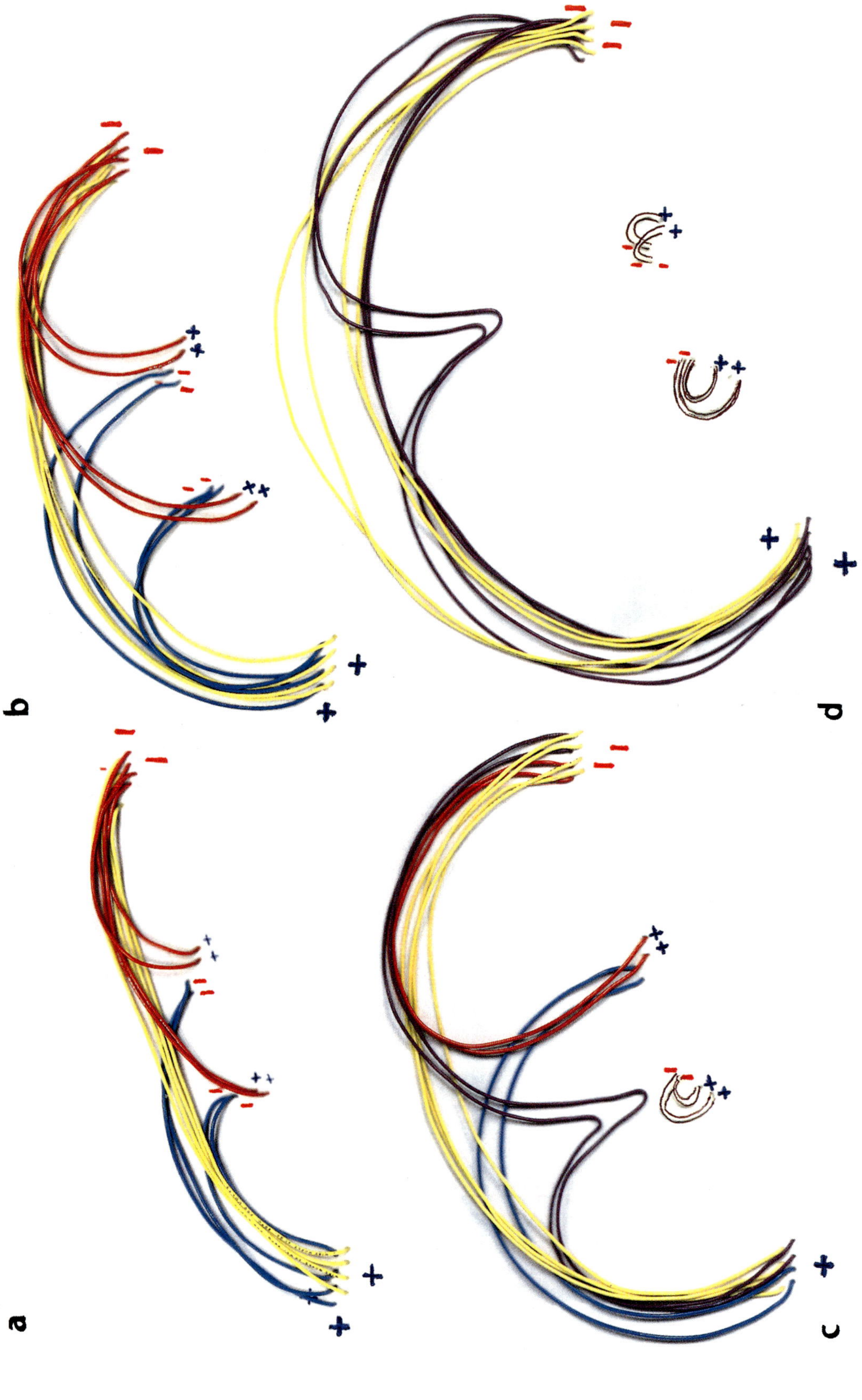

134 MAGNETIC HELICITY IN ERUPTING FILAMENTS AND CMES

Plate 3. (a) A simplified representation of the model in Plate 2 if it erupted. (b) ascent of the filament magnetic field during the early stages of eruption. (c) further ascent after magnetic reconnection has detached the left pair of barbs from their photospheric roots on each side of the filament. (d) filament axial field continuing to ascend after reconnection detached the right pair of barbs from the Sun. There is no assumption of detachment of the more massive ends of the filament.

full length of the observed filament linked to negative and positive network magnetic fields at its two extreme ends;

3. the observed coincidence of the major barb ends with patches of minority polarity on each side of the filament as seen in overlays of the Hα filtergrams and photospheric magnetograms; all barbs, major and minor, are depicted as having the same relationship as observed for major barbs because there is no evidence that minor barbs are different from major barbs;

4. mass motions from the filament axis into the barbs and vice versa; this provides the basis for the joining of the barbs to the filament axis in the simplest conceived configuration.

A key relationship, implied in (2) and (4) above, and employed in combining the above observed features into the model is that observable fine structures, and mass motions along fine structures are aligned with the local magnetic field. This relationship is substantiated by numerous observations of photospheric magnetic fields in relation to structural patterns in chromospheric fibrils, in filaments, and in the corona [*Smith and Ramsey*, 1967; *Foukal*, 1971; *Poletto*, 1975; *Martin*, 1990; *Martin, Bilimoria, and Tracadas*, 1994].

3.2 Changes Predicted in Filament Magnetic Fields During Eruption

The applicability of the empirical model can be tested by seeing if it helps to predict other features or behavior that are independent of the information used in constructing the model. In the present context of CMEs, we ask, "What changes would take place as the filament axis rises during the early stage of an erupting filament?"

To illustrate how the model must be dynamically altered during a typical eruption, we first simplify the model by limiting it to two barbs on each side of the filament as shown in Plate 3a. The first change, shown in Plate 3b, is the rising of the axis such that the middle of the axis ascends while the ends remain rooted. The effect of this change is that the barbs become more vertical. (We note that for quiescent filaments, the rising of the center of the filament is consistent with the rooting of the barbs in small pockets of magnetic flux in which the flux density is much less than at the ends of the filament.) The increasing vertical components of the stretched barb magnetic fields set up an ideal configuration for magnetic reconnection to take place between the oppositely-directed barb fields beneath the filament. In Plate 3b, the magnetic field of the left pair of barbs on opposite sides of the filament are represented immediately before reconnection and in Plate 3c immediately after reconnection. The right pair of barbs are represented prior to reconnection in Plate 3c and immediately after reconnection in Plate 3d. Additional reconnections between the barb fields and adjacent magnetic fields are not specifically included in the model but are not excluded either. They would lead to barb displacement rather than barb detachment.

There are several consequences of the proposed magnetic reconnection: (1) the formation of loops beneath reconnection site, (2) the linking of the upper part of the barb fields above the reconnection site but below the primary axial fields of the filament, and (3) the introduction of helical structure of approximately 180 degrees (1/2 turn) below the filament axis due to the linking of barb fields that are initially on opposite sides of the filament axis. While these changes in the magnetic field are not yet accessible by direct observation, visible mass motions in Hα are expected to accompany and correspond to these magnetic changes as follows:

(1a) barb mass in Hα would fall to the chromosphere along loops that would be formed beneath the reconnection site;

(1b) if the loop density and lifetime is sufficient, visible structure would be expected in Hα along these loops but not necessarily in soft X-rays because the falling mass also could be too low in temperature;

(2) barb mass above the reconnection site would be accelerated upwards because magnetic tension in the field lines would cause the magnetic field to tend to straighten; this would concurrently pull barb mass upward and into near alignment with the rising axial fields;

(3) the development of a small degree of helical structure (180 degrees or less) in the filament as the eruption continues; (3a) a dextral filament should develop right-helical structure, (3b) a sinistral filament should develop left-helical structure. The upper axial fields running the full length of the filament (bundle of yellow wires in the model in Plate 2) would not necessarily share in the formation of the helical structure unless secondary reconnections took place between the upper part of the reconnected barb magnetic fields and the higher axial fields.

The above predictions should be tested by future observations of structural changes in prominences and concurrent Doppler shifts observable during the early stages of their eruption. However, some evidence might be found by searching the archives of solar observatories. An outstanding example of the apparent detachment of barbs during an erupting prominence was photographed in Hα at the Sacramento Peak Observatory on 5 February 1959. Figure 4a shows the prominence prior to eruption. Figure 4b reveals the two middle barbs (legs) as the filament axis ascends. Figures 4c and

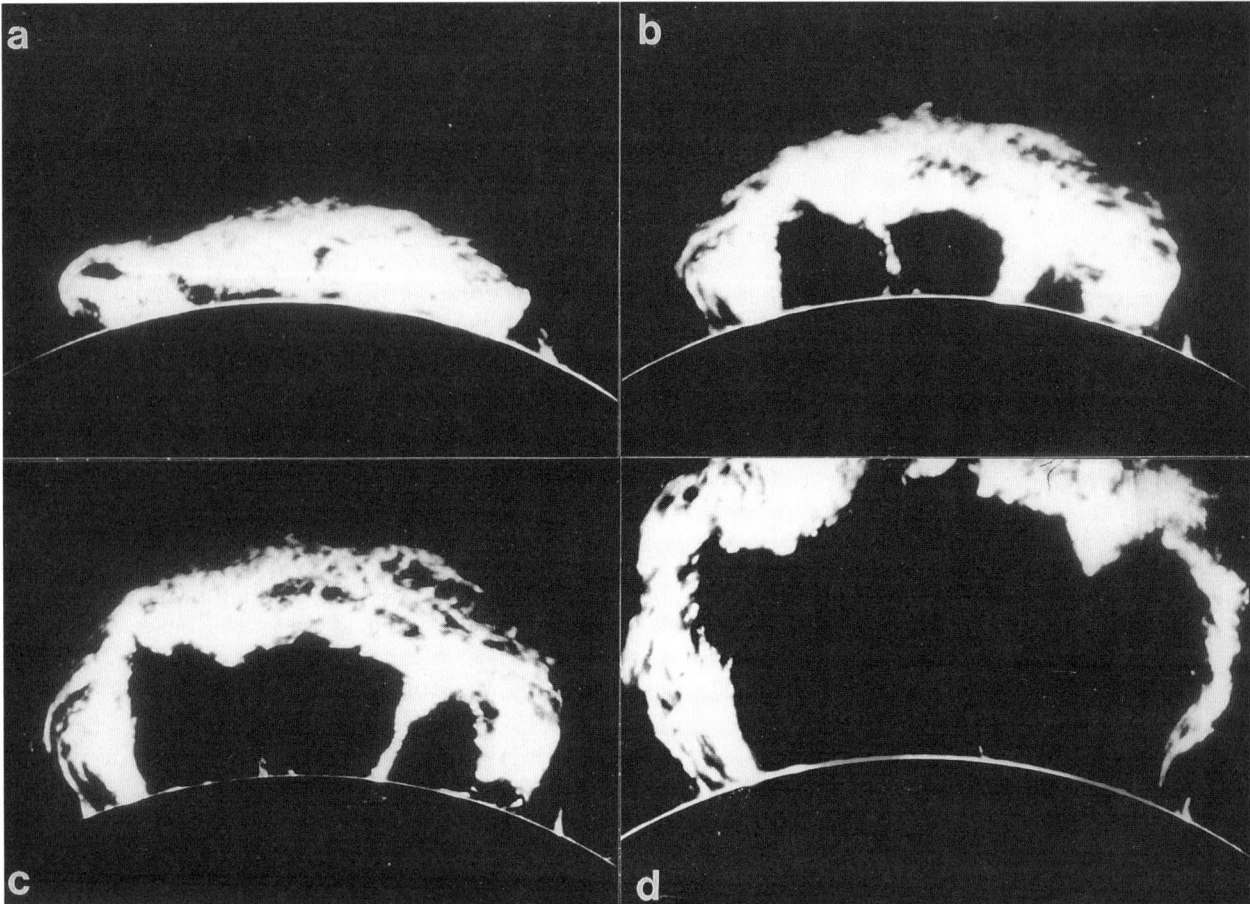

Figure 4. A symmetric erupting filament on 5 Feb 1959 at Sacramento Peak Observatory. (a) prior to eruption, 21:17 UT (b) beginning of the rapid ascent, 22:18 UT (c) detachment of the left of the two major barbs accompanied by upward acceleration of the barb mass toward the filament axis, 22:23 UT (d) detachment of and upward acceleration of the second major barb, 22:38 UT, probably due to magnetic reconnection.

4d respectively show the apparent detachment of the middle barbs upon further ascent of the main body of the prominence with the outer legs remaining rooted to the chromosphere. This event is suggestive of a number of features represented in the model erupting magnetic field in Plate 3: (1) the stretching of the legs (barbs) of the prominence as the filament ascends, (2) the apparent disconnection of the two intermediate legs in the corona (3) the falling of mass below the disconnection zone, and (4) the apparent upward acceleration of the mass of the left leg above the disconnection zone until that mass joins the upper axial structure of the prominence.

The changing of the inferred magnetic field geometry of the filament axis, due both to magnetic reconnection and to the ascent of the axis, could be the cause of the downward draining of mass at the ends of erupting filaments.

3.3 Discussion of Predicted Geometric Changes in the Magnetic Fields of Erupting Filaments

The sense of helical structure introduced by barb reconnection is opposite to that predicted (and ascertained) for reconnecting coronal arcades. The amount of helical structure initiated by the proposed barb reconnection is about 1/2 turn. The acquired helical structure is limited in comparison to the many full turns of helical structure that would result from reconnection within large scale arcades if the models cited in Section 3 are correct. If helicity is conserved, the reconnected fields, left behind the rising filament, have acquired an equal amount of helicity of opposite sign [*Ruzmaikin*,

1996; *Ruzmaikin, Feynman and Martin*, presented at this meeting].

It is not clear whether recognizable helical structure of erupting filament mass would be retained within a CME (with helical structure dominantly of the opposite sign) as the CME is transported into the interplanetary medium. The changing geometry of the whole erupting structure with time and possible magnetic reconnection within an interplanetary cloud could alter the field configurations characteristic of the early phase of the erupting filament and its associated CME. Currently, there are no definitive identifications of filament mass within interplanetary clouds. However, it is well known that helical structures within erupting filaments are detectable near the sun while the mass is still within one or two solar radii. Unfortunately, in most existing observations, the sense of helicity in erupting filaments is indeterminate due to the lack of data from which to measure or infer their 3-dimensional properties.

To definitely test this dynamic model of filament eruption, we propose new magnetic field and Hα Doppler observations of filaments be made before and during the eruption of filaments. To test the initial configuration of the Martin model, we suggest direct observations of filament magnetic fields such as those made recently by Lin [*Haosheng Lin*, submitted to Astrophys. J. 1996]. Doppler observations of the fine structure of filaments are suggested as another means of learning whether filaments have helical structure prior to eruption. The sign of helicity in filament structure and some of the 3-dimensional properties of filaments can be interpreted by learning which structures within a filament (or erupting filament) are closest and furthest from the observer. This can be learned from time series of images showing Doppler and plane-of-sky motions in filaments.

The sense of helical field predicted from this filament model, is opposite to the sense of helical field proposed in flux tube models of filaments [*Rust and Kumar*, 1994; *Low*, 1994]. This difference arises because the flux tube models assume that the barb structure of filaments represents mass in the lower part of an assumed flux tube that is left-helical for dextral filaments and right-helical for sinistral filaments. The Martin model (Plate 2) does not assume an initial helical configuration; as depicted in Plate 3, helical structure is a consequence of magnetic reconnection between the fields of the filament barbs and this necessarily adds right helical structure to dextral filaments and left-helical structure to sinistral filaments. Ascertaining the magnetic structure directly or the sense of helical structure in portions of erupting filaments linked to barbs is a test of the Martin model versus twisted flux tube models.

In applying these suggested tests, it should be noted that helical structure in filaments might be introduced by other means than those suggested here. For example, some observations of erupting filaments [*A. Bhatnagar*, personal communication 1995; *S. F. Martin*, personal information] would be consistent with bodily rolling of the axial fields of filaments as they erupt. While there is no mechanism yet known to initiate such motions, attention should be paid to possible structural properties that might exist in addition to those suggested by current models.

Using the Rust and Kumar filament model [1994] and by assuming that filament magnetic fields become the fields of interplanetary clouds, Rust and Kumar anticipated the sign of interplanetary clouds to be the same sign as we predict from interpreting the interplanetary clouds as arising from coronal arcades overlying filaments.

4. SUMMARY

From the previously established relationship of left-skewed coronal arcades overlying dextral filaments and right-skewed coronal arcades overlying sinistral filaments, together with empirical models of the geometry of coronal magnetic fields and filaments as they erupt, we predict that:

1a. During eruption, right-skewed arcades will result in right-helical CMEs and hence in right-helical interplanetary magnetic clouds; left-skewed arcades will result in left-helical CMEs and left-helical interplanetary magnetic clouds.

1b. Due to the dominance of right-skewed arcades in the northern hemisphere and left-skewed arcades in the southern hemisphere, a corresponding percentage of right- and left-helical interplanetary clouds would be expected north and south of the plane of the ecliptic. This prediction of the sign of helicity in interplanetary clouds and hemisphere of origin is the same as shown by Rust and Kumar [1994] and Bothmer and Rust [this volume] although the model, on which our predictions are based, is different from Rust and Kumar [1994].

2. During eruption, dextral filaments will develop right-helical structure if the barbs are detached from their photospheric roots by magnetic reconnection beneath the filament axis; similarly sinistral filaments are expected to develop left-helical structure as they erupt. The magnitude of helical structure developed along the length of a filament by this mechanism is of the order of 180 degrees or less. This prediction does not exclude the development of helical structure of the same or opposite sign by other processes.

3. From (1) and (2), it is concluded that the sign of helical structure of filaments can be opposite in sign to

that of the coronal mass ejections within which they are embedded. Secondly, the helical structure developed in erupting filaments by the proposed reconnection of barb magnetic fields is much smaller in magnitude than in the surrounding CME magnetic fields. Proposed observations for testing these predictions and the models on which they are based are (1) combined time series of Doppler and plane-of-sky motions in filaments and (2) direct magnetic field measurements in filaments.

Acknowledgments. The authors appreciate numerous discussions about the modelling of filament magnetic fields with B.C. Low. SFM was supported in this research by California State University, Northridge through a subcontract with NASA JPL. AHM completed his contribution to the written paper under NASA grant NAGW-5073 to Helio Research.

REFERENCES

Antiochos, S. K., R. B. Dahlburg, and J. A. Klimchuk, The magnetic field of solar prominences, *Astrophys. J.*, 420, L41–L44, 1994.
Bothmer, V. and D. M. Rust, The field configuration of magnetic clouds and the solar cycle, this volume, 1997.
Bothmer, V., and R. Schwenn, Eruptive prominences as sources of magnetic clouds in the solar wind, *Space Sci. Rev.*, 70, 215, 1994.
Foukal, P., Morphological relationships in the chromospheric Hα fine structure, *Solar Phys.*, 19, 59, 1971.
Gosling, J. T., Coronal mass ejections and magnetic flux ropes in interplanetary space, in *Physics of Magnetic Flux Ropes*, edited by C. T. Russell, E. R. Priest, and L. C. Lee, Geophys. Monogr. Ser. vol. 58, p. 343, 1990.
Hiei, E., A. J. Hundhausen, and D. G. Sime, Reformation of a coronal helmet streamer by magnetic reconnection after a coronal mass ejection, *Geophys. Res. Lett.*, 20(24), 2785–2788, 1993.
Hundhausen, A. J., Coronal mass ejections, in *Cosmic Winds and The Heliosphere*, edited by J. R. Jokipii, C. P. Sonett, and M. S. Giampapa, (University of Arizona Press), in press, 1996.
Low, B. C., Magnetohydrodynamic processes in the solar corona: Flares, coronal mass ejections, and magnetic helicity, *Phys. Plasmas*, 1(5), 1684, 1994.

Martin, S. F., Conditions for the formation of prominences as inferred from optical observations, in *Dynamics of Quiescent Prominences*, edited by V. Ruzdjak and E. Tandberg-Hanssen, (Springer-Verlag), p.1–44, 1990.
Martin, S. F., Footpoints of solar filaments (abstract), *B. A. A. S.*, 26(4), 185th meeting, (paper no. 123.03), 1522, 1994.
Martin, S. F. and A. H. McAllister, The skew of X-ray coronal loops overlying Hα filaments, in *Magnetohydrodynamic Phenomena in The Solar Atmosphere, Prototypes of Stellar Magnetic Activity*, edited by Y. Uchida and H. S. Hudson and T. Kosugi, (Kluwer Academic Pub., Dordrecht), 497–498, 1996.
Martin, S. F., R. Bilimoria, and P. W. Tracadas, Magnetic field configurations basic to filament channels and filaments, in *Solar Surface Magnetism*, edited by R. Rutten and C. J. Schrijver, (Kluwer Academic Pub., Dordrecht), p. 303–338, 1994.
Martin, S. F., and C. R. Echols, An observational and conceptual model of the magnetic field of a filament, in *Solar Surface Magnetism*, edited by R. Rutten and C. J. Schrijver, (Kluwer Academic Pub., Dordrecht), p. 339–346, 1994.
Marubashi, K., Structure of the interplanetary magnetic clouds and their solar origins, *Adv. Space Res.*, 6, 335–338, 1986.
Marubashi, K., Interplanetary flux ropes and solar filaments, this volume, 1997.
Poletto, G., G. S. Vianna, M. V. Zombeck, A. S. Krieger, and A. F. Timothy, A comparison of coronal X-ray structures of active regions with magnetic fields computed from photospheric observations, *Solar Phys.*, 44, 83, 1975.
Rust, K., and A. Kumar, Helical magnetic fields in filaments, *Solar Phys.*, 155, 69–97, 1994.
Ruzmaikin, A., Redistribution of magnetic helicity at the Sun, *Geophys. Res. Lett.*, 23, 2649–2552, 1996.
Smith, S. F. and H. R. Ramsey, Flare positions relative to photospheric magnetic fields, *Solar Phys.*, 2, 158–170, 1967.
van Ballegooijen, A. A., and P. C. H. Martens, Formation and eruption of solar prominences, *Astrophys. J.*, 343, 971–984, 1989.

S. F. Martin, Helio Research, 5212 Maryland Ave., La Crescenta, CA 91214 (e-mail: sara@primenet.com)

A. H. McAllister, High Altitude Observatory, National Center for Atmospheric Research, P.O. Box 3000, Boulder, CO 80307-3000.

The Field Configuration of Magnetic Clouds and the Solar Cycle

V. Bothmer[1] and D.M. Rust[2]

[1]Institut für Reine und Angewandte Kernphysik, University of Kiel, 24118 Kiel, Germany
[2]Applied Physics Laboratory, The John Hopkins University, Laurel, Maryland 20723, USA

By means of solar wind measurements from near-Earth spacecraft, the field structure of magnetic clouds (MCs) for the years 1965-1993 was investigated. A solar cycle variation appears in magnetic field orientation in the clouds. From about the time of sunspot maximum in 1981, for example, most MCs had northward-directed fields leading southward-directed ones. Since the sunspot maximum of 1991, southward-directed fields have led northward fields in MCs. Their axial fields are equally distributed between eastward-directed and westward-directed, and there are as many MCs with right-hand rotating fields as there are left-hand ones regardless of the phase of the cycle. The MCs are found to be associated with Hα filament disappearances at the Sun, and their magnetic field properties are interpreted in terms of the known properties of filament fields, namely, that the predominant axial field direction in each solar hemisphere reverses at sunspot maximum as does the direction of the fields overlying filament channels.

1. INTRODUCTION

The interplanetary counterparts of coronal mass ejections (CMEs) that exhibit the topology of helical magnetic flux ropes are commonly called magnetic clouds (MCs) [*Burlaga et al.*, 1981; *Klein and Burlaga*, 1982; *Gosling*, 1990; *Bothmer and Schwenn*, 1996a]. MCs make up about one-third of all solar wind streams that are identified as interplanetary consequences of CMEs [*Gosling et al.*, 1990; *Bothmer and Schwenn*, 1996a].

In a recent analysis *Bothmer and Schwenn* [1996b] found that 34 of 46 MCs (74%) observed by the Helios spacecraft during the years 1974-1981 had a southward-directed (with respect to the ecliptic) magnetic field in the leading portion and a northward-directed field in the trailing portion (see also *Bothmer* [1993]). Only 12 MCs had a northward-directed field in the leading portion and a southward-directed field in the trailing portion. In this paper we call the MCs South-North (SN) and North-South (NS) clouds to emphasize the sequence of out-of-ecliptic field deflections that a magnetometer would record as the cloud passed. Several important distinctions between these two types of clouds were pointed out by *Zhang and Burlaga* [1988] who studied 19 clouds from near the 1980 peak of solar cycle 21. SN clouds outnumbered NS clouds 13 to 6, and they were associated with large increases in solar wind speed. Solar wind speed was essentially constant at about 400 km/sec inside and outside the NS clouds. SN clouds caused larger magnetic storms at Earth: the average minimum D_{st} was -125 γ for SN clouds and -91 γ for NS clouds.

MCs have been associated with Hα filament disappearances at the Sun [*Burlaga et al.*, 1982; *Wilson and Hildner* 1986]. *Rust* [1994] and *Bothmer and Schwenn* [1994] found links between the inferred magnetic structures in and around the filaments and those of MCs. *Rust* emphasized the agreement between the inferred sense of twist (chirality) of associated filament fields and MC fields while *Bothmer and Schwenn* emphasized the agreement between MC fields and the pattern of photospheric fields near the filaments. MCs have also been associated with other features of solar activity, e.g., with flares [*Burlaga et al.*, 1981]. *Bothmer and Schwenn* [1996b] suggested that the

Coronal Mass Ejections
Geophysical Monograph 99
Copyright 1997 by the American Geophysical Union

Table 1.

MC [Year:Month:Day]	MC-Speed [km/s]	Associated DB [Month:Day]	Solar Position of DB	Magnetic Polarity of DB	Inferred DB Type	Inferred MC Type
65:11:04/5	>400	10:31	N20E16-N31E13	-/+	NWS	NWS
66:12:13/14	>450	12:09	N02E79-N02E59	∓	NWS	NWS
67:01:14	450-500	01:12	N68E83-N43E06	-\+	SEN	SEN
67:05:02/3	400-500	04:29	S40E19-S31W02	-/+	NES	NES
67:12:30/31	400-500	12:27	S25E13-S20E07	-/+	NES	NES
68:02:27/28	300	02:21	S20E53-S10E43	+/-	SWN	SWN
68:09:07/8	400-500	09:04	N29E45-N29E16	∓	NWS	NWS
68:10:30/31	>600	10:27	S30E12-(S36E03)	-/+, SGD	NES	NES
69:02:11/12	450-550	02:07	N43E17-N45E02	-/+	NWS	SEN
69:08:26/27	350-450	08:21	N05E13-N07W01	-/+	NWS	NWS
70:01:15/16	350-400	01:10	S40W03-S40W15	∓	NES	NWS
71:06:23	350	no				SEN
72:03:27/28	400-450	03:23	N44E20-N18E02	-\+	SEN	SEN
72:06:18/19	?	06:16	N30W38-N18W52	-\+	SEN	SEN
72:11:01	400-650	10:30	S33E14-S35W01	-\+	SWN	SW?N
73:01:20	350-500	no				NES
73:03/04:31/01	400-450	03:27	S02E31-N03E20	-/+	NW?S	NWS
73:05:21/22	>650	?				NE?S
73:09:26/27	350-450	no				SEN
74:10:13/14	400-500	10:11	N30E12-N38E10	+/-, SGD	SEN	SEN
75:04:20	350-400	no				SEN
75:05:25/26	?	05:20	N41E07-N40W02	-\+, SGD	SEN	SEN
75:11:17/18	350-400	11:10	S35E74-S27E39	+/-	SWN	SW?N
76:01:10/11	350-400	?				SWN
77:09:22	650-700	09:19	S50E03-S47W17	+/-, SGD	SWN	SWN
77:09:27	400	no				SWN
78:01:04/5	450-650	?				SWN
78:01:17/18	300-350	01:11*	S37W01-S24W10	-/+	NES	NES
78:04:03/4	450-500	03:30	S60E39-S45E09	+/-	SWN	SWN
78:08:27/28	400-500	08:23	N16E18-N11E04	+\-	NWS	NWS
78:09:29	>750	?				SEN
78:10:30	350-400	10:26	N42E61-N25E51	-\+	SEN	SEN
79:04:03/4	450-600	03:30	S44W10-S22W20	+/-	SWN	SWN
79:04:25/26	500-600	04:22	S40E29-S36E07	+/-	SWN	SWN
79:09:18	350-400	no				SEN
79:12:03/4	350-400	11:27	N13E04-S02W02	+\-	NWS	NE?S
80:02:16/17	350-450	02:11	N19E02-N17W22	-\+	SEN	SEN

Table 1. (continued)

MC [Year:Month:Day]	MC-Speed [km/s]	Associated DB [Month:Day]	Solar Position of DB	Magnetic Polarity of DB	Inferred DB Type	Inferred MC Type
80:03:19/20	300-400	?				NES
80:12:19/20	450-550	?				SEN
81:03:05/6	550-650		DB-catalog ends in 1980			SWN
81:07:23/24	450-500					SEN
81:07:25/26	550->750					SWN?
81:10:22	450-600					NES
82:02:12/13	450-600					NWS
82:02:23	350-500					NW?S
82:09:21/22	>750					NES
82:09:26	450-550					NES
82:12:15-17	?					NES
83:07:23/24	400-450					NWS
85:07:12/13	400-500					SEN
86:11:24	450-500					NWS
88:01:14/15	500-750					NWS
88:02:21/22	?					NWS
88:06:13	300-400					NES
89:06:13/14	450-550					NWS
89:08:28/29	450-550					NWS
90:06:14/15	500-700					SWN
90:07:10/11	400-500					NWS
91:04:25/26	350-450					NW?S
91:07:09/10	600-700					SEN
92:02:08-10	400-600					SWN
92:02:21/22	400-500					SE?N
92:05:07/8	300-400					NES
92:07:20/21	300-350					NWS
92:09:09/10	400-600					NES
92:11:09	400-450					SEN
93:12:08	400-500					SEN

Table 1. Magnetic clouds (MCs) observed by near-Earth spacecraft (no duty cycle consideration) and associated filament disappearances ("disparition brusques" (DBs)) at the Sun. The parameters from left to right are: date of the MC, speed interval for the MC, date of associated DB, solar position of the filament, magnetic polarities on both sides of the filament's axis (+: field lines point away from the solar surface, -: field lines point towards the solar surface), magnetic flux rope type of the filament as inferred from the given magnetic polarities under assumption of left-handed (right-handed) helicity dominance in the northern (southern) solar hemisphere, and inferred flux rope type of the MC in interplanetary space. SGD: Magnetic polarities according to synoptic charts published in *Solar Geophysical Data* (SGD). *:Several filaments disappeared on this day, the selected DB was closest to central meridian.

dominance of SN MCs during 1974-1981 could indicate a relationship between the magnetic structure of MCs and the cyclic reversal of the Sun's magnetic field. Here we pursue this idea by investigating MCs observed by spacecraft near Earth. The source of data is the OMNI-database at NSSDC [*King*, 1991]. The observations cover more than a 22-year magnetic solar cycle.

2. DATA AND SELECTION CRITERIA

The OMNI-database provides hourly averaged solar wind data from the end of 1963 to the present. To identify MCs we inspected Carrington-plots of plasma and magnetic field data, looking for events with the characteristic signatures of MCs as prescribed by *Burlaga* [1991]. These signatures are: a) a smooth coherent rotation of the magnetic field vector over a time interval of the order of 1 day, b) a higher than average solar wind magnetic field strength (we selected only events with B>10 nT), c) a lower than average solar wind temperature. Events with complex field variations were excluded from the present study. In all, we identified 67 MCs during the years 1965-1993 (see Table 1). Note that we did not take the duty cycle of the spacecraft into account, so we can say nothing about the number of clouds per solar cycle.

Data from a typical cloud, which was also identified by *Klein and Burlaga* [1982], are shown in Figure 1. The MC is characterized by a smooth rotation in the latitudinal angle θ of the magnetic field from North (N) to South (S) with respect to the ecliptic (θ=90°≡N, θ=-90°≡S) and an azimuthal angle φ of ~270° (φ=0°≡sunward direction, φ=90°≡East (E), φ=270°≡West (W)).

Every MC was classified according to its field rotation in the direction normal to the ecliptic and according to the azimuth direction of its field vector near cloud center with respect to the ecliptic. We will use the notation introduced by *Bothmer and Schwenn* [1996b]: In SEN (SWN) clouds the field vector turns from South to East (West) near the cloud's center and finally to the North in the cloud's trailing portion. Similarly in NES (NWS) clouds the field vector rotates from North to East to South (North to West to South). According to this classification the MC represented in Figure 1 is of type NWS. All identified MCs and their types are listed in Table 1. More generally one recognizes that the angular variation or twist of MC fields is right-handed for SWN and NES clouds and left-handed for NWS and SEN clouds (see Figure 2). This classification helps one think in terms of magnetic helicity [*Burlaga*, 1988; *Lepping et al.*, 1990; *Rust*, 1994], which can be related to filament and CME models and solar dynamo models.

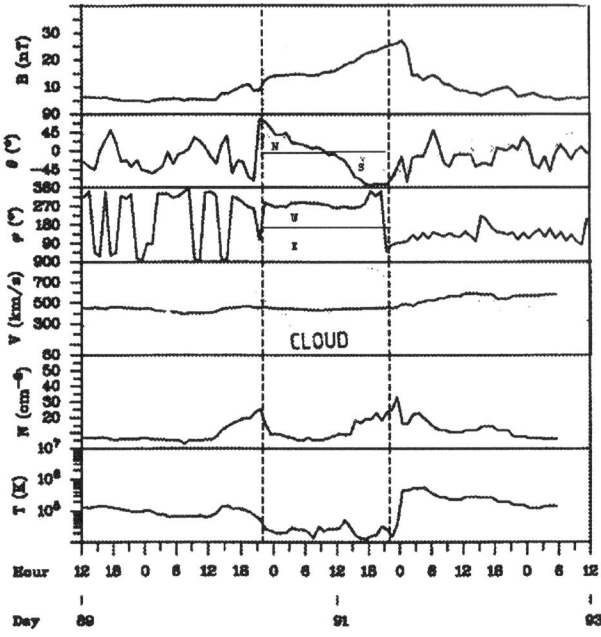

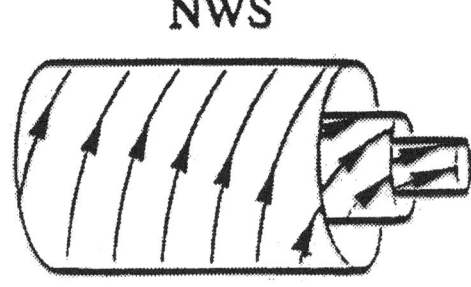

Figure 1. Top: Solar wind data for a magnetic cloud of type NWS observed at 1 AU in 1973 by the IMP s/c. The top panels show: Magnetic field magnitude, latitude (θ) and azimuthal (φ) angle of the field, solar wind speed, density and temperature. North (N), South (S), East (E), West (W) mark sectors of the field direction with respect to the ecliptic. Bottom: Sketch of the inferred flux rope type of the MC. The MC is of type NWS, i.e. during the passage of the cloud the field direction changes from North (N) to South (S) with a westward (W) direction in field azimuth near its center.

3. ASSOCIATIONS OF MAGNETIC CLOUDS WITH DISAPPEARING FILAMENTS

Using the observed plasma speeds of the MCs to estimate their start-times at the Sun, we searched the "Catalog of solar filament disappearances 1964-1980" (*C.S. Wright*, World Data Center A for Solar-Terrestrial Physics, 1991) for Hα filament disappearances (disparitions brusques (DBs)) recorded near these times. Table 1 lists the associations found between MCs and DBs. Note that it is always difficult to associate interplanetary and solar events uniquely. This is also reflected in Table 1 by the number of cases in which the associated DBs were observed quite far away from the Sun's central meridian (CM) and in some cases for which multiple DBs were observed. However, many associations are unique in the sense that at the estimated onset-time of the MC at the Sun, a single DB was observed which was not preceded or followed by other DBs in close proximity in either space or time.

4. MAGNETIC STRUCTURE OF FILAMENTS AND MAGNETIC CLOUDS

Recent observations reveal that, independently of the phase of the 22-year solar cycle, mid- to high-latitude filaments in the Sun's northern hemisphere have a characteristic fine structure and orientation with respect to the surrounding fields that *Martin et al.* [1994] termed 'dextral', whereas filaments in the Sun's southern hemisphere are like mirror images of the dextral ones and so are called 'sinistral'. From several lines of evidence, *Rust and Kumar* [1994] inferred that the twist of the magnetic fields in dextral filaments is left-handed and that it is right-handed in sinistral filaments. This point is discussed in detail in the paper by *Rust* [1997] in this volume.

As sketched in Figure 2, we inferred the field direction in filaments from the expected axial direction and from the polarity of the photospheric fields on either side. Near sunspot minimum, a family of filaments forms near latitude 30 degrees in each hemisphere. The locus of these filaments drifts toward the pole until solar maximum, when it has reached latitude about 55 degrees. Filaments remain at this latitude, forming the so-called polar crowns, until the next solar minimum. Then the band of filaments resumes its poleward march. This 'rush to the poles' takes about four years, until the solar maximum. Filaments in the northern polar crown between 1981 and 1990 most often had negative fields on their northern side and positive fields on their southern side. The sense of the fields was reversed in the south.

The prevailing direction of the axial fields in mid- to high-latitude filaments also reverses every eleven years [*Rust*, 1967; *Leroy*, 1989; *Martin et al.*, 1994] according to well known rules. From these rules and from the records of magnetic polarities near the filaments, we inferred the flux rope type of each filament and compared these inferences with the observed flux rope types of the associated MCs in interplanetary space (see Table 1). We assumed, as is usually done (e.g., *Priest* [1989]), that magnetic arcades span the filament channel more or less perpendicular to the filament axis and, that the arcade fields can be identified more or less with the fields just inside the MC boundaries (see Figure 2).

Underlying our study was the supposition that magnetic helicity [*Kumar and Rust*, 1996] should be conserved in filaments, even as they erupt and expand into interplanetary space. We note here that *Martin and McAllister* [this volume] have proposed a different interpretation for the helicity of filaments relative to MCs, and we refer the interested reader to their paper for more details.

Twenty-seven of the 39 MCs that we identified during the years 1965-1980 could be associated with a DB, and in 24 of the 27 cases, the inferred structure of the filament fields agreed with that for the associated MC, i.e. MCs with left-handed fields (types SEN, NWS) were associated with left-handed filaments from the Sun's northern hemisphere, and MCs with right-handed fields (SWN, NES) were associated with right-handed filaments from the southern hemisphere. Unfortunately, the DB-catalog ends in 1980 so we could not complete the study of all MCs.

Most filaments associated with MCs were at heliographic latitudes of 30-40 degrees, i.e. outside active regions. This places an important restriction on our conclusions, since the magnetic topology of those CMEs that stem from active solar regions and are frequently accompanied by flares could be different from our flux rope type CMEs.

5. FREQUENCY OF SN AND NS MAGNETIC CLOUDS

Working with measurements from the Helios probes, *Bothmer* [1993] and *Bothmer and Schwenn* [1996b] found that during the years 1974-1981 MCs of type SN were observed three times more frequently than MCs of type NS. They also found that the number of MCs with left-handed magnetic helicity was equal to the number of MCs with right-handed helicity. Their unexpected finding of a preponderance of SN clouds for this time-interval led us to study the OMNI data set in order to establish whether the dominant magnetic helicity of MCs as a class varies in time, maybe in relationship to the solar cycle.

The SN and NS clouds identified in the OMNI data during the years 1965 to 1993 are listed in Table 1. The data coverage was rather different for the individual years, and the rate of typical MCs per year (~ 6-12) observed at 1 AU is small [*Klein and Burlaga*, 1982]. Nevertheless, we found a dominance of SN clouds for the years 1974 to 1981 and a dominance of NS MCs before ~1970 and after ~1981. Figure 3 shows the number of SN and NS MCs identified for the years 1974-1981 and 1982-1991 (the time intervals have been selected primarily for comparison with the results from the Helios observations). Of all 67 identified MCs during 1965-1993, 34 were SN and 33 were NS clouds. Thirty-seven MCs had left-handed magnetic helicity and 30 MCs had right-handed magnetic helicity. The results found from the Helios observations are hereby supported [*Bothmer and Schwenn*, 1996]. Note that during years of high solar activity, when the Sun's global magnetic field reverses and the upper tier of

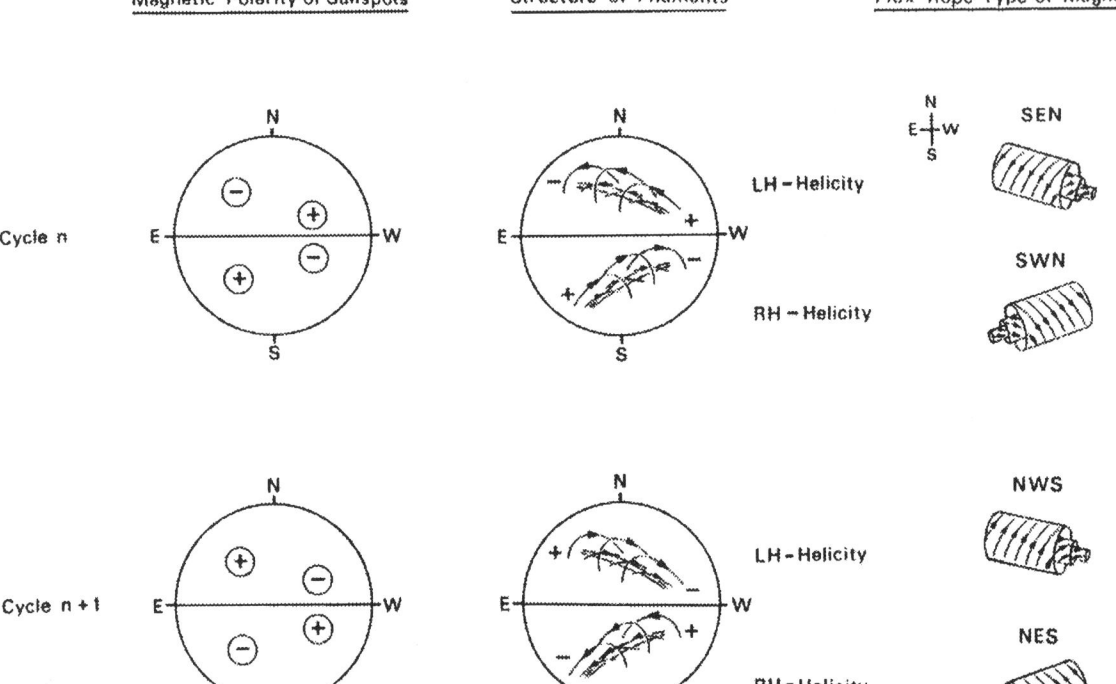

Figure 2. An extended association between the magnetic field structure of solar filaments and interplanetary magnetic clouds (MCs) that can explain the observed solar cycle variation of the field structure of MCs.

filaments is being replaced by the lower one, both SN and NS clouds were observed frequently (e.g. around 1970, 1980, 1992).

6. SUMMARY AND CONCLUSIONS

From an investigation of the structure of MCs identified in near-Earth solar wind data for the years 1965-1993, we find evidence for a dependence of the magnetic structure of MCs on the phase of the solar cycle. MCs with a SN rotation of the magnetic field direction with respect to the ecliptic were preferentially observed during 1974-1981, whereas during 1982-1991 the observations show a dominance of MCs with a NS rotation. Over the years studied, the total number of SN-type MCs equaled the total number of NS-type MCs and the number with right-hand magnetic fields was roughly equal to the number of left-hand fields.

We also found a convincing association of MCs with filament disappearances, in agreement with *Rust* [1994] and *Bothmer and Schwenn* [1994]. It was not difficult to identify filament disappearances for each expected MC launch time, based on solar wind speed. From a comparison of the magnetic structure of the MCs with that inferred for the associated filaments we found agreement of chirality (handedness) in 24 of 27 cases. Note that our results do not necessarily imply that the filament itself evolves into the MC in interplanetary space. From a comparison of the spatial sizes of the features seen in Hα filtergrams and white-light coronagraph observations it seems likely that the filament may be either the bottom part of a large flux tube or just an indicator for the much larger scale structure of the overlying coronal fields.

Figure 2 summarizes our view of the solar origins of the cyclic variation of the magnetic structure of MCs. As sketched in the figure's left column, the magnetic polarity of sunspots and all other fields reverses each cycle. Hα filaments commonly occur in solar regions where the sunspots themselves are no longer visible. The magnetic arcades over the filaments arise in the large-scale magnetic polarity regions on either side (as sketched in the center column of Figure 2). The magnetic structure of the filaments illustrated in Figure 2 is an interpretation based on results from *Bothmer and Schwenn* [1994] and the helicity rule of *Rust and Kumar* [1994]. The right column of Figure 2 finally shows the expected types of MCs in interplanetary space. Not every MC in the interplanetary medium is necessarily related to a

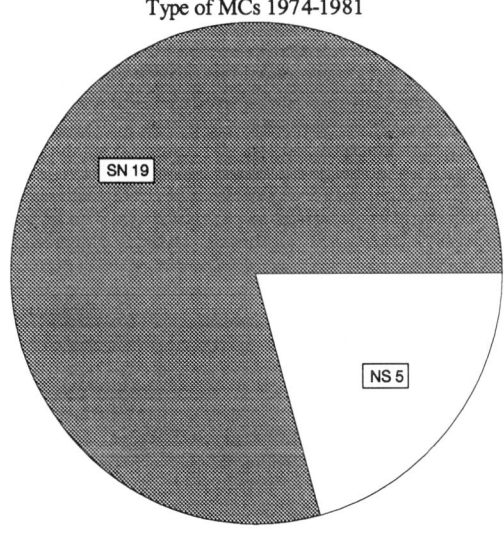

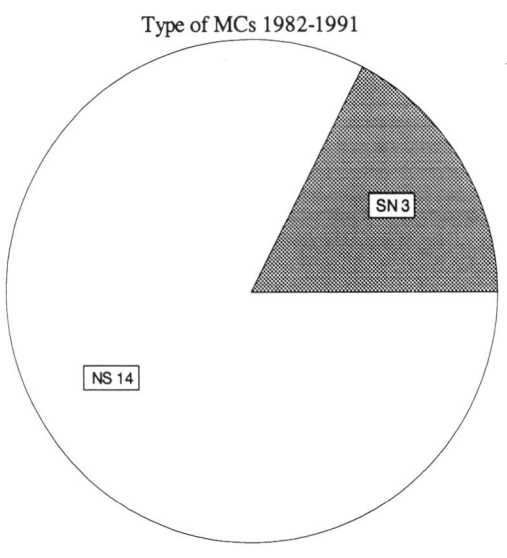

Figure 3. Top: Frequency of SN and NS magnetic clouds during 1974-1981. Bottom: Frequency of SN and NS clouds during 1982-1991.

filament at the Sun. Other solar activity features, like flares or instabilities of the coronal streamer belt may also give rise to MCs, but we have no evidence of it.

As a consequence of this mechanism we expect that generally MCs of SN-type should occur predominantly in the eleven years from shortly after the peak of even-numbered cycles whereas NS-type MCs should dominate in the eleven years after the peak of odd-numbered cycles. The frequency of MCs with left- and right-handed magnetic helicity would be the same in each cycle, assuming that the general level of activity is the same in the northern and southern hemispheres.

Our results are important for space weather forecasts: It is well known that CMEs are the causes of the strongest geomagnetic storms and that MCs form an important subset of all CMEs. We predict that the majority of MCs up to the peak of the next cycle (cycle number 23, years 1997 - 2003) will be SN. According to the work of *Zhang and Burlaga* [1988], this means that geomagnetic storms due to MCs will be more intense, as measured by the average minimum in D_{st}, than during the most recent cycle. The magnetic storm of January 10, 1997, which resulted from Earth's collision with a SN MC, was an example of what we expect.

Another important aspect is the concept of magnetic helicity in astrophysical plasmas: magnetic fields maybe expelled from the Sun outward into interplanetary space without loosing the intrinsic helicity that has been created by solar mechanisms which might even be linked with the solar dynamo (see *Rust* [Science, Vol. 269, p. 1517, 1994]).

Acknowledgments. We acknowledge the use of the OMNI-database at NSSDC, Greenbelt, Maryland, USA. We thank both referees for their assistance in evaluating this paper.

REFERENCES

Bothmer, V., Die Struktur magnetischer Wolken im Sonnenwind - Zusammenhang mit eruptiven Protuberanzen und Einfluß auf die Magnetosphäre der Erde -, ph.d. thesis, University of Göttingen, 1993.

Bothmer, V., and R. Schwenn, Signatures of fast CMEs in interplanetary space, Adv. Space Res., 17, 319-322, 1996a.

Bothmer, V., and R. Schwenn, The structure of magnetic clouds in the solar wind, submitted to Ann. Geophys., 1996b.

Bothmer, V., and R. Schwenn, Eruptive Prominences as Sources of Magnetic Clouds in the Solar Wind, Proceedings of the II Soho Workshop at Elba, Italy, 1993, Space Sci. Rev., 70, 215-220, 1994.

Burlaga, L.F., Magnetic Clouds, in Physics of the Inner Heliosphere Vol. II, ed. by R. Schwenn and E. Marsch, Springer, Berlin/Heidelberg, 1-22, 1991.

Burlaga, L.F., Magnetic clouds and force-free fields with constant alpha, J. Geophys. Res,. 93, 7217-7224, 1988.

Burlaga, L.F., L. Klein, N.R. Sheeley Jr., D.J. Michels, R.A. Howard, M.J. Koomen, R. Schwenn, and H. Rosenbauer, A magnetic cloud and a coronal mass ejection, Geophys. Res. Lett., 9, 1317-1320, 1982.

Burlaga, L.F., E. Sittler, F. Mariani, and R. Schwenn, Magnetic loop behind an interplanetary shock: Voyager, Helios, and IMP 8 observations, J. Geophys. Res., 86, 6673-6684, 1981.

Gosling, J.T., Coronal mass ejections and magnetic flux ropes in interplanetary space, in Physics of Magnetic Flux Ropes, ed. by E.R. Priest, L.C. Lee, and C.T. Russell, AGU Geophysical Monograph, 58, 343-364, 1990.

King, J.H., Long-Term Solar Wind Variations and Associated Data Sources, J.Geomag. Geolelectr., 43, 865-880, 1991.

Klein, L.W., and L.F. Burlaga, Interplanetary Magnetic Clouds At 1 AU, J. Geophys. Res., 87, 613-624, 1982.

Kumar, A., and D.M. Rust, Interplanetary magnetic clouds, helicity conservation, and intrinsic-scale flux ropes, J. Geophys. Res., 101, 15667-15684, 1996.

Lepping, R.P., J.A. Jones, and L.F. Burlaga, Magnetic field structure of interplanetary magnetic clouds at 1 AU, J. Geophys. Res., 95, 11957-11965, 1990.

Leroy, J. L., Observations of prominence magnetic fields, in Dynamics and Structure of Quiescent Solar Prominences, edited by E. Priest, Kluwer, Dordrecht, p. 77, 1989.

Martin, S.F., and A.H. McAllister, Predicting the sign of helicity in erupting filaments and coronal mass ejections, this volume.

Martin, S.F., R. Bilimoria, P.W. Tracadas, Magnetic field configurations basic to filament channels and filaments, in Solar Surface Magnetism, ed. by R.J. Rutten and C.J. Schrijver, NATO ASI Series C, Vol. 433, Kluwer Academic Publ., Dordrecht, 303-338, 1994.

Priest, E.R., Ed.: Dynamics and Structure of Quiescent Solar Prominences, Astrophysics and Space Science Library, Kluwer Academic Publishers, 101 Phillip Drive, Norwell, MA 02061, USA, 1989.

Rust, D. M., Helicity Conservation, this volume, 1997.

Rust, D.M., Spawning and shedding of helical magnetic fields in the solar atmosphere, Geophys. Res. Lett., 21, 241-244, 1994.

Rust, D.M., Magnetic Fields in Quiescent Solar Prominences I. Observations, Astrophys. J., 150, 313, 1967.

Rust, D.M., and A. Kumar, Helical Magnetic Fields in Filaments, Solar Phys., 155, 69-97, 1994.

Wilson, R.M., and E. Hildner, On the Association of Magnetic Clouds With Disappearing Filaments, J. Geophys. Res., 91, 5867-5872, 1986.

Zhang, G., and L.F. Burlaga, Magnetic Clouds, Geomagnetic Disturbances, and Cosmic Ray Decreases, J. Geophys. Res., 93, 2511-2518, 1988.

V. Bothmer, Institut für Reine und Angewandte Kernphysik, University of Kiel, 24118 Kiel, Germany, e-mail: Bothmer@ifkki.kernphysik.uni-kiel.de.

D.M. Rust, Applied Physics Laboratory, The John Hopkins University, Laurel, Maryland 20723, USA, e-mail: david.rust@jhuapl.edu.

Interplanetary Magnetic Flux Ropes and Solar Filaments

K. Marubashi

Communications Research Laboratory, Tokyo, Japan

This paper aims at clarifying the physical relationship between interplanetary magnetic clouds and solar magnetic fields. For this purpose, we analyzed twelve magnetic clouds whose magnetic field variations are well explained by a flux rope model. Attempts were made to determine the geometry of flux ropes that fit the observed magnetic field variations and to identify disappearances of solar filaments that can be associated with the generation of those magnetic structures. A comparison of the magnetic field structures of flux ropes fitted to the interplanetary observations and the structures of solar magnetic fields suggests a model for the generation of magnetic clouds, or interplanetary magnetic flux ropes. We propose that the solar magnetic field surrounding a disappearing filament already has a flux rope structure at the time of eruption and that the structure extends through interplanetary space to be observed as an interplanetary flux rope. In addition, a model is proposed for a possible three-dimensional structure of an interplanetary magnetic flux rope, and observational support is presented.

1. INTRODUCTION

A magnetic flux rope is a cylindrical body of magnetized plasma with twisted fields [*Priest*, 1990]. An interplanetary magnetic cloud was proposed by *Klein and Burlaga* [1982] for interpreting peculiar magnetic field structures in the solar wind that are characterized by the smooth rotation of the observed magnetic field vectors typically over the course of a day. Subsequent studies showed that the magnetic field rotation can be well explained as a field variation observed when a spacecraft traversed a flux rope structure [*Goldstein*, 1983; *Marubashi*, 1984; 1986; *Burlaga*, 1988; *Lepping et al.*, 1990; *Vandas et al.*, 1991]. In view of these subsequent works, I use the terminology, "magnetic flux rope" instead of "magnetic cloud" throughout this paper. Special attention has been paid to the interplanetary magnetic flux ropes as possible counterparts of coronal mass ejections (CMEs) since the first identification of a magnetic cloud by *Burlaga et al.* [1981]. Here, and throughout this paper, the term CME is used in a strictly solar coronal sense (See Kahler [1987] for a review of CMEs). Statistical studies [*Wilson and Hildner*, 1984; 1986] showed that certain types of magnetic clouds are associated with flare-related type II radio bursts and with disappearances of solar filaments.

Among various forms of solar activity, prominence eruptions or filament disappearances are known to be most strongly correlated with CMEs [*Munro et al.*, 1979; *Webb and Hundhausen*, 1987]. *Marubashi* [1986] pointed out close relationships between magnetic field configurations of interplanetary flux ropes and solar magnetic fields around the solar filaments for two cases. More recently, *Bothmer and Schwenn* [1994] and *Rust* [1994] have extended the analysis on this relationship by using more flux rope events. It is thus expected that we can get further insight into the generation mechanism of interplanetary flux ropes by studying their structures and comparing them with structures of solar magnetic fields surrounding associated disappearing filaments.

The purpose of this paper is to examine this view by analyzing more examples. For this purpose we selected twelve well-defined flux ropes from data during January 1965 through October 1978. The criteria for selection are mentioned in the next section. The data sets used in this study were taken from the hourly solar wind data which were compiled at the National Space Science Data Center, Goddard Space Flight Center. The original data sources are the solar wind field and plasma measurements by the IMP, VELA, HEOS, ISEE 3 satellites. As a result of the analysis of these flux rope events, a model is proposed on the relationship between the structures of CMEs observed by coronagraphs and interplanetary magnetic flux ropes, and on the three-dimensional configuration of flux ropes in interplanetary space. Finally, a possibility is suggested that various types of magnetic field changes can be explained by the proposed geometry of the interplanetary magnetic flux rope.

2. GEOMETRY OF INTERPLANETARY MAGNETIC FLUX ROPES

Table 1 presents 12 flux rope events selected for detailed analysis. The selection is essentially based on identifying the rotation of magnetic field vectors of about 180° nearly in a plane. The actual selection procedure consists of the following three steps. (1) Search the time intervals where the magnetic fields are relatively strong (≥ 10 nT) and a rotation of the field vector is seen over large latitude angles.

(2) Confirm the rotation of magnetic field vectors with a vector plot such as the bottom diagram of Figure 1. (3) Perform minimum variance analysis [*Sonnerup and Cahill*, 1967], and select the start and end times of the event. The start and end times are subsequently revised, when appropriate, after the model fitting. Results of minimum variance analysis are also presented in Table 1: the ratio of the intermediate to minimum eigenvalues λ_2/λ_3, and the minimum variance direction (θ_n and ϕ_n). These values indicate that the axes of these selected flux ropes are directed at large angles from the Sun-Earth line.

The flux rope type is defined by the direction of the field-aligned electric current flowing in the flux rope [*Marubashi*, 1986] and presented in the table: parallel type (P) and anti-parallel type (A). These two types are also called right-handed and left-handed, respectively, in accordance with the handedness of the magnetic helicity [*Burlaga*, 1988; *Lepping et al.*, 1990; *Bothmer and Schwenn*, 1994; *Rust*, 1994]. It is easy to determine the type of an observed flux rope when the rotation of field vectors is apparent. The parallel type is identified by anti-clockwise magnetic field rotation, and the anti-parallel type by clockwise rotation, with respect to the trajectory of the spacecraft traversing the flux rope. The last two columns of Table 1 provide the dates and times of the shocks that preceded the flux ropes, and their velocities (V_s). The shock velocities were calculated by the Rankine-Hugoniot relation [e.g., *Abraham-Shrauner and Yun*, 1976]:

$$V_s = (n_2 U_2 - n_1 U_1) / (n_2 - n_1), \quad (1)$$

TABLE 1. Twelve Flux Rope Events Selected for Detailed Analysis

No.	Flux Rope Event						Associated Shock	
	Start Day/Time	End Day/Time	λ_2/λ_3	θ_n (deg)	ϕ_n (deg)	Type (P/A)	Day/Time	V_s (km/s)
1	65/11/04/13	65/11/05/18	47.8	11.7	346.2	A		
2	66/11/17/19	66/11/18/13	16.3	32.9	20.5	A	66/11/17/0017	407.9
3	67/01/06/13	67/01/07/08	21.1	31.0	2.8	P	67/01/06/0714	
4	67/05/02/12	67/05/03/14	38.0	-7.6	355.0	P	67/05/01/1907	549.4
5	67/12/30/18	67/01/01/09	22.4	-21.9	355.6	P	67/12/29/2227	549.3
6	69/02/11/09	69/02/12/18	8.7	-15.1	322.5	P	69/02/10/2026	531.7
7	69/08/26/14	69/08/27/04	41.5	-16.9	13.9	A	69/08/26/0435	413.4
8	71/06/23/10	71/06/24/13	23.2	-0.4	356.5	A		
9	73/03/31/23	73/04/02/00	9.5	-0.8	358.2	A		
10	75/08/01/07	75/08/02/03	8.6	18.8	352.9	A		
11	76/01/10/15	76/01/11/13	41.7	-19.2	2.1	P	76/01/10/0621	396.0
12	78/10/30/00	78/10/31/12	23.4	-4.9	5.2	A		

Results of minimum variance analysis is presented; λ_2/λ_3 is the ratio of the intermediate to minimum eigen values; θ_n, ϕ_n are the latitude and longitude angles of the minimum variance direction in GSE coordinates.

Here, the proton densities (n_1, n_2) and velocities (U_1, U_2) are 1-hour averages in the intervals nearest to and excluding the shock transition. In this sense, the estimation is rather inaccurate, but both the shock speeds and the cloud speeds are used to estimate the time of the solar event.

In examining the geometry of the flux rope, we use a constant-α force-free cylindrical model including self-similar expansion, in view of the success of previous modelling [*Burlaga*, 1988; *Farrugia et al.*, 1992; 1993], though a spheroid model is also proposed for magnetic clouds [*Vandas et al.*, 1991; 1993]. Suppose that R_0 is the radius of a flux rope at the time of the first encounter with a spacecraft and that the expansion had proceeded for the time duration t_0. Then the flux rope radius, R, the expansion velocity, v, and the magnetic field, B, at time t_0+t are given by the following expressions. (Cylindrical coordinates, ρ, φ, ζ, are used with ζ-axis along the axial magnetic field of the flux rope.)

$$R = R_0 (1 + t / t_0), \qquad (2)$$

$$\boldsymbol{v} = v_\rho\, \boldsymbol{e}_\rho, \qquad (3)$$

$$v_\rho = \rho / (t + t_0) \qquad (\rho \leq R), \qquad (4)$$

$$\boldsymbol{B} = B_\varphi\, \boldsymbol{e}_\varphi + B_\zeta\, \boldsymbol{e}_\zeta, \qquad (5)$$

$$B_\varphi = s\, B_0\, J_1(a\rho) / (1 + t / t_0), \qquad (6)$$

$$B_\zeta = B_0\, J_0(a\rho) / (1 + t / t_0)^2, \qquad (7)$$

Here J_0 and J_1 are Bessel functions of the first kind of order 0 and 1, respectively, B_0 is the magnetic field intensity at the cylinder axis at time t_0, a is chosen so that aR gives the first zero of J_0, and $s=1$ and $s=-1$ correspond to the parallel and anti-parallel type of flux rope, respectively. In this model, the observed solar wind bulk velocity is taken as the vector sum of the bulk flow velocity, U_{sw}, in the $-X$ direction, and the expansion velocity $\boldsymbol{v}$.

In order to calculate magnetic field and velocity changes as the flux rope is traversed by a spacecraft, we need to specify the geometric relation between the flux rope and the spacecraft trajectory. The relation is determined by three parameters: the latitude and longitude angles of the ζ-axis, θ_a and ϕ_a, and the distance between the ζ-axis and the spacecraft trajectory at the closest approach. We define this quantity, the impact parameter p, as a distance measured along the vector $\boldsymbol{e}_x \times \boldsymbol{e}_\zeta$. It should be specially noted that p can take both positive and negative values.

Now we can determine the model parameters of a flux rope, R_0, t_0, and B_0, the geometric parameters, θ_a and ϕ_a, and the bulk velocity of the flux rope, U_{sw}, by minimizing the difference between the observed and calculated solar wind velocity and magnetic fields with a nonlinear least squares fitting technique. Figure 1 presents the results for a typical flux rope event observed on February 11, 1969, which was first analyzed by *Klein and Burlaga* [1982]. Plotted from top to bottom are the magnetic field strength, the X, Y, and Z components in GSE coordinates, the ratio of standard deviations of field intensities to the average intensities, the bulk flow speed, the proton density, the temperature, the plasma beta (as derived from the proton parameters), and the magnetic field vectors projected on X-Y, X-Z, and Y-Z planes. The dashed curve plotted together with the proton temperature shows the temperature expected from the correlation between the solar wind speed and the proton temperature [*Lopez*, 1987] to identify the abnormally depressed proton temperature [*Richardson and Cane*, 1993]. This event exhibits many signatures characteristic of ejecta [*Gosling*, 1990], namely, the small variance of magnetic fields, the low plasma beta, and the abnormally depressed proton temperature. In addition to the magnetic field rotation in the flux rope, whose end points are indicated by two vertical lines, a similar rotation is also seen in the sheath region after the shock indicated by the dashed line. This feature may be interpreted as the draping of interplanetary magnetic fields around the ejecta [*McComas et al.*, 1988; *Fainberg et al.*, 1996]. The thick lines for the magnetic field and the bulk flow speed show the result of the nonlinear least squares fitting, which reveals good agreement with the observations.

We carried out the nonlinear least squares fitting for the twelve flux rope events listed in Table 1; the results are summarized in Table 2; the meaning of fitted parameters is annotated below the table. The agreement between observed and fitted quantities is generally satisfactory, as can be inferred from the rms deviations of calculated fields from observed fields, $\sigma(\Delta B)$, in the last column. In the actual calculations, some modifications were needed depending on the characteristics and availability of the data. For event No. 2, the solar wind velocity data are not available. Therefore the fitting was done under the assumption of $U_{sw}=400$. In event No. 9, the velocity change and the magnetic field intensities are strongly affected by the interaction with the overtaking fast stream [*Gosling*, 1990]. Therefore we used only the first 10 data points for the solar wind velocity, where the velocity decreases with time, and the fitting to the magnetic field was made between the observed and calculated unit vectors, $\boldsymbol{B}(t)/B$.

One problem which must be pointed out about the present model from the results in Table 2 is that the t_0 values are too large compared with the typical transit times from Sun to Earth. A possible reason is that the velocity profiles are

150 INTERPLANETARY FLUX ROPES AND SOLAR FILAMENTS

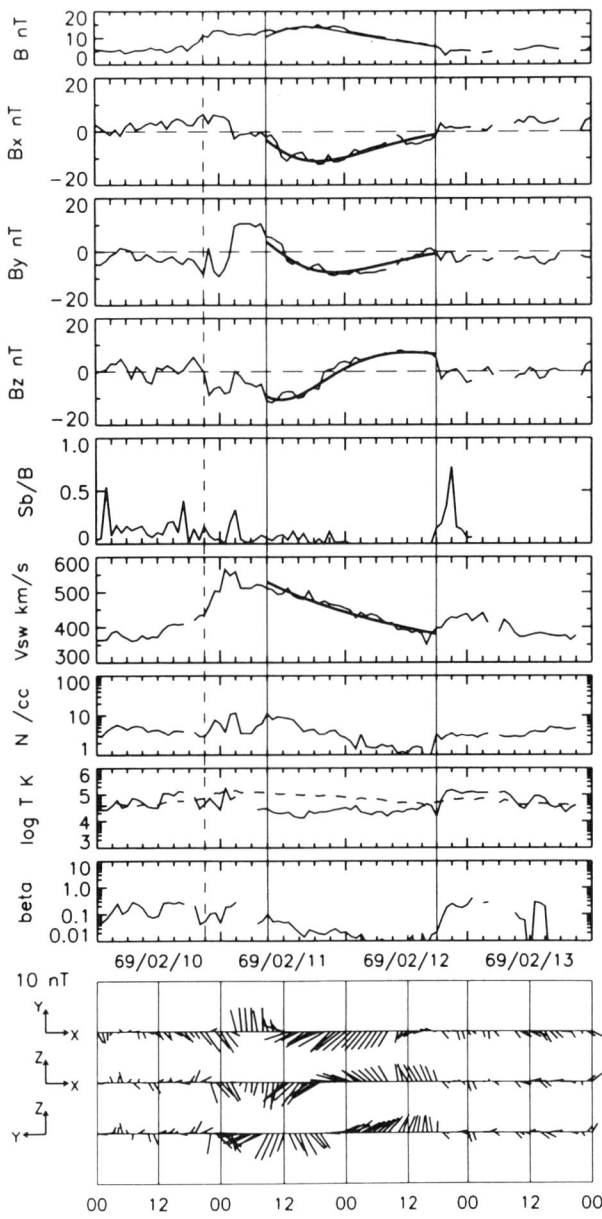

Figure 1. Solar wind parameters surrounding the February 11, 1969 flux rope event. An associated shock is indicated by a dashed line. Thick curves present the results of nonlinear least squares fitting (see text).

strongly affected through interactions with the surrounding solar wind plasma, so that the self-similar expansion model is much too simplified in describing the dynamics of the flux rope. However, it is reasonable to suppose that the exact nature of the model does not change the results of this paper, particularly those on the correspondence with disappearing filaments.

3. RELATIONS TO FILAMENTS DISAPPEARANCES

In an effort to find solar phenomena which can be related to the generation of the flux ropes, we searched for filament disappearances [*Joselyn and McIntosh*, 1981] and solar flares accompanying strong type II/IV radio bursts at metric wavelengths [*Cook and McCue*, 1979] in reasonable time windows. Though flares may not generate large-scale solar wind disturbances [*Gosling*, 1993], flares are often associated with CMEs, and the metric type II/IV bursts are evidence for disturbances propagating through the corona [*Kahler*, 1992]. Table 3 summarizes the result. The time window was defined as 2-day interval centered at the time of the launch of each magnetic flux rope. The launch times are estimated under the assumption that the flux ropes traveled with constant speeds U_{sw}, the bulk speed of the flux rope at the time of encounter and thereafter, as listed in Table 2. For the shock-associated events, the shock velocities provide another estimate for the launch time.

We can see that one or more disappearances of quiescent filaments took place near the central meridian within each time window in 9 cases out of 12. The flare association is seen only in one case (No. 5). For each flux rope, we chose one event that took place nearest to the central meridian. Then each of the 9 corresponding flux ropes were imposed on the chosen filaments in Figure 2. The flux rope axis was first projected onto the Y-Z plane by using θ_a and ϕ_a in Table 2, and then the resultant orientation was mapped on the relevant filament, with a correction for the angle between the ecliptic and equatorial planes. The straight and curved arrows indicate the direction of the axial field and the direction of rotation of the helical field of each flux rope thus mapped. This mapping method gives the approximate orientation of the flux rope at the launch site if the flux rope was created in the solar atmosphere and if its orientation was conserved during propagation. It seems reasonable to assume conservation of orientation for those cases in which the flux ropes were encountered in interplanetary space near the top part of the extended loop (see Figure 3). The relevant filaments are indicated by circles, and the dashed lines indicate the approximate directions of the long axes of the filaments. The flux rope type, parallel (P) or antiparallel (A), is indicated at the bottom of each map.

The following features are evident in Figure 2. (1) The flux ropes of parallel type (right-handed) correspond to the filament disappearances in the southern hemisphere and those of anti-parallel type (left-handed) to the filaments in the northern hemisphere [*Marubashi*, 1986, *Rust*, 1994]. (2) The directions of the helical magnetic fields in the outer portion of the flux ropes (or the azimuthal fields referred to in the cylindrical structure) are consistent with the directions of magnetic field loops above the inversion lines of

TABLE 2. Flux Rope Parameters Determined by the Least Squares Fitting

No.	U_{sw}	B_0	θ_a	ϕ_a	R_0	p	t_0	$\sigma(\Delta B)$
1	435.8	27.0	33.9	219.9	0..081	-0.154	32.3	3.03
2	(400)	25.5	75.2	63.2	0.091	0.532	107.3	4.11
3	358.6	18.0	81.1	51.1	0.096	-0.571	501.4	3.03
4	442.1	26.9	-59.4	115.2	0.124	0.310	101.5	3.85
5	410.9	22.2	-44.3	155.5	0.129	0.389	97.1	3.23
6	453.7	20.3	-10.9	238.0	0.122	-0.318	56.0	2.83
7	406.9	16.7	17.4	289.1	0.062	-0.108	125.2	3.83
8	340.5	12.0	63.1	100.7	0.101	0.054	144.8	3.30
9	(431)	22.1	-20.0	272.9	0.122	-0.040	217.9	17.41
10	368.0	18.4	55.6	3.5	0.085	0.617	165.2	3.00
11	382.1	21.5	-22.3	271.9	0.099	0.234	230.2	6.26
12	384.9	14.6	-57.7	110.3	0.149	0.073	187.3	4.19

U_{sw} is the bulk speed of the flux rope (km/s); B_0 is the axial field intensity at the center of the flux rope (nT); θ_a and ϕ_a are the latitude and longitude angles of the axial magnetic field (deg); R_0 is the radius of the flux rope (AU); p is the impact parameter, in unit relative to R_0; t_0 is the duration of self-similar expansion (hours); $\sigma(\Delta B)$ is the rms deviation (nT).

TABLE 3. Solar Events Found in the Time Windows

No.	Estimated Launch Time		Candidate Events[a]			
	from U_{sw}	from Vs	Date	Long	Lat	Events[b]
1	65/10/31/14		10/31	195	N25	FD (E05)
2	66/11/13/11	66/11/12/18				None
3	67/01/01/17		01/02	255	N09	FD (East Limb)
4	67/04/28/14	67/04/28/15	04/29	252	S32	FD (W40)
			04/29	205	S34	FD (Center)
5	67/12/26/12	67/12/26/19	12/26	230	S30	D (E40)
			12/27	249	S30	FD (Center)
			12/27	265	S55	FD (W10)
			12/27	195	S17	1N Flare (E60)
6	69/02/07/13	69/02/07/14	02/08	290	S29	FD (W22)[c]
7	69/08/22/08	69/08/21/20	08/21	215	N04	FD (Center)
8	71/06/18/18		06/18	59	N09	FD (E05)
9	73/03/27/22		03/27	120	N25	FD (E30)
			03/28	179	S15	DBP (27-28) (W40)
10	75/07/27/14					None
11	76/01/06/02	76/01/05/21	01/05	85	S26	DB1 (4-5) (E10)
12	78/10/25/12		10/25	228	N26	FD (E20)

[a] Solar events that could be associated with the 12 flux rope events are presented; dates of their occurrence and Carrington longitudes, and heliographic latitudes of event centers.

[b] FD indicates events taken from *McIntosh* [1979], and *Solar Geophysical Data*, NOAA; and DB indicates events taken from *Cartes Synoptiques de la Chromosphere Solaire et Catalogues des Filaments et des Centres d'Activite*, Observatoire de Paris, Meudon.

[c] The disappearance occurred during 0000-0150 UT, on February 8, 1969. [*Bulletin of Solar Phenomena, Tokyo Astronomical Observatory, Vol. 21, No. 1*]

152 INTERPLANETARY FLUX ROPES AND SOLAR FILAMENTS

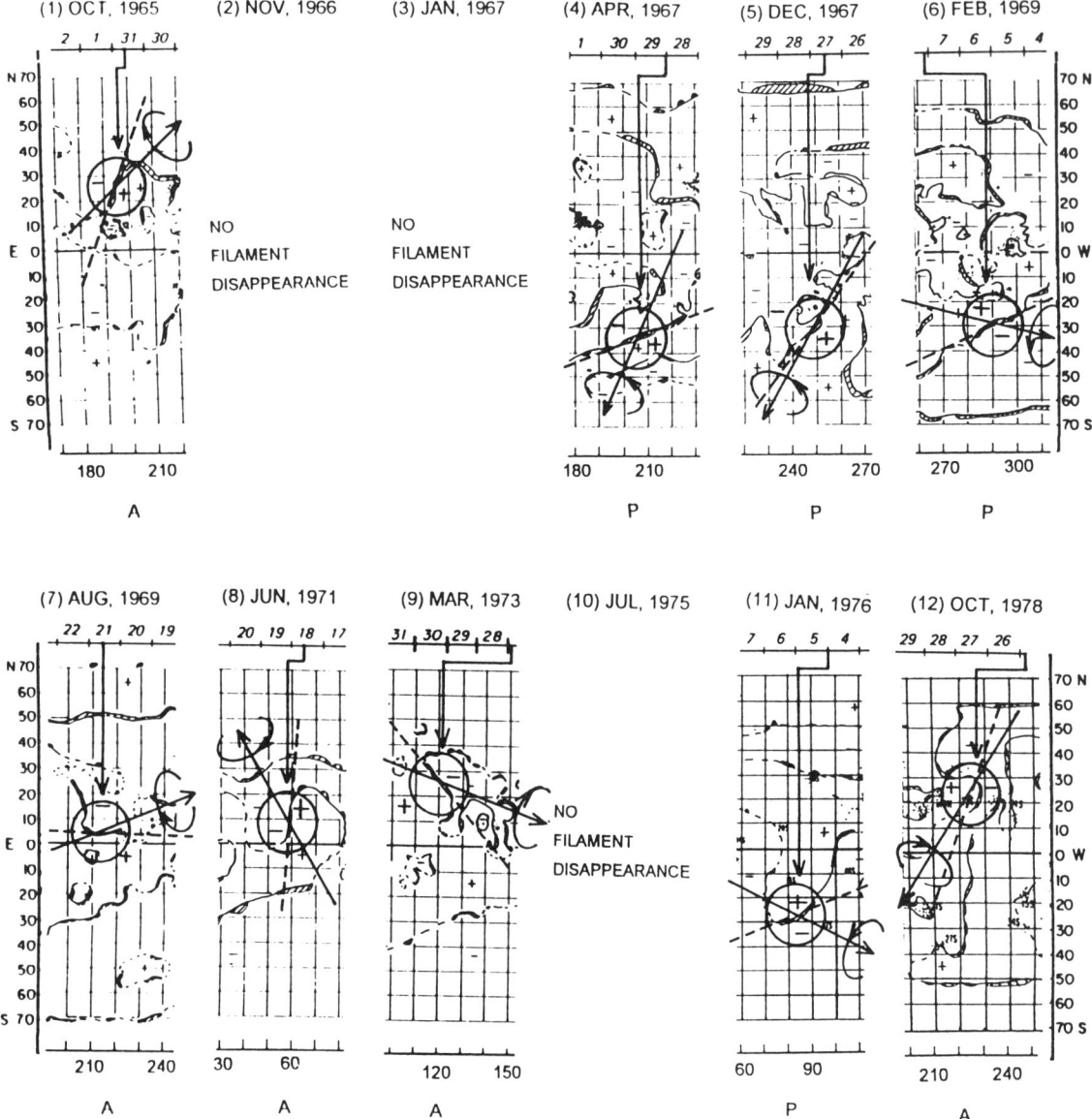

Figure 2. Relations between magnetic field structures of nine interplanetary flux ropes and the corresponding solar filaments on Hα synoptic charts showing photospheric magnetic polarities (outward +) and inversion lines. Magnetic configurations of projected flux ropes are indicated by directions of axial fields and helical fields. Arrows from the top indicate the dates when the circled filaments disappeared. The flux rope type is indicated at the bottom.

vertical solar magnetic fields. The loops are expected to be approximately potential fields in the region high above the filaments [*Martin et al.*, 1994]. (3) The direction of the flux rope axis seems to be aligned with, rather than to be perpendicular to, the filament axis. On the other hand, there is evidence for the existence of solar magnetic fields in the vicinity of the filament which are more or less parallel to filaments, indicating strongly sheared magnetic field structures [*Leroy et al.*, 1984; *Martin et al.*, 1994]. Thus it seems possible that this sheared magnetic field configuration above the filament develops into the structure similar to the projected flux rope through the combined effects of upward motion and magnetic field reconnection [*Gosling et al.*, 1995]. The results suggest that the magnetic field overlying an eruptive prominence already has a flux rope structure at the time of eruption and that the structure extends or propagates through interplanetary space with its orientation maintained and is observed as an interplanetary magnetic flux rope. Many CMEs commonly have a three-part structure: a bright frontal loop followed by a dark cavity surrounding a bright core [*Illing and Hundhausen*, 1986]. It seems reasonable to suppose that the dark cavity could be the main body of the flux rope at the time of launch, in view of the above-mentioned possible interpretation for the generation of magnetic flux ropes.

Some problems should be pointed out about the present analysis that could affect the geometrical relation between the interplanetary magnetic flux rope and the filament. First of all, the determination of flux rope geometry is completely model-dependent. Secondly, determining the boundaries of flux ropes is a problem. However, it is expected that effects of these problems would not change the orientation of flux rope axes significantly. Finally, the cylindrical structure is simply an approximation for describing the local magnetic field configuration of the flux ropes. A more precise mapping method would require knowledge about the effect of propagation and the resultant global structure. The inclusion of these effects could significantly affect the analysis result, particularly when the spacecraft traverses the flux rope away from the central part.

Figure 3 depicts the concept of an expanding flux rope proposed from the present study: (a) an event associated with eruption of an east-west oriented filament viewed from the north, (b) an event associated with a north-south oriented filament projected onto a meridional plane by rotation, and (c) the internal magnetic field configuration for the antiparallel type in an east-west view. It should be noted that the leg parts in Figures 3a and 3b are curved nearly along the Archimedian spiral when viewed from the north. The longitudinal extent in (a) and the latitudinal extent in (b) are taken to be about ±30°, based on the fact that the flux ropes were associated with disappearing filaments within about ±30° of the central meridian, though the spatial size of the interplanetary flux rope is not known yet. Radial expansion in the central part is assumed in drawing the shape of the expanding flux ropes. The connection of both ends of the loop is drawn by dashed lines, because it is still unknown whether they are connected to the Sun or disconnected from the Sun [*Burlaga et al.*, 1990; *Gosling*, 1990]. It should be mentioned here that the actual shape of a flux rope can be a combination of case (a) and case (b), as is evident from various directions of the cylinder axes listed in Table 2.

4. NEW ASPECTS ON THREE-DIMENSIONAL STRUCTURE

If the interplanetary flux ropes really have the shapes depicted in Figure 3, there should be occasions when a

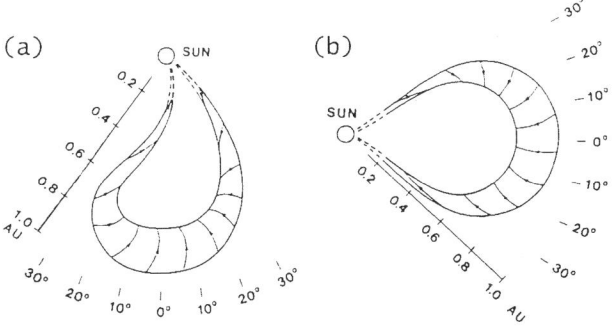

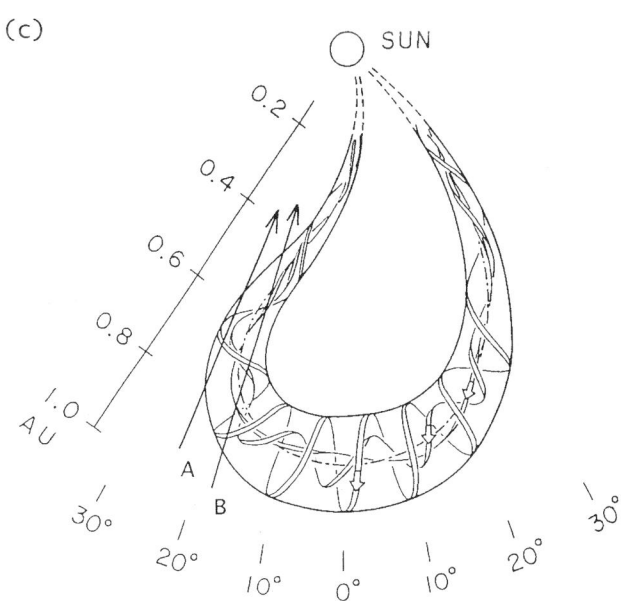

Figure 3. A schematic illustrating possible geometry of interplanetary flux ropes.

spacecraft encounters them at the curved portions, as indicated by pass A and pass B in Figure 3. Figure 4 presents one such case, in which the observed magnetic field variations can be interpreted by the passage of a spacecraft at the proximity of the outer boundary, as indicated by pass A in Figure 3. We can see a time interval of very stable magnetic fields from 1200 UT on October 4, 1979, through 0700 UT on October 5, 1979. This interval coincides with a bidirectional electron heat flow event [*Gosling et al.*, 1987], and exhibits other ejecta signatures [*Gosling*, 1990]. The magnetic field in the ecliptic plane is parallel to the Archimedian spiral. However, the large Z-component precludes an explanation in terms of simple spiral fields with negative polarity. An attempt was made to see if the magnetic field variation can be explained by the flux rope structure. For

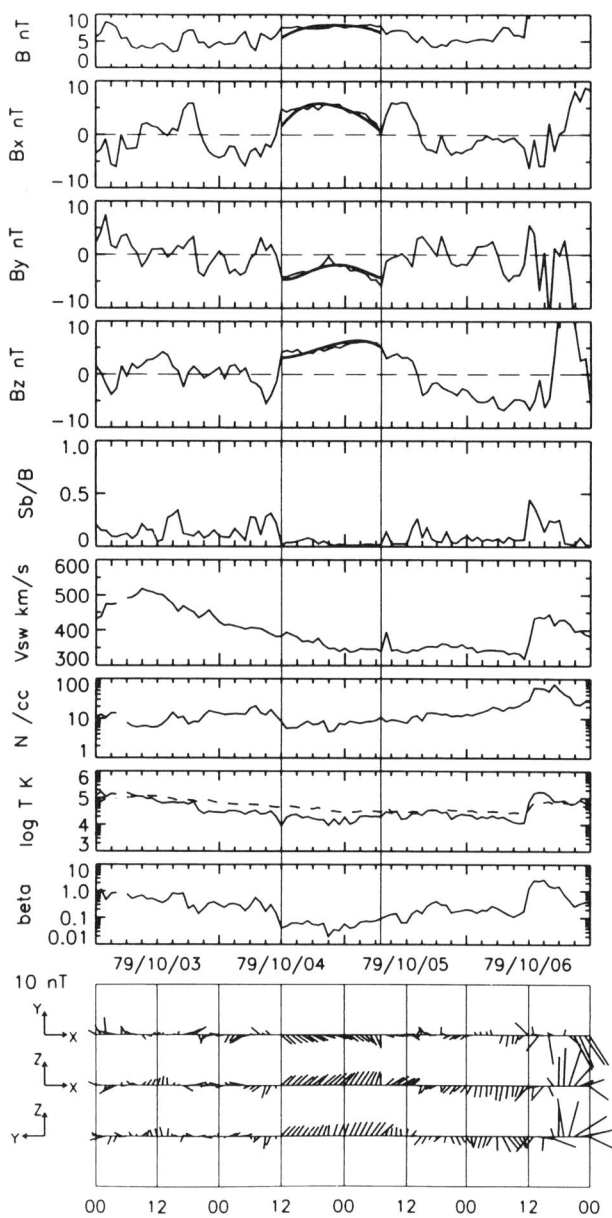

Figure 4. Same as Figure 1 except for the October 4, 1979 event. Only magnetic filed observations were used for the model fitting.

this event, it is clear that we cannot use the cylindrical model, because the direction of the observed magnetic field remains roughly unchanged. The only possible geometry for the cylinder model that provides the constant field direction is the cylinder oriented along the Sun-Earth line (X direction). Then the spacecraft would observe the constant field with its X-directed motion. In this case, however, the spacecraft can never get away from the flux rope. Thus we need a model that takes into account the curvature. Accord-

ingly, we developed a least squares fitting method applicable to a torus-shaped flux rope as an approximation for the global curved shape of flux ropes. The result of the fit is presented by thick lines. This fit was obtained for a torus with a major radius of 0.17 AU nearly vertical to the ecliptic plane and with a cross-sectional radius of 0.037 AU. This flux rope was identified as an anti-parallel type by the fitting process.

Another example is the November 17, 1975, event shown in Figure 5. This event was presented by *Klein and Burlaga* [1982] as an example showing a magnetic field rotation of nearly 360°. Again the solar wind parameters in the region between the two vertical lines exhibit signatures characteristic of ejecta [*Gosling*, 1990], namely, small magnetic field variances, low temperatures, and low proton beta. The observed magnetic field variation can be well explained by a torus model, as indicated by thick lines. The fit in this case was obtained for a torus with a major radius of 0.09 AU and with a cross-sectional radius of 0.057 AU, tilted about 34° from the ecliptic plane. The spacecraft crosses the axis twice, as indicated by pass B in Figure 3, and thus the magnetic field vectors rotate by 360°. This torus-shaped flux rope is also classified as an anti-parallel type, as is evident by the clockwise rotation in the Y-Z plane.

One question which arises here is why the flux rope sizes in these two events are much smaller than sizes of the flux ropes in Table 1. This question must be addressed in future studies, in which it is expected that we can get more details about the three-dimensional shapes of interplanetary magnetic flux ropes.

5. CONCLUSIONS AND DISCUSSIONS

We have explored the relationship between interplanetary magnetic flux ropes and solar filaments with twelve selected examples. The main findings are: (1) Interplanetary magnetic flux ropes are associated with disappearing filaments, in 9 cases out of 12; (2) The magnetic field structure of interplanetary flux ropes is consistent with the structure of solar magnetic fields in the regions overlying the disappearing filaments [*Bothmer and Schwenn*, 1994]; (3) Interplanetary flux ropes of parallel type (right-handed helicity) are associated with disappearing filaments in the southern hemisphere of the Sun, and those of anti-parallel type (left-handed) with filaments in the northern hemisphere. As a result of this study, a possible explanation for the generation of interplanetary magnetic flux ropes has been proposed. The proposed explanation is that solar magnetic fields overlying a disappearing filament already have a flux rope structure at the time of eruption, and that this structure is extended through interplanetary space to be observed as an inter-

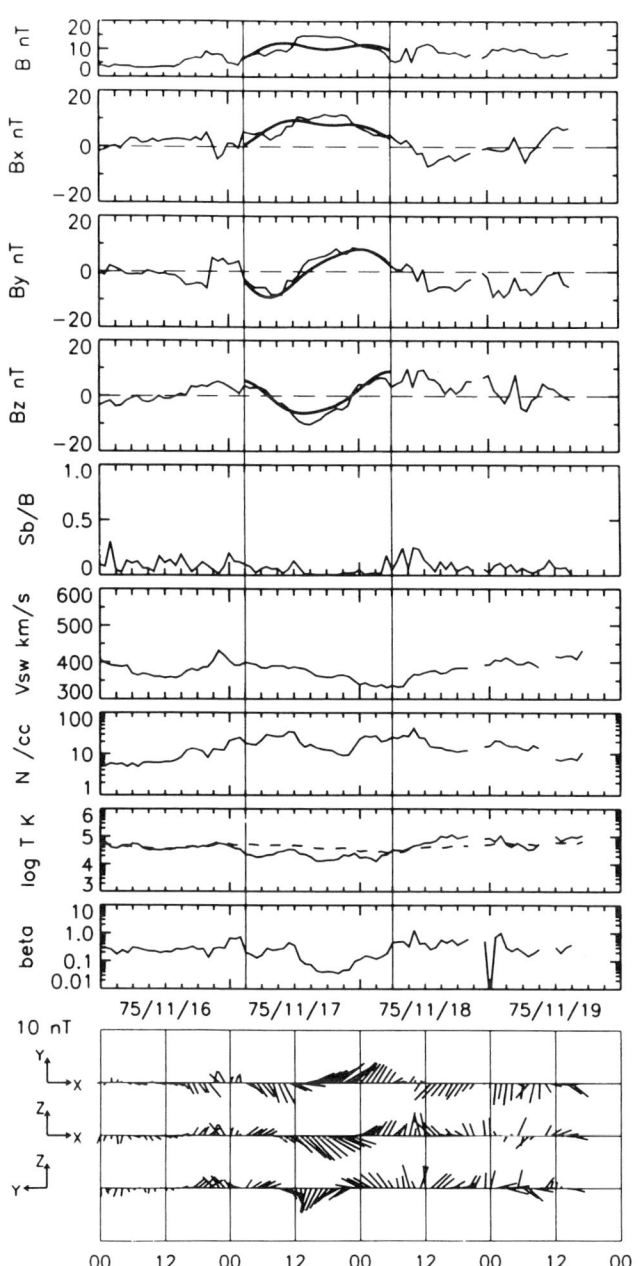

Figure 5. Same as Figure 4 except for the November 17, 1975.

planetary flux rope. Further, we have found interplanetary magnetic field variations that can be explained as spacecraft passage through a curved portion of a flux rope structure, though in these cases the flux ropes were much smaller than those treated in the earlier section.

For further clarification of the relationship between interplanetary magnetic flux ropes and solar filaments, more studies are needed to determine the global configuration of the flux ropes, such as MHD simulations and multi-spacecraft observations.

Acknowledgments. The solar wind data used in this study were supplied from the US National Space Science Data Center, NASA, Goddard Space Flight Center, through the Space Observation Data Center, ISAS, Japan. The author wishes to thank people at both of these centers for their support.

REFERENCES

Abraham-Shrauner, B. and S. H. Yun, Interplanetary shocks seen by Ames plasma probe on Pioneer 6 and 7, *J. Geophys. Res., 81,* 2097-2102, 1976.

Bothmer, V. and R. Schwenn, Eruptive prominences as sources of magnetic clouds in the solar wind, *Space Sci. Rev., 70,* 215-220, 1994.

Burlaga, L. F., Magnetic clouds and force-free fields with constant alpha, *J. Geophys. Res., 93,* 7217-7224, 1988.

Burlaga, L. F., E. Sittler, F. Mariani, and R. Schwenn, Magnetic loop behind an interplanetary shock: Voyager, Helios, and IMP 8 observations, *J. Geophys. Res., 86,* 6673-6684, 1981.

Burlaga, L. F., R. P. Lepping, and J. A. Jones, Global configuration of a magnetic cloud, in *Physics of Magnetic Flux Ropes, Geophys. Monogr. Ser.,* vol.58, edited by C. T. Russell, E. R. Priest, and L. C. Lee, pp. 373-377, AGU, Washington, D. C., 1990.

Cook, F. E. and C. G. McCue, Solar-terrestrial relations and short-term ionospheric forecasting, *Radio Electronic Engineer, 45,* 11-30, 1975.

Fainberg, J., V. A. Osherovich, R. G. Stone, R. J. MacDowall, and A. Balogh, Ulysses observations of electron and proton components in a magnetic cloud and related wave activities, in *Solar Wind Eight, AIP Conference Proc., 382,* edited by D. Winterhalter, J. T. Gosling, S. H. Habbal, W. S. Kurth, and M. Neugebauer, pp. 554-557, 1996.

Farrugia, C. J., L. F. Burlaga, V. A. Osherovich, and R. P. Lepping, A comparative study of dynamically expanding force-free, constant-alpha magnetic configurations with applications to magnetic clouds, in *Solar Wind Seven,* edited by E. Marsch and R. Schwenn, *COSPAR, vol. 3,* pp. 611-614, Pergamon, New York, 1992.

Farrugia, C. J., L. F. Burlaga, V. A. Osherovich, I. G. Richardson, M. P. Freeman, R. P. Lepping, and A. J. Lazarus, A study of an expanding interplanetary magnetic cloud and its interaction with the Earth's magnetosphere: The interplanetary aspect, *J. Geophys. Res., 98,* 7621-7632, 1993.

Goldstein, H., On the field configuration in magnetic clouds, in *Solar Wind Five, NASA Conf. Publ., 2280,* 731-733, 1983.

Gosling, J.T., Coronal mass ejections and magnetic flux ropes in interplanetary space, in *Physics of Magnetic Flux Ropes, Geophys. Monogr. Ser.,* vol.58, edited by C. T. Russell, E. R. Priest, and L. C. Lee, pp. 330-364, AGU, Washington, D. C., 1990.

Gosling, J.T., The solar flare myth, *J. Geophys. Res., 98,* 18,937-18,949, 1993.

Gosling, J.T., D. N. Baker, S. J. Bame, W. C. Feldman, and R. D.

Zwickl, Bidirectional solar wind electron heat flux events, *J. Geophys. Res., 92*, 8519-8535, 1987.

Gosling, J. T., J. Birn, and M. Hesse, Three-dimensional magnetic reconnection and the magnetic topology of coronal mass ejection events, *Geophys. Res. Lett., 22*, 869-872, 1995.

Illing, R. M. E. and A. J. Hundhausen, Disruption of a coronal streamer by an eruptive prominence and coronal mass ejection, *J. Geophys. Res., 91*, 10,951-10,960, 1986.

Kahler, S., Coronal mass ejections, *Rev. Geophys., 25*, 663-675, 1987.

Kahler, S. W., Solar flares and coronal mass ejections, Annu. *Rev. Astron. Astrophys., 30*, 113-141, 1992.

Klein, L. W. and L. F. Burlaga, Interplanetary magnetic clouds at 1 AU, *J. Geophys. Res., 87*, 613-624, 1982.

Lepping, R. P., J. A. Jones, and L. F. Burlaga, Magnetic field structure of interplanetary magnetic clouds at 1 AU, *J. Geophys. Res., 95*, 11,957-11,965, 1990.

Leroy, J. L., V. Bommier, and S. Sahal-Bréchot, New data on the magnetic structure of quiescent prominences, *Astron. Astrophys., 131*, 33-44, 1984.

Lopez, R., Solar cycle invariance in solar wind proton temperature relationships, *J. Geophys. Res., 92*, 11,189-11,194, 1987.

Martin, S. F., R. Bilimoria, an P. W. Tracadas, Magnetic field configurations basic to filament channels and filaments, in *Solar Surface Magnetism*, R. J. Rutten and C. J. Schrijver (Eds.), *NATO ASI Series C*, Vol. 433, Kluwer Academic Publ., Dordrecht, pp. 303-338, 1994.

Marubashi, K., Dynamics of solar wind plasma clouds produced by disappearing filaments, paper presented at the Joint US-Japan Seminar on Recent Advances in the Understanding of Structure and Dynamics of the Heliosphere during the Current Maximum and Declining Phase of Solar Activity, Kyoto, Japan, Nov. 5-9, 1984.

Marubashi, K., Structure of the interplanetary magnetic clouds and their origins, *Adv. Space Res., 6(6)*, 335-338, 1986.

McComas, D. J., J. T. Gosling, D. Winterhalter, and E. J. Smith, Interplanetary magnetic field draping about fast coronal mass ejecta in the outer heliosphere, *J. Geophys. Res., 93*, 2519-2526, 1988.

McIntosh, P. S., Annotated atlas of Hα synoptic charts for solar cycle 20 (1964-1974), Carrington solar rotations 1487-1616, *Report UAG-70*, WDC A for Solar-Terrestrial Physics, NOAA, Boulder, Colo., 1979.

Munro, R. H., J. T. Gosling, E. Hildner, R. M. MacQueen, A. Poland, and C. L. Ross, The association of coronal mass ejection transients with other forms of solar activity, *Sol. Phys., 61*, 201-215, 1979.

Priest, E. R., The equilibrium of magnetic flux ropes (Tutorial lecture), in *Physics of Magnetic Flux Ropes, Geophys. Monogr. Ser.*, vol.58, edited by C. T. Russell, E. R. Priest, and L. C. Lee, pp. 1-31, AGU, Washington, D. C., 1990.

Richardson, I. G. and H. V. Cane, Signatures of shock drivers in the solar wind and their dependence on the solar source location, *J. Geophys. Res., 98*, 15,295-15,304, 1993.

Rust, D. M., Spawning and shedding helical magnetic fields in the solar atmosphere, *Geophys. Res. Lett., 21*, 241-244, 1994.

Sonnerup, B. U. O. and L. J. Cahill, Jr., Magnetopause structure and attitude from Explorer 12 observations, *J. Geophys. Res., 72*, 171-183, 1967.

Vandas, M., S. Fischer, and A. Geranios, Spherical and cylindrical models of magnetic clouds and their comparison with spacecraft data, *Planet. Space Sci., 39*, 1147-1154, 1991.

Vandas, M., S. Fischer, P. Pelant, and A. Geranios, Evidence for a spherical structure of magnetic clouds, *J. Geophys. Res., 98*, 21,061-21,069, 1993.

Webb, D. F. and A. J. Hundhausen, Activity associated with the solar origin of coronal mass ejections, *Sol. Phys., 108*, 383-401, 1987.

Wilson R. M. and E. Hildner, Are interplanetary magnetic clouds manifestations of coronal transients at 1 AU?, *Sol. Phys., 91*, 169-180, 1984.

Wilson R. M. and E. Hildner, On the association of magnetic clouds with disappearing filaments, *J. Geophys. Res., 91*, 5867-5872, 1986.

K. Marubashi, Space Science Division, Communications Research Laboratory, Koganei, Tokyo 184, Japan

Magnetic Clouds

Vladimir Osherovich

Hughes STX Corporation, Lanham Maryland

L. F. Burlaga

Laboratory for Extraterrestrial Physics, NASA Goddard Space Flight Center, Greenbelt, Maryland

An interplanetary magnetic cloud is defined as a type of ejecta characterized by strong magnetic fields, a large rotation in the magnetic field direction as the cloud moves past a spacecraft, and low proton temperatures. To first approximation, a magnetic cloud can be modeled locally as a constant-α, static, force-free flux rope with cylindrical geometry. Magnetic clouds expand, and a self-similar model for an expanding cylindrical flux tube, based on an exact class of MHD solutions, describes this expansion. The model explains the observed asymmetry of the magnetic field strength profile as a consequence of the expansion (and the related decrease of the magnetic field strength with time t, $B_{max} \sim t^{-2}$) while the magnetic cloud passes the spacecraft. Asymptotically, assuming constant speed of propagation, $B_{max} \sim r^{-1/\gamma}$, where r is the distance from the sun and γ is the polytropic index, found to be close to 0.5. The prediction of the self-similar model concerning the dependence of B_{max} on distance from the sun was verified using observations of a magnetic cloud at 1 AU and at 2 AU. Since the model strongly suggests $\gamma < 1$, one expects to observe an anticorrelation between density and the temperature and the Ulysses data confirm it. We identify three plasmas, based on the relative value of the electron temperature T_e and the proton temperature T_p: 1) a magnetic cloud, where $T_e \gg T_p$ ($\gamma_e \approx 0.4 - 0.5$, and $\gamma_p \approx 1.1$); 2) the sheath that surrounds a magnetic cloud, where $T_e \ll T_p$ ($\gamma_e \approx 0.7 - 0.8$, and $\gamma_p \approx 5/3$); and 3) the typical solar wind, where T_e is comparable to T_p. A 1-fluid MHD model can approximate any one of these states, but not all three. A self-similar model with $\gamma_e \approx 0.5$ is applicable only to a magnetic cloud.

1. INTRODUCTION

The definition of a magnetic cloud includes both magnetic and thermodynamic properties [Burlaga et al., 1981]. According to this definition, the observational signature of a magnetic cloud consists of a) enhanced magnetic field strength, b) a large and smooth rotation of the magnetic field direction (a bipolar signature of one or two components) and c) low proton temperature. Other interplanetary ejecta may have one or two of these signatures, but they are not magnetic clouds, by definition. In this sense, magnetic clouds represent a subset of the broad variety of interplanetary disturbances.

A typical magnetic cloud at 1 AU is shown in Figure 1. The total magnetic field strength is significantly enhanced relative to the background. One of the two angles, the

158 MAGNETIC CLOUDS

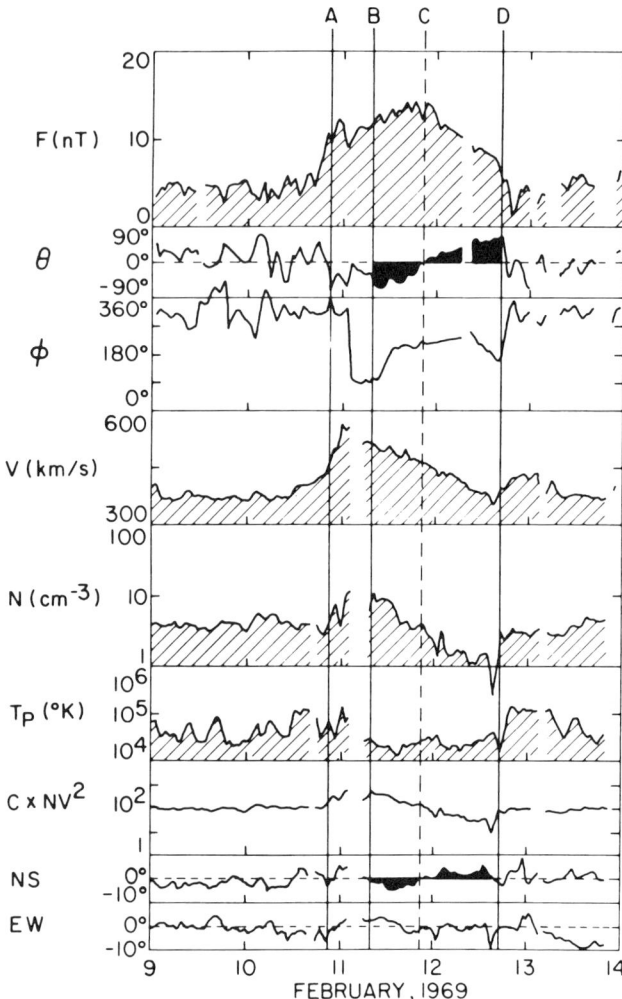

Figure 1. This figure, from Klein and Burlaga [1982] shows a magnetic cloud observed at 1 AU on February 11 - 12, 1969. The upper panel shows the magnetic field strength. The forward shock is denoted by "A", the magnetic cloud is located between B and D, and the sheath in front of the magnetic cloud is between A and B. The maximum magnetic field strength at C coincides with the change in the sign of the elevation angle θ (in solar ecliptic coordinates). The azimuthal angle φ is shown in the panel below. The bulk speed V, the proton density N, and the proton temperature T_p respectively are presented in the three middle panels. The next lower panel shows the quantity CNV^2. The final panels show the elevation angle and the azimuthal angle for the flow vector. The bipolar signature of the velocity elevation angle suggests rotation of the magnetic cloud.

elevation angle θ, changes sign as it rotates smoothly from south to north. The stronger magnetic field is accompanied by a lower proton temperature T_p. Other panels of Figure 1 illustrate a related density depletion and a monotonic

decrease of the bulk speed inside the magnetic cloud (a signature of expansion, Klein and Burlaga, 1982). In the frame of reference related to the center of the magnetic cloud, the decreasing speed profile turns into a profile of outflow from the center. The lower three panels will be discussed below. Magnetic clouds have been interpreted as loop-like configurations which are locally cylindrical [Burlaga et al., 1981; Suess, 1988; Burlaga, Lepping and Jones, 1990], and observation of the magnetic field direction as a magnetic cloud moves past a spacecraft has been interpreted as signature of a flux rope with a component B_z along the axis of the tube and with an azimuthal component B_ϕ changing it sign (Figure 2, from Goldstein, 1983). We believe 15 years of magnetic cloud research has firmly established the local approximately cylindrical geometry of magnetic clouds. Those who argue in favor of alternative geometries have not reproduced consistently the observed magnetic, dynamic, and thermodynamic signatures of a magnetic cloud. A comparison of the magnetic flux-rope model and the spheromak model has been discussed elsewhere [Farrugia, Burlaga and Osherovich, 1995a; and Farrugia, this volume]. Below we adopt locally cylindrical geometry for magnetic clouds. The early work on magnetic clouds was reviewed by Burlaga [1984, 1991, and 1995] and Farrugia, Burlaga and Lepping [1996]. A discussion of magnetic clouds is included in the review of ejecta by Gosling [1990].

There are at least two major reasons to study magnetic clouds. The first reason is that magnetic clouds are more regular than the other ejecta, and thus are more amenable to analysis. A magnetic cloud can be identified with confidence, and owing to its simplicity it can be studied analytically and modeled numerically. In other words, research on magnetic clouds, as we again demonstrate in this paper, reveals fundamental physical processes (which are not contained in the original definition of a magnetic cloud, but which have been verified through a systematic study of this object). The second reason to study magnetic clouds is their importance for Sun-Earth relations. Whereas the magnetic field in the ambient solar wind is weaker than that in the magnetic cloud and changing direction often, the magnetic field in a magnetic cloud is strong and its direction can remain southward for of the order of 12 hours. Disturbances with large southward magnetic fields allow reconnection with the magnetospheric magnetic fields [Dungey, 1961]. Thus magnetic clouds are geomagnetically effective [Burlaga et al., 1981], particularly when the magnetic field strength is enhanced by a shock passing through it or by the interaction with a solar wind stream or ejecta [Burlaga and Behannon, 1987].

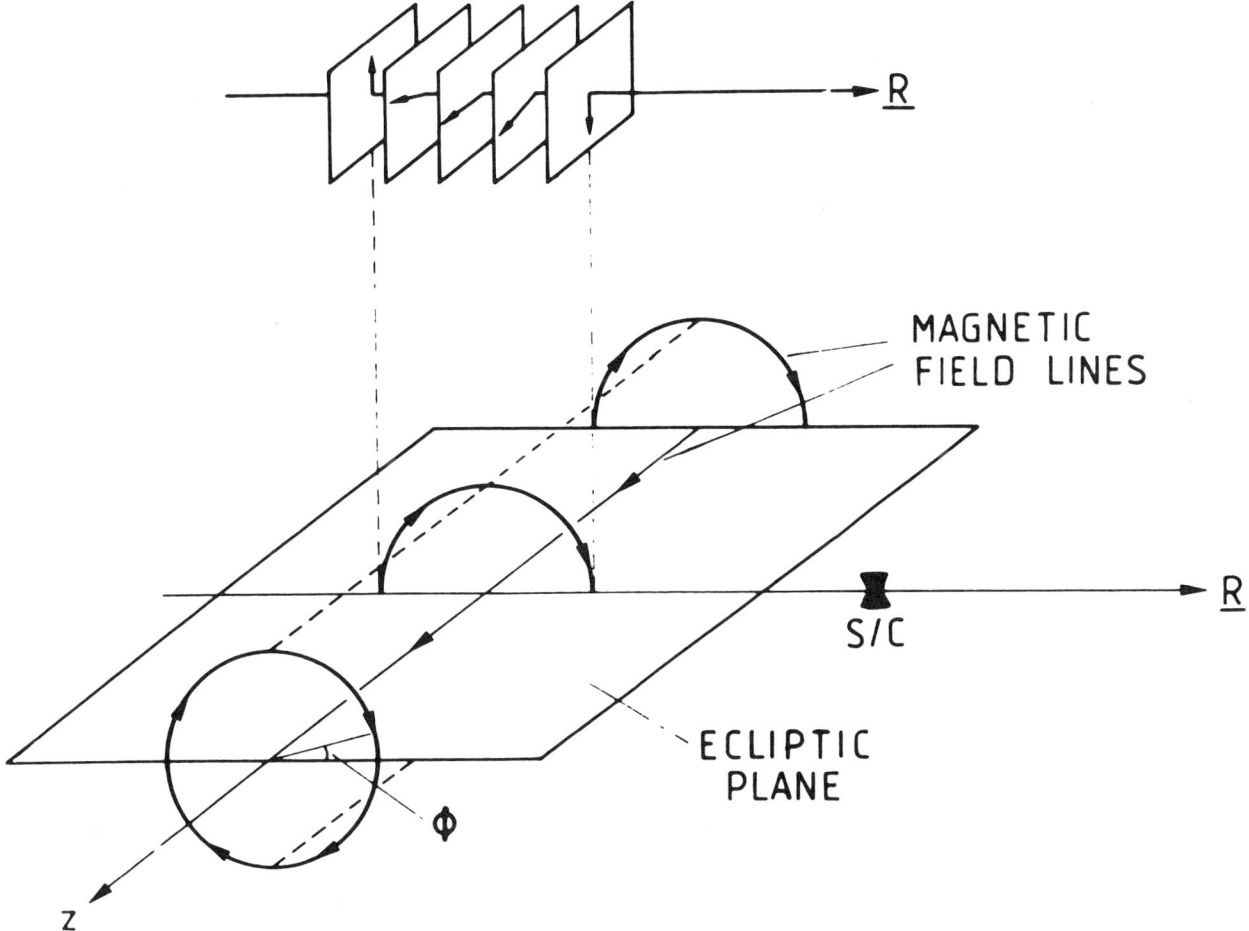

Figure 2. A cylindrical magnetic cloud with its axis in the equatorial plane perpendicular to the radial direction **r**. The magnetic field lines are circular far outside and straight lines parallel to the z-axis inside. When this configuration passes a spacecraft (S/C) a rotation of the field vector in a plane perpendicular to **r** is observed. This is illustrated in the upper part of the figure (compare with the observations sketched in Burlaga and Behannon [1982]). **r** is the direction of minimum variance, and the axis is oriented parallel to the direction of medium variance (from Goldstein, 1983).

A magnetic cloud is a polarized signal which is ideal for probing the magnetosphere [Burlaga, Lepping and Jones, 1990]. The difference between the coherent structure of a magnetic cloud and the irregular interplanetary magnetic field is similar to the difference between polarized and unpolarized light.

The outline of our paper is as follows. The next section describes a self-similar model of a magnetic flux-rope. The section establishes the relation between the early static force-free cylindrical models and the new time-dependent self-similar MHD model, and eight predictions of the new model are formulated. The third and fourth sections are devoted to the verification of the predicted properties of magnetic clouds. In the third section we present multi-spacecraft observations of a single magnetic cloud and compare the observed changes in the magnitude of the magnetic field and the shape of the magnetic field profile with the theoretical results. The fourth section contains results of the Ulysses observations of a magnetic cloud outside the ecliptic. In this section we discuss the differences between thermodynamic parameters of the magnetic cloud and the surrounding sheath, verifying the prediction of the model concerning ion-acoustic waves in the magnetic cloud. The summary and brief discussion outline the strength and limitations of the self-similar magnetic flux rope model.

2. SELF-SIMILAR MODEL OF A MAGNETIC FLUX-ROPE: A NATURAL EXTENSION OF EARLY STATIC FORCE-FREE CYLINDRICAL MODELS.

The notion that magnetic clouds have low β ($\beta \approx 0.1$ or less) lead to successful force-free modeling [Goldstein, 1983; Marubashi, 1991; Burlaga 1988]. General solutions of the force-free equation

$$\mathbf{J} \times \mathbf{B} = 0 \qquad (1)$$

or equivalently,

$$\nabla \times \mathbf{B} = \alpha \mathbf{B} \qquad (2)$$

$$\mathbf{B} \cdot \nabla \alpha = 0 \qquad (3)$$

are not available. For cylindrical geometry Schluter [1957] found a general solution, using the generating function method. For constant-α (or linear) force-free fields, the problem can be reduced to the Helmholtz equation, for which general solutions are known in different system of coordinates corresponding to different symmetries [Chandrasekhar and Kendall, 1957]. The simplicity of constant-α solutions allows one to fit the observations of the magnetic field components to the analytical solution. Good fits to the data are obtained, indicating that the constant-α, force-free cylindrical flux rope geometry is a fair description of the magnetic field geometry [Burlaga, 1988; Lepping, Jones and Burlaga, 1990]. Through such a fit, one obtains both the local orientation of a magnetic cloud and the target parameter for the trajectory of the spacecraft relative to the center of the magnetic cloud.

It is understood that force-free field models describe static magnetic field configurations, decoupled from the thermodynamic and dynamic factors. The next step in the study of magnetic clouds is to find a full nonlinear MHD solution which incorporates flows (such as expansion) and thermodynamics, while preserving the simplicity of the approach. This step has been achieved by a self-similar model of an expanding flux-rope.

Combining the only space variable R (the distance from the axis of the flux rope) and the time t, following a long-stand tradition of the self-similar approach, we introduce a self-similar parameter

$$\eta = R / y(t) \qquad (4)$$

where y(t) is an evolution function for which a dynamic equation is sought. One finds that y(t) is comparable to the radius of the flux rope. For a magnetic flux rope

$$\mathbf{B} = B_z(R,t)\mathbf{e}_z + B_\phi(R,t)\mathbf{e}_\phi \qquad (5)$$

where the components B_z and B_ϕ can be found from the vector potential

$$\mathbf{A} = A_z(R,t)\mathbf{e}_z + A_\phi(R,t)\mathbf{e}_\phi \qquad (6)$$

Assuming that A_z and the magnetic flux function $\Psi = RA_\phi$ are functions of only one variable η, we find that

$$\mathbf{B} = \nabla \times \mathbf{A} = B_o(\eta)y^{-1}(t)\mathbf{e}_\phi + B_1(\eta)y^{-2}(t)\mathbf{e}_z \qquad (7)$$

where

$$B_1(\eta) = \frac{dA_z}{d\eta} \quad \text{and} \quad B_o(\eta) = \eta^{-1}\frac{d\psi}{d\eta} \qquad (8)$$

The above formulation introduces a self-similar flux rope. It implies the important result that for any expansion, when y(t) grows monotonically, the rate of decrease with increasing time of the B_z component is faster than the rate of change of B_ϕ. A particularly simple and important case is a linear expansion of the radius of the flux tube, for which $y = y_o + V_o t$, where y_o and V_o are constants. In order to separate the variables η and t we impose an additional condition on $B_1(\eta)$ and $B_o(\eta)$, viz.,

$$\chi B_o \frac{dB_o}{d\eta} = -(B_1 / \eta)\frac{d}{d\eta}(\eta B_1) \qquad (9)$$

where χ is a constant. For this class of separable solutions, relation (9) for the static case (y(t) = y_o = constant and χ = 1) coincides with the force-free condition (1). This relation with χ = 1 allows to treat the evolution of all initially force-free fields with cylindrical geometry. Relation (9) with $\chi \neq 1$ includes configurations with gas pressure (finite β) which carry a relation between the poloidal and toroidal parts of the field that is similar to but not coincident with that for force-free fields. The complete set of MHD equations is reduced to a second order nonlinear differential equation for the evolution function [Osherovich, Farrugia and Burlaga, 1993b]

$$\frac{d^2y}{dt^2} = Sy^{-3} - Qy^{-1} + Ky^{(-2\gamma+1)} - \nu\frac{dy}{dt} \qquad (10)$$

where S, Q, K, ν and the polytropic index γ are constants. The first term Sy^{-3} describes the gradient of magnetic field energy of the component along the tube (i.e. B_z); the opposing pinch force is represented by Qy^{-1}; the gradient of gas pressure corresponds to the term $Ky^{(-2\gamma+1)}$; and the drag force in the last term is assumed to be proportional to the plasma density and the bulk speed. For K = ν = 0 there is no expansion; the magnetic flux rope oscillates around the force-free state (in which the gradient of $B_z^2/8\pi$ is exactly balanced by the pinch force). Expansion may occur only when the gas pressure is included, and only when $\gamma < 1$ [Osherovich, Farrugia and Burlaga, 1993b, 1995].

The asymptotic solution of equation (10) for $t \to \infty$ is

$$y = (2\gamma K / \nu) t^{1/2\gamma} \quad (11)$$

From this solution, assuming an unvarying average bulk speed of the center of the magnetic cloud, Osherovich, Farrugia and Burlaga [1993b] found that the variation of the maximum magnetic field strength in the magnetic cloud B_{max} with distance r from the sun is

$$B_{max} \propto r^{-1/\gamma} \quad (12)$$

For the case $\gamma = \frac{1}{2}$ (to be justified later) the time variation of the magnetic field components is

$$B_z \propto t^{-2} \quad (13)$$

$$B_\phi \propto t^{-1} \quad (14)$$

and the corresponding dependence of the maximum field strength on r is

$$B_{max} \propto t^{-2} \propto r^{-2} \quad (15)$$

Asymptotically, the velocity of expansion for the case of $\gamma = \frac{1}{2}$ is

$$\mathbf{V} = R \frac{y'}{y} \mathbf{e}_r = \frac{R}{2\gamma t} \mathbf{e}_r = \frac{R}{t} \mathbf{e}_r \quad (16)$$

which is the same as that for free linear expansion. Our recent research demonstrated that inside magnetic clouds, in which electrons are the dominant thermal component, the polytropic component is indeed below unity. At 1 AU the pressure of the electrons is 6 - 7 times greater than the pressure of the protons, and the polytropic index of the electrons is close to ½ [Osherovich et al., 1993c]. For a polytropic gas in the magnetic flux rope the relation between the gas pressure p and the plasma density ρ involves a dependence on the magnetic flux function ψ

$$p \propto F(\psi) \rho^\gamma \quad (17)$$

Assuming an ideal gas, $p \propto \rho T$, and plotting T versus ρ on a log-log scale one can recover the polytropic index only when $F(\psi)$ is a slowly varying function in comparison with the power law ρ^γ across the magnetic cloud. The importance of this for the ambient solar wind electrons was previously discussed by Sittler and Scudder [1980]. Figure 3 shows that

$$T \propto \rho^{\gamma-1} \quad (18)$$

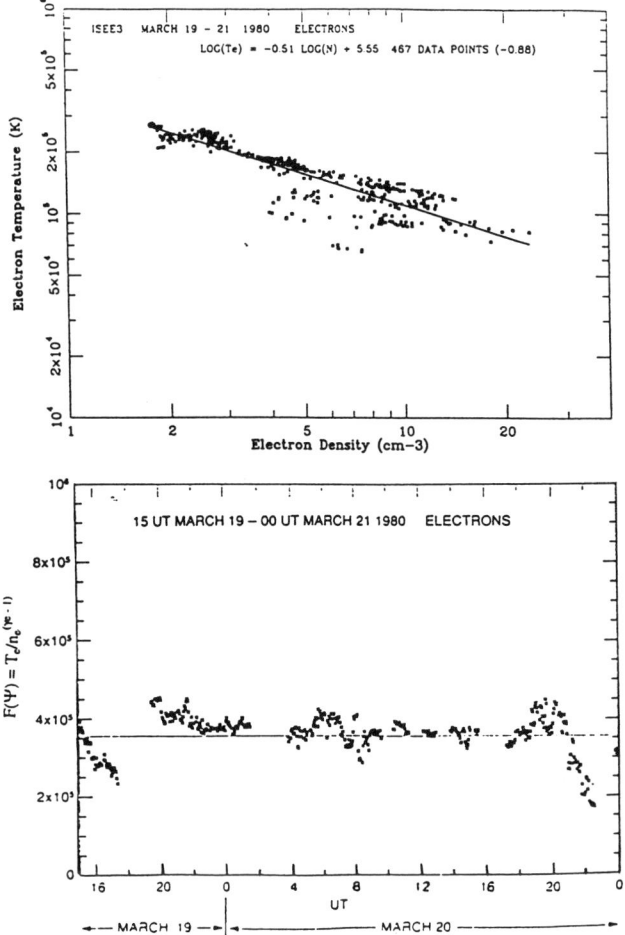

Figure 3. A scatter plot of the electron temperature T_e versus the electron density n_e for the magnetic cloud observed by ISEE-3 on March 19 - 20, showing that $\gamma_e \approx 1/2$ (upper panel). The lower panel shows that the quantity $F(\psi) = T_e/n_e^{(\gamma_e - 1)}$ has little scatter about the fitted constant value across the magnetic cloud. (Both figures are adopted from Osherovich et al., 1993).

where $\gamma \approx 1/2$ and that $F(\psi)$ stays constant across the magnetic cloud.

In the self-similar model for expansion of a magnetic flux rope, treating the most favorable case of gas pressure picked on the axis of the cylindrical tube, we concluded that the tube will not expand unless the polytropic index γ is less than 1. Since the observations of γ_e are below unity (and therefore T_e and ρ are anti-correlated) in all available magnetic clouds, the self-similar model of an expanding magnetic flux rope passes its first important test. The dependence of B_{max} on distance from the sun given by (15) is the second prediction of this model. When equation (15) predicts the dependence of the magnetic field strength on distance from the sun, two previous asymptotic relations, (13) and (14) carry information concerning the shape of the magnetic field profile. Since the total magnetic field in the flux rope is

$$B = \left(B_z^2 + B_\phi^2\right)^{1/2} \quad (19)$$

where $B_z \sim t^{-2}$ and $B_\phi \sim t^{-1}$, not only does B_{max} decrease with time but the magnetic field across the entire magnetic cloud is a monotonically decreasing function of time. This process is termed "aging" of the magnetic cloud [Osherovich, Farrugia and Burlaga, 1993b]. The effect of aging is illustrated in Figure 4 (top panel) from Farrugia, Osherovich and Burlaga [1995b], which shows snapshots of the magnetic field strength profile from (19) for a series of times at equally spaced intervals. When a spacecraft at 1 AU enters the magnetic cloud, the magnetic cloud is young (after traveling approximately 60 hours from the sun) and the magnetic field is relatively strong. When the spacecraft is leaving the magnetic cloud (after approximately 30 hours inside the magnetic cloud) the magnetic cloud is 50% older; therefore on the same magnetic surface at the end of the magnetic cloud the magnetic field is significantly weaker. As Figure 4 shows, sliding from one snapshot profile to another (with a weaker field) leads to an asymmetry of the magnetic field strength profile with an enhancement in front of the magnetic cloud. It is understood that the interaction of the magnetic cloud with the ambient solar wind might also contribute to the asymmetry, as suggested by Klein and Burlaga, [1982]. However, to a large degree the asymmetry can be accounted for by the aging process [Osherovich, Farrugia and Burlaga, 1993b; Farrugia, Osherovich and Burlaga, 1995b]. The asymmetry in the magnetic field strength profile at a given distance from the sun is the third prediction of our model.

The ratio of the time that a spacecraft is inside the magnetic cloud to the lifetime of the magnetic cloud is much smaller at large distances than at 1 AU. The magnetic field strength profiles of magnetic clouds should be more symmetric at 4 - 5 AU than at 1 AU (the fourth prediction of the model).

According to (19), the magnetic field strength will decrease with increasing time. However, the component along the flux rope, B_z, asymptotically decreases as t^{-2}, while the azimuthal component decreases only as t^{-1}. For young clouds, $B_{\phi max}$ can be as small as $(1/2) B_{z max}$ at 1 AU, but B_z decreases faster than B_ϕ. Since B_ϕ contributes to the total magnetic field strength profile mostly in the wings of the B distribution, for older magnetic clouds one should

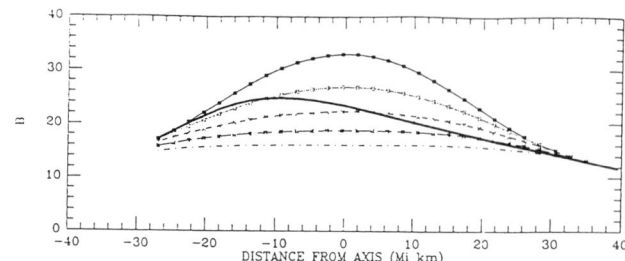

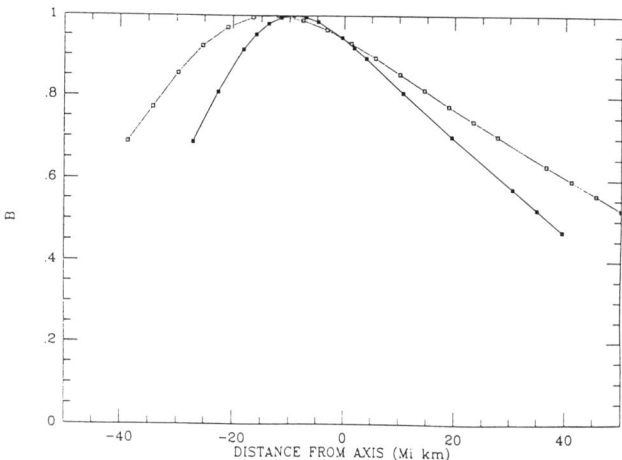

Figure. 4. (Top panel) Snapshots of (symmetric) magnetic field profiles at equal intervals of time, and the resultant asymmetric magnetic profile seen by a spacecraft passing through the magnetic cloud (heavy trace). (Lower panel) Normalized magnetic field strength profiles for the same magnetic cloud at two different times approximately one day apart. The flatter profile (open squares) corresponds to the later interval. (After Farrugia, Osherovich and Burlaga, 1995b).

expect a flattening of the magnetic field strength profile with increasing time as illustrated by the two curves in the lower panel of Figure 4 (see the Figure caption). The flattening of the magnetic field strength profile is our fifth prediction.

To the list of unusual properties of magnetic clouds which the self-similar model suggests, we add high electron temperatures and ion-acoustic waves. Expansion for ions ($\gamma_p > 1$) causes cooling, whereas expansion for electrons ($\gamma_e < 1$) implies an increase of the electron temperature T_e. For T_e/T_p of the order of unity, Landau damping is effective for the ion acoustic waves. At 1 AU, T_e/T_p is approximately 6 to 7 in a magnetic cloud [Osherovich et al, 1993c], and at a few AU the ratio T_e/T_p should be even greater (the sixth prediction). Therefore, one should expect to observe ion-acoustic waves in magnetic clouds, especially at a few AU (the seventh prediction).

So far we have outlined magnetic and thermodynamic properties of magnetic clouds suggested by the self-similar model. Dynamic properties showed up in the speed profile (16). As simple as it looks, this theoretical speed profile, when fitted to the speed observations, should give the age of the magnetic cloud (the time from the moment when self-similar expansion started) and a snapshot of the size of the magnetic cloud at the moment when the spacecraft enters the magnetic cloud. Goodness of such fits and compatibility of the age of the magnetic cloud with the estimate of the time needed to travel from the sun to the spacecraft provide the eighth test of the model [Farrugia, Osherovich, and Burlaga, 1995b]. All except the last prediction will be tested in the next two sections. The fit to the speed profile and related discussion can be found in the paper by Farrugia [this volume].

Rotation of some magnetic clouds was suggested by Klein and Burlaga [1982], as illustrated in the second panel from the bottom of Figure 1 from that paper. The rotation of plasma around a flux rope axis can be incorporated in a self-similar model. The dynamic equation for the self-similar model of a rotating magnetic cloud differs slightly from equation (10) [Farrugia, Osherovich and Burlaga, 1995b].

3. MULTISPACECRAFT OBSERVATIONS OF ONE MAGNETIC CLOUD. THE BEST TEST OF THE MODEL.

The predictions of our self-similar model concerning the magnitude of B and the shape of the magnetic field strength profile are specific. Given that $\gamma \approx 1/2$, according to (15) $B_{max} \sim t^{-2} \sim r^{-2}$. Figure 5 shows the total magnetic field strength profiles of a magnetic cloud observed first by Helios 2 at 1 AU and later by Voyager 2 at 2 AU [Burlaga et al., 1981]. The maximum magnetic field strength decreased from 20 nT at 1 AU to 5 nT (the solid line in Figure 5), as expected according to equation (15). Indeed, the B profile is more flattened and the asymmetry of the B profile is less prominent at 2AU than at 1 AU. Recently, our general formula for B_{max} (12) has been confirmed by Vandas et al. [1995]. Starting their numerical code with $\gamma = 5/3$ this group arrived at the following result:

$$B_{max} \sim r^{-3/2} \sim r^{-1/0.7} \qquad (20)$$

However, neither the case of $\gamma = 5/3$ (the dashed line in Figure 5) nor the case of an effective polytropic index $\gamma = 0.7$ suggested by Vandas et al. [1995, 1996] (the dot-

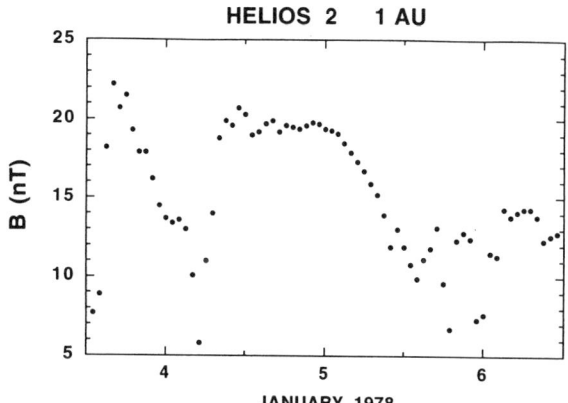

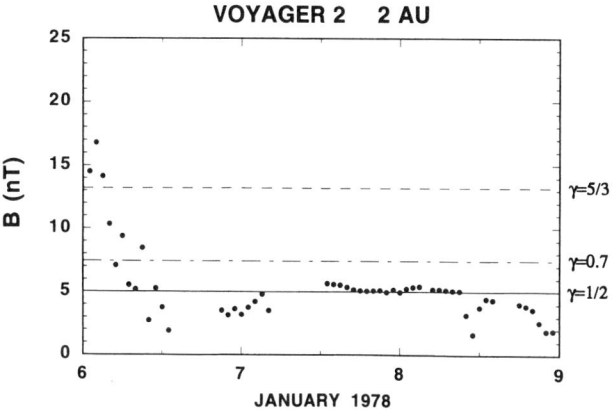

Figure 5. The decrease of B_{max} for the magnetic cloud seen at 1 AU (by Helios-2) and at 2 AU (by Voyager 2) follows formula (12) with $\gamma = 1/2$. (solid line). Neither the case $\gamma = 5/3$ (dashed line) nor the case $\gamma = 0.7$ (dot-dashed line) fits the observations. Adapted from Burlaga et al., 1981)

dashed line in Figure 5) fits the observations. In our model, γ is not a free parameter that can be chosen to fit the magnetic field observations. The polytropic index γ has to be determined from the observations of the density ρ and the temperature T for the dominant component. For those who, starting with $\gamma > 1$ and varying a number of parameters in their model, finally obtain satisfactory results for the magnetic field with some effective $\gamma < 1$, the observed anticorrelation between T and ρ remains a challenge. For the self-similar flux rope model, the anticorrelation between T and ρ results from expansion, which is possible only for a real polytropic gas with $\gamma < 1$.

4. THERMODYNAMIC AND MAGNETIC STRUCTURES OF OLD MAGNETIC CLOUDS OBSERVED AT 4.64 AU AND S. 32.5° BY ULYSSES. ION ACOUSTIC WAVES IN THE MAGNETIC CLOUD

Since the evolution of magnetic and thermodynamic fields provides the ultimate test of the theory, we turn to an "old" magnetic cloud, which after traveling from the sun for almost 12 days was observed by the Ulysses spacecraft at 4.64 AU from June 10 to June 13, 1993 [Armstrong et al., 1994; Gosling et al., 1994; Fainberg et al., 1996]. Properties of this enormous structure are depicted in Figure 6. With the arrival of a shock on June 9, the spacecraft entered the sheath. The magnetic field strength increases by a factor of two across the shock. The magnetic field vector rotates as the sheath moves past the spacecraft, but a smooth rotation of the magnetic field vector begins only after passage of the front boundary of the magnetic cloud on June 10. After three days of smooth rotation of **B**, the spacecraft enters the sheath which ends a day later with the passage of a reverse shock immediately after June 14. The smoothness of the magnetic field allowed us to identify the boundaries of the magnetic cloud (the dashed lines on June 10 and 13). As predicted, the magnetic field strength profile is flat in this old magnetic cloud. As predicted, this profile is rather symmetric compared to the profile for a young magnetic cloud shown in Figure 1.

The rotation of **B** in Figure 6 (top panel) is shown in the **n-t** plane, where the **n** and **t** directions are perpendicular to the direction **r** pointed radially away from the sun. As shown in Figure 6 (middle panel) the normal component of **B**, B_n, changes sign in the middle of the magnetic cloud (like B_φ in Figure 2). The other components B_r and B_t do peak in the central part of the magnetic cloud without changing signs in the magnetic cloud. Therefore, we combine them in one component $(B_r^2 + B_t^2)^{1/2}$ (Figure 6, middle panel) which presumably is along the flux tube (like B_z in Figure 2). In this way we see that it is the large azimuthal component (with a bipolar signature) which, contributing greatly in the wings of the B-distribution, makes the whole B profile flat, as we described in Section 2.

It is clear that the magnetically defined boundaries of the magnetic cloud are confirmed by a sharp decrease in T_p and a steep rise in T_e (Figure 6, lower panel). Since $T_e \gg T_p$ in the magnetic cloud, the electron component is the major contributor to the total gas pressure (by a margin of 10 - 20 through most of the magnetic cloud). A strong anticorrelation between T_e and ρ in the magnetic cloud is illustrated in Figure 7. The striking detailed anticorrelation observed is a consequence of a polytropic relation with γ_e

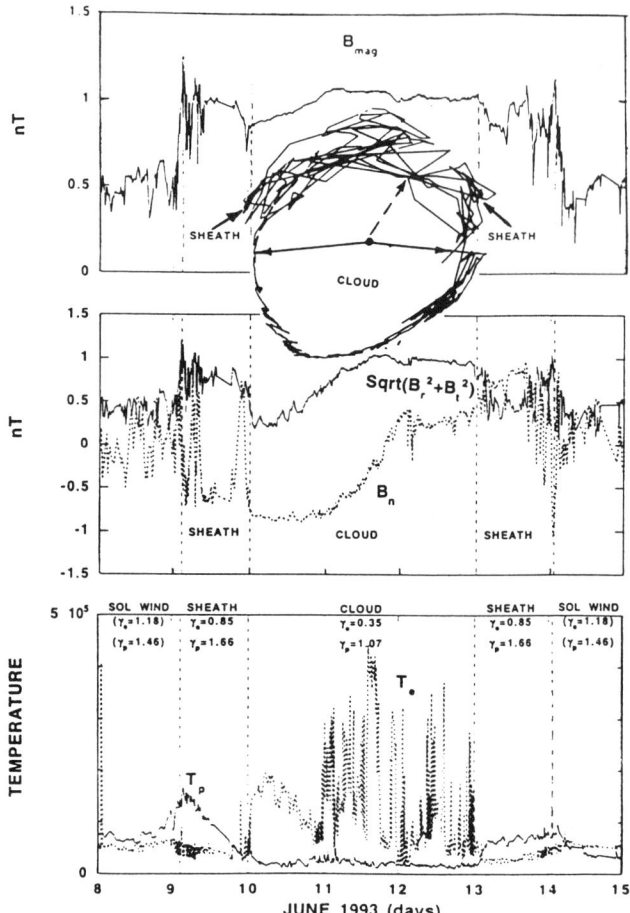

Figure 6. (Top panel) A magnetic cloud observed at 4.64 AU at 32.5° by Ulysses. The magnetic field strength profile is flat. The rotation of the magnetic field vector in the **n-t** plane in the leading sheath and the trailing sheath (separated by a dashed arrow) is not smooth; the rotation inside the magnetic cloud is smooth. (Lower panel). T_e and T_p shown in the magnetic cloud ($T_e \gg T_p$), in the sheath ($T_e \ll T_p$) and in the solar wind (T_e and T_p are comparable). The polytropic indices in these three plasmas differ significantly. The values of γ_e and γ_p shown in the solar wind (in brackets) are taken from the literature.

< 1 between the electron pressure and density illustrated by the open circles in Figure 8. For comparison, a polytropic law for protons with $\gamma_p = 1.03$ is shown by the solid circles in Figure 8. In fact, Fainberg et al. [1996] have shown that a magnetic separatrix divides this magnetic cloud into two non-equal parts with slightly different γ_e, suggesting a topology which is more complex than a single flux tube. The polytropic indices presented in Figure 6 (bottom panel) are averages over these two regions, which are presumably two flux ropes. After averaging between the

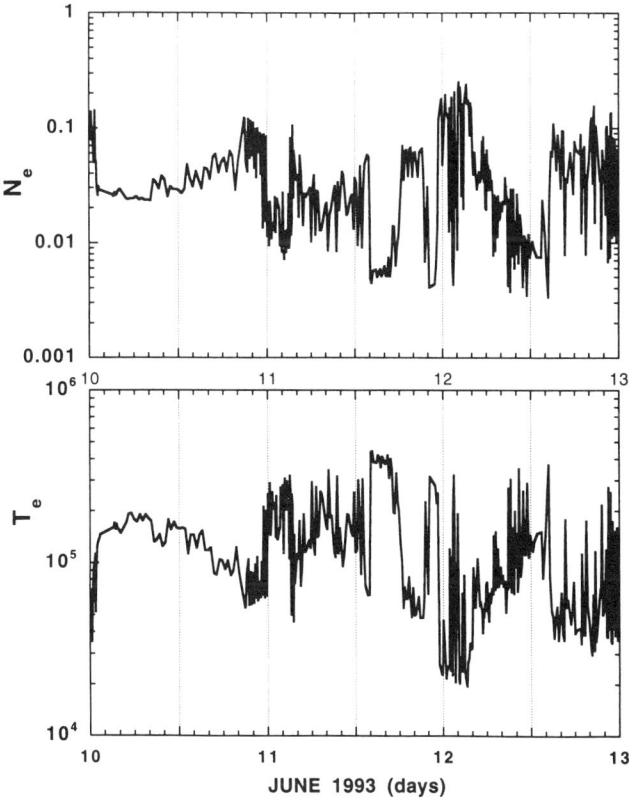

Figure 7. The anti-correlation between T_e and n_e in the magnetic cloud is in compliance with the prediction of the self-similar model of a magnetic flux rope. (After Fainberg et al., 1996).

two sheath regions γ_p is found to be close to 5/3 and γ_e = 0.85. Thus, the dense sheath with a dominant proton component is adiabatic. Outside the sheath T_e and T_p in the ambient solar wind are comparable (but different). Figure 9 (top panel) shows that T_e/T_p is below unity in the sheath and rises to 10 - 20 at its maximum on the boundary of the magnetic cloud. As discussed in Section 2, Landau damping is not effective in such plasma and these conditions are favorable for ion-acoustic waves, which indeed have been observed in this magnetic cloud (Figure 9, lower panel, after Stone et al., 1995). Note that in the sheath, $T_e/T_p \ll 1$ and Landau damping completely suppresses wave activity.

5. SUMMARY AND DISCUSSION

We have analyzed a number of predictions of a self-similar model of an expanding cylindrical magnetic flux rope applied to magnetic clouds. The evolution of B_{max} and the profile of B predicted by the model are in good agreement with multispacecraft observations at 1 AU and 2 AU. The unusual prediction of the model that magnetic clouds are polytropic flux tubes with $\gamma < 1$ has been verified at 1 AU and at 4.6 AU. The resulting anticorrelation between T_e and N_e has been found in every magnetic cloud that has been analyzed. An analysis of the profile of T_e across a magnetic cloud relative to the density profile [Osherovich et al., 1993c] established an anticorrelation between T_e and N_e corresponding to a polytropic law with $\gamma \approx \frac{1}{2}$.

The electron temperature T_e is much greater than the proton temperature in magnetic clouds [Burlaga et al., 1981]. Thus, the polytropic exponent below unity in magnetic clouds reflects the properties of the electrons. To first approximation, the gas in a magnetic cloud can be regarded as a single fluid, where the electron component is the main contributor to the gas pressure, and this component makes this fluid polytropic with $\gamma < 1$.

Ion acoustic waves are found in magnetic clouds. In retrospect, the ion acoustic waves observed by Helios 1 on December 1, 1977 [Burlaga et al, 1980] were observed in a magnetic cloud. Ion acoustic waves were also found in the June 1993 magnetic cloud. The presence of ion acoustic waves in magnetic clouds is attributed to the fact that $T_e/T_p \gg 1$ in magnetic clouds, which implies that Landau damping of the waves is ineffective there.

The observations of the June 1993 magnetic cloud illustrate the important fact that three different plasma have to be considered in modeling a magnetic cloud moving through the solar wind: 1) the plasma inside the magnetic cloud itself, where $T_e/T_p \gg 1$ ($\approx$ an electron fluid with $\gamma_e <$ 1); 2) the plasma in the sheath surrounding the magnetic cloud, where $T_e/T_p \ll 1$ ($\approx$ a proton fluid with $\gamma_p \approx 5/3$); and the ambient solar wind plasma, where T_e and T_p are different but comparable ($\approx$ a 'single' fluid with effective temperature $T_e + T_p$). For the ambient solar wind, Sittler and Scudder [1980] obtained the empirical value $\gamma_e \approx 1.18$ and Trotten, Freeman and Arya [1995] found $\gamma_p \approx 1.46$ for the protons. (See also the other references on polytropic exponents in this paper). Presumably, more studies of the polytropic laws in the solar wind plasma will be forthcoming. We expect differences between fast and slow solar wind flows as well as a dependence of γ_e and γ_p on distance from the sun and on heliospheric latitude. However, with the data available now we can definitely say that the polytropic indices obtained in the solar wind are different from those in a magnetic cloud and in the sheath (Figure 6, bottom panel). In other words, any attempt to use one fluid MHD models with a single polytropic index γ for all three plasmas is not physically realistic.

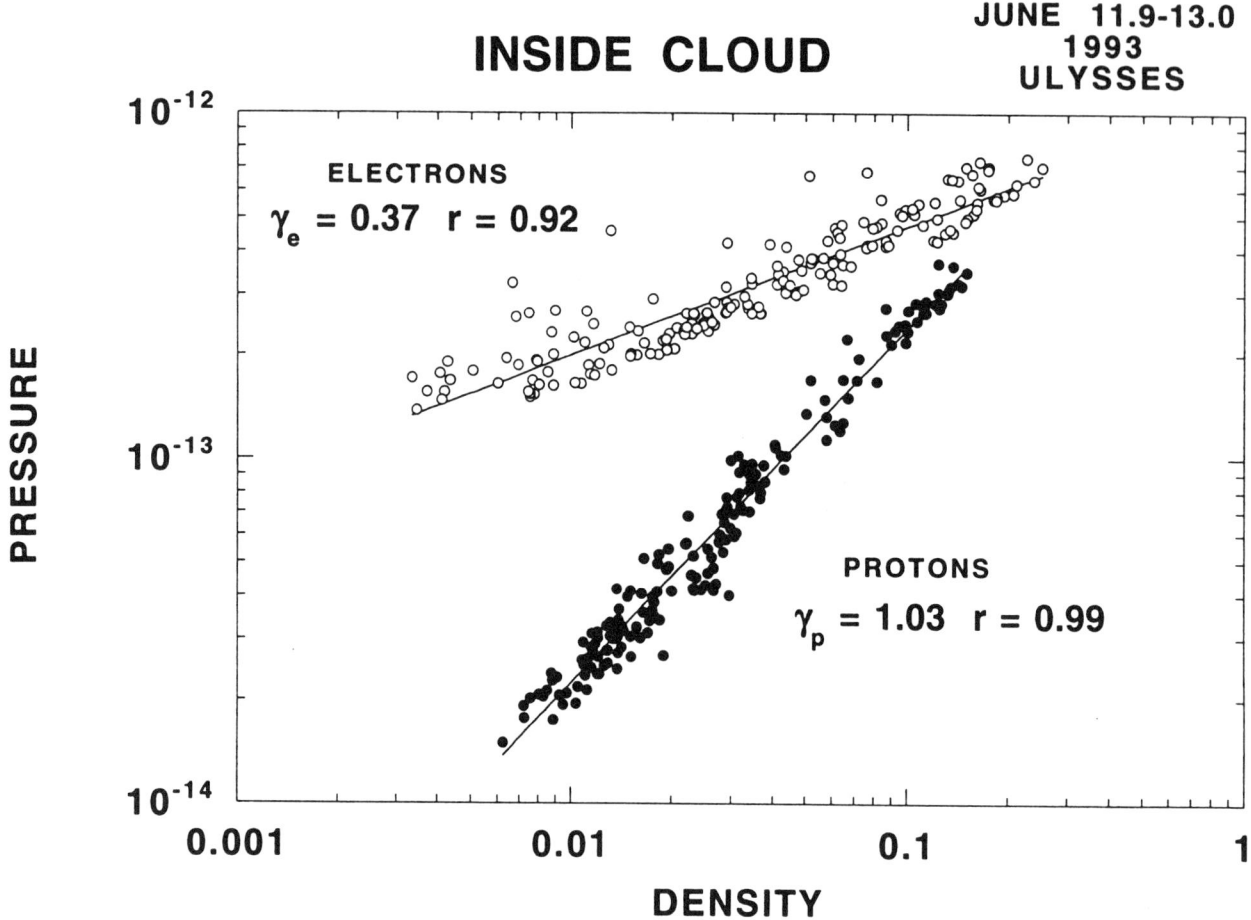

Figure 8. Polytropic relations between the electron pressure and the electron density (open circles) and between the proton pressure and proton density solid circles) for one of two parts of the magnetic cloud. For the second part of the magnetic cloud the relation is similar, with a slightly different γ_e but with significantly different $F(\Psi)$.

Gosling et al. [1994] modeled the forward-reverse shock pair for the June 1993 magnetic cloud assuming a single fluid with $\gamma = 5/3$. As shown by Chen and Garren [1993], the adiabatic expansion of magnetic clouds with $\gamma = 5/3$ leads to magnetic cloud temperatures of approximately 3°K at 1 AU, while the observed temperatures are $10^4 - 10^5$ °K. From our point of view, such modeling is valid only in the sheath, where protons dominate and $\gamma_p = 5/3$. However, such a 1-fluid approach with $\gamma_p = 5/3$ does not model the magnetic cloud. By the same token, the self-similar model with $\gamma < 1$ is applicable only to the magnetic cloud, unless it is modified to match the conditions outside the magnetic cloud. It has been known for five decades that self-similarity is limited to within a certain radius, beyond which self-similarity breaks down and a shock is formed [see, e.g., the discussion in Osherovich, Farrugia and Burlaga, 1993a]. The ways to match the self-similar solution with a non self-similar one at the contact surface have been described in the literature [Bernstein and Kulsrud, 1965; Low, 1984]. Further development of our model shall follow this tradition.

We depicted a magnetic cloud as a self-similar expanding magnetic flux rope, adiabatically isolated from the ambient solar wind by a dense sheath region. Since there is no heat transfer through the sheath, the only way of heating the plasma in a magnetic cloud is along the flux rope. There is enough observational evidence to support the idea that at least one "foot" of the magnetic loop is connected to the sun [Kahler and Reames, 1991; Richardson et al., 1991; Farrugia et al., 1993 a, b]. All of the above means that in observing magnetic clouds we are sampling expanding coronal loops with unusual

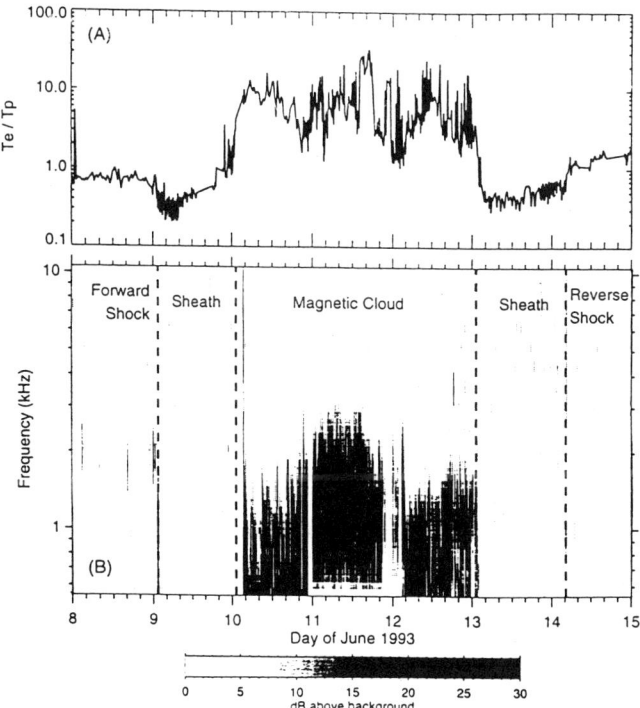

Figure 9. Wave activity is observed in and only in the magnetic cloud, where $T_e/T_p \sim 10 - 20$ and Landau damping is not efficient. There are no waves in the sheath, where T_e/T_p drops below unity. The observed frequencies correspond to those of Doppler shifted ion-acoustic waves.

thermodynamic properties ($\gamma_e < 1$). In this case the suggestion of Scudder [1992] that the low-density non-maxwellian plasma in the inhomogeneous corona may have a polytropic index below unity should be applicable to magnetic clouds. Indeed, Fainberg et al. [1996] verified that, in the June 1993 magnetic cloud observed by Ulysses, γ was indeed less than 1 and the contribution of halo electrons (the non-maxwellian fraction) to the total gas pressure was comparable to the contribution of the core electrons (the maxwellian part). Outside the magnetic cloud, the core electrons dominated. The ability to sample in situ non-maxwellian coronal plasma confined by the magnetic fields of a flux rope provides a third strong reason to study interplanetary magnetic clouds.

Acknowledgments. We acknowledge valuable discussions with J. Fainberg, C. Farrugia and R. Stone. Fainberg and Farrugia also helped in preparing some of the figures. One of us (VO) received partial support from a NASA Space Physics Theory Program grant to the Goddard Space Flight Center.

REFERENCES

Armstrong, T. P., et al., Observations by Ulysses of hot (~270 keV) coronal particles at 32° south heliolatitude and 4.6 AU, *Geophys. Res. Lett., 21*, 17, 1747, 1994.

Bernstein, I. B., and R. M. Kulsrud, On the explosion of a supernova into the interstellar magnetic field, 1, *Ap. J., 142*, 479, 1965.

Burlaga, L. F., MHD processes in the outer heliosphere, *Space Sci. Rev., 39*, 255, 1984.

Burlaga, L. F., Magnetic clouds and force-free fields with constant alpha, *J. Geophys. Res., 93*, 7217-7224, 1988.

Burlaga, L. F., Magnetic clouds, Chapter 6 in *Physics of the Inner Heliosphere*, Vol. 2, edited by R. Schwenn and E. Marsch, p. 1, Springer-Verlag, Berlin-Heidelberg, 1991.

Burlaga, L. G., *Interplanetary Magnetohydrodynamics*, Oxford University Press, New York, 1995.

Burlaga, L. F. and K. W. Behannon, Magnetic clouds: Voyager observations between 2 and 4 AU, *Solar Physics, 81*, 181, 1982.

Burlaga, L. F., and K. W. Behannon, Compound streams, magnetic clouds and major geomagnetic storms, *J. Geophys. Res., 92*, 5725, 1987.

Burlaga, L. F., R. Lepping, and J. Jones, Global configuration of a magnetic cloud, *Physics of Flux Ropes*, p. 373, Edited by C. T. Russell, E. R. Priest and L. C. Lee, AGU Geophysical Monograph 58, American Geophysical Union, Washington D. C., 1990.

Burlaga, L. F., R. Lepping, R. Weber, T. Armstrong, C. Goodrich, J. Sullivan, D. Gurnett, P. Kellogg, E. Keppler, F. Mariani, F. Neubauer, H. Rosenbauer, and R. Schwenn, Interplanetary particles and fields, November 22 to December 6, 1977: Helios, Voyager and Imp observations between 0.6 AU and 1.6 AU, *J. Geophys. Res., 85*, 2227, 1980.

Burlaga, L. F., E. Sittler, F. Mariani, and R. Schwenn, Magnetic loop behind an interplanetary shock: Voyager, Helios and IMP-8 observations, *J. Geophys. Res., 86*, 6673, 1981.

Chandrasekhar, S., and P. C. Kendall, On force-free magnetic fields, *Astrophys. J., 126*, 457-460, 1957.

Chen, J., and D. A. Garren, Interplanetary magnetic clouds: Topology and driving mechanism, *Geophys. Res. Lett., 20*, 2319, 1993.

Dungey, J. W., Interplanetary magnetic field and the auroral zones, *Phys. Rev. Lett., 6*, 47, 1961.

Fainberg, J., V. A. Osherovich, R. G. Stone, R. J. MacDowall, and A. Balogh, Ulysses observations of electron and proton components in a magnetic cloud and related wave activity, in *Solar Wind Eight*, to appear, 1996.

Farrugia, C. J., Recent work on modeling the global field line topology of interplanetary magnetic clouds, this volume.

Farrugia, C. J., I. G. Richardson, L. F. Burlaga, R. P. Lepping, and V. A. Osherovich, Simultaneous observations of solar MeV particles in a magnetic cloud and in the earth's northern tail lobe: Implications for the global field line topology of magnetic clouds, and for the entry of solar particles into the magnetosphere during cloud passage, *J. Geophys. Res., 98*, 15,497, 1993a.

Farrugia, C. J., L. F. Burlaga, V. A. Osherovich, I. G. Richardson, M. P. Freeman, R. P. Lepping, and A. J. Lazarus, A study of an expanding interplanetary cloud and its interaction with the Earth's magnetosphere: The interplanetary aspect, *J. Geophys. Res., 98,* 7621, 1993b.

Farrugia, C. J., V. A. Osherovich, and L. F. Burlaga, The self-similar, nonlinear evolution of rotating magnetic flux ropes, *Annales Geophysicae, 13,* 815, 1995a.

Farrugia, C. J., V. A. Osherovich, and L. F. Burlaga, The magnetic flux rope versus the spheromak as models for interplanetary magnetic clouds, *J. Geophys. Res., 100,* 12, 293, 1995b.

Farrugia, C. J., L. F. Burlaga, and R. P. Lepping, Magnetic clouds and the quiet-storm effect at earth, in *Proceedings of the Chapman Conference on Magnetic Storms,* Jet Propulsion Laboratory, Pasadena, February 12 - 16, 1996, to appear, 1996.

Goldstein, H., On the field configuration in magnetic clouds, in *Solar Wind Five,* edited by M. Neugebauer, *NASA Conf. Publ., NASA CP-2280,* 731-733, 1983.

Gosling, J. T., Coronal mass ejections and magnetic flux ropes in interplanetary space, in *Physics of Magnetic Flux Ropes, Geophys. Monogr. Ser.,* vol. 58, edited by C.T. Russell, E.R. Priest, and L.C. Lee, p. 343, AGU, Washington, D. C., 1990.

Gosling J. T., S. J. Bame, D. J. McComas, J. L. Phillips, E. E. Sciane, V. J. Pizzo, B. E. Goldstein, and A. Balogh, A forward-reverse shock pair in the solar wind driven by over-expansion of a coronal mass ejection: Ulysses observations, *Geophys. Res. Lett.* 21,237, 1994.

Klein, L. W., and L. F. Burlaga, Interplanetary magnetic clouds at 1 AU, *J. Geophys. Res.,*87, 613, 1982.

Kahler, S. W. and D. V. Reames, Probing the magnetic topologies of magnetic clouds by means of solar energetic particles, *J. Geophys. Res., 96,* 9419, 1991.

Low, B. C., Self-similar magnetohydrodynamics. III. The subset of spherically symmetric gasdynamic flows, *Ap. J., 281,* 381, 1984.

Marubashi, K., Interplanetary magnetic flux ropes observed by the Pioneer-Venus Orbiter, *Adv. Space Res., 11(l),* 57-60, 1991.

Osherovich, V. A., C. J. Farrugia, and L. F. Burlaga, The non-linear evolution of magnetic flux ropes: 1. The low beta limit, *J. Geophys. Res., 98,* 1325 1993a.

Osherovich, V. I., Farrugia, C. J., and L. F. Burlaga, Dynamics of aging magnetic clouds, *Adv. Space Res., 13,* 6 (6), 57, Pergamon Press, 1993b.

Osherovich, V. A., C. J. Farrugia, L. F. Burlaga, R. P. Lepping, J. Fainberg, and R. G. Stone, Polytropic relationship in interplanetary magnetic clouds, *J. Geophys. Res., 98,* 15,331, 1993c.

Osherovich, V. A., C. J. Farrugia, and L. F. Burlaga, The non-linear evolution of magnetic flux ropes: 2. Finite beta plasma, *J. Geophys. Res., 100,* 12307, 1995.

Richardson, I. G., C. J. Farrugia, and L. F. Burlaga, Energetic ion observations in the magnetic cloud of 14-15 January 1988 and their implications for the magnetic field topology, in *Proceedings of the 22nd International Cosmic Ray Conference (Dublin), Vol. 3,* SH7.8, p 597, Dublin Institute for Advanced Studies, 1991.

Schluter, A., Kraftfreie Magnetfelder II, *Z. Naturforschung, 12a,* 855, 1957.

Scudder, J. D., On the cause of temperature change in inhomogeneous low density astrophysical plasmas, *Astrophys. J., 398,* 299 -318, 1992.

Shafranov, V. D., Plasma equilibrium in a magnetic field, *Rev. Plasma Phys., 2* 113, 1966.

Sittler, E. C., Jr., and J. D. Scudder, An empirical polytrope law for solar wind thermal electrons between 0.45 and 4.67 AU: Voyager 2 and Mariner 10, *J. Geophys. Res., 85,* 5131, 1980.

Stone, R. G., et al., Ulysses radio and plasma wave observations at high southern heliographic latitudes, *Science, 268,* 1026, 1995.

Trotten, T. L., and J. W. Freeman, An empirical determination of the polytropic index for the free-streaming solar wind using Helios 1 data, *J. Geophys. Res., 100,* 13, 1995.

Vandas M., S. Fischer, M. Dryer, Z. Smith and T. Detman, Simulation of magnetic cloud propagation in the inner heliosphere in two dimensions, 1, A loop perpendicular to the ecliptic plane, *J. Geophys. Res., 100,* 12285, 1995.

Vandas, M. and S. Fisher, Parametric study of loop-like magnetic cloud propagation, *J. Geophys. Res., 101,* 15,645, 1996.

L. F. Burlaga, Laboratory for Extraterrestrial Physics, NASA Goddard Space Flight Center, Greenbelt, MD 20771 (e-mail: u2leb@lepvax.gsfc.nasa.gov).

V. Osherovich, Hughes STX Corporation, 4400 Forbes Blvd., Lanham, MD 20706

Flux Ropes and Spheromaks: A Numerical Study

M. Vandas, and S. Fischer

Astronomical Institute, Academy of Sciences, Prague, Czech Republic

D. Odstrčil

Astronomical Institute, Academy of Sciences, Ondřejov, Czech Republic

M. Dryer, Z. Smith, and T. Detman

NOAA Space Environment Center, Boulder, Colorado

Propagation of magnetic clouds as flux ropes or spheromaks is studied with use of a self-consistent MHD numerical approach. Flux ropes are approximated by cylinders in a 2.5-D treatment, while spheromaks (plasmoids) described initially as spherical clouds are treated in 3-D. We deal with the reliability of 2.5-D solutions, comparing results of different codes, different mesh resolutions and different conditions for the cloud's injection. Despite this variety, we find that the results of the cloud's propagation are very similar and consistent. Indeed, we find that the results are independent of code mesh size and the numerical technique used. Cylindrical clouds are deflected from straight, radial propagation to the side where the inner and outer magnetic fields have the opposite sense. The magnitude of the deflection depends on the magnetic field strength and the gradient of the ambient magnetic field near the cloud's boundary. Spherical clouds with a poloidal field and axes aligned along the background magnetic field are not deflected. But they evolve into toroids, closed flux ropes. These results were obtained using a unipolar background magnetic field, that is, we do not consider interactions with the heliospheric current sheet.

INTRODUCTION

Magnetic clouds are a specific subset of coronal mass ejections [*Wilson and Hildner*, 1986] and originally have been defined from single spacecraft measurements as regions in the solar wind having the following properties: (1) the magnetic field magnitude is higher than that in the surrounding plasma; (2) the magnetic field vector rotates smoothly through a large angle; and (3) the proton temperature is lower than that of the surrounding plasma [*Burlaga*, 1988]. The three-dimensional shape of magnetic clouds is usually represented as flux ropes (loops) carried away by the solar wind from the Sun and magnetically connected to it [*Goldstein*, 1983; *Marubashi*, 1986; *Burlaga*, 1988]. This model has been confirmed in many cases through studies of magnetic field component profiles and suprathermal and energetic particle behavior (see other papers in this monograph). However, one cannot exclude the possibility that some magnetic clouds are fully disconnected from the Sun, forming a

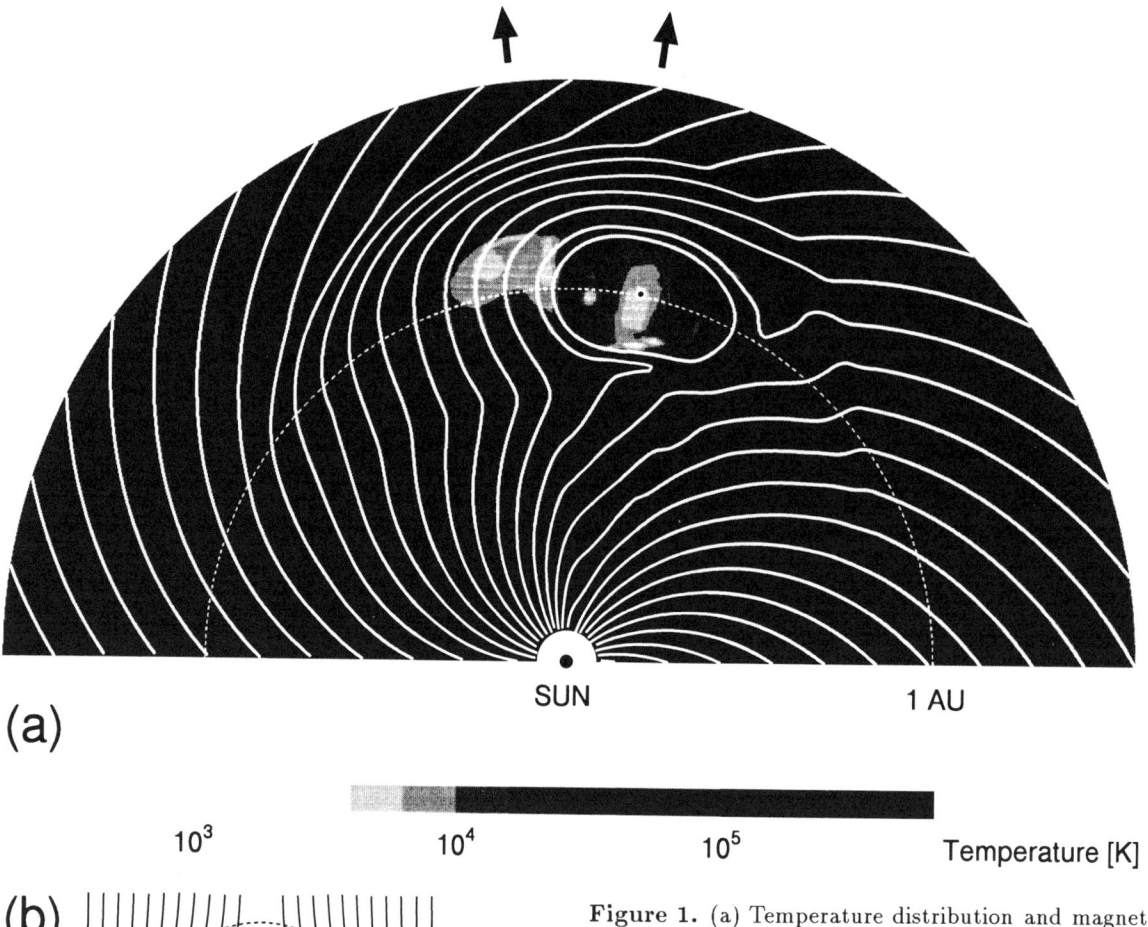

Figure 1. (a) Temperature distribution and magnetic field lines in the ecliptic plane at the time when a cylindrical magnetic cloud reached 1 AU. (b) Starting magnetic field configuration of the magnetic cloud.

plasmoid (spheromak) [*Ivanov and Harshiladze*, 1985; *Vandas et al.*, 1993].

In the present paper, propagation and evolution of magnetic clouds are studied with use of a self-consistent MHD numerical approach, as we have done in previous works [*Vandas et al.*, 1995, 1996a, b]. Evolution of cylindrical magnetic clouds through MHD simulations has been also studied by *Cargill et al.* [1996] but under simplified external conditions. Analytical approaches to this problem can be found in *Osherovich et al.* [1993] or *Chen and Garren* [1993].

METHOD

The propagation of magnetic clouds was studied with use of a self-consistent, time-dependent, single-fluid MHD numerical code [*Han et al.*, 1988]. Two models of magnetic clouds were treated: a flux rope and a spheromak. Flux ropes were approximated by cylinders with a force-free magnetic field configuration [*Lundquist*, 1950] in a 2.5-D treatment. Spheromaks (plasmoids) were initially described as spherical clouds with a force-free poloidal magnetic field configuration [*Chandrasekhar and Kendall*, 1950] and treated in 3-D. We used the two-step Lax-Wendroff numerical scheme in spherical coordinates. The grid resolution was $0.5\,R_\odot$ ($R_\odot$ is the solar radius) and $0.75°$ in 2.5-D cases, and $0.25\,R_\odot$ and $0.375°$ in 3-D cases. The inner boundary was situated at $18\,R_\odot$. Detailed descriptions of our method are given in *Vandas et al.* [1995, 1996a] and in a paper submitted to the Jour-

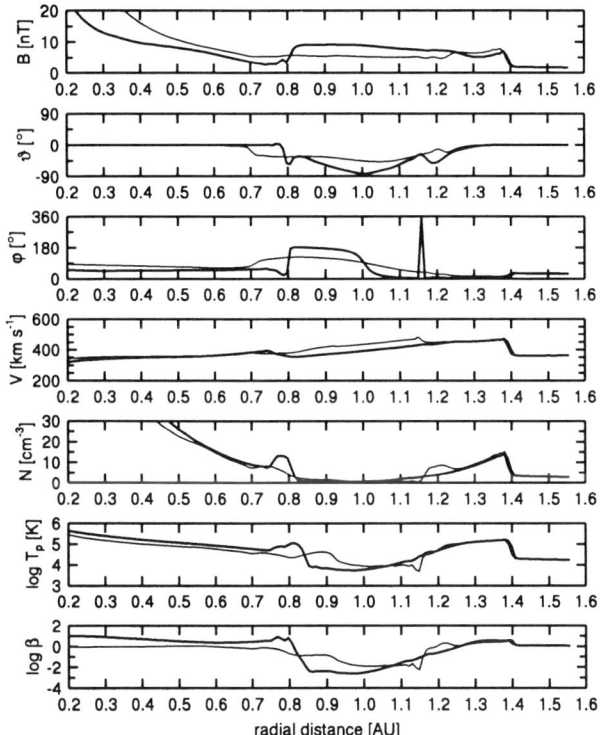

Figure 2. Radial profiles of magnetic field and plasma quantities for the magnetic cloud (thick lines) and its surroundings (thin lines) from Figure 1.

nal of Geophysical Research (M. Vandas, S. Fischer, P. Pelant, M. Dryer, Z. Smith, and T. Detman, Propagation of a spheromak, 1, Some comparisons of cylindrical and spherical magnetic clouds, 1996). We do not discuss the reconnection in the paper because "numerical" reconnection is an artifact within the context of our ideal fluid approximation. We wish to suppress this effect as much as possible, therefore a small grid size was chosen in the present work.

FLUX ROPES

Figure 1a shows the temperature distribution and magnetic field lines in the ecliptic plane for a cylindrical cloud with the axis perpendicular to the ecliptic plane. The initial structure of the cloud and its surroundings are shown in Figure 1b. The inner dashed circle is the cloud's boundary (its radius is 3.9 Gm), and the outer dashed circle indicates a region where the ambient magnetic field is mostly affected. We call this outer circle "the envelope" and choose its radius to be 2 R_c, where R_c is the radius of the cloud. The cloud was injected with a velocity of 750 km s^{-1}, which was 3× higher than the ambient velocity, and is shown after 92 h, when it reached 1 AU. The temperature inside the cloud is decreased, but it is highly inhomogeneous. Note also a large, low temperature region outside (left side) of the cloud. The shock wave driven by the cloud accelerates and compresses the plasma and leaves behind itself a rarefied region. Because the plasma behaves adiabatically, a low temperature region (compression and heating at the shock, expansion and cooling farther behind it) is formed. This low density and low temperature region is also seen in Figure 2 (thin lines).

Figure 2 gives two radial profiles of plasma and magnetic field quantities for the case shown in Figure 1. The thick lines are for a radial cut through the cloud's axis, while the thin lines are for a cut that passes near, but outside of, the left border of the cloud. These directions are shown by arrows in Figure 1a. The temperature and β are decreased in both cases, but the magnetic field enhancement and field rotation is exhibited only inside the magnetic cloud, which stretches from approximately 0.8 to 1.2 AU. The magnetic field magnitude inside the cloud is flat, and the velocity exhibits a smooth decrease towards the rear. This is caused by the cloud's expansion (as discussed by *Osherovich et al.* [1993]), which also results in a decrease of the density and temperature in the cloud.

Figure 1a also shows the cloud's deflection. The cloud was initially injected in the central radial direction that bisects the computational region. The sense of deflection depends on the mutual orientation of the internal and external magnetic fields. Figure 3 summarizes results for two different cloud's magnetic field rotations and axes' orientations (the axis is perpendicular to the ecliptic plane for cases 1a and 1b, and it is parallel to the ecliptic plane for cases 2a and 2b). The deflection is to the side where the internal and external magnetic fields are oppositely directed [*Vandas et al.*, 1996a]. It is caused by the Lorentz magnetic force which results from a high level of the value rot **B** here. Figure 4 shows unit magnetic field vectors in a meridional plane, two magnetic field lines draping over the cloud from both sides, and the cloud itself at 1 AU. An arc, drawn at 1 AU through the cloud's center, represents a zero level of polar forces (i.e., the θ-components of various forces). The pressure gradient is displayed around the arc in the thin line; its positive values (a poleward direction of the force, downward in the figure) are plotted antisunward of the arc. The second force is the Lorentz magnetic force (not shown here). The total force, as a sum of these two, is given by the thick line. We see

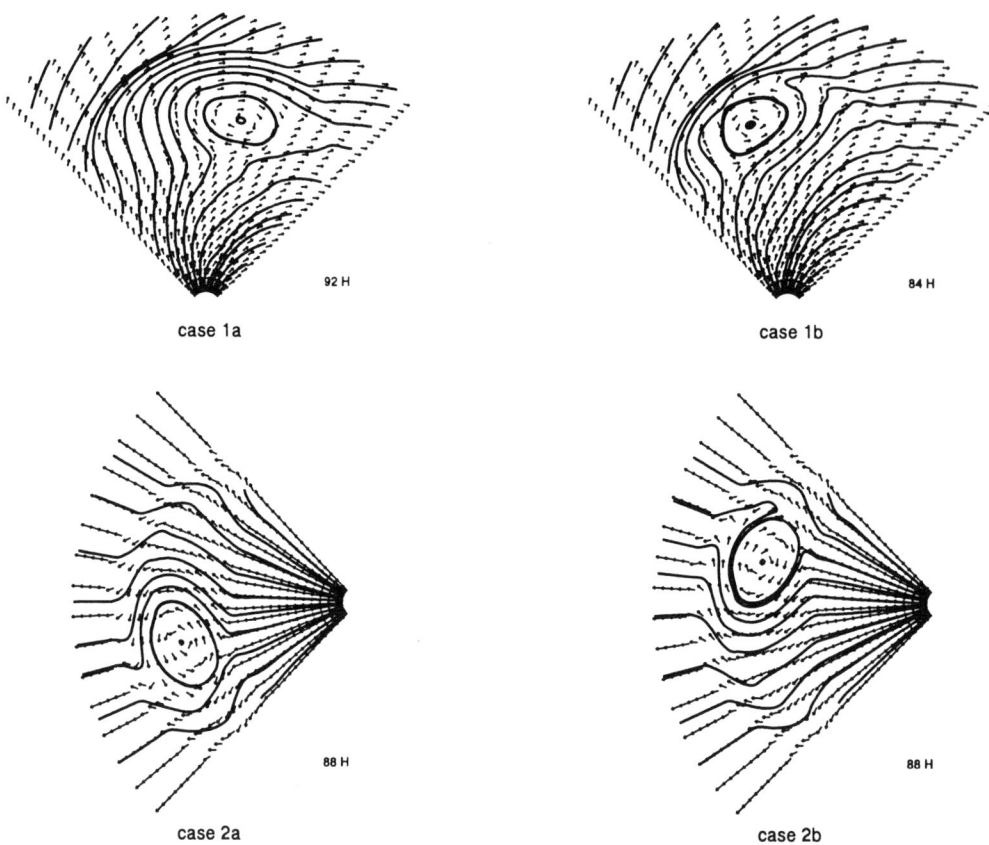

Figure 3. The cloud's deflection depends on the mutual orientation of the internal and external magnetic fields [from *Vandas et al.*, 1996a].

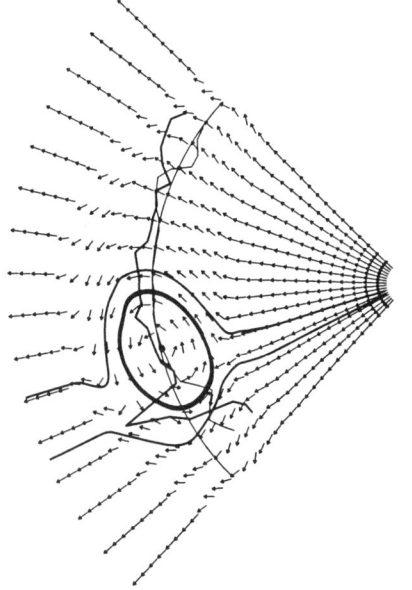

Figure 4. The polar forces acting at 1 AU on the magnetic cloud with the axis parallel to the ecliptic plane (case 2a in Figure 3) [from *Vandas et al.*, 1996a].

a large force at the poleward edge of the cloud, which pulls it poleward. The main contribution here is the Lorentz magnetic force, which means that the deflection also depends on the magnetic field strength [*Vandas et al.*, 1996a, b]. Figure 5 demonstrates that the deflection is larger near the Sun, where the magnetic field is higher. Magnetic clouds with lower injection velocities are deflected more at 1 AU because they spend more time in the region near the Sun with the stronger ambient field. This interpretation is supported by another finding that the deflection depends on the magnetic field gradient near the cloud's boundary. Figure 1b shows an envelope with a radius of $2\,R_c$. If we make the envelope smaller, the cloud is deflected even more. Figure 6 gives a superposition of the results with the cloud envelope with a radius of $2\,R_c$ (thick lines, repeated from Figure 1a) and of $1.2\,R_c$ (thin lines). The smaller envelope means a higher level of the value rot **B**, hence a larger deflection.

To exclude the possibility that the deflection is a numerical effect of the code, we tested our simulations and found that the deflection is not affected by the grid size

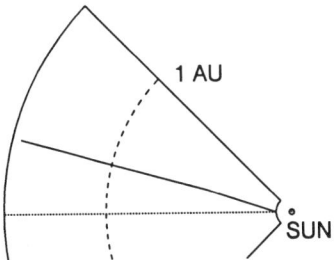

Figure 5. The trajectory of the cloud axis for the magnetic cloud with the axis parallel to the ecliptic plane (case 2b in Figure 3). A part of the meridional plane is shown. The magnetic cloud was injected along the dotted line which is also the projection of the ecliptic plane [from *Vandas et al.*, 1996b].

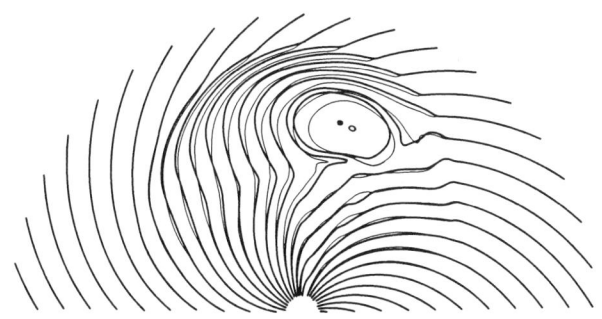

Figure 6. Comparison of simulations with two different envelope sizes shows a change in deflection. A smaller envelope (thin lines) causes a larger deflection.

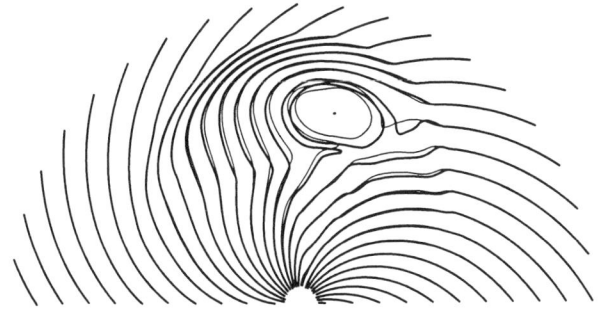

Figure 7. Comparison of simulations with standard (thick lines) and coarse (thin lines) grid sizes.

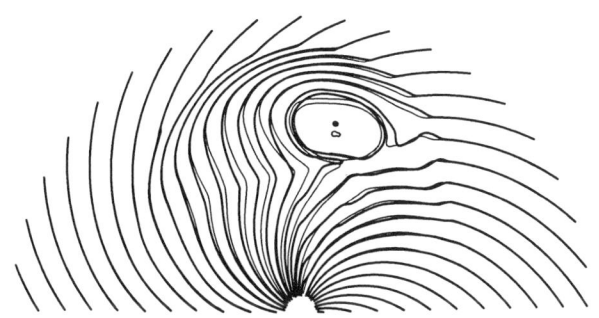

Figure 8. Comparison of simulations with two different numerical codes, the Lax-Wendroff scheme (thick lines) and the total variation diminishing Lax-Friedrich algorithm (thin lines).

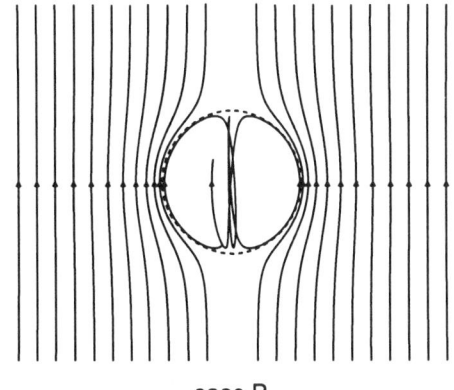

case P

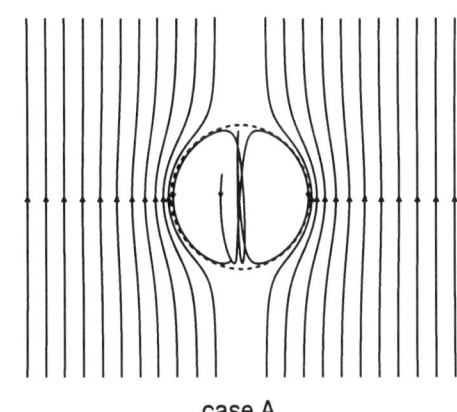

case A

Figure 9. The starting magnetic field configurations of the magnetic clouds for the parallel (case P) and antiparallel (case A) orientation of the cloud polar axis.

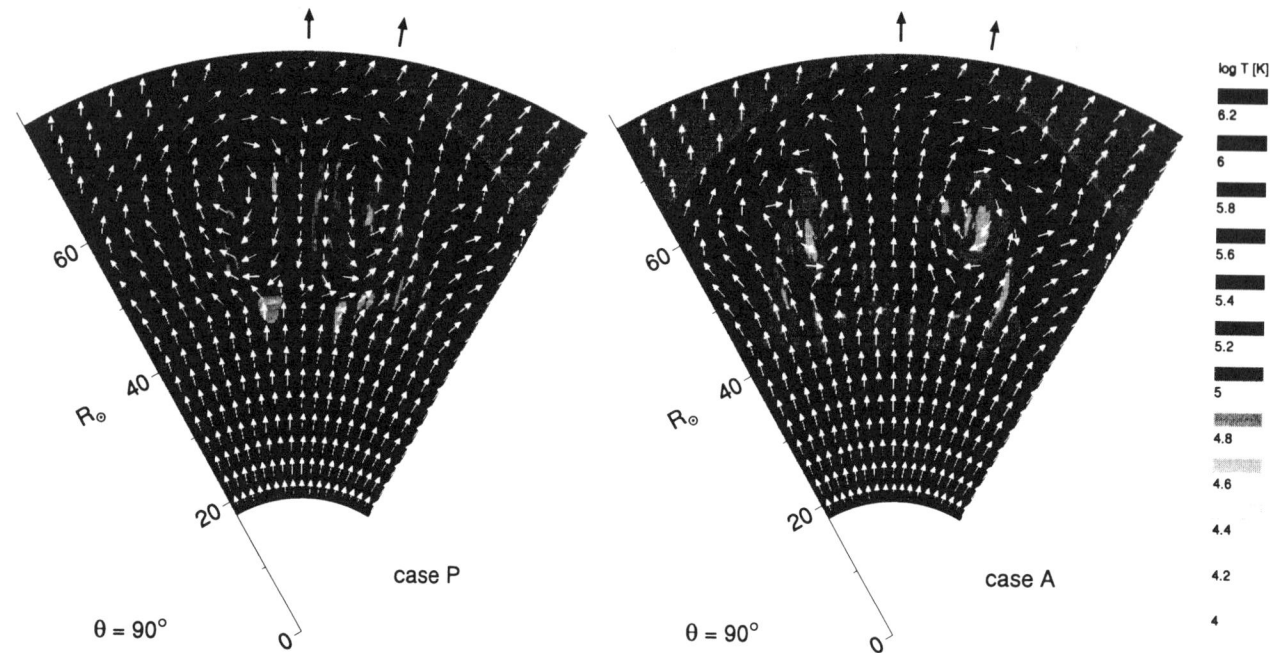

Figure 10. Temperature distribution and magnetic field vectors in the ecliptic plane at the time when the spherical magnetic clouds reached about 60 $R_\odot$.

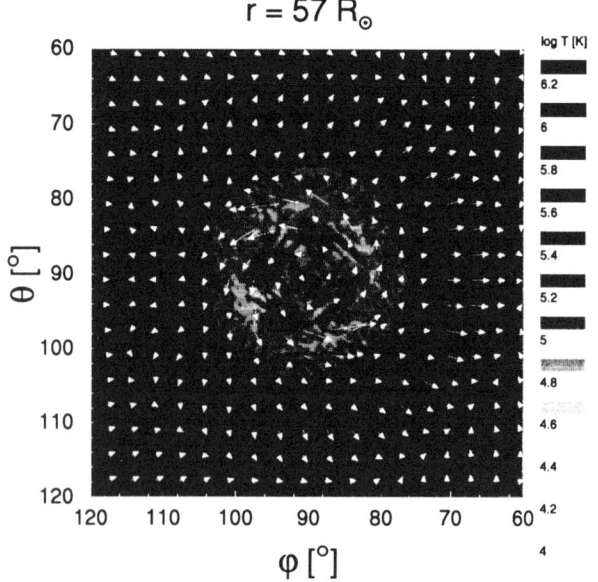

Figure 11. Temperature distribution and magnetic field vectors on a sphere with the radius going through the central parts of the spherical magnetic cloud with the parallel axis orientation (case P) from Figure 10.

(i.e., by the numerical reconnection). Figure 7 is a superposition of two simulations: the thick lines are for the grid resolution of Figure 1 (standard grid), the thin lines are for a 2× coarser grid resolution. The size of the cloud is different (due to numerical reconnection), but the positions of the clouds' axes coincide well. We also compared our simulations with a simulation using a different numerical code based on the total variation diminishing Lax-Friedrich (TVD-LF) numerical scheme [*Tóth and Odstrčil*, 1996]. Figure 8 shows resuls for the simulation with the standard grid and described above (thick lines) compared with the TVD-LF simulation (thin lines). The agreement is rather good.

The described drift mechanism has been used by *Smith et al.* [1997] to explain an observation of a magnetic cloud near the Earth. Its source was in the southern hemisphere, and it drifted towards the ecliptic plane.

SPHEROMAKS

We have performed runs for two cases of spherical clouds in the computational region extending to 0.3 AU. The clouds were again injected at the inner boundary at 18 $R_\odot$ with a velocity of 750 km s^{-1}. The starting

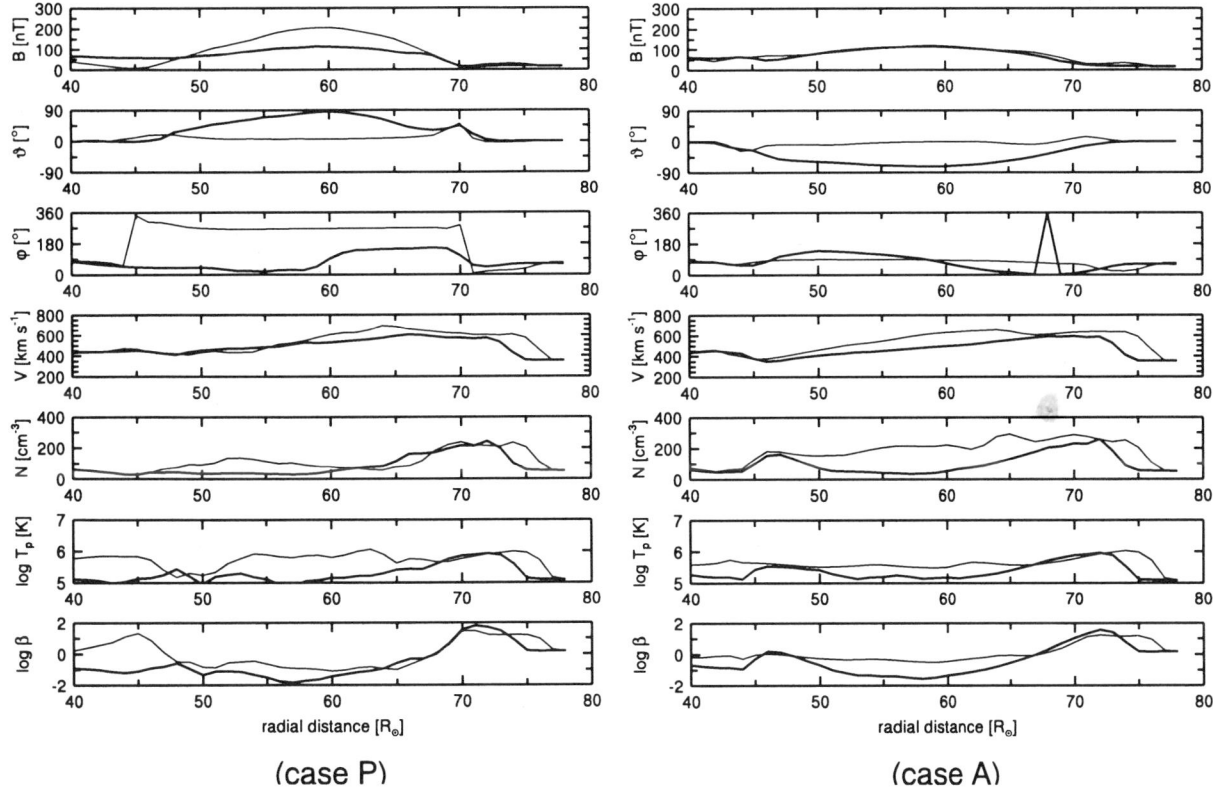

Figure 12. Radial profiles of magnetic field and plasma quantities at the magnetic cloud polar axis (thin lines) and through the toroidal core (thick lines).

magnetic field configurations of the magnetic clouds for these two cases are shown in Figure 9. The spheromaks have the same poloidal magnetic field and differ only in the orientation of the polar axis. In the parallel orientation (case P), the boundary field initially has the same orientation as the ambient field, while at the polar axis the field has an opposite direction. In the antiparallel orientation (case A), the boundary field initially opposes the ambient field, while at the polar axis the field has the direction of the ambient field.

Figure 10 shows the temperature distribution and magnetic field vectors in the ecliptic plane for the spheromak with the parallel orientation (case P) at time 16.9 h and for the spheromak with the antiparallel orientation (case A) at time 16.4 h. Both cases give a similar evolution during propagation. Shock waves form, and the spheromaks expand, evolving into a toroidal structure, although the cloud in case A expands more rapidly. The coldest plasma is inside the toroid. As the toroidal structure expands, there is a compressed region at the cloud polar axis with a higher temperature and magnetic field.

The toroidal structure of the cloud is clearly seen in Figure 11, which shows the temperature distribution and magnetic field vectors at a sphere with a heliospheric radius $r = 57\ R_\odot$. The simulations indicate that spheromaks with a poloidal field do not retain their original configuration and evolve into toroids, which are in fact closed magnetic flux ropes. This evolution is more pronounced for the antiparallel axis orientation when the directions of the cloud magnetic field at the axis and the ambient field are the same. The numerical MHD simulations give different results from analytical solutions for spheromak expansion [*Farrugia et al.*, 1995] where a self-similar condition retains the spheromak original configuration. In contrast to the cylindrical clouds, for both cases, parallel and antiparallel orientations, the clouds are not deflected.

Figure 12 shows the radial profiles of plasma and magnetic field quantities for both axis orientations. The thin lines are for the radial line going near the polar axis, and the thick lines are for the radial line going through the toroid's inner part (the directions of the radial cuts are shown by arrows in Figure 10). The thick lines exhibit signatures of a magnetic cloud (higher magnetic field, low density and temperature) extending approximately from 50 to 70 $R_\odot$, in contrast to the thin lines (high magnetic field, but hotter and denser plasma).

CONCLUSIONS

MHD simulations of propagation of initially force-free cylinders or spheres in the expanding solar wind reproduce signatures of magnetic clouds: an increase and rotation of the magnetic field together with a decrease of (proton) temperature. Cylindrical magnetic clouds are deflected from the radial direction (the direction of the solar wind), and the magnitude of the deflection depends on the magnetic field strength and its gradient near the cloud's boundary. Spherical clouds with a poloidal field and boundary magnetic field aligned with the ambient field are not deflected. They evolve into toroids, i.e., into closed flux ropes.

Acknowledgments. This publication is based on work sponsored by the U.S.-Czechoslovak Science and Technology Joint Fund in cooperation with NOAA and MŠMT ČR under Project numbers 920 30 and 94082. We were also supported by grant 205/96/1575 (M.V. and S.F.) from the Grant Agency of the Czech Republic and by grant 303402 (D.O.) and project RA K1042603 (M.V., S.F. and D.O.) from the Academy of Sciences of the Czech Republic. Supercomputing resources for this project were supported, in part, through a grant from the Westinghouse Electric Corporation.

REFERENCES

Burlaga, L. F., Magnetic clouds and force-free fields with constant alpha, *J. Geophys. Res., 93*, 7217-7224, 1988.

Cargill, P. J., J. Chen, D. S. Spicer, and S. T. Zalesak, Magnetohydrodynamic simulations of the motion of magnetic flux tubes through a magnetized plasma, *J. Geophys. Res., 101*, 4855-4870, 1996.

Chandrasekhar, S., and P. C. Kendall, On force-free magnetic fields, *Astrophys. J., 126*, 457-460, 1957.

Chen, J., and D. A. Garren, Interplanetary magnetic clouds: Topology and driving mechanism, *Geophys. Res. Lett., 20*, 2319-2322, 1993.

Farrugia, C. J., V. A. Osherovich, and L. F. Burlaga, Magnetic flux rope versus the spheromak as models for interplanetary magnetic clouds, *J. Geophys. Res., 100*, 12293-12306, 1995.

Goldstein, H., On the field configuration in magnetic clouds, *Solar Wind Five*, ed. M. Neugebauer, *NASA Conf. Publ., NASA CP-2280*, 731-733, 1983.

Han, S. M., S. T. Wu, and M. Dryer, A three-dimensional, time-dependent numerical modeling of supersonic, super-alfvenic MHD flow, *Comput. Fluids, 16*, 81-103, 1988.

Ivanov, K. G., and A. F. Harshiladze, Interplanetary hydromagnetic clouds as flare-generated spheromaks, *Sol. Phys., 98*, 379-386, 1985.

Lundquist, S., Magnetohydrostatic fields, *Ark. Fys., 2*, 361-365, 1950.

Marubashi, K., Structure of the interplanetary magnetic clouds and their solar origins, *Adv. Space Res., 6*(6), 335-338, 1986.

Osherovich, V. A., C. J. Farrugia, and L. F. Burlaga, Dynamics of aging magnetic clouds, *Adv. Space Res., 13*(6), 57-62, 1993.

Smith, Z., S. Watari, M. Dryer, P. Manoharan, and P. McIntosh, Identification of the solar source for the 18 October 1995 magnetic cloud, *Solar Phys.*, in press, 1997.

Tóth, G., and D. Odstrčil, Comparison of some flux corrected transport and total variation diminishing numerical schemes for hydrodynamic and magnetohydrodynamic problems, *J. Comput. Phys., 128*, 82-100, 1996.

Vandas, M., S. Fischer, P. Pelant, and A. Geranios, Evidence for a spheroidal structure of magnetic clouds, *J. Geophys. Res., 98*, 21,061-21,069, 1993.

Vandas, M., S. Fischer, M. Dryer, Z. Smith, and T. Detman, Simulation of magnetic cloud propagation in the inner heliosphere in two-dimensions, 1, A loop perpendicular to the ecliptic plane, *J. Geophys. Res., 12*, 285-12, 292, 1995.

Vandas, M., S. Fischer, M. Dryer, Z. Smith, and T. Detman, Simulation of magnetic cloud propagation in the inner heliosphere in two-dimensions, 2, A loop parallel to the ecliptic plane and the role of of helicity, *J. Geophys. Res., 101*, 2505-2510, 1996a.

Vandas, M., S. Fischer, M. Dryer, Z. Smith, and T. Detman, Parametric study of loop-like magnetic cloud propagation, *J. Geophys. Res., 101*, 15,645-15,652, 1996b.

Wilson, R. M., and E. Hildner, Are interplanetary magnetic clouds manifestations of coronal transients at 1 AU? *Solar Phys., 91*, 169-180, 1984.

T. Detman, M. Dryer, and Z. Smith, Space Environment Center, NOAA, R/E/SE, 325 Broadway, Boulder, CO 80303

S. Fischer, and M. Vandas, Astronomical Institute, Academy of Sciences, Boční II 1401, 141 31 Praha 4, Czech Republic

D. Odstrčil, Astronomical Institute, Academy of Sciences, 251 65 Ondřejov, Czech Republic

Recent Work on Modelling the Global Field Line Topology of Interplanetary Magnetic Clouds

C. J. Farrugia

Institute for the Study of Earth, Oceans, and Space, University of New Hampshire, Durham

Magnetic clouds form a distinct and important subset of coronal mass ejections ("CME's") distinguished by a few simple properties. In recent years there has been lively interest in modelling the global magnetic field line topology of magnetic clouds. Are they closed magnetic configurations disconnected from the Sun (magnetic "bubbles"), or are they better described as magnetic loops or "tongues" or flux ropes (magnetic "bottles"), perhaps even retaining magnetic connection to the Sun ? In this selective review we discuss the arguments advanced in favour of one or other of two models developed enough to allow meaningful comparison of theoretical predictions with in situ spacecraft data: on the one hand, a magnetic flux rope and, on the other, a spheromak. The experimental datum that magnetic clouds are strongly evolving configurations will be central to the discussion. Our comparisons will be based on exact solutions of the ideal MHD equations describing the radial, self-similar expansion of cylindrical magnetic flux ropes and spheromaks. On present evidence, this comparison favours the magnetic flux rope topology. We shall also briefly discuss recent MHD simulations of the propagation of magnetic clouds into interplanetary space and how the results thus obtained compare with analytical conclusions. A short discussion of magnetic field line draping as a potential means of distinguishing the two topologies follows. Finally, we point out the importance in further analytical elaboration of the models of incorporating the interaction of magnetic clouds with surrounding flows.

1. INTRODUCTION

Much attention focusses on the global magnetic field line topology of an important subclass of solar ejecta (coronal mass ejections, CME's) known as magnetic clouds. In this review we treat two models, locally-straight, cylindrical magnetic flux ropes, and spheromaks, and assume a specific evolution in time: self-similar, radial expansion. (Some authors refer to a magnetic configuration detached from the Sun and representing a magnetic cloud as a plasmoid but we prefer the much more exact terminology spheromak.) Admittedly, these models are but two of several candidate topologies which have been proposed [see, e.g., *Chen* [this volume]). However, the two models (a) are exact solutions of the (ideal) MHD equations, which (b) contain specific predictions on the temporal behavior of the magnetic, kinematic, and thermodynamic structures of these ejecta. Furthermore, numerical MHD simulations of the propagation in interplanetary space of these two model configurations are advanced enough to allow comparison of simulation work with theory.

Magnetic clouds are characterized by (i) enhanced magnetic field strengths, (ii) a large and smooth rotation of the magnetic field vector as seen by a stationary heliospheric observer, and (iii) low proton temperatures and low proton beta [*Burlaga et al.*, 1981; see also reviews by *Burlaga*, 1990, 1995, and by *Osherovich* and *Burlaga*, this volume]. Magnetic clouds are expanding configurations [*Klein* and *Burlaga*, 1982] and should thus be considered as strongly non-stationary objects. Spacecraft observe magnetic clouds which have left the Sun at different times in the past (i.e., are of different ages. By age we mean the time the configuration has been expanding self-similarly prior to observation, see further below.). Furthermore, at 1 AU the flight time of a spacecraft through a magnetic cloud, of order 1 day, is a substantial fraction of the age of the configuration. The possibility which thus exists of confronting predictions of the two models with in situ data on clouds at different stages of their evolution further helps to discriminate between these two major proposed topologies.

A much-studied magnetic cloud, which was observed at 1 AU on 14 - 15 January, 1988, is shown in Figure 1 [after *Farrugia et al.*, 1992]. From top to bottom, the panels show the bulk flow speed, density, and temperature of the protons; the GSE B_x, B_y, B_z components of the magnetic field, and the total field. The plasma data are at 1 min while the field data are at ~15 s resolution. According to the definition above, the data between the vertical guidelines define the magnetic cloud interval. A shock preceeds the cloud by ~5.5 h. The monotonically decreasing bulk speed profile implies that the cloud is expanding radially. The bulk speed profile is an evolutionary signature central to our discussion. Besides expanding into, the cloud is also overtaking the ambient medium: its average bulk speed (613 km/s) is higher than pre-shock values. However, the cloud is not being overtaken by faster flows from the rear. Another dual signature of evolution is evident in the total field profile, which exhibits two asymmetries: (i) the fields are stronger towards the front than towards the rear edge, and (ii) maximum field strength is reached well before the middle of the time interval shown. These aspects of the signature will occupy us further below (see also the review by *Osherovich* and *Burlaga*, this volume).

2. TWO MODELS OF MAGNETIC CLOUDS

Goldstein [1983] considered magnetic clouds as equilibrium force-free configurations of straight cylindrical

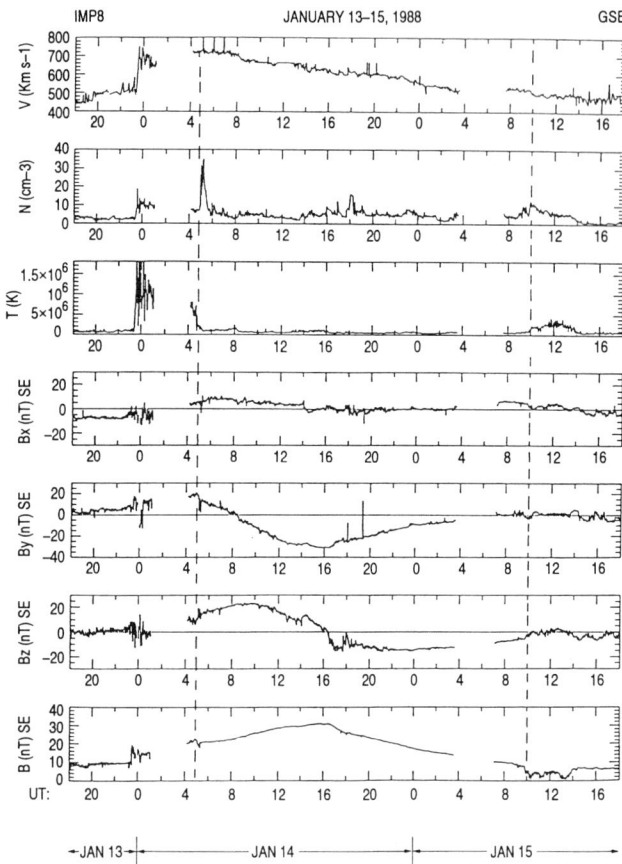

Figure 1. Plasma and field data for the much-studied January, 1988 magnetic cloud (after *Farrugia et al.*, 1992). The panels show, from top to bottom, the proton bulk flow speed, density and temperature; and the GSE magnetic field components and total field strength. The cloud interval is shown between vertical guidelines.

geometry, i.e., as solutions in cylindrical coordinates (R, Φ, Z) of $curl\mathbf{B} = \alpha(\mathbf{R})\mathbf{B}$, where $\alpha(\mathbf{R})$ is an arbitrary function of position $\mathbf{R}$, which is constant along magnetic lines. A specific $\alpha(\mathbf{R})$ variation was considered by *Marubashi* [1986], while *Burlaga* [1988] and *Lepping et al.* [1990] studied the case $\alpha = const.$, for which a solution had been derived by *Lundquist* [1950]. Good agreement with magnetic cloud data at 1 AU was obtained, particularly concerning the direction of the field. The flux tube's symmetry axis is simultaneously the magnetic axis, and the magnetic surfaces are surfaces of coaxial cylinders. The field lines form a set of nested helices whose pitch decreases with distance from the axis. The magnetic field strength on the axis is twice that on the periphery, as usually defined, and the total field is symmetric about closest approach on a line intersecting the tube.

Solutions of the force-free equation in spherical polar coordinates (r, ϕ, θ) are also known in closed form [*Chandrasekhar* and *Kendall*, 1957; *Rosenbluth* and *Bussac*, 1979], the spheromak being one such solution. That a spheromak might be a good model for magnetic clouds was first suggested by *Ivanov* and *Harshiladze* [1985]. Spheroidal models, including spheromaks, were least-squares fitted to magnetic field profiles by *Vandas et al.* [1991, 1993], and the quality of the fits were compared with those obtained from the flux rope model. The field lines of the spheromak wind around tight tori whose cross-section is not circular. The magnetic surfaces are centred around a circular magnetic axis along which the field is azimuthal. (The relevant static formulae and schematics of the two configurations may be found in *Farrugia et al.* [1995]).

Discussion based on these two static configurations centred around: (a) observed asymmetries in total field profiles and (b) the amplitude of variation of the total field. The static Lundquist tube was unable to reproduce the first feature and seemed constrained by the second. These objections need to be reassessed when expansion is taken into account [*Farrugia et al.*, 1992]. Comparisons of the two static models were often hampered by imprecise knowledge of cloud boundaries, which were either stipulated or obtained through least-squares fitting, with neither procedure being wholly satisfactory (see reviews by *Burlaga* [1991, 1995]). The determination of cloud boundaries remains a problem to this day, though some progress has been made [*Fainberg et al.*, 1995; *Osherovich* and *Burlaga*, this volume; *Farrugia et al.*, 1997a]

3. RADIAL, SELF-SIMILAR EXPANSION AND ITS CONSEQUENCES; EXACT MHD SOLUTIONS

We now consider spheromaks and cylindrical flux tubes expanding self-similarly. The temporal and spatial variation is then characterized by a so-called "self-similar" parameter (denoted by η and ξ, for the flux rope and spheromak, respectively), defined as the ratio of the radial coordinate to the "evolution function" $(y(t), F(t)$, respectively), the latter quantity representing the radius of the configuration. For the cylindrical flux rope and the spheromak we have, respectively,

$$\eta = \frac{R}{y(t)}; \quad \xi = \frac{r}{F(t)} \quad (1)$$

For radial flow, the self-similarity assumption allows us to extend the static force-free solutions to include time. The derivation is given elsewhere [*Osherovich et al.*, 1993a, 1995; *Farrugia et al.*, 1997b]. Studies of self-similarly evolving structures with spherical flow have been done among others by *Bernstein* and *Kulsrud* [1965], *Kulsrud et al.* [1965] and *Low* [1982]. The resulting magnetic fields are, for the spheromak

$$B_r = [2B_0/F^2(t)\alpha\xi]j_1(\alpha\xi)\cos\theta \quad (2)$$
$$B_\phi = (B_0/F^2)j_1(\alpha\xi)\sin\theta \quad (3)$$
$$B_\theta = -(B_0\sin\theta/\alpha F^2\xi)[\sin(\alpha\xi) - j_1(\alpha\xi)], \quad (4)$$

where B_0 is the field strength at the center, and $j_1(x)$ is the spherical Bessel function of order 1. For the magnetic flux rope we have:

$$B_R = 0 \quad (5)$$
$$B_\Phi = (B_0/y)J_1(\alpha\eta) \quad (6)$$
$$B_z = (B_0/y^2)J_0(\alpha\eta), \quad (7)$$

where J_0 and J_1 are Bessel functions of order 0 and 1, respectively. To determine the differential equations satisfied by $y(t)$ and $F(t)$ (the "evolution equations"), we solve the full set of ideal MHD equations for a polytrope and describing radial, self-similar expansion. In the momentum equation we include a drag force proportional to the product of the density and velocity of radial expansion. The energy transport is described by a polytrope of index, γ. In the present application, we assume a configuration where the gas and magnetic pressures both maximize on the symmetry axis.

The analysis is aimed at separating variables in the respective momentum equations, thus leading to the required evolution equations describing the nonlinear dynamics of the corresponding configurations. This is achieved by working with "separable" magnetic fields, defined by imposing a relation between the amplitudes of the field components in self-similar form [*Osherovich et al.*, 1993a, 1995]. Further analysis [*Farrugia et al.*, 1995] leads to the evolution equations, which are

$$d^2y/dt^2 = Sy^{-3} - Qy^{-1} + Ky^{-2\gamma+1} - \nu dy/dt \quad (8)$$

for the flux rope, and

$$d^2F/dt^2 = aF^{-2} - \nu dF/dt \quad (9)$$

for the spheromak.

Quantities S, Q, K, ν and a in equations (8) and (9) are constants. Equation (9) differs from the corresponding equation in *Low* [1982] by the drag term $\nu dF/dt$.

The terms on the right-hand-side of (8) are from left to right: the magnetic pressure gradient due the axial field component; the pinch force; the gas pressure gradient; and the drag force representing mechanical dissipation in the system. From (9), the gradient of the poloidal and toroidal components of the spheromak field, as well as the gradient of gas pressure follow the same variation in F, and are lumped together in one term, aF^{-2}. The different way the forces evolve in the two configurations has important observational consequences.

Some points may be made. It was shown [*Osherovich et al.*, 1995] that separable solutions can be found for all γ; but for the cylindrical flux rope to expand self-similarly, the polytropic index in (8) has to be ≤ 1. For the spheromak, the only known separable solution has a polytopic index equal to 4/3 [*Low*, 1982; *Farrugia et al.*, 1995]. While for the spheromak, the magnetic (both poloidal and toroidal) and gas pressure terms all scale in the same way (equation (9)), for the expanding flux rope, each scales to a different power of y. In the absence of drag, an expanding flux rope is exactly force-free only initially. By contrast, an initially force-free spheromak remains so thereafter (a=0 and $\nu = 0$ in (9)). The magnetic fields of both models decrease with time (but differently from the ambient solar wind [*Osherovich et al.*, 1993b]). In addition, however, the flux rope's field flattens in time because the axial component decreases faster than the azimuthal component, i.e., as time progresses, the ratio of the axial field strength to the field strength at the periphery (the azimuthal field) decreases. Physically, this means that in the magnetic flux rope, magnetic energy is being transferred from the axial to the azimuthal component, whereas in the spheromak this energy exchange does not take place. This circumstance can be used to distinguish the two configurations: if observations should indicate that the poloidal component gains energy at the expense of the toroidal component (as may be tested by examining the same cloud at different phases in its evolution), then this would constitute a strong argument against the configuration in question being a self-similarly evolving spheromak. It would also strongly support the magnetic flux rope if the magnetic energy transfer is quantitatively as predicted by the theory.

Flattening of the magnetic field strength profile may occur even at 1 AU if the magnetic cloud is sufficiently slow. Thus, in the cloud seen at IMP 8 on the dawnside on August 27 - 29, 1989 (Figure 2), the flatter B-profile with respect to that of the January 1988 cloud is likely to be due in part to the August 1989 cloud being older, consistent with its slower average speed (440 vs. 613

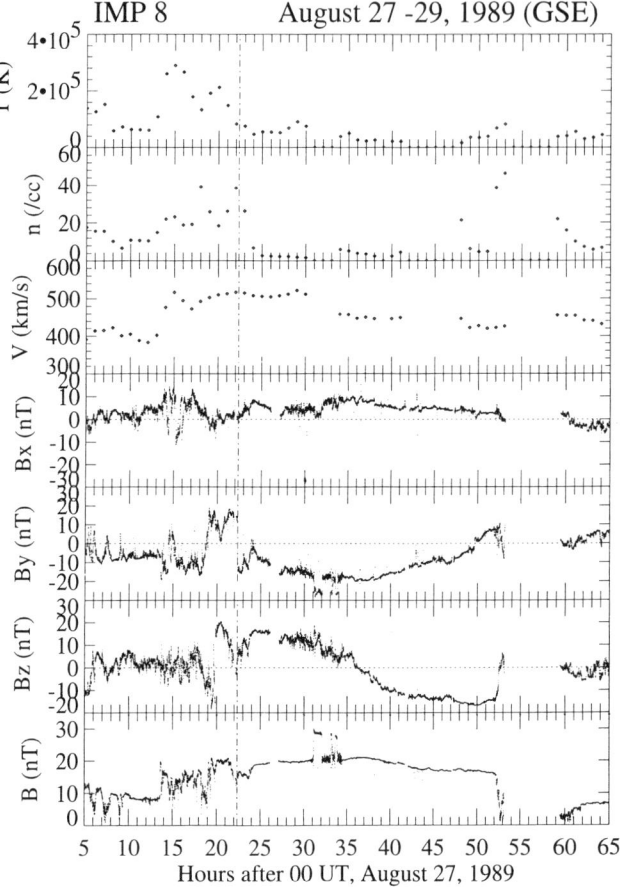

Figure 2. Plasma and magnetic field data for the magnetic cloud observed by IMP 8 on the dawnside. The figure shows hour-averages of the temperature, density and bulk speed of the protons, and the magnetic field in GSE coordinates. We argue that the bulk speed profile of this cloud is flatter than that of the January 1988 cloud because this cloud is older (lower speed).

km s^{-1}). (The large, sudden jumps in the magnetic field are due to crossings of the bow shock. Note also the clear $T_p - n$ structure in the sheath, with two clear humps in T_p coinciding with dips in n.)

In comparing the two models with data, we shall use the asymptotic expressions for y and F (and hence for the velocity) instead of the full equations (8) and (9). The use of asymptotic expressions for data comparisons is an approximation. It can be checked a posteriori from the inner consistency and the reasonableness of the results obtained. It can be shown [*Osherovich et al.*, 1993b; *Farrugia et al.*, 1995, 1997b] that the asymptotic radial expansion velocities are:

$$v = \frac{1}{2\gamma}\left(\frac{R}{t+t_0}\right) \quad (10)$$

and

$$v = \frac{r}{3(t+t_0)} \quad (11)$$

for the flux rope and the spheromak, respectively. In the case of the flux rope, the asymptotic radial velocity with $\gamma = 0.5$ corresponds to free expansion, studied in *Farrugia et al.* [1992, 1993]. Equations (10) and (11) yield different estimates of the ages and radii of the configuration, which may be checked for their plausibility.

4. COMPARISON OF ANALYTICAL RESULTS WITH DATA

4.1 Asymmetry in the total field profile

Expansion is one major contributor to the twofold asymmetry of the total field profile of a cloud not being overtaken by succeeding flows (see Figure 1). By the time the fields at the rear end of the cloud are sampled, they have had enough time to decrease relative to those near the forward end. Furthermore, the maximum (on the axis) of an initially spatially symmetric magnetic field profile will shift towards the leading edge once the self-similar field decrease overtakes the positive spatial gradient. Figure 3 illustrates the situation for the spheromak. As mentioned earlier, for the flux rope we have also a flattening of the profile. These asymmetries are kinematic in nature. Other asymmetries are expected to be induced through momentum exchange with the ambient medium, discussed briefly below.

4.2 The bulk speed profile

A sensitive way of assessing the relative merits of the spheromak and magnetic flux rope as models of magnetic clouds is to least-squares fit their predicted asymptotic bulk flow speed to data. In Figure 4 we show least-squares fit of the asymptotic bulk speed formulas of the flux rope (formula 10; solid line, with $\gamma = 0.5$) and the spheromak (formula (11); dashed line) to the bulk flow speed data for the January 1988 magnetic cloud in Figure 1. (The dot-dashed line fit is explained below.) In these fits the average bulk speed of the cloud has been folded into the expressions for the radial expansion speeds [*Farrugia et al.*, 1993]. The two free parameters, labeled r_0 and t_0, stand for the radial size of the cloud and the time the cloud has been expanding when the observations start. Both the modelled asymptotic velocities of the flux rope and the spheromak yield

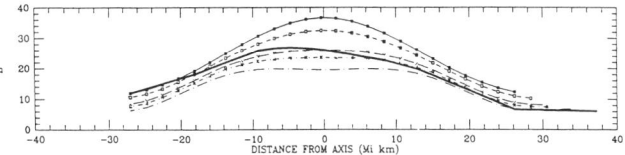

Figure 3. Five snapshots of the magnetic field strength along the symmetry axis of a spheromak (symbols joined by lines), taken at equal intervals of time. The heavy trace shows the field strength observed by a spacecraft travelling relative to the structure. (After *Farrugia et al.*, 1995)

very good fits to the data. However, considering that the average bulk flow speed of the cloud was 613 km/s, the flux rope fit gives a reasonable estimate for the age (65.4 h) while the spheromak fit (just 16.1 h) does not.

4.3 The magnetic field profile

Both model fields have also been fitted to the January 1988 magnetic field data. The algorithm is described in *Farrugia et al.* [1992]: we fit first the v-profile to obtain r_0 and t_0, and, using these, we fit the magnetic field data to obtain the field strength on the axis at the start of observations, the orientation of the cloud in space, and the distance of closest approach of the spacecraft to the cloud axis. This procedure is iterated if necessary. The resulting least-squares fits for the two models are shown in Figure 5a, b. From the point of view of the algorithm, both models give good fits of comparable quality, the sum of residues being about equal. Inspecting Figures 5a, b, it is clear that both models fit the center portion of the cloud well, with the maximum shifted towards the front edge in both models and data, as anticipated. In both models, the amount of field decrease due to self-similar expansion yields peak field values which are lower than the measurements. This is probably indicative of further compression of the cloud's field by the ambient medium it is overtaking. The flanks of the plot (i.e., the data segments near the front and rear boundaries of the cloud) are clearly better fitted by the straight cylindrical tube model, since the spheromak fit has changes in polarity in B_y and B_z which are not present in the data. When these points are taken into account, we conclude that the flux rope model gives a better fit to the data.

4.4 The thermodynamic structure

To expand self-similarly, cylindrical flux ropes require a polytropic index $\gamma \leq 1$, while spheromaks need a $\gamma = 4/3$. A study of the proton and electron com-

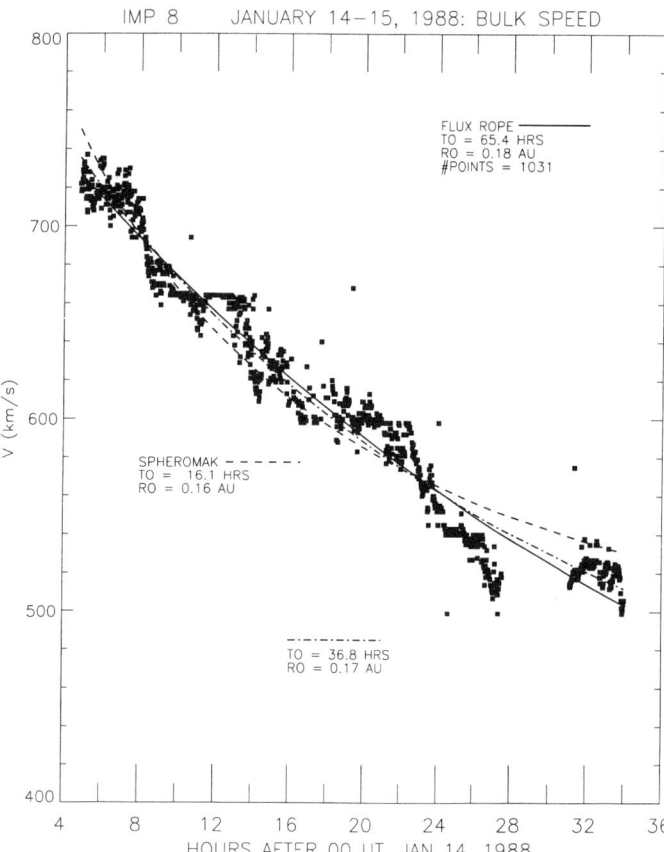

Figure 4. Two theoretical and one simulation asymptotic bulk speed profiles are shown fitted to the bulk speed measurements for the January, 1988 magnetic cloud. The solid line represents the fit for the magnetic flux rope; the dashed line, that for the spheromak, and the dot-dashed line, that derived from MHD simulations.

ponents in magnetic clouds using case examples [*Osherovich et al.*, 1993c; *Farrugia* and *Burlaga*, 1994] yielded these results: (a) inside the cloud the electron temperature is higher than the proton temperature by about an order of magnitude, and (b) both electron and proton components satisfy polytropic laws. The appropriate polytropic indices are 0.4 - 0.5 for the electrons and 1.1 - 1.3. for the protons. (See, also, *Osherovich* and *Burlaga*, this volume.) The major component in magnetic clouds is thus the electrons. These results are consistent with a self-similarly expanding flux rope model but are not consistent with a self-similarly expanding spheromak model. (One should not confuse the thermodynamics of magnetic clouds with that of the solar wind since magnetic clouds are distinct structures in the solar wind and their heat transfer need not be the same. Both T_e and n_e depend on heliospheric distance. The behavior of T_e and n_e in the solar wind as a function of heliospheric distance was studied by *Sittler* and *Scudder* [1980]. The behavior of T_e and n_e in magnetic clouds as a function of heliospheric distance, and for the same period, was studied by *Sittler* and *Burlaga* [1996]. The results of the two studies are different.)

We close this section with a Gedankenexperiment (*J. D. Scudder*, Private Communication, 1996). The possibility of deducing a polytropic index from a set of (T, n) measurements made along a spacecraft orbit intersecting the cloud is sometimes denied on the grounds that these measurements are not being made on the same streamline, or, equivalently, using the infinite conductivity assumption, on the same magnetic field line. This denial is a mistake. As argued by *Osherovich et al.* [1993], the axisymmetry relaxes this requirement and demands only that measurements be made on the same magnetic surface. Furthermore, during a crossing, a given magnetic surface is crossed twice in the case of a flux rope (and four times in the case of a spheromak, but we remain with the flux rope for clarity). Thus each (T, n) measurement made along a given orbit can be paired off (for sufficiently high resolution data) with another (T, n) measurement made as the same magnetic surface is crossed a second time. One can use the magnetic field model to determine when the second crossing of the magnetic surface occurs and hence determine the (T, n) measurement made there. A set of polytropic indices can be determined from each such data pairs and, if they are close enough to each other, a polytropic index can be deduced. *Osherovich et al.* [1993] used an alternative, but equivalent, way of deducing the polytropic indices: they made internal checks to verify that the magnetic-surface dependent quantity to which $T/\rho^{(\gamma-1)}$ was proportional indeed remained approximately constant across the tube [*Osherovich* and *Burlaga*, this volume]. A physical mechanism for obtaining a polytropic index less than unity has been given by *Scudder* [1992], who showed this may be the result of deviations from Maxwellian distributions in a low density plasma.

In summary, we have applied one line of theoretical and numerical investigation (MHD model for self-similar, radial expansion), in conjunction with in situ field and plasma data, to study the suitability of the magnetic flux rope and the spheromak as models of magnetic clouds. Within this approach, it appears that it is unlikely that magnetic clouds are self-similarly expanding spheromaks. On the other hand, the self-similarly expanding magnetic flux rope gives good agreement with a wide variety of data.

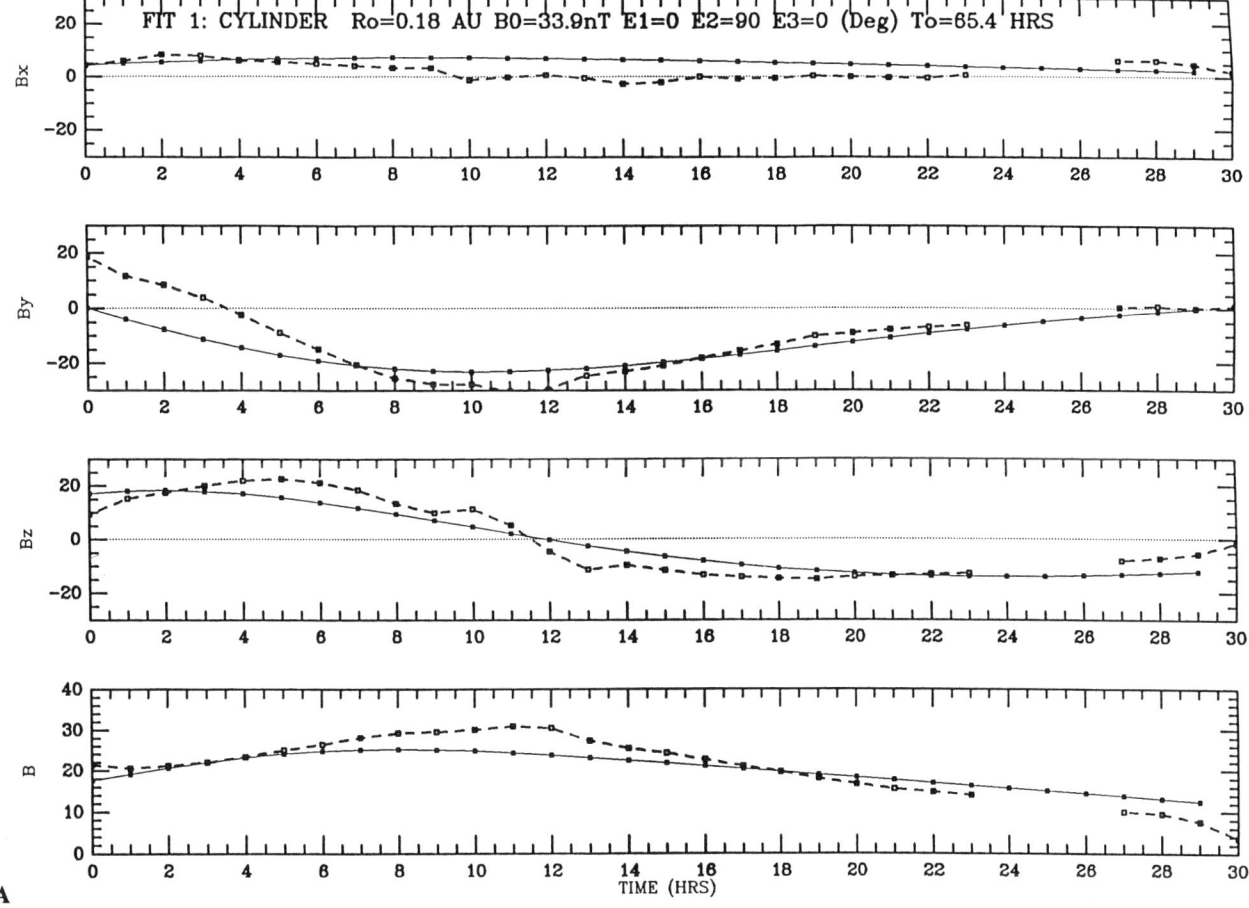

Figure 5. Fits of the flux rope (5a) and the spheromak magnetic fields (5b) to that of the January 1988 magnetic cloud. (After *Farrugia et al.*, 1995)

5. SIMULATION RESULTS; COMPARISON OF SIMULATION AND ANALYTICAL CONCLUSIONS

The propagation of cylindrical magnetic flux ropes from the vicinity of the Sun to beyond 1 AU was simulated by *Vandas et al.* [1995a, 1996] using a 2.5-dim MHD code, and by *Vandas et al.* [this volume]. (For simulation results on the propagation of spheromaks, the reader is referred to *Vandas et al.*, this volume, and also to *Detman et al.* [1991].) In these simulations, the initial configuration is a Lundquist flux tube. Simulations were made for flux ropes ejected radially with their axes either parallel or perpendicular to the ecliptic plane. A further work by *Vandas et al.* [1996b] examines the effect of various parameters on the evolution and undertakes a comparison of simulation and analytical results.

In this body of work, a number of interesting, new results are presented. We cite some of these. Magnetic clouds are found to expand faster near the Sun and, in the cases studied, to reach an asymptotic bulk speed independent of the details of release. The magnetic flux ropes are deformed as they propagate antisunward, and the authors find that at 1 AU the azimuthal dimension is of order 1.5-2.0 larger than the radial dimension. (A deformation in the same sense of the spheromak was also found by *Detman et al.* [1991], with the initially spherical shape becoming flattened in the direction perpendicular to the ecliptic.) The influence of the inclination of the flux rope's axis to the ecliptic is found not to affect the evolution of cloud parameters significantly. A further, interesting result is that magnetic cloud flux ropes deflect from a central radial path toward the side where the magnetic fields external and

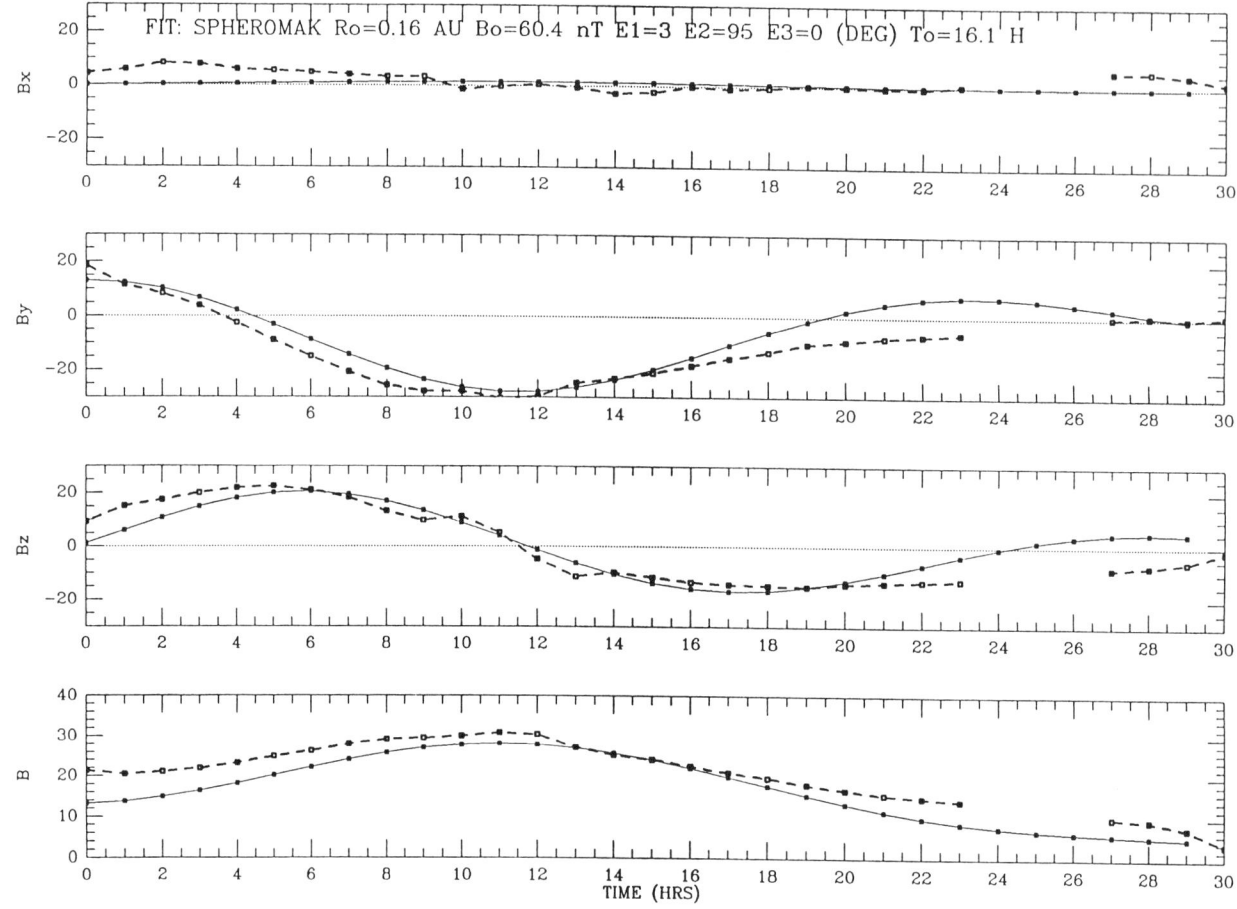

Figure 5 (continued).

internal to the cloud are oppositely directed. The authors find that the magnetic and plasma signature of a simulated cylindrical flux rope is in good qualitative agreement with observations.

One emphasis of the work by *Vandas et al.* [1996b] is the asymptotic behavior of simulated flux ropes, and the authors draw some points of comparison with corresponding results from the analytical theory reviewed above. In the simulation work, the polytropic index is 5/3 (as opposed to ~ 0.5 in the theory). Nonetheless the authors arrive at formulae for the asymptotic behavior of various physical quantities which are formally similar to those obtained in the theory (*Osherovich et al.* [1993b], see above). We shall discuss here the simulation result for the asymptotic expansion speed (formula (2), *Vandas et al.* [1996b]) and compare it with expression (10). The simulation of a cloud projected initially at 1000 km s^{-1} into the nominal solar wind (defined in *Vandas et al.* [1996b]) gives $v \approx (1/1.6)(r/t)$ (*Vandas et al.* [1996b], their Figure 8). We fit this formula (with the average bulk speed folded in) to the bulk speed data for the January 1988 magnetic cloud. The result is shown by the dot-dashed line in Figure 4. The fit to the data is good, and the inferred radius of the tube at the start of observations (r_0) is also reasonable and in agreement with those derived from the two analytical models. The inferred age, however, (36.8 h) appears too short and would require a huge or prolonged acceleration. Other simulations reported by *Vandas et al.* [1996b] give $v \approx (1/1.4)(r/t)$. Fitted to the January 1988 data (but not shown), this formula yields an age of 43.5 h and a radial size of 0.18 AU. In short, it seems important to examine under what conditions (if any) the simulations predict an asymptotic bulk speed which is essentially $v = r/t$, as this example shows that any factor on the right hand substantially different from unity leads to what appear to be underestimates of the cloud's age.

6. CONSIDERATIONS OF DRAPING

In principle, the draping of the interplanetary magnetic field (IMF) around the two model structures should help to distinguish between them because draping depends on the shape of the obstacle. Let us consider first a straight cylindrical flux rope. MHD theory of flow around a cylindrical flux rope predicts in this case that the IMF component orthogonal to both the symmetry axis of the tube and the flow velocity of the solar wind should be the one to be enhanced most, while the other components are just convected gasdynamically in the sheath region [Erkaev et al., 1995]. Thus, e.g., for a flux tube aligned along the east-west direction (a common orientation of magnetic cloud axes) propagating antisunward, it is the IMF component perpendicular to the ecliptic which determines the structure of the sheath and its depletion layer (width, and variation of parameters.) On the other hand, if the flux rope is orthogonal to the ecliptic, then it is rather the east-west component of the IMF which drapes most and thus influences the sheath structure most.

The situation of the expanding spheromak is different. Even without deformation, both IMF components perpendicular to the bulk flow velocity of the solar wind are enhanced and are important in determining the structure of the sheath and its depletion layer, when present. In this case the situation is analogous to that of the terrestrial magnetosheath.

We have seen above that the study of Detman et al. [1991] predicts that a spheromak should get deformed in such a way that the azimuthal dimension increases relative to the radial. If we idealize somewhat and consider the spheromak to have developed an elliptical cross-section described by two radii of curvature (a smaller one in a plane through the "nose" perpendicular to the ecliptic, and a larger one in a plane through the "nose" along the ecliptic), then we have to consider the effect of the non-axisymmetry on the sheath flow. An analogous situation has been studied in connection with planets Jupiter and Saturn, whose magnetospheres depart strongly from axisymmetry in the same sense as spheromaks (i.e., they are broadened in the rotational equator [Erkaev et al., 1996; Farrugia et al., 1997c]). When the obstacle is non-axisymmetric, the thickness of the magnetosheath and the width and structure of the depletion layer are strong functions of the orientation of the interplanetary field. In particular, the effect is strongest (weakest) when the IMF is perpendicular (parallel) to the plane of largest curvature radius. In addition, the magnetic field in the sheath undergoes a smooth rotation towards the normal to the ecliptic as the obstacle is approached.

The last point is illustrated in the hodographs of B_y vs. B_z (normalized to the free-stream magnetic field) along the stagnation line shown in Figure 6. Each plot extends from the bow shock (lower left) to the magnetopause. The various plots are parametrised by the co-latitude of the interplanetary magnetic field. The starred symbols indicate equal intervals of time. Parameter q is a measure of the flattening, which is more pronounced for Jupiter. The top panel refers to the terrestrial magnetosheath, and would be applicable to a spheromak which is not deformed. Here both IMF components orthogonal to the solar wind velocity are enhanced by the same amount. The second and third panels refer to Saturn and Jupiter, respectively, and would in turn be applicable to a spheromak which is deformed as reported. In addition to the enhancement, the field rotates as described, the rotation being fastest close to the obstacle. This effect is not present in the terrestrial magnetosheath (equal radii of curvature of magnetosphere) or in the case of flow around straight cylindrical flux ropes (one radius of curvature is infinite). This effect is worth studying in simulations of spheromak propagation. Admittedly, as other studies of ejecta sheaths have shown, the effect might be hard to find, not because it is small (see Figure 6) but because the fluctuations in the sheath ahead of magnetic clouds make it hard to study systematic changes there.

7. DISCUSSION AND CONCLUSIONS

Determination of the global field topology of magnetic clouds is a large issue which, in the context of earlier so-called "plasma clouds", has engaged the attention of many scientists for a long time. It has not been our intention to settle this issue here in this brief review. Our aim has been much more circumscribed: to report on the results of an exact MHD theory in a specific approach (radial self-similar expansion) as applied to two models, the magnetic flux rope of locally straight, cylindrical topology, and the spheromak. Restricting the scope, however, also carries an advantage: the analytical models are developed enough to contain clear predictions which can be checked against data. Predicting power is always a sign of a good theory. Furthermore, these predictions can be checked against sophisticated MHD simulations. Within the approach we adopted, the conclusion is clear: where it has been tested, the magnetic flux rope accounts for the data better than does the spheromak. We had nothing to

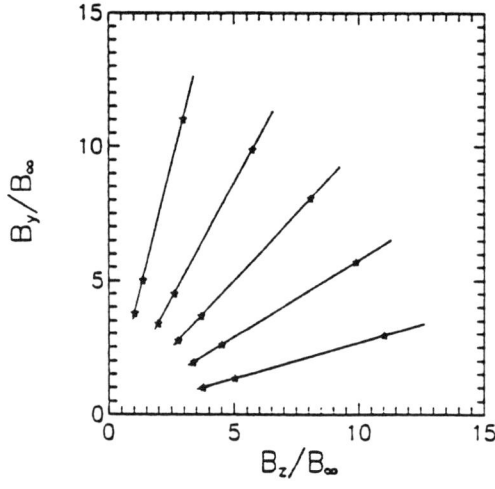
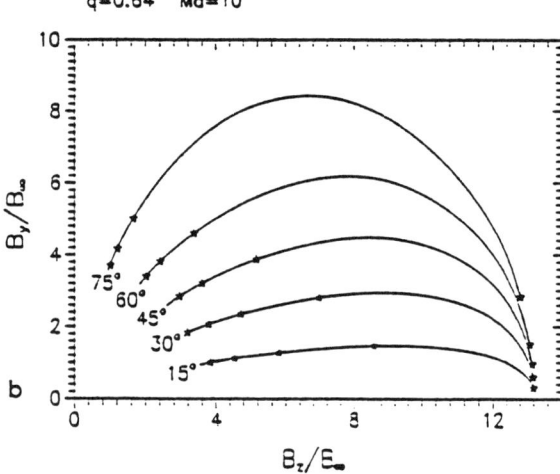
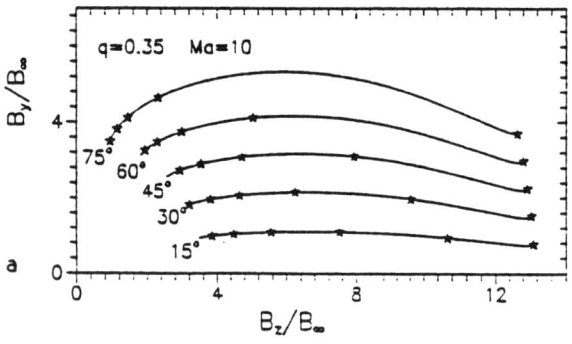

Figure 6. Hodographs of B_y vs. B_z along the stagnation streamline for (a) Earth, (b) Saturn, and (c) Jupiter. The various plots are parametrized by the colatitude angle of the IMF. The curves start at the bow shock (lower left) and end at the magnetopause. The starred symbols indicate equal times along the stagnation streamline.

say about other possible topologies and other possible approaches.

Aside from our specific modelling, there is abundant circumstantial evidence favouring the flux rope topology. By definition, the spheromak is a structure detached from the Sun, whereas the magnetic flux rope may still be magnetically connected to the Sun. A lot of data on the directional properties and intensity variations of solar energetic (few MeV) particles in magnetic clouds can all be consistently interpreted in terms of a magnetic flux rope which on a large scale is bent and connected to the Sun (see review by *Richardson* [this volume]). (The inference that magnetic cloud flux ropes are bent on a large scale was drawn by *Burlaga et al.* [1990] in a study of data from several widely-separated spacecraft.) Again, reviewing coronal mass ejections, *Gosling* [1990] considered also that subset whose members possess a coherent and large magnetic field rotation (many of which are magnetic clouds) and concluded that the evidence for an underlying flux rope topology is very strong.

In future elaboration of the flux rope model, there is need for more detailed study of the interaction of the magnetic cloud with surrounding flows. The January 1988 cloud was not being overtaken by faster flows but was running into a slower stream (Figure 1). It is quite possible that part of the compression of the forward edge of the cloud, that part which was not accounted for by the expanding Lundquist field (see Figure 4), is due to this interaction. When, in addition, magnetic clouds are followed by faster flows, the ensuing interaction may on occasion result in substantial changes to the magnetic and plasma profiles, as can be seen in many magnetic cloud observations made by the WIND spacecraft. In Figure 7 we show an extreme case. Here a magnetic cloud driving a shock (shown by the first vertical guideline from the left, the magnetic cloud interval itself being approximately within the other two guidelines) is clearly being overtaken by a faster stream. The signature is affected by the interaction. In particular, the bulk speed profile does not have the usual expansion signature and would, *prima facie*, be interpreted rather as a signature of contraction. Nevertheless, aside from such extreme cases, when one considers the detailed and quantitative predictions made on the basis of the analytical model for self-similar expansion of flux ropes on a variety of issues, and the general agreement with observations, one has to conclude that the approach is scientifically mature and should be pursued further.

Acknowledgments. I would like to thank L. F. Burlaga and V. A. Osherovich, J. D. Scudder, and N. V. Erkaev for

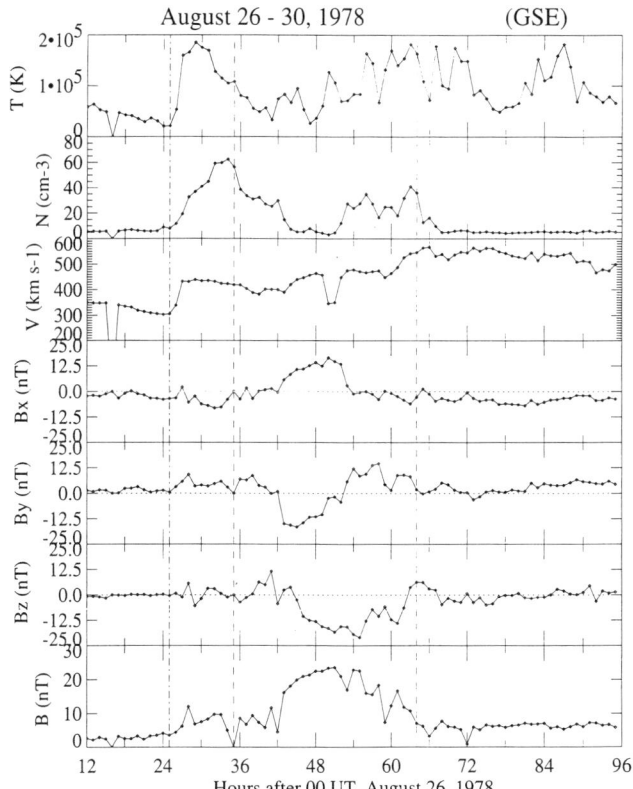

Figure 7. Plasma and magnetic field data for the magnetic cloud for August 26-30, 1978

useful discussions. Many thanks to Laurence Janoo and D. J. Ryan for technical help. This work was supported in part by NASA Grant NAG 5-2834 and DARA grant 50-OC-8911-0.

REFERENCES

Bernstein, I. B., and R. M. Kulsrud, On the explosion of a supernova into the interstellar magnetic field, *Astrophys. J.*, *142*, 479, 1965.

Burlaga, L. F., Magnetic Clouds: Constant alpha force-free configurations, *J. Geophys. Res.*, *93*, 7217, 1988.

Burlaga, L. F., Magnetic Clouds, Chapter 5 in *Physics of the Inner Heliosphere*, Vol. 2, edited by R. Schwenn and E. Marsch, p.1, Springer-Verlag, Berlin-Heidelberg, 1991.

Burlaga, L. F., *Interplanetary Magnetohydrodynamics*, Oxford University Press, New York, 1995.

Burlaga, L. F., E. Sittler, F. Mariani, and R. Schwenn, Magnetic loop behind an interplanetary shock: Voyager, Helios and IMP 8 observations, *J. Geophys. Res.*, *86*, 6673, 1981.

Burlaga, L. F., R. P. Lepping, and J. A. Jones, Global configuration of a magnetic cloud, in *Physics of Magnetic Flux Ropes*, edited by C. T. Russell, E. R. Priest, and L. C. Lee, AGU Geophysical Monograph 58, American Geophysical Union, Washington, DC, p. 373, 1990.

Chandrasekhar, S, and P. C. Kendall, On force-free magnetic fields, *Astrophys. J.*, *126*, 457, 1957.

Chen, J., Coronal mass ejections: Causes and consequences, a theoretical view, this volume.

Detman, T. R., M. Dryer, T. Yeh, S. M. Han, S. T. Wu, and D. J. McComas, A time-dependent, three-dimensional MHD numerical study of interplanetary magnetic field draping around plasmoids in the solar wind, *J. Geophys. Res.*, *96*, 9531, 1991.

Erkaev, N. V., C. J. Farrugia, H. K. Biernat, L. F. Burlaga, and G. A. Bachmaier, Ideal MHD flow behind interplanetary shocks driven by magnetic clouds, *J. Geophys. Res.*, *100*, 19,919, 1995.

Erkaev, N. V., C. J. Farrugia, and H. K. Biernat, Effects on the Jovian magnetosheath arising from solar wind flow around non-axisymmetric bodies, *J. Geophys. Res.*, *101*, 10,665, 1996.

Fainberg, J., V. A. Osherovich, R. G. Stone, R. J. MacDowall, and A. Balogh, Ulysses observations of electron and proton components in a magnetic cloud and related wave activity, in *Solar Wind Eight*, AIP Conference Proceedings 382, edited by D. Winterhalter, J. Gosling, S. R. Habbal, W. S. Kurth, and M. Neugebauer, p. 554, 1995.

Farrugia, C. J., L. F. Burlaga, V. A. Osherovich, and R. P. Lepping, A comparative study of dynamically expanding force-free, constant-alpha magnetic configurations with applications to magnetic clouds, in *Solar Wind Seven*, edited by E. Marsch and R. Schwenn, Pergamon Press, p. 611, 1992.

Farrugia, C. J., L. F. Burlaga, V. A. Osherovich, I. G. Richardson, M. P. Freeman, R. P. Lepping, and A. J. Lazarus, A study of an expanding interplanetary magnetic cloud and its interaction with Earth: The interplanetary aspect, *J. Geophys. Res.*, *98*, 7621, 1993.

Farrugia, C. J., and L. F. Burlaga, A fast-moving magnetic cloud and features of its interaction with the dayside magnetosheath, in *The Solar Wind-Magnetosphere System*, edited by H. K. Biernat, G. A. Bachmaier, S. Bauer, and R. P. Rijnbeek, Austrian Academy of Sciences Press, Vienna, p. 33, 1994.

Farrugia, C. J., V. A. Osherovich, and L. F. Burlaga, Magnetic flux rope versus the spheromak as models for interplanetary magnetic clouds, *J. Geophys. Res.*, *100*, 12,293, 1995.

Farrugia, C. J., L. F. Burlaga, and R. P. Lepping, Magnetic Clouds and the quiet-storm effect at Earth, in *Magnetic Storms*, edited by B. T. Tsurutani, W. D. Gonzales, and Y. Kamide, AGU Monograph, Washington, D. C, in press, 1997a.

Farrugia, C. J., V. A. Osherovich, and L. F. Burlaga, The non-linear evolution of magnetic flux ropes: 3. Effects of Dissipation, *Annales Geophys.*, *15*, 152, 1997b.

Farrugia, C. J., H. K. Biernat, and N. V. Erkaev, Numerical modelling of solar wind flow past non-axisymmetric magnetospheres: Planets Jupiter and Saturn, *Adv. Space Res.*, in press, 1997c.

Goldstein, H., On the field configuration in magnetic clouds, in *Solar Wind Five*, edited by M. Neugebauer, p.731, NASA Conf. Publ. 2280, Washington, D. C., 1983.

Gosling, J. T., Coronal mass ejections and magnetic flux ropes in interplanetary space, in *Physics of Magnetic Flux Ropes*, edited by C. T. Russell, E. R. Priest, and L. C. Lee, p. 344, AGU Geophysical Monograph 58, American Geophysical Union, Washington, D. C., 1990.

Ivanov, K. G., and A. F. Harshiladze, Interplanetary hydromagnetic clouds as flare-generated spheromaks, *Solar Phys.*, *98*, 379, 1985.

Klein, L. W., and L. F. Burlaga, Interplanetary magnetic clouds at 1 AU, *J. Geophys. Res.*, *87*, 613, 1982.

Kulsrud, R. M, I. B. Bernstein, M. Kruskal, J. Fanucci, and N. Ness, On the explosion of a supernova into the interstellar magnetic field, II, *Astrophys. J.*, *142*, 491, 1965.

Lepping, R. P., J. A. Jones, and L. F. Burlaga, Magnetic field structure of interplanetary magnetic clouds at 1 AU, *J. Geophys. Res.*, *95*, 11,957, 1990.

Low, B. C., Self-similar magnetohydrodynamics, I, The $\gamma = 4/3$ polytrope and the coronal transient, *Astrophys. J.*, *254*, 796, 1982.

Lundquist, S., Magnetohydrostatic fields, *Ark. Fys.*, *2*, 361, 1950.

Marubashi, K., Structure of the interplanetary magnetic clouds and their origins, *Adv. Space Sci.*, *6(6)*, 335, 1986.

Osherovich, V. A., C. J. Farrugia, and L. F. Burlaga, Nonlinear evolution of magnetic flux ropes: 1. The low beta limit, *J. Geophys. Res.*, *98*, 13,225, 1993a.

Osherovich, V. A., C. J. Farrugia, and L. F. Burlaga, Dynamics of aging magnetic clouds, *Adv. Space Res.*, *13*, *6(6)*, 57, 1993b.

Osherovich, V. A., C. J. Farrugia, L. F. Burlaga, R. P. Lepping, J. Fainberg, and R. G. Stone, Polytropic relationship in interplanetary magnetic clouds, *J. Geophys. Res.*, *98*, 15,331, 1993c.

Osherovich, V. A., C. J. Farrugia, and L. F. Burlaga, The non-linear evolution of magnetic flux ropes: 2. Finite beta plasma, *J. Geophys. Res.*, *100*, 12,307, 1995.

Osherovich, V. A., and L. F. Burlaga, Magnetic Clouds, this volume.

Richardson, I. G., Using energetic particles to probe the magnetic topology of ejecta, this volume.

Rosenbluth, M. N., and M. N. Bussac, MHD stability of spheromak, *Nucl. Fusion*, *19*, 489, 1979.

Scudder, J. D., On the causes of temperature change in inhomogeneous low-density astrophysical plasmas, *Astrophys. J.*, *398*, 299, 1992.

Sittler, E. C., Jr., and J. D. Scudder, An empirical polytropic law for solar wind thermal electrons between 0.45 and 4.76 AU: Voyager 2 and Mariner 10, *J. Geophys. Res.*, *85*, 5131, 1980.

Sittler, E. C., and L. F. Burlaga, Electron temperatures within magnetic clouds between 2 and 4 AU: Voyager observations (abstract), *EOS Trans. AGU*, *77(46)*, Fall Meet. Suppl., F589, 1996.

Vandas, M., S. Fischer, and A. Geranios, Spherical and cylindrical models of magnetic clouds and their comparison with spacecraft data, *Planet. Space Sci.*, *39*, 1147, 1991.

Vandas, M., S. Fischer, P. Pelant, and A. Geranios, Spheroidal models of magnetic clouds and their comparison with spacecraft measurements, *J. Geophys. Res.*, *98*, 11,467, 1993.

Vandas, M., S. Fischer, M. Dryer, Z. Smith, and T. Detman, Simulation of magnetic cloud propagation in the inner heliosphere in two dimensions, 1. A loop perpendicular to the ecliptic plane, *J. Geophys. Res.*, *100*, 12,285, 1995a.

Vandas, M., S. Fischer, M. Dryer, Z. Smith, and T. Detman, Self-consistent simulation of cylindrical magnetic cloud propagation in the heliosphere with its axis both perpendicular to, and lying within, the ecliptic plane, *Adv. Space Res.*, *17(4/5)*, 327, 1995b.

Vandas, M., S. Fischer, M. Dryer, Z. Smith, and T. Detman, Simulations of magnetic cloud propagation in the inner heliosphere in two dimensions, 2. A loop parallel to the ecliptic plane and the role of helicity, *J. Geophys. Res.*, *101*, 2505, 1996a.

Vandas, M., S. Fischer, M. Dryer, Z. Smith, and T. Detman, Parametric study of loop-like magnetic cloud propagation, *J. Geophys. Res.*, *101*, 15,645, 1996b.

Vandas, M., S. Fischer, D. Odstrcil, S. Fischer, M. Dryer, Z. Smith, and T. Detman, Flux ropes and spheromaks: A numerical study, this volume.

C. J. Farrugia, Institute for the Study of Earth, Oceans, and Space, University of New Hampshire, Durham, N. H. 03824. (e-mail: farrugia@monet.sr.unh.edu)

Using Energetic Particles to Probe the Magnetic Topology of Ejecta

I.G. Richardson[1]

Laboratory for High Energy Astrophysics, NASA Goddard Space Flight Center, Greenbelt, Maryland

Observations of energetic particles over a wide range of rigidities can provide information on the presence of ejecta, and their structure. A large fraction of ejecta produce a local, few percent depression of the cosmic ray density. The depression typically extends over the ejecta region as determined from a range of ejecta signatures, in particular plasma with abnormally low proton temperatures. At lower rigidities (less than 0.5 GV, E_{proton}<100 MeV), the depression can be greater than 70%. These observations suggest that ejecta are predominantly closed magnetic structures. The bidirectional particle flows present in regions of some ejecta are consistent with particle circulation/reflection within closed structures. The observation of solar particle event onsets inside some ejecta suggests that field lines in these ejecta are rooted at the Sun. These particles arrive from a range of directions, including from the east. The observations are consistent with the presence of looped magnetic field lines rooted at the Sun in many ejecta, although recent observations from the WIND spacecraft suggest that ejecta may be more complex and include regions of field lines both connected to and disconnected from the Sun. At high heliolatitudes, Ulysses observations suggest that energetic particle signatures of ejecta are slightly different. In particular, low energy (~MeV) particle enhancements, rather than depressions, are found in ejecta, apparently because ejecta guide particles to high latitudes where the ambient particle intensity is lower than near the ecliptic.

1. INTRODUCTION

Energetic particles are valuable probes of ejecta structure. Their gyroradii (e.g. 4×10^{-5} AU for a 1-MeV proton in a 20-nT field) are small compared to ejecta scale sizes while their speeds (1.4×10^4 km s^{-1} for a 1-MeV proton) are much greater than typical ejecta speeds. In addition, ejecta often contain low levels of magnetic field turbulence, resulting in long (>> 1 AU [*Tranquille et al.*, 1987]) particle scattering mean free paths parallel to the magnetic field which allow particles to travel large distances along ejecta field lines with little scattering. Thus energetic particles can provide information on the large-scale magnetic field topology of ejecta including the presence of open or closed field lines, and can indicate when field lines are connected to particle sources such as the Sun, interplanetary shocks and galactic cosmic rays. They also provide evidence for the propagation of local structures, such as ejecta boundaries, past the spacecraft. In this paper, various ways in which energetic particle observations can provide clues to the presence of eject, and to their structure, are reviewed. Several examples will be discussed using data principally from the Goddard Space Flight Center (GSFC) instruments on the ISEE 3, IMP 8 and Helios 1 and 2 spacecraft, and the University of Kiel instruments on Helios 1 and 2. ISEE 3 and IMP 8 were in the vicinity of the Earth, while Helios 1 and 2 were in heliocentric orbits extending 0.3-1 AU from the Sun.

2. PARTICLE DEPRESSIONS IN EJECTA

Entry into and exit from an ejecta is typically accompanied by a decrease and a recovery, respectively, in the particle density over a range of rigidities. Figure 1 shows an ejecta-

[1]Also at Department of Astronomy, University of Maryland, College Park.

190 ENERGETIC FIELD PARTICLES AND EJECTA TOPOLOGY

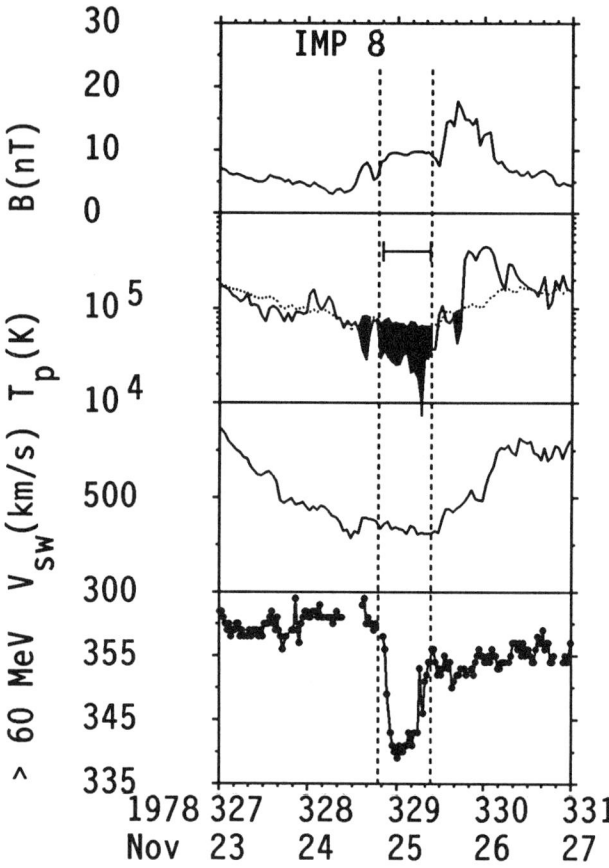

Figure 1. A cosmic ray depression (bottom panel), indicated by the counting rate (counts s^{-1}) of the anti-coincidence guard of the GSFC instrument on IMP 8, which is associated with an ejecta. One signature of the ejecta is the region of plasma with abnormally low proton temperatures in panel 2. The shading indicates the difference between the observed proton temperature T_p (solid line) and the temperature expected for normal solar wind expansion (T_{ex}, dotted line), which is based on the solar wind speed in the third panel, when $T_p < 0.5\ T_{ex}$. Another ejecta signature is the bidirectional solar wind electron heat flux (BDE) indicated by the horizontal line in panel 2 [*Gosling et al., 1987*]. The top panel shows the magnetic field intensity.

associated, short duration, cosmic ray decrease at IMP 8 in November 1978. A typical signature of an ejecta is the extended region of abnormally low proton temperature (T_p), identified in panel 2 by comparing T_p (solid line) with the temperature expected for normally expanding solar wind (T_{ex}, dotted line) [*Richardson and Cane, 1995*]. Black shading indicates when $T_p < 0.5\ T_{ex}$. Another signature is the interval of bidirectional solar wind electron flows (BDE) indicated by the horizontal line in panel 2 [*Gosling et al., 1987*]. The bottom panel shows count rates from the anticoincidence guard of the GSFC IMP 8 instrument. The guard detects greater than 60 MeV amu^{-1} particles and at quiet times indicates the cosmic ray density at a median rigidity of ~2 GV. The ejecta produced a local, ~6% decrease in the

cosmic ray density. A similar event, associated with a magnetic cloud (i.e. an ejecta region including an enhanced magnetic field with a smooth rotation, and low plasma β (ratio of the plasma and magnetic field pressures) [*Burlaga et al.* [1981]) is shown in Figure 10 of *Richardson and Cane*, [1995].

Figure 2 shows a more typical situation where a fast ejecta generates a shock ahead of it which accelerates ions to MeV energies. The shock (solid vertical line) and ejecta (between the dashed vertical lines) passed Helios 2 (Figure 2(a)), located 0.4 AU from the Sun and 5° west of the Sun-Earth line, around one and a half days before encountering IMP 8 (Figure 2(b)). The three panels for each spacecraft show T_p and T_{ex} (with low temperature plasma associated with the ejecta shaded), the anticoincidence guard rate, and ~4-MeV ion intensity. The anticoincidence guards of the University of Kiel Helios and GSFC IMP 8 instruments have similar energy responses, but in the figure, the IMP 8 counting rate has been normalized to the Helios 2 rate to allow them to be intercompared. Although there are short data gaps, the guard rates at both spacecraft show evidence of a "two-step" cosmic ray depression. The first decrease (caused by the post-shock turbulence) commences close to shock passage and is followed by an additional abrupt decrease on entry to the ejecta. The total decrease is 14% at Helios 2 and 11% at IMP 8. The recovery is more gradual, with a change in the recovery rate at the trailing edge of the ejecta which marks the return from ejecta field lines to field lines connected to the shock on which the cosmic ray density is still depressed by the postshock turbulence. Thus there is a local cosmic ray decrease in the ejecta commencing at the leading edge of the ejecta (as inferred from T_p and other ejecta signatures) together with an extended depression associated with the postshock turbulence. This interpretation of two-step cosmic ray decreases was discussed by *Barnden* [1972] and is supported by the multispacecraft study of *Cane et al.* [1994] which used anticoincidence guard data to confirm the separate contributions of the local ejecta-associated decrease and the extended depression in the postshock region. Overall, the vast majority (~90%) of ejecta observed by the Helios 1 or 2 spacecraft were accompanied by significant depressions in the cosmic ray density (of which ~65% were ≥4%), suggesting that such depressions are a robust signature of ejecta [*Cane et al.*, 1997].

At both spacecraft in Figure 2, there is an abrupt decrease in the ~4-MeV ion intensity by greater than 90% at the leading edge of the ejecta. This decrease marks a transition from field lines which are connected to the shock and loaded with accelerated particles, to field lines inside the ejecta which do not connect to the shock. (The October 18-20, 1995 magnetic cloud shows this type of decrease particularly clearly (T. Sanderson, private communication, 1996).) Often, there is a corresponding increase in the shock-accelerated particle intensity on exit from the ejecta as the spacecraft returns to field lines connected to the shock [e.g. *Cane et al.*, 1988], but there is no evidence of such an increase at the trailing edge of this particular ejecta. The shock-accelerated particle intensity usually falls off with distance from the shock in the postshock region so presumably, the particle intensities inside and outside the ejecta trailing edge were similar in this event.

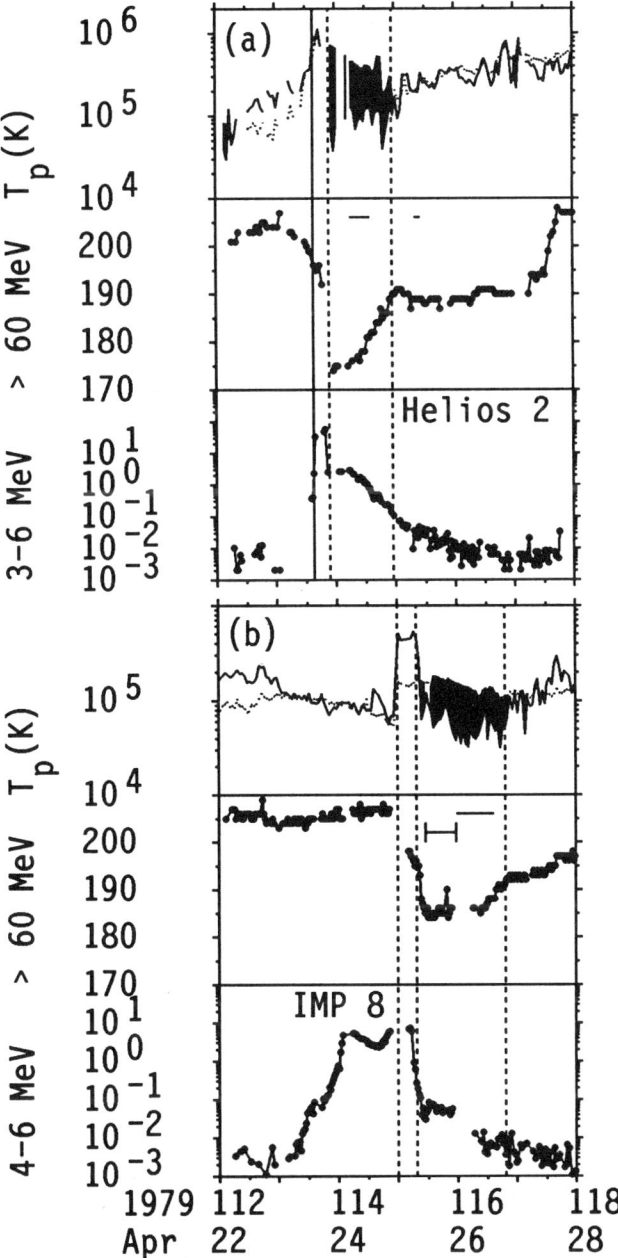

Figure 2. A "two-step" cosmic ray depression produced by a shock (solid vertical line) and ejecta (the region between the dashed vertical lines) passing (**a**) Helios 2 at 0.4 AU, 5°W and then (**b**) the Earth. As in Figure 1, the ejecta includes abnormally low temperature plasma (shaded regions in the top panel of each figure). The counting rates from the anticoincidence guards of the Kiel (Helios 2) or GSFC instruments (IMP 8) in the middle panels show a step down in the cosmic ray density commencing in the vicinity of the shock and which extends into the region after ejecta passage, together with a local depression in the ejecta which commences abruptly on entry to the ejecta. The bottom panel in each figure indicates the large (greater than 90%) fall in the shock-accelerated ~4-MeV ion intensity ((MeV s cm^2 sr)$^{-1}$) on entry to the ejecta. The horizontal lines without bars at the ends in the middle panel of each figure indicate intervals of bidirectional MeV ion flows (BIFs) while that with bars indicates an interval of bidirectional solar wind electron heat fluxes (BDEs).

Figure 2 illustrates that the depth of the particle depression is dependent on particle rigidity. To examine this further, Figure 3 shows the percent depression versus particle rigidity (P) for several ejecta. The lowest and intermediate rigidity data are from instruments on IMP 8 and ISEE 3, while higher rigidity data are from the University of Tasmania neutron monitor network and the Cambridge muon telescope. The depressions are >70% at P<0.5 GV (E_{proton}<100 MeV) and fall off as ~P$^{-0.6}$ at higher rigidity. There is little difference in the response of magnetic clouds (unfilled circles) and other ejecta. The local cosmic ray depressions in ejecta may be modelled by assuming that cosmic rays enter the interior of the ejecta (initially empty of cosmic rays) by perpendicular diffusion [*Cane et al.*, 1995; *Vanhoefer*, 1996]. Qualitatively, the rigidity dependence in Figure 3 arises because higher rigidity particles are able to diffuse more rapidly across field lines. This model is supported by observations of the same ejecta at different distances from the Sun (e.g. Figure 2) which suggest that the cosmic ray depression at nearly the same heliolongitude is deeper closer to the Sun [*Cane et al.*, 1994].

Studies of the interaction of cosmic rays with ejecta have also been made using data from multiple neutron monitors. For example *Belov et al.*, [1995] combined hourly-averaged data from around 40 neutron monitors in order to infer the cosmic ray distribution in the vicinity of Earth during particle depressions. Their results suggest that significant changes in the cosmic ray distribution occur which may be indicative of particle flows within ejecta. Such studies have only recently been attempted now that data from neutron monitors covering a wide range of geomagnetic longitudes and latitudes are readily available, but they have the potential for providing new insights into ejecta structure.

The close association between the boundaries of ejecta (as inferred from T_p depressions and other ejecta signatures [*Gosling*, 1990]) and energetic particle depressions suggests that ejecta are predominantly closed magnetic structures. Otherwise, if energetic particles were to enter the interior of ejecta along open magnetic field lines, the particle depression would not extend for the duration of the ejecta, and regions would be observed within ejecta in which the cosmic ray density is less depressed. Also, the abrupt decrease of shock-accelerated particles at the leading edge of many ejecta implies that field lines in the leading edge of these ejecta are not connected to field lines in the post-shock solar wind which are likely to intercept the shock.

The observations described above summarize the situation at ~1 AU near the ecliptic plane. The recent Ulysses mission has provided information on the relationship between energetic particles and ejecta at high heliographic latitudes. One major difference is that at high latitudes, the ~MeV ion intensity may be enhanced inside ejecta relative to the intensity outside rather than depressed as in the events discussed above [*Bothmer et al.*, 1995]. The reasons for this difference appear to be (1) there is a much weaker enhancement of shock-accelerated ions outside the ejecta at high latitudes, and (2) the ejecta either carries particles contained within it to high latitudes, or ejecta field lines guide particles to high latitudes.

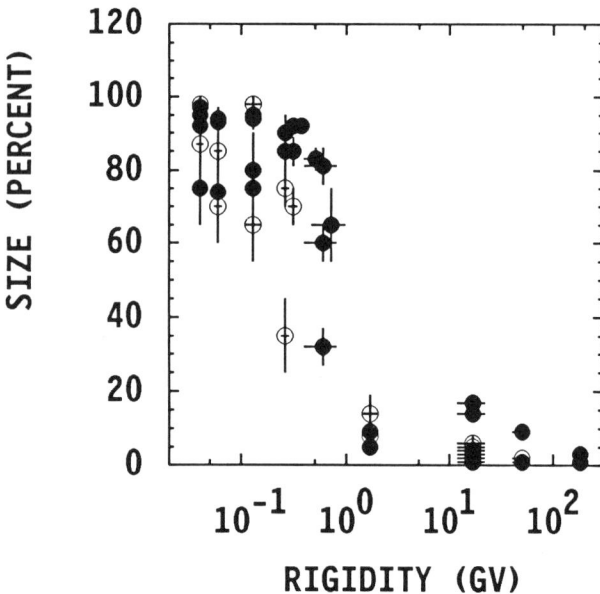

Figure 3. The rigidity dependence of the particle decrease for several ejecta with (unfilled circles) and without magnetic cloud signatures (after *Cane et al.* [1995]).

3. LONGITUDINAL EXTENT OF EJECTA AND IDENTIFICATION OF THE RELATED SOLAR EVENT

Two interesting aspects of ejecta can be examined using energetic particle data: (1) the longitudinal extent of ejecta, and (2) the identification of the solar event associated with the initiation of the ejecta. Both issues are addressed in a recent study [*Cane et al.*, 1996] in which particle signatures (as above) are used to infer the solar wind structures (e.g. shock or shock plus ejecta) responsible for all large (≥4%) cosmic ray depressions at the Mount Wellington or Deep River neutron monitors in 1964-1994. This study provides a list of all energetic ejecta which passed the Earth during this 30 year period. Such a list cannot be compiled using near-Earth solar wind observations alone because of data gaps whereas the energetic particle data are nearly complete. The time of the related solar event can often be inferred from the onset of the associated particle enhancement. Observations of e.g. Hα flare or filament activity can then be used to identify the event location. The particle intensity-time profile also provides a clue to the event longitude relative to the observer [*Cane et al.*, 1988]. The list gives the solar events associated with these energetic ejecta if an unambiguous association can be made. Most studies attempting to compare the structures of magnetic clouds and the related solar events such as erupting filaments [e.g. *Bothmer*, 1993; *Rust*, 1994] have not taken energetic particle observations into account when identifying the time of the solar event, but have used the questionable assumption that ejecta travel from the Sun at constant speed. Such observations should be considered in order to relate solar wind transient structures more reliably with solar events.

The *Cane et al.* [1996] study also provides information on the longitudinal extent of ejecta. In Figure 4, the cosmic ray depression is plotted versus the longitude of the solar event relative to central meridian. Filled (unfilled) circles indicate depressions in which the ejecta is (is not) encountered. The largest depressions occur when ejecta from near central meridian are encountered, presumably because the Earth is more likely to penetrate deep into the ejecta. Note that ejecta are only observed following shocks from events within ~50° of central meridian. A similar conclusion was obtained from studying ejecta signatures following shocks with well-established solar sources [*Richardson and Cane*, 1993]. However, the longitudinal extent of ejecta is probably less than the 100° suggested by these results since (1) the Hα flare may occur anywhere between the footpoints of the ejecta [*Harrison et al.*, 1990] and so may not indicate the exact radial propagation direction of the centre of the ejecta, and (2) only a few ejecta observed by one Helios spacecraft were detected at the other spacecraft, even though these spacecraft were generally separated by less than 40° [*Cane et al.*, 1997].

4. BIDIRECTIONAL PARTICLE FLOWS

Field-aligned bidirectional energetic (~MeV) ion flows (BIFs) are another feature of many ejecta (but also of some other solar wind regions) [e.g. *Rao et al.*, 1967; *Palmer et al.*, 1978; *Sanderson et al.*, 1983; *Marsden et al.*, 1987; *Richardson and Reames*, 1993; *Richardson*, 1994, and references therein]. In Figure 2(a) and (b), BIFs at ~1 MeV detected by the GSFC instruments on each spacecraft [*Richardson and Reames*, 1993; *Richardson*, 1994] are indicated in the middle panels by horizontal lines without bars at the ends. BIFs are generally interpreted in terms of particles circulating within a plasmoid-like region disconnected from the Sun, or being reflected within the legs of looped magnetic field lines rooted at the Sun. The expansion of these structures reduces the particle pitch-angles, creating the field-aligned flows [*Palmer et al.*, 1978]. The origin of these particles is unclear. They may be the small fraction of shock-accelerated particles which penetrate the ejecta, be accelerated at the time when the ejecta leaves the Sun, or, as discussed below, be injected into the ejecta by unrelated solar events. Studies of the ion composition and charge states may help to elucidate their origin.

Rao et al. [1967] and *Kahler and Reames* [1991] suggested an alternative interpretation of BIFs, that they develop on open field lines from particle reflection between the post-shock magnetic field enhancement and the strengthening field close to the Sun. This interpretation is unlikely to explain BIFs in ejecta since the depressions in the shock-accelerated ion intensity inside ejecta suggest that ejecta field lines are not directly connected to the post-shock region, as discussed above.

Note that the BIFs in Figure 2 do not occur throughout the ejecta. This is typical of most ejecta, so that as a result, BIF durations are often much less than those of complete ejecta [*Richardson and Reames*, 1993]. One possible explanation is that if the ions travel several AU along field lines between reflections, they may encounter conditions (e.g. enhanced turbulence) on certain ejecta field lines which cause the beam to be disrupted. In addition, in some regions of an ejecta, the ion intensity may be too low for the flows to be examined.

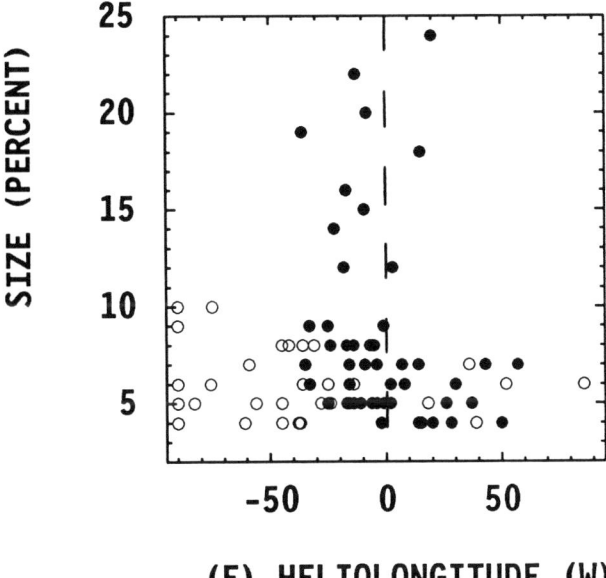

Figure 4. Cosmic ray depression at Mount Wellington or Deep River vs. solar event longitude for ≥4% depressions in 1964-1994 [after *Cane et al.,* 1996]. Filled circles: ejecta plus shock encountered; unfilled circles: shock only. The largest depressions are produced when the ejecta is encountered, and this occurs when the solar event is less than 50° from central meridian.

Low ion intensities are also a reason why there are no reported BIFs in some ejecta.

The horizontal line with bars at the ends in the middle panel of Figure 2(b) indicates a BDE interval [*Gosling et al.,* 1987]. Note that the BDE and BIF periods show little overlap in this ejecta but that together, they indicate nearly the complete ejecta region as inferred from T_p and energetic particle data. *Crooker et al.* [1990] studied the relationship between various ejecta signatures in one ejecta and found that in that case, BIFs preceded the BDEs, the opposite situation to that in Figure 2(b). Overall, there appears to be no universal relationship between the different ejecta signatures, nor one signature which uniquely defines an ejecta. Thus a range of signatures should ideally be examined to determine the presence and extent of an ejecta.

Gosling et al. [1995] suggested that the cessation of BDEs inside an ejecta indicates a change from a region of looped magnetic field lines rooted at the Sun to a region of field lines which have become open through reconnection with field lines in the ambient solar wind. However, the cosmic ray depression in Figure 2(b) clearly extends into the ejecta region following the BDE. This suggests that field lines in this region were still closed since if they were open, we might have expected cosmic rays to have rapidly populated this region of the ejecta. A possible explanation for the cosmic ray observations, assuming the interpretation of the absence of a BDE is correct, is that if reconnection occurred close to the Sun where cosmic ray densities are likely to be low, few cosmic rays might have entered through the base of the ejecta. In any case, the cosmic ray depression evidently gives a better indication of the complete ejecta region at 1 AU than the BDE.

5. SOLAR PARTICLE EVENTS OBSERVED INSIDE EJECTA

Occasionally, the onset of a solar particle event is observed by a spacecraft located inside an ejecta [*Kahler and Reames,* 1991; *Richardson et al.,* 1991; *Farrugia et al.,* 1993a, b; *Richardson and Cane,* 1996]. Such observations suggest that field lines inside these ejecta are connected to the Sun, rather than detached. The onsets discussed in these studies were observed in the ecliptic within 1 AU from the Sun. A similar example was detected by Ulysses at 32°S and 4.6 AU [*Armstrong et al.,* 1994], suggesting that ejecta field lines at high latitudes and large distances from the Sun may continue to be rooted at the Sun.

An energetic electron event observed by GSFC instrument on ISEE 3 inside a magnetic cloud is shown in Figure 5. The top three plots in the main panel show the enhanced magnetic field strength and smooth rotation of the magnetic field direction characteristic of a magnetic cloud. The horizontal bars in the azimuthal (ø) angle panel indicate intervals of BIFs observed at two energies (35-1000 keV and 1-4 MeV) by instruments on ISEE 3 [*Marsden et al.,* 1987; *Richardson and Reames,* 1993]. Again, the BIFs are evidently associated with the magnetic cloud, but do not extend throughout the complete structure. The bottom panel shows the 0.2-2-MeV electron intensity (in $(MeV\,s\,cm^2\,sr)^{-1}$). The prompt solar particle event commencing late on February 12 (Day of year (DOY) 43), 1982 [*Kahler and Reames,* 1991] is clearly evident. The pie plots at the top of the figure show examples of >0.2-MeV electron counting rates in eight azimuthal sectors plotted versus viewing direction with the Sun to the top of the figure. The arrows indicate the magnetic field direction and the numbers give the count rate (s^{-1}) in the maximum sector. The solar particles had a unidirectional flow at onset (distribution for DOY 43; 22 UT) and subsequently became bidirectional (e.g. distribution for DOY 44; 04 UT). This change from unidirectional to bidirectional flow is consistent with the reflection of the particles inside the legs of looped magnetic fields rooted at the Sun or by the subsequent injection of particles into the opposite leg of the looped field lines. Another clue that the magnetic field structure in this ejecta is unusual is that the pie plot for event onset shows maximum intensities in the left-hand sectors. This means that the first particles arrived from the *east* of the Sun, nearly opposite to the direction expected for particles arriving from the Sun along Parker-spiral field lines. Evidently field lines near the axis of this magnetic cloud and extending *eastward* from the spacecraft were connected to the Sun. To examine this further, flows in 39 solar particle events observed by Helios 1 and 2 or ISEE 3 inside ejecta (identified using a comprehensive set of ejecta signatures), and in 60 events observed outside ejecta, have been studied by *Richardson and Cane* [1996]. Figure 6 summarizes ~2-MeV proton flows during the first 6 hours after the onsets of events observed outside ejecta, inside magnetic clouds, and inside other ejecta. Vectors (defined by the first harmonic of a Fourier series fit to the sectored particle counts [*Richardson and*

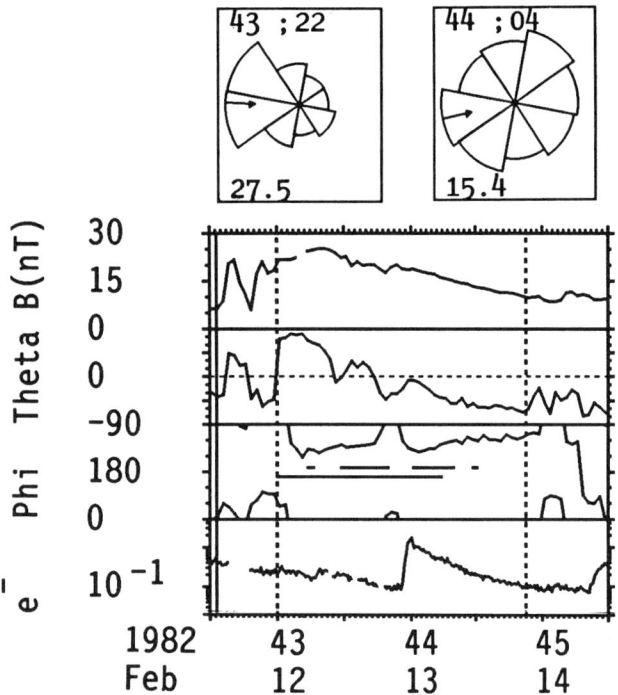

Figure 5. Example of a solar electron injection (the bottom panel shows the 0.2-2.0-MeV electron intensity in units of (MeV s cm^2 sr)$^{-1}$) observed by ISEE 3 when inside a magnetic cloud (note the enhanced magnetic field intensity and rotation in the polar angle (θ)). The pie plots show particle counts in eight azimuthal sectors, plotted vs. viewing direction with the Sun to the top of the page, at event onset and ~6 hours later (times given as DOY; UT). The arrows indicate the magnetic field direction. The electrons arrived from the east of the Sun suggesting that field lines extending east from the spacecraft were connected to the Sun. The pie plot for ~6 hours after onset shows the development of bidirectional field-aligned flows.

Reames, 1993]) giving the direction and magnitude of the particle flow during successive 15-minute data accumulation periods are plotted tip-to-tip, starting at the origin of the panel with the Sun to the top of the page. The vectors for virtually all non-ejecta events (left-hand panel) move to the lower-left, indicating flows away from the Sun along near-Archimedean field lines. The flows for events in ejecta, both with (middle panel) and without (right-hand panel) magnetic cloud signatures, are more varied. In ~40% of these events (including three of the five magnetic cloud events), the vectors move to the right, indicating flow from the east of the Sun (as in Figure 5) while flows from the west are found in other events. Thus flows from the west or from the east are observed at the onset of particle events detected inside ejecta. Another difference is that inside ejecta, 62% of the events developed bidirectional flows compared with 12% outside. The observations are consistent with the presence of looped field lines, rooted at the Sun, in ejecta whereas outside ejecta, field lines are predominantly open and near-spiral. This study suggests that unusual particle flow directions at the start of a solar particle event indicate that the spacecraft was probably within an ejecta. Multispacecraft observations of the same particle event made both inside and outside ejecta support these conclusions. For example, Figure 7 shows pie plots of particle distributions at the onset of an event late on December 11 (DOY 345), 1978 at Helios 1 and 2 and IMP 8. At Helios 2, outside ejecta material, the flow was away from the Sun along spiral field lines. Helios 1 was in an ejecta and observed a sunward flow. At ISEE 3, a BDE was occurring at the time of event onset [*Gosling et al.*, 1987], suggesting the presence of an ejecta, and flows from the east were observed.

This type of analysis can only be made when a solar event happens to occur while a spacecraft is inside an ejecta on field lines connected to the Sun. We cannot also exclude the possibility that some ejecta are totally disconnected from the Sun when detected at ~1 AU, or that ejecta contain regions of disconnected magnetic field, since prompt solar particle events are unlikely to be observed inside these regions. Initial

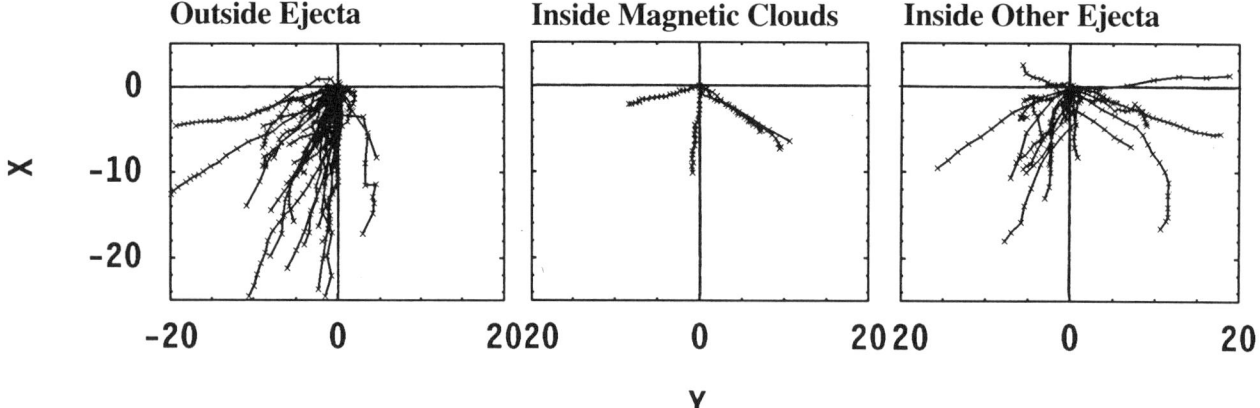

Figure 6. Summary of ~2-MeV ion flow vectors for the first 6 hours of solar particle events observed outside ejecta, and inside ejecta with and without magnetic cloud signatures. Outside ejecta (left-hand panel), particle flows are generally away from the Sun and close to the Parker spiral direction. Inside ejecta both with (middle panel) and without magnetic cloud signatures (right-hand panel), particles arrive from a wide range of azimuths, including from the east of the Sun. The observations suggest that looped magnetic field lines are present in ejecta, but are rare outside ejecta where spiral field lines predominate.

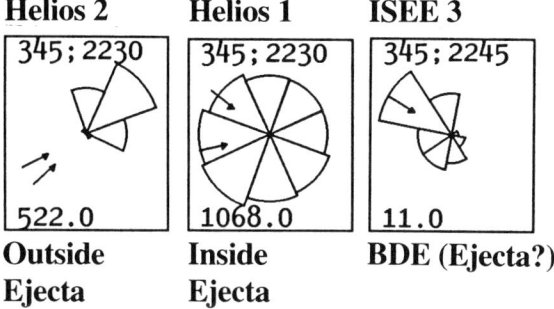

Figure 7. Azimuthal distributions of ~1-MeV ions following the onset of a solar event on December 11 (DOY 345), 1978, observed in the ambient solar wind at Helios 2, inside an ejecta at Helios 1, and during a BDE (suggesting the presence of an ejecta) at ISEE 3,. The flows are consistent with those in Figure 6 in that the particles were streaming away from the Sun along spiral field lines only at Helios 2.

results suggest that observations from the 3-D plasma experiment on the WIND spacecraft, which detects electrons in the energy interval between the ~200-keV data discussed above and solar wind energies, will provide important information on these aspects. For example, several solar electron bursts were detected when WIND was located within the magnetic cloud of October 18-20, 1995 [Larson et al., 1996]. Two features of these observations are particularly interesting. Intermittent drop-outs of the electrons at all energies suggest that the spacecraft encountered regions of field lines disconnected from the Sun. Also, larger travel distances from the Sun were inferred from velocity dispersion for events observed nearer to the edges of the magnetic cloud. This latter observation is consistent with the flux-rope like magnetic field structure suggested by *Burlaga et al.* [1981] in which the path length from the Sun is longer along helical field lines nearer the edge of the magnetic cloud. A similar conclusion was reached by *Farrugia et al.* [1993b] who considered observations on February 15, 1980 at ISEE 3 and at IMP 8, which was only 0.012 AU further from the Sun. A solar particle event commenced just before the leading edge of a magnetic cloud passed ISEE 3 and then IMP 8. The arrival of low energy particles (<7 MeV amu^{-1} ions) was delayed at ISEE 3 inside the magnetic cloud, but not at IMP 8 outside the magnetic cloud. At higher particle speeds, both spacecraft saw similar intensity-time profiles. The delay in the arrival of the slower particles at ISEE 3 inside the magnetic cloud was shown to be consistent with the ~3.5 times longer path length from the Sun along helical field lines near the boundary of the magnetic cloud inferred from fitting the magnetic field observations to a flux-rope model.

High-energy solar particles observed as "ground-level events" (GLEs) by neutron monitors can also provide information on the presence of unusual magnetic field configurations associated with ejecta. For example *Cramp et al.* [1995] found that GLE particles arrived from opposite directions along the magnetic field in the October 12, 1981 event which occurred when Earth was inside an ejecta [*Richardson et al.*, 1991]. They interpreted this as evidence for looped magnetic field lines, consistent with the conclusions of *Richardson et al.* [1991] for this event based on lower energy spacecraft data. In fact, *Meyer et al.* [1956], by examining a GLE in order to infer the particle propagation characteristics of a region in the yet-to-be-discovered solar wind which produced a cosmic ray depression, were perhaps the first to use energetic particles to probe an ejecta.

6. SUMMARY

This paper has reviewed several ways in which energetic particle observations provide information on ejecta structure. Near the ecliptic, the particle decreases in ejecta over a wide range of rigidities, which can attain greater than 70% at rigidities equivalent to proton energies of less than 100 MeV, suggest that ejecta contain predominantly closed magnetic field lines, in particular at the ejecta leading edge where the start of the particle decrease is generally located close to the onset of abnormally cool plasma. At high latitudes, ejecta are characterized by MeV particle intensity enhancements rather than by the depressions observed in the ecliptic because ejecta apparently transport or guide MeV particles from lower latitudes.

A large fraction of ejecta (~90%) produce a significant, local depression in the cosmic ray density, so this is a useful signature of the presence of an ejecta. Such depressions can be simply explained in terms of particle entry into the ejecta by perpendicular diffusion. Evidence for the reconnection of ejecta field lines with those in the ambient solar wind during intervals without BDEs is not particularly clear in the cosmic ray observations since there are intervals in ejecta without BDEs in which the cosmic ray density remains depressed. On the other hand, WIND electron observations do show evidence for such field lines. Thus it is possible that the different gyroradii of the particles which probe structures of different scale sizes determine whether signatures of reconnected field lines are observed. There is evidence from solar particle flows that ejecta can include magnetic fields which deviate significantly from the Parker spiral configuration. In particular, they may include looped field lines rooted at the Sun. Finally, although frequently neglected, energetic particle data can make a valuable contribution to studies relating solar events, such as flares and filament eruptions, to the related structures subsequently observed in the solar wind, and can be used to infer the presence of ejecta in the solar wind even when in-situ solar wind plasma observations are not available.

Acknowledgments. H. V. Cane, C. J. Farrugia, T. T. von Rosenvinge and G. Wibberenz are thanked for their contributions to the work reviewed here. H. Kunow (University of Kiel) is also thanked for providing data from the University of Kiel Helios experiments. The use of data from the Helios plasma experiments (P.I. F. M. Neubauer) provided by the National Space Science Data Center

(NSSDC) is gratefully acknowledged. The near-Earth solar wind parameters are from the NSSDC OMNI data base. This work was supported by NASA grant NGR 21-002-316.

REFERENCES

Armstrong, T. P., D. Haggerty, L. J. Lanzerotti, C. G. Maclennan, E. C. Roelof, M. Pick, G. M. Simnett, R. E. Gold, S. M. Krimigis, K. A. Anderson, R. P. Lin, E. T. Sarris, R. Forsyth, and A. Balogh, Observation by Ulysses of hot (~270 keV) coronal particles at 32° south heliolatitude and 4.6 AU, *Geophys. Res. Lett., 21,* 1747-1750, 1994.

Barnden, L. R., The large-scale magnetic configuration associated with Forbush decreases, *Proc. 13th Int. Cosmic Ray Conf., 2,* 1277-1282, 1972.

Belov, A. V., L. I. Dorman, E. A. Eroshenko, N. Iucci, G. Villoresi, and V. G. Yanke, Anisotropy of cosmic rays and Forbush decreases in 1991, *Proc. 24th Int. Cosmic Ray Conf., 4,* 912-915, 1995.

Bothmer, V., Die Struktur magnetischer Wolken im Sonnenwind, *Ph.D. thesis, University of Göttingen,* 1993.

Bothmer, V., R. G. Marsden, T. R. Sanderson, K. J. Trattner, K.-P. Wenzel, A. Balogh, R. J. Forsyth, and B. E. Goldstein, The Ulysses south polar pass: Transient fluxes of energetic ions, *Geophys. Res. Lett., 22,* 3369-3372, 1995.

Burlaga, L. F., E. Sittler, F. Mariani, and R. Schwenn, Magnetic loop behind an interplanetary shock: Voyager, Helios, and IMP 8 observations, *J. Geophys Res., 86,* 6673-6684, 1981.

Cane, H. V., D. V. Reames, and T. T. von Rosenvinge, The role of interplanetary shocks in the longitude distribution of solar energetic particles, *J. Geophys. Res., 93,* 9555-9567, 1988.

Cane, H. V., I. G. Richardson, T. T. von Rosenvinge, and G. Wibberenz, Cosmic Ray decreases and shock structure: a multispacecraft study, *J. Geophys. Res., 99,* 21,429-21,441, 1994.

Cane, H. V., I. G. Richardson, and G. Wibberenz, The response of energetic particles to the presence of ejecta material, *Proc. 24th Int. Cosmic Ray Conf., 4,* 377-380, 1995.

Cane, H. V., I. G. Richardson, and T. T. von Rosenvinge, Cosmic Ray decreases: 1964-1994, *J. Geophys. Res., 101,* 21,561-21,572, 1996.

Cane, H. V., I. G. Richardson, and G. Wibberenz, Helios 1 and 2 observations of particle decreases, ejecta, and magnetic clouds, *J. Geophys. Res., in press,* 1997.

Cramp, J. L., M. L. Duldig, and J. E. Humble, The effect of near-Earth IMF structure on the modelling of ground level enhancements, *Proc 24th Int. Cosmic Ray Conf., 4,* 289-292, 1995.

Crooker, N. U., J. T. Gosling, E. J. Smith, and C. T. Russell, A bubblelike coronal mass ejection flux rope in the solar wind, in *Physics of Magnetic Flux Ropes, Geophys. Monogr. Ser.,* vol. 58, edited by C. T. Russell, E. R. Priest, and L. C. Lee, p. 365-371, AGU, Washington, 1990.

Farrugia, C. J., L. F. Burlaga, V. A. Osherovich, I. G. Richardson, M. P. Freeman, R. P. Lepping, and A. J. Lazarus, A study of an expanding interplanetary magnetic cloud and its interaction with the Earth's magnetosphere: The interplanetary aspect, *J. Geophys. Res., 98,* 7621-7632, 1993a.

Farrugia, C. J., I. G. Richardson, L. F. Burlaga, V. A. Osherovich, and R. P. Lepping, Simultaneous observations of solar MeV particles in a magnetic cloud and in the Earth's northern tail lobe, *J. Geophys. Res., 98,* 15,497-15,507, 1993b.

Gosling, J. T., Coronal mass ejections and magnetic flux ropes in interplanetary space, in *Physics of Magnetic Flux Ropes, Geophys. Monogr. Ser.,* vol. 58, edited by C. T. Russell, E. R. Priest, and L. C. Lee, p. 343-364, AGU, Washington, 1990.

Gosling, J. T., D. N. Baker, S. J. Bame, W. C. Feldman, R. D. Zwickl, and E. J. Smith, Bidirectional solar wind electron heat flux events, *J. Geophys. Res., 92,* 8519-8535, 1987.

Gosling, J. T., J. Birn, and M. Hesse, Three-dimensional magnetic reconnection and the magnetic topology of coronal mass ejection events, *Geophys. Res. Lett., 22,* 869-872, 1995.

Harrison, R. A., E. Hildner, A. J. Hundhausen, D. G. Sime, and G. M. Simnett, The launch of solar coronal mass ejections: Results from the coronal mass ejection onset program, *J. Geophys. Res., 95,* 917-937, 1990.

Kahler, S. W., and D. V. Reames, Probing the magnetic topologies of magnetic clouds by means of solar energetic particles, *J. Geophys. Res., 96,* 9419-9424, 1991.

Larson, D. E., R. E. Ergun, J. M. McTiernan, R. P. Lin, J. P. McFadden, C. W. Carlson, K. A. Anderson, M. McCarthy, G. Parks, H. Reme, J. M. Bosqued, K.-P. Wenzel, and T. R. Sanderson, Wind spacecraft observations of energetic particles in the October 1995 magnetic cloud, *Abstracts, AGU Chapman Conf. on Coronal Mass Ejections: Causes and Consequences,* p. 29, 1996.

Marsden, R. G., T. R. Sanderson, C. Tranquille, K.-P. Wenzel, and E. J. Smith, ISEE 3 observations of low-energy proton bidirectional events and their relation to isolated interplanetary structures, *J. Geophys. Res., 92,* 11,009-11,019, 1987.

Meyer, P., E. N. Parker, and J. A. Simpson, Solar cosmic rays of February 1956 and their propagation through interplanetary space, *Physical Rev., 104,* 768-783, 1956.

Palmer, I. D., F. R. Allum, and S. Singer, Bidirectional anisotropies in solar cosmic ray events: Evidence for magnetic bottles, *J. Geophys. Res., 83,* 75-90, 1978.

Rao, U. R., K. G. McCracken, and R. P. Bukata, Cosmic ray propagation processes, 2, The energetic storm particle event, *J. Geophys. Res., 72,* 4325-4341, 1967.

Richardson, I. G., A survey of bidirectional ≥ 1 MeV ion flows during the Helios 1 and Helios 2 missions: observations from the Goddard Space Flight Center instruments, *Astrophys. J., 420,* 926-942, 1994.

Richardson, I. G., and H. V. Cane, Signatures of shock drivers in the solar wind and their dependence on the solar source location, *J. Geophys. Res., 98,* 15,295-15,304, 1993.

Richardson, I. G. and H. V. Cane, Regions of abnormally low proton temperature in the solar wind (1965-1991) and their association with ejecta, *J. Geophys. Res., 100,* 23,397-23,412, 1995.

Richardson, I. G. and H. V. Cane, Particle flows observed in ejecta during solar event onsets and their implication for the magnetic field topology, *J. Geophys. Res., 101,* 27,521-27,532, 1996.

Richardson, I. G., and D. V. Reames, Bidirectional ~1 MeV/amu ion intervals in 1973-1991 observed by the Goddard Space Flight Center instruments on IMP 8 and ISEE 3/ICE, *Astrophys. J. (Suppl.), 85,* 411-432, 1993.

Richardson, I. G., H. V. Cane, and T. T. von Rosenvinge, Prompt arrival of solar energetic particles from far eastern events: The role of large-scale interplanetary magnetic field structure, *J. Geophys. Res., 96,* 7853-7860, 1991.

Rust, D., Spawning and shedding helical magnetic fields in the solar atmosphere, *Geophys. Res. Lett., 21,* 241-244, 1994.

Sanderson, T. R., R. G. Marsden, R. Reinhard, K.-P. Wenzel, and E. J. Smith, Correlated particle and magnetic field observations of a large-scale magnetic loop structure behind an interplanetary shock, *Geophys. Res. Lett., 10,* 916-919, 1983.

Tranquille, C., T. R. Sanderson, R. G. Marsden, and K.-P. Wenzel, Properties of a large-scale interplanetary loop structure as deduced from low energy proton anisotropy and magnetic field measurements, *J. Geophys. Res., 92,* 6-14, 1987.

Vanhoefer, O., Auswirkungen Interplanetarer Plasmawolken auf die Kosmische Strahlung, *Diploma Thesis,* University of Kiel, Kiel, Germany, 1996.

I. G. Richardson, Code 661, Goddard Space Flight Center, Greenbelt, MD 20771.

Using Charged Particles to Trace Interplanetary Magnetic Field Topology

S. W. Kahler

Phillips Laboratory Geophysics Directorate, Hanscom Air Force Base, Massachusetts

The topology of interplanetary magnetic field lines relative to the Sun has traditionally been determined from the directions of the fields observed in space. By assuming that charged particles originate in the solar corona and follow field lines as they propagate away from the Sun, it is possible to infer field topologies by comparing the local particle flow directions with the field directions. Recent results with E > 2 keV electrons to study magnetic polarities at sector boundaries show that polarity changes do not correlate well with magnitudes of changes in field directions. Work with heat-flux electrons has shown that intrasector field reversals arise from (1) folds in the unipolar fields, or (2) injections of closed fields into the sector structure, presumably from coronal mass ejections. We consider additional candidate techniques for inferring the magnetic polarity of the interplanetary field.

1. INTRODUCTION

It is well known that interplanetary magnetic fields lie predominately along an Archimedean spiral pattern and are organized into large sectors of unipolar fields [*Wilcox and Ness*, 1965]. The sector structure results from a solar field pattern in which positive and negative fields are separated by a warped, global heliospheric current sheet [e.g., *Behannon et al*, 1989]. The interplanetary field polarities have been determined on the basis of time averages of the measured field directions. Fields with longitudinal directions Φ centered around 135° are assumed to be of positive, outward-pointing polarity and those centered around 315° to be of negative, inward-pointing polarity.

This scheme of simple spiral fields organized into the two polarities often breaks down. When the fields lie nearly perpendicular to the normal spiral angle, their polarities become uncertain. Even when the fields lie along the Archimedean spiral angle, their directions may indicate polarities opposite to the true situation. Figure 1 shows several such situations. In (1a) the field is all positive polarity but kinked in one region. As those fields are convected past the Earth, the signature of the field directions will match that of the very different situation in (1b), where a small negative sector will be convected past the Earth. Let us also consider the closed loop fields of (1c) and (1d), which represent dynamic situations. In (1c) there is no change in the field directions at the Earth as the closed loops follow the open loops, but there is a significant change in the field topology. In (1d) the transition from inward pointing to outward pointing field regions will be obvious, but the change in topology will not. Clearly, a technique better than simple field direction is needed to determine the polarities of the interplanetary field. Here we define the term polarity to mean the polarity of the field line at the solar source region.

A simple technique to use solar electrons to determine the interplanetary field polarity was introduced by Kahler and Lin [1994]. The basic idea is shown in Figure 2. When solar electrons propagate away from the Sun on a positive polarity field, they will always flow parallel to the direction of the magnetic field, even when the field kinks backward toward the Sun. In case of a negative polarity

Coronal Mass Ejections
Geophysical Monograph 99
This paper is not subject to U.S. copyright.
Published in 1997 by the American Geophysical Union

198 CHARGED PARTICLES TO TRACE INTERPLANETARY FIELD TOPOLOGY

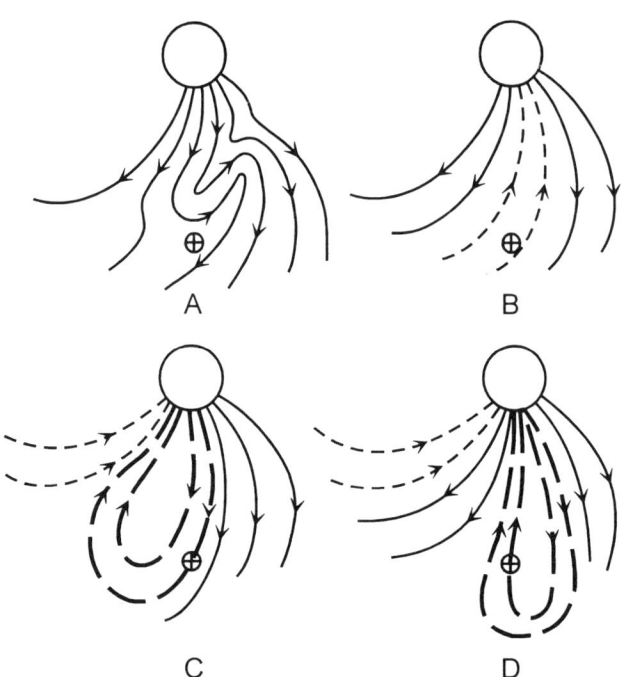

Figure 1. Four basic interplanetary field topologies which can be probed by solar electrons. (a) a kinked field in a positive polarity sector; (b) a small region of negative polarity field in a positive polarity sector; (c) a closed loop (heavy dashed lines) at a sector boundary; (d) a closed loop (heavy dashed lines) within a positive sector. An observer at Earth (⊕) sees a normal field direction in (c), but an intrasector field reversal (IFR) in (a), (b), and (d). From Kahler et al. [1996].

field, they will always flow antiparallel to the field. One must always be careful that the electrons are solar in origin and not flowing sunward from the Earth's bowshock or an interplanetary shock. There is furthermore the question of pitch-angle isotropy; when the distribution tends toward isotropy, the direction of flow and the inferred polarity is less certain.

2. DETERMINING SECTOR BOUNDARIES

The E > 2 keV electrons observed with the UC Berkeley detector on ISEE-3 were used in a short study to determine polarities in periods around apparent sector boundaries [*Kahler & Lin*, 1994] when the electron fluxes were sufficiently high. Electron data were summed over periods of 20 minutes to 2 hours when the magnetic field directions remained relatively steady in direction. An example from that work is shown in Figure 3. What appears as a sector crossing at 0300 UT on 13 September is not a change in polarity in terms of the solar source

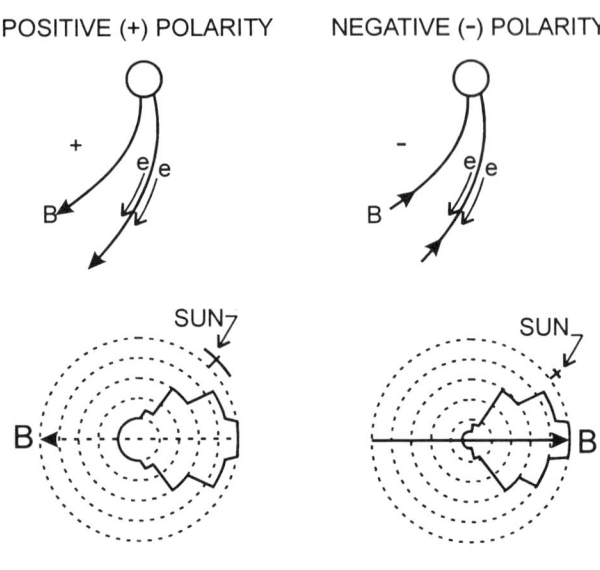

Figure 2. In positive polarity magnetic fields the solar electrons flow parallel to the field direction (left). For a negative polarity field they flow antiparallel (right). The bottom panels show azimuthal plots of solar electrons observed in the ecliptic plane with the UC Berkeley electron detector on ISEE-3. Note that the detector view direction is 180° opposite that of the electron flow direction. Adapted from Kahler and Lin [1994].

fields. The electron data further show that the multiple fluctuations in field direction during the period 1200 to 2400 UT all lie in a negative polarity region. This period would usually be described as "mixed polarity" based on field directions only.

In a survey of the electron data around several selected sector boundaries Kahler & Lin [1995] compared the rotational angles, ω, across polarity changes with the rotational angles across large rotations with no polarity changes. Their basic result, shown in Figure 4, was that ω is a poor guide to polarity reversals. Although statistics were limited, they found that about half the polarity reversals were characterized by $\omega < 90°$, and about half the $\omega > 120°$ discontinuities were not polarity reversals. Their results further emphasized how unreliable the field directions are for locating sector boundaries or determining field polarities in the vicinity of sector boundaries.

3. STUDY OF CORONAL MASS EJECTIONS

At energies more than a decade below 2 keV there is also an interplanetary electron population, the heat-flux electrons [*Feldman et al.*, 1975], which propagate away

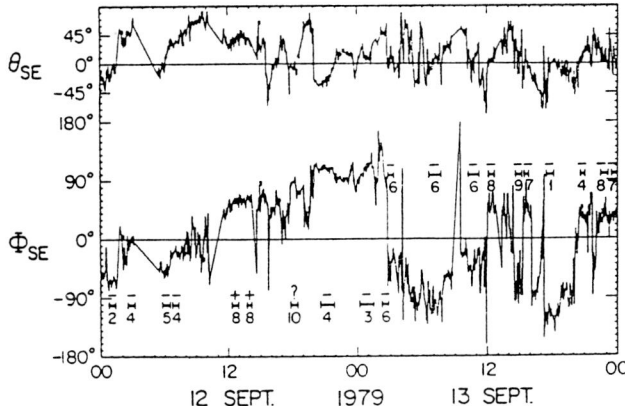

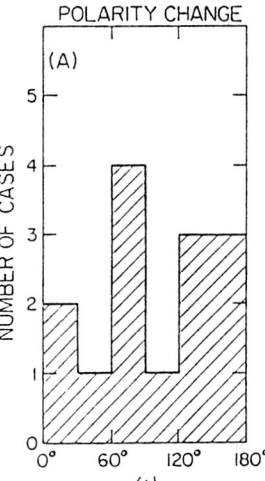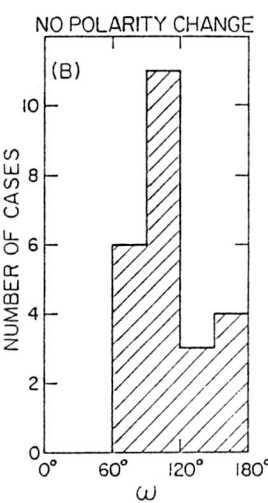

Figure 3. Top: the interplanetary field latitude measured on ISEE-3 during a two-day period in September 1979. Bottom: the longitudinal direction of the interplanetary field. The normal positive polarity field direction is 135° and the negative field is -- 45°. Horizontal bars mark the periods when the electron anisotropies were measured. The field polarities marked above the bars were determined from the electron anisotropies, assuming solar field sources. The degree of anisotropy (0 the highest, 10 the least) is given under the bars. The polarity was negative throughout the large directional fluctuations on 13 September. From Kahler and Lin [1994].

Figure 4. (a): the distribution of cases of polarity change as a function of ω, the angular change across the field discontinuities. (b): the distribution of cases of no polarity change. This distribution is limited to cases of $\omega > 60°$. From Kahler and Lin [1995].

from the Sun along magnetic field lines with little scattering. While less reliable as field line tracers than the E > 2 keV electrons [Lin and Kahler, 1992], the heat-flux electrons are nearly always present in the solar wind at flux levels sufficient to measure their directions continuously. Kahler et al. [1996] have used those electrons to probe the topology of intrasector field reversals (IFRs), defined to be short periods of opposite-polarity fields lying well within a sector of a given polarity (Figure 5). The basic idea was to determine whether IFRs are due to kinks in the field (Figure 1a) or to isolated regions of opposite polarity (Figure 1b). They concluded that kinks in the field are rather common occurrences, but the isolated regions of opposite polarity are rare or nonexistent.

An interesting result of the study of IFRs is that the IFRs with polarity reversals are almost always associated with periods of bidirectional electron fluxes (BDEs). The BDEs have been widely accepted as signatures of coronal mass ejections (CMEs) [Gosling, 1990]. A reverse survey, starting with the BDEs of the table of Gosling et al. [1987] showed that almost all BDEs occur within about 10 hours of an IFR, usually an IFR of reverse polarity. Thus the origin of regions of reverse polarity in a sector is not through small "island" regions on the solar source surface, but through an injection of closed fields via a CME. This shows that the BDEs define only part of a CME in the interplanetary medium, and that some of the transient fields are open.

A natural interpretation of this result is provided by Plate 1, which shows a schematic configuration of the topology of an erupting CME field proposed by Gosling et al. [1995]. Let us assume that the field configuration moves outward to an interplanetary observer and ask what he will see. The observed sequence will be the blue, red, and green field lines, respectively. An observer to the right side of the figure will interpret the blue and red lines as negative polarity, based on the dominant component of the bidirectional heat flux arising from the inward pointing leg of the closed fields [Kahler et al., 1996]. The green lines will be interpreted as positive polarity fields. For an observer to the left of the axis of the flux rope, the blue, red and green lines will all appear as positive polarities. In a negative sector field the observer to the right will see no IFRs, but in a positive sector field all three field lines of Plate 1 will appear as IFRs. The situation is reversed for the observer to the left side of the flux rope. Of course, the BDEs will indicate to both observers that the blue and red fields are closed.

It is of interest to ask what changes in field directions or polarities observed at 1 AU can be interpreted as evidence of CMEs. The various possibilities are listed in Table 1. We assume in each case that a BDE is observed and ask what nearby structures may also be a part of the CME.

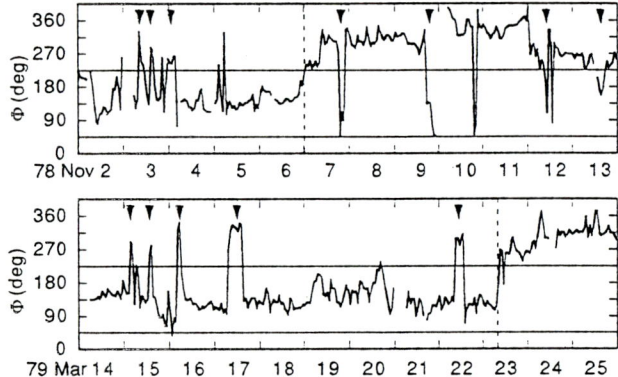

Figure 5. Plots of longitudinal angles Φ of interplanetary magnetic fields from ISEE-3. Vertical arrows designate intra-sector field reversals (IFRs), and the two solid horizontal lines at 45° and 225° mark the nominal boundaries between positive and negative polarity fields. The dashed vertical line in each panel shows the assumed sector boundaries. From Kahler et al. [1996].

Structure 1 has the same polarity as the surrounding sector, so it could be either a part of the sector or part of the CME. Structures 2 and 4 are reversed polarities, which we have seen are only associated with BDEs and are therefore, CMEs. Structure 4, the reversed polarity IFR, was discussed in detail in Kahler et al. [1996], but structure 2 is a relatively unknown feature. Its direction appears consistent with the sector structure, but the electrons flow back to the sun, indicating that the polarity is reversed. At least one such feature was encountered in the study of Kahler et al. [1996], although they were not the subject of the study. Structure 3 is most often found without accompanying BDEs, but when it does occur in association with a BDE, we assume that it is most likely a CME rather than a chance occurrence. A full CME in space is assumed to consist of the BDE(s) and associated structures 2, 3, and/or 4 and perhaps contributions of structure 1.

4. OTHER CANDIDATE TECHNIQUES TO MEASURE MAGNETIC FIELD POLARITIES

We have seen that energetic electrons provide a powerful tool for determining the polarity of the interplanetary magnetic field. However, there are times when such measurements are not available or when heat flux dropouts (HFDs) occur, and the field polarities can not be determined. Are there other techniques which can be used to infer polarities? In general, we look for any phenomenon which, like energetic particles, follows the magnetic field lines away from the Sun. In this section we discuss the

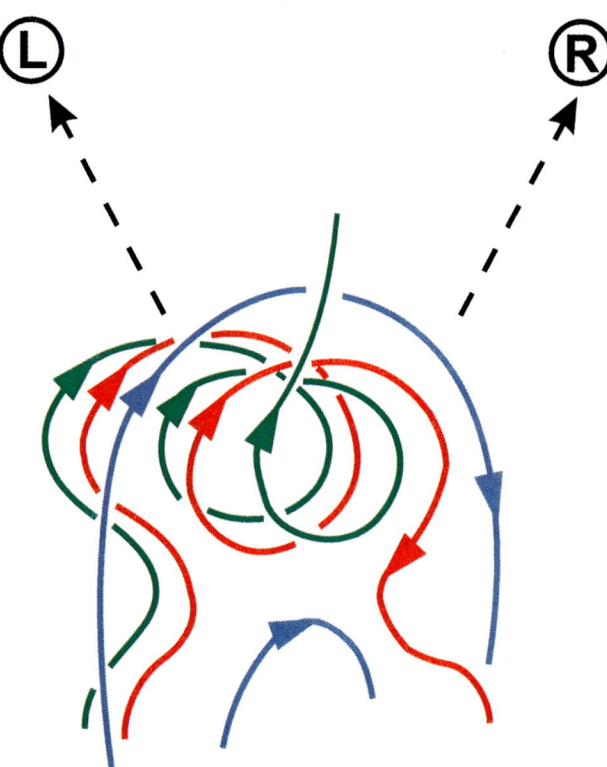

Plate 1. Schematic showing several possible magnetic field line topologies in a CME which has undergone some magnetic reconnection. The green line is an open field of positive polarity, and the blue and red lines are closed fields. As the expanding structure is convected past observers at the left (L) and right (R), those observers detect the blue, red, and green field lines, in that order. Adapted from Gosling et al. [1995].

advantages of and difficulties with several possible alternative techniques.

4.1. Polar Rain

We have determined IMF polarities by using in situ measurements of the electron heat-flux flow directions. The flow directions can also be determined indirectly from measurements of polar rain, the smoothly varying electron heat-fluxes of 100 to 500 eV that precipitate into the Earth's polar cap [*Gosling et al.*, 1986; *Gussenhoven and Madden*, 1990]. Figure 6 from Gosling et al. [1986] illustrates schematically how the heat-flux electrons follow interplanetary field lines which are connected to the open field lines of the polar caps. Positive (negative) polarity IMFs result in preferential precipitation in the north (south) polar caps. Thus by measuring the difference in the polar rain fluxes between the north and south polar caps one can

Table 1. Rules for Interpreting Open Field Structures[a]

#	Field Direction	Polarity	Sector
1	Normal (sector)	Normal	Sector or CME
2	Normal (sector)	Reversed	CME
3	Reversed (IFR)	Normal	CME
4	Reversed (IFR)	Reversed	CME

[a]Presence of intrasector BDE (closed field of CME) assumed.

infer the polarity of the IMF. Gosling et al. [1986] have pointed out that the unusual periods of hemispherically symmetric rain correspond to periods of bidirectional electron heat fluxes. Note that if the IMF is kinked by nearly 180° at the Earth, the polar rain still yields the correct IMF polarity. In that case an away sector would then be described in Figure 6 by the lower ("toward") cartoon with the labels "solar wind heat flux" and "to outer heliosphere" reversed, with the result that the heat flux would still enter the north polar cap, as shown in the top cartoon for a normal spiral field direction. Thus while a positive polarity field might have a direction indicating a negative polarity, the polar rain flux would still correctly indicate the positive polarity.

A daily polar rain index has been devised by Gussenhoven and Madden [1990] from electron measurements made on polar passes of DMSP satellites. It was necessary to make corrections for seasonal variations and the day to night variations within a given polar cap. A comparison of the daily index with the IMF polarities inferred from their directions yielded an 84% agreement over a 12-year period [Gussenhoven and Madden, 1990]. Since some days were characterized by sector boundaries and intrasector reversals, the agreement should be considered excellent. The daily polar rain indices are inadequate for studies of IMF polarities on shorter time scales, but the polar rain appears to provide a useful alternative method for deducing IMF polarities.

4.2. Lunar Shadowing

Another technique to measure the IMF polarity, similar in principle to the use of polar rain, is that of lunar shadowing [Lin, 1968]. The Moon has no magnetic interaction with the solar wind and acts simply as a large body to block the flow of solar electrons along field lines of the IMF. Lin [1968] observed decreases in E > 22 keV and > 45 keV anisotropic solar electron event fluxes with an experiment on the lunar-orbiting satellite AIMP-2. Most decreases were observed when the AIMP-2 was behind the Moon relative to the Sun, but some were observed in front of the Moon. Lin [1968] suggested that the latter events

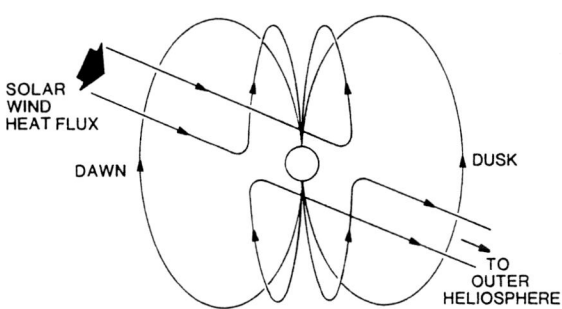

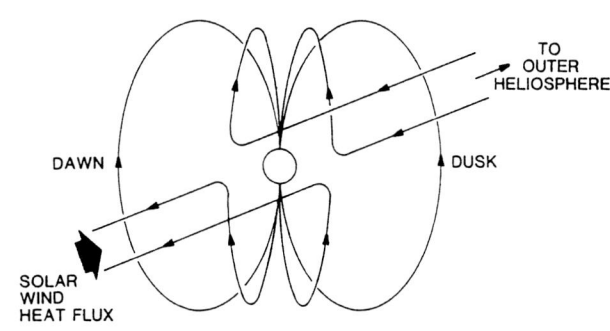

Figure 6. Schematic view from the sun showing the electron heat flux into the geomagnetic polar caps by a connection of a positive polarity field to the northern polar cap (top) and of a negative polarity field to the southern polar cap (bottom). From Gosling et al. [1986].

resulted from kinks or loops in the IMF lines. In this case the Moon simply provided a means for determining the flow direction of the electrons relative to the magnetic field direction for an electron detector with no directional discrimination.

4.3. The Svalgaard-Mansurov Effect

A method of inferring the polarity of the interplanetary magnetic field from diurnal variations in the polar geomagnetic field was discovered by Svalgaard [1968] and Mansurov [1969]. For an outward polarity field the vertical component of the field of a station near the northern pole is decreased during several hours near noon. For an inward polarity field the vertical component is enhanced at that time. The effects are reversed for a station near the southern pole. Figure 7 shows the

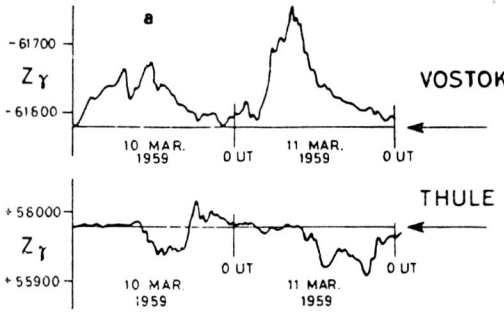
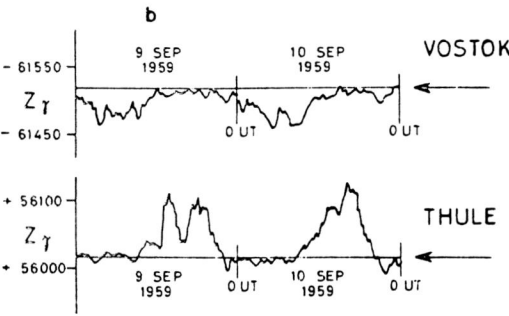

Figure 7. Sample magnetograms of the polar geomagnetic field at a north polar station (Thule) and a south polar station (Vostok) during an away sector (a) and a toward sector (b). From Wilcox [1972].

contrasting patterns for away sectors (a) and toward sectors (b). A comparison of the daily inferred polarities with those observed by spacecraft over a three-year period indicated agreement on 82% of the days [Wilcox et al., 1975]. For many years the inferred polarities have been published in Solar-Geophysical Data, using observations from the Thule station at geomagnetic latitude 86.8°N for half the day and observations from the Vostok station at geomagnetic latitude 84.9°S for the other half [Wilcox et al., 1975].

While the detailed physics of the variations of the polar geomagnetic fields may be unclear, the direction of the azimuthal component of the interplanetary field has been identified as the cause of the effect [Wilcox, 1972]. Since the interplanetary field usually lies close to the spiral direction, the azimuthal direction matches well with its associated polarity. This indicates, however, that the Svalgaard-Mansurov technique measures a directional component of the interplanetary field and not the field polarity in terms of its solar origin.

4.4. Alfven Waves

If Alfven waves are generated at the Sun and flow outward along the interplanetary field lines, as originally proposed by Belcher and Davis [1971], then the flow direction of the waves relative to the background field should also indicate the field polarity. The question then is whether such waves exist among the MHD fluctuations present in the solar wind and can be measured easily in space experiments. Observationally, the normalized cross helicity is calculated from the product of the solar wind velocity fluctuations and the magnetic field fluctuations, with the amplitude giving the degree of Alfvenicity and the sign the direction of the wave propagation [Roberts et al., 1987]. The MHD fluctuations are measured over a range of wavenumbers, so the appropriate wavenumber range must also be determined.

An important discovery of the Ulysses mission was the presence of large-amplitude Alfven waves in the high-latitude solar wind [e.g., Tsurutani et al., 1996]. The waves are outwardly propagating with a normalized wave power about a factor of 2 higher than that measured near the ecliptic plane. This favorable situation at high latitudes and large heliocentric distances (2.5 to 5 AU) is mitigated, however, by the lack of either routine observations or much doubt about the field polarities in those regions.

The more important question is what one can observe in the ecliptic plane. Roberts et al. [1987] studied the origin and evolution of low frequency ($< 10^{-4}$ Hz) interplanetary fluctuations from 0.3 to 20 AU. While these fluctuations in the inner solar system are quite Alfvenic and outward travelling, they evolve rapidly to become substantially less Alfvenic and less outward travelling at 1 AU and beyond [Roberts et al., 1987]. This seems to be due to two effects, a second source of fluctuations from interplanetary stream shears and a turbulent cascade sending the fluctuations to higher wave numbers. Figure 8 shows how the distributions of measured cross helicities become more symmetrical with distance from the Sun. Thus near 1 AU one might expect that it would usually be difficult to determine the field polarity from Alfven waves. On the other hand, the cleanest detection of Alfven waves appears in the 10^{-4} to 10^{-3} Hz range [Marsch, 1991], convenient for looking at features with 1-hour time scales. In the absence of an electron heat-flux signal the propagation direction inferred from the sign of a sufficiently Alfvenic cross helicity could yield the probable polarity of the field.

4.5. The Differential Flow $\Delta V = V_a - V_p$

In the solar wind the average speed of alpha particles exceeds that of the protons. The maximum velocity

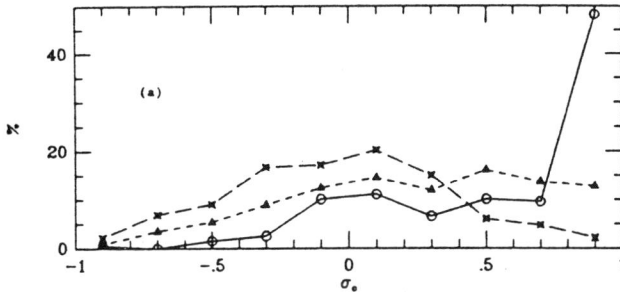

Figure 8. Plot of the distributions of cross helicity σ_c observed from Helios 1 (circles) at 0.3 AU, from Voyager 1 (triangles) at 2 AU, and from Voyager 2 (crosses) at 20 AU. The highly positive values indicate a high Alfvenicity of outward traveling waves. The distributions are calculated for a 3-hr timescale. From Roberts et al. [1987].

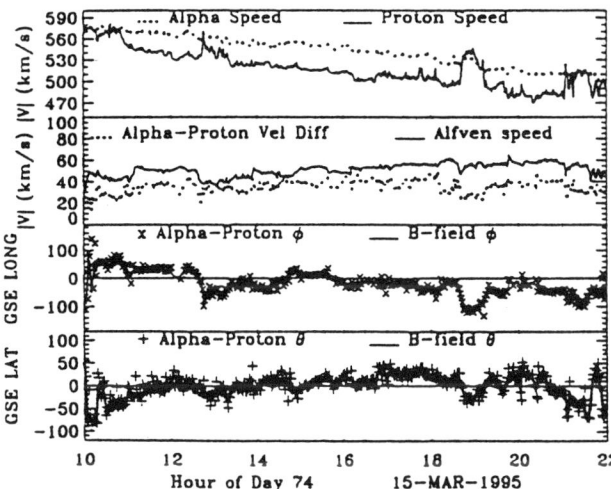

Figure 9. Plot of the alpha particle and proton parameters measured by the Solar Wind Experiment on the WIND spacecraft. Top panel: the alpha particle drift speed usually exceeds the proton drift speed. Second panel: ΔV usually is less than the Alfven speed. Third and fourth panels: the GSE vector components of the alpha-proton velocity difference align very closely to the interplanetary magnetic field direction. From Steinberg et al. [1996].

difference, ΔV, which is usually limited to the Alfven speed, generally increases with the bulk solar wind speed and decreases with solar distance [*Neugebauer et al.*, 1996]. It appears that the protons participate in the Alfven wave motion while the alpha particles stream freely and disregard the Alfvenic turbulence [*Marsch*, 1991]. The effect is most obvious in high-speed streams, where Alfven waves are also most abundant, and least obvious in low-speed, dense streams where it may be limited by Coulomb collisions [*Neugebauer et al.*, 1996].

It is well known [e.g., *Neugebauer et al.*, 1996] that the direction of ΔV tracks the direction of the magnetic field. If interplanetary magnetic field lines were kinked back toward the Sun as in Figure 1a, we might expect that in the kink ΔV would become negative. Negative values of ΔV have in fact often been observed. In some cases they result from changes in the electric potential across shocks [*Ogilvie et al.*, 1982; *Neugebauer et al.*, 1996]. However, Ogilvie et al. [1982] reported an observation on ISEE-3 in which a negative ΔV was accompanied by a 180° turn in the electron heat flux which they interpreted as a kink in the magnetic field. Recently, Steinberg et al. [1996] presented a clear case in which $|V_p| > |V_a|$ during a large azimuthal change in the field direction, as shown in Figure 9. Their interpretation that the alpha particles were still outrunning the protons, but in a magnetic kink directed back toward the Sun, was supported by a corresponding change in the electron heat-flux flow direction. This and the earlier cases suggest that away from shocks the direction of the vector ΔV can be used as a proxy for the electron heat flux direction to determine the polarity of the interplanetary magnetic field. The differential velocity is similar to the electron heat flux in being defined by a single continuously varying vector direction. This is an advantage over the use of Alfven waves, for which an appropriate time scale must be determined for the cross-helicity calculation.

REFERENCES

Behannon, K.W., L.F. Burlaga, J.T. Hoeksema, and L.W. Klein, Spatial variation and evolution of heliospheric sector structure, *J. Geophys. Res.*, 94, 1245, 1989.

Belcher, J.W., and L. Davis, Large-amplitude Alfven waves in the interplanetary medium, 2, *J. Geophys. Res.*, 76, 3534, 1971.

Feldman, W.C., J.R. Asbridge, S.J. Bame, M.D. Montgomery, and S.P. Gary, Solar wind electrons, *J. Geophys. Res.*, 80, 4181, 1975.

Gosling, J.T., Coronal mass ejections and magnetic flux ropes in interplanetary space, in *Physics of Magnetic Flux Ropes*, edited by C.T. Russell, E.R. Priest, and L.C. Lee, p.343, American Geophysical Union, Washington, 1990.

Gosling, J.T., D.N. Baker, S.J. Bame, and R.D. Zwickl, Bidirectional solar wind electron heat flux and hemispherically symmetric polar rain, *J. Geophys. Res.*, 91, 11352, 1986.

Gosling, J.T., D.N. Baker, S.J. Bame, W.C. Feldman, R.D. Zwickl, and E.J. Smith, *Bidirectional solar wind electron heat flux events*, 92, 8519, 1987.

Gosling, J.T., J. Birn, and M. Hesse, Three-dimensional magnetic reconnection and the magnetic topology of coronal mass ejection events, *Geophys. Res. Let., 22,* 869, 1995.

Gussenhoven, M.S., and D. Madden, Monitoring the polar rain over a solar cycle: a polar rain index, *J. Geophys. Res., 95,* 10399, 1990.

Kahler, S., and R.P. Lin, The determination of interplanetary magnetic field polarities around sector boundaries using E > 2 keV electrons, *J. Geophys. Res., 21,* 1575, 1994.

Kahler, S.W., and R.P. Lin, An examination of directional discontinuities and magnetic polarity changes around interplanetary sector boundaries using E > 2 keV electrons, *Solar Phys., 161,* 183, 1995.

Kahler, S.W., N.U. Crooker, and J.T. Gosling, The topology of intrasector reversals of the interplanetary magnetic field, *J. Geophys. Res., 101,* 24373, 1996.

Lin, R.P., Observations of lunar shadowing of energetic particles, *J. Geophys. Res., 73,* 3066, 1968.

Lin, R.P., and S.W. Kahler, Interplanetary magnetic field connection to the sun during electron heat flux dropouts in the solar wind, *J. Geophys. Res., 97,* 8203, 1992.

Mansurov, S.M., New evidence of a relationship between magnetic fields in space and on earth, *Geomagn. Aeron., 9,* 622, 1969.

Marsch, E., Kinetic physics of the solar wind plasma, in *Physics of the Inner Heliosphere, 2,* edited by R. Schwenn and E. Marsch, p.45, Springer-Verlag, Berlin, 1991.

Neugebauer, M., B.E. Goldstein, E.J. Smith, and W.C. Feldman, Ulysses observations of differential alpha-proton streaming in the solar wind, *J. Geophys. Res., 101,* 17047, 1996.

Ogilvie, K.W., M.A. Coplan, and R.D. Zwickl, Helium, hydrogen, and oxygen velocities observed on ISEE-3, *J. Geophys. Res., 87,* 7363, 1982.

Roberts, D.A., M.L. Goldstein, L.W. Klein, and W.H. Matthaeus, Origin and evolution of fluctuations in the solar wind: Helios observations and Helios-Voyager comparisons, *J. Geophys. Res., 92,* 12023, 1987.

Steinberg, J.T., A.J. Lazarus, K.W. Ogilvie, R. Lepping, and J. Byrnes, Differential flow between solar wind protons and alpha particles: first WIND detection, *J. Geophys. Res., 23,* 1183, 1996.

Svalgaard, L., Sector structure of the interplanetary magnetic field and daily variation of the geomagnetic field at high latitudes, *Dan. Meteorol. Inst. Geophys. Pap. R-6,* 1, 1968.

Tsurutani, B.T., C.M. Ho, J.K. Arballo, E.J. Smith, B.E. Goldstein, M. Neugebauer, A. Balogh, and W.C. Feldman, Interplanetary discontinuities and Alfven waves at high heliographic latitudes: Ulysses, *J. Geophys. Res., 101,* 11027, 1996.

Wilcox, J.M., Inferring the interplanetary magnetic field by observing the polar geomagnetic field, *Rev. Geophys. Space Phys., 10,* 1003, 1972.

Wilcox, J.M., and N.F. Ness, Quasi-stationary corotating structure in the interplanetary medium, *J. Geophys. Res., 70,* 5793, 1965.

Wilcox, J.M., L. Svalgaard, and P.C. Hedgecock, Comparison of inferred and observed interplanetary magnetic field polarities, 1970-1972, *J. Geophys. Res., 80,* 3685, 1975.

S.W. Kahler, PL/GPSG, Hanscom AFB, MA 01731-3010.

The Current Status in Our Understanding of Energetic Particles, Coronal Mass Ejections, and Flares

H. V. Cane

Laboratory for High Energy Astrophysics, Code 661, NASA/GSFC, Greenbelt, Maryland
and
Physics Department, University of Tasmania, Hobart, Australia

This paper presents a summary of observations of coronal mass ejections, flares, shocks, and energetic particles with the aim of clarifying the relationship between these phenomena. The paper points out areas where our understanding is incomplete, with particular emphasis on particle acceleration. It is argued that flares result from the same magnetic instability that creates mass ejections and that flares are not responsible for interplanetary disturbances or for major proton events. It is probable that coronal shocks are directly related to flares but interplanetary shocks are driven by coronal mass ejections. Particle acceleration by shocks is a very important process and accounts for the majority of major proton increases, although the physics is not yet fully understood.

INTRODUCTION

This paper reviews the observations which have led researchers to conclude that solar flares do not cause effects detected in the near-Earth space environment, apart from ionospheric disturbances and some particle events. Rather, interplanetary disturbances and some flares are both the result of a more fundamental process, that of mass ejection. These mass ejections, commonly called CMEs (coronal mass ejections), are observed in white light and were first imaged with space-borne coronagraphs in the early 1970s [*Tousey*, 1973; *Gosling et al.*, 1974]. Although various aspects of this "new paradigm" have been proposed since the early 1980s, the controversy became a heated debate in 1993 after J. T. Gosling published a paper in the *Journal of Geophysical Research* entitled "The Solar Flare Myth" [*Gosling*, 1993] (see also *Kahler* [1992]). This paper has led to a "flares versus CMEs" controversy in the solar-terrestrial community (e.g., *Hudson et al.*, 1995; *Miller*, 1995). In the afore mentioned work, and the present paper, the term "flare" implies a sudden chromospheric brightening. (For a discussion of the evolution of solar flare nomenclature, see *Cliver* [1995]).

In the simplest terms, the old understanding was that a flare generated a shock which caused a geomagnetic storm when it impacted the Earth's magnetosphere. The shock was detected when it was close to the Sun by virtue of the meter wavelength radio emission (type II burst) it generated. The shock appeared to be generated ahead of a plasma cloud (also called the piston or driver gas) indicated by the presence of type IV emission. (For a more detailed description of the early ideas, see pages 576-580 in *Kundu* [1965]). Energetic particles either came from flares or possibly were accelerated by the shock, since a good correlation between particle events and type II bursts was found [*Svestka and Fritzova-Svestkova*, 1974]. However, since type II bursts typically last 5-10 minutes [*Kundu*, 1965], impulsive particle injection was implied. For the largest events, the four phenomena, flares, interplanetary shocks, energetic particles, and type II/type IV bursts, are well correlated. Initially the incorporation of CMEs into the picture required no modification since CMEs

Coronal Mass Ejections
Geophysical Monograph 99
Copyright 1997 by the American Geophysical Union

were thought to be the plasma clouds emitted by flares. But various inconsistencies became apparent, and these are the subject of the first part of the paper where the relationships between CMEs and a) interplanetary shocks, b) flares and c) type II bursts are examined. Energetic particle observations are then reviewed with an emphasis on those observations that link particle acceleration with CMEs/interplanetary shocks rather than flares. The paper also provides a summary of the present ideas and points out areas in which our understanding is incomplete.

CMEs AND INTERPLANETARY DISTURBANCES

The earliest observations suggesting a relationship between solar activity and geomagnetic activity date from the previous century. *Sabine* [1852] was first to note that geomagnetic activity tracks the solar cycle. The first flare ever observed, in white light, was followed about 18 hours later by a geomagnetic storm [*Carrington*, 1860]. Later, *Newton* [1943] found a significant correlation between large flares and subsequent geomagnetic storms.

Gold [1955] was the first to suggest that the sudden commencements of geomagnetic storms were caused by the impact of shocks driven by plasma clouds ejected from the Sun. The idea of plasma clouds had been suggested earlier by *Lindeman* [1919] and *Chapman and Ferraro* [1929]. The existence of such clouds and the partial exclusion of cosmic rays from the interior of the clouds was invoked to explain the sudden decreases in cosmic rays [*Forbush*, 1938] associated with geomagnetic storms. *Morrison* [1956] proposed that magnetic fields in plasma clouds were turbulent whereas others [e.g., *Cocconi et al.,* 1958] suggested they were smooth. There were also conflicting ideas about whether the plasma clouds were closed, plasmoid-like structures disconnected from the Sun [*Piddington,* 1958] or whether they remained connected back to the Sun leading to magnetic tongues [*Cocconi et al.,* 1958]. This debate is still not totally resolved. There is evidence for the existence of magnetic tongues [e.g., *Richardson and Cane,* 1996], but the existence of plasmoids cannot be ruled out. Note that Parker's theory of the solar wind was not developed until 1958. *Parker* [1961] introduced the idea of a blast wave created by a widespread elevation in the solar wind flow speed. This could cause a cosmic ray decrease via a shell of turbulent field behind the shock. This, of course, was a different proposal to that of *Gold* [1955] whose suggestion involved the introduction of closed magnetic fields from a region of the corona which previously did not contribute to the solar wind.

By the early 1970s, less than 10 years after the first spacecraft observations of interplanetary shocks [*Sonnett et al.,* 1964], a fairly reasonable idea had been formed of the geometry of shocks and plasma clouds [*Hundhausen,* 1972].

Hundhausen [1972] used the term "flare ejecta" for the plasma cloud and in the rest of this paper the term "ejecta" will be used instead of plasma cloud or driver gas. Subsequently, *Barnden* [1973] suggested that Hundhausen's two-component model of ejecta plus swept up, compressed ambient field with shock, could explain the observed two-step structure seen in many Forbush decreases.

CMEs were first observed in the early 1970s. From studies of the events observed by Skylab [*Gosling et al.,* 1974] it was obvious that these ejections were capable of producing the high speed flows required to produce interplanetary shocks because a number of the events had outward speeds above 1000 km/sec. Since these were associated with flares, it was assumed that CMEs were caused by flares. Second generation coronagraphs were flown in the 1980s. During cycle 21, the Solwind coronagraph [*Howard et al.,* 1985] observed many CMEs and showed without a doubt that CMEs are the drivers of interplanetary shocks [*Sheeley et al.,* 1985; *Cane et al.,* 1987]. The Solar Maximum Mission (SMM) coronagraph [*MacQueen et al.,* 1980] observed over 1000 CMEs [*Burkepile and St. Cyr,* 1993] during two solar maxima and also during solar minimum. The SOHO spacecraft was successfully launched in December 1995, and onboard was a new coronagraph [*Brueckner et al.,* 1995]. Already new and exciting results have been obtained [*Brueckner et al.,* 1996; *Howard et al.,* this volume].

Although there is a clear association between flares, shocks and CMEs, several observations suggest that flares are not responsible for CMEs and shocks. Thus, whereas most major interplanetary shocks are preceded by a flare and meter wavelength type II and type IV bursts [*Hundhausen,* 1972; *Cane,* 1985], not all shocks can be associated with a flare, and most flares cannot be associated with an interplanetary shock [*Hundhausen,* 1972]. Furthermore, it is unlikely that the high speed flow following shocks and lasting for a day or so can come from a flare which lasts typically a few hours.

CMEs AND FLARES

It was the coronagraph on SMM that enabled a detailed study of the relationship between flares and CMEs. There are six basic results from the studies which should dispel any belief that flares can cause CMEs:

1. Statistics: Only a minority of CMEs (<40%) are associated with flares, with the most common association (~70%) being with eruptive prominences [*Munro et al.,* 1979; *Webb and Hundhausen,* 1987; *St. Cyr and Webb,* 1991].

2. Spatial Sizes: The characteristic angular sizes of CMEs imply disruptions of the low corona on spatial scales much larger than those of flares and active regions [*Hundhausen,* 1988; *Kahler et al.,* 1989; *Harrison et al.,* 1990]. From the

SMM data base [*Burkepile and St Cyr*, 1993], the average angular width of all CMEs was 48°.

3. Latitude distributions: The latitudes of flaring regions drift equatorward as solar activty progresses from minimum to maximum. In contrast CMEs, like helmut streamers, originate from a broader latitude range extending to high latitudes as maximum solar activity is approached. These patterns are shown in Figure 1, based on a figure by *Hundhausen* [1993].

4. Timing: CME onsets tend to occur before flares [*Wagner*, 1982; *Harrison*, 1994, and references therein].

5. Locations: Flares can occur anywhere under a CME and are not centered underneath, as was originally believed and anticipated if they cause CMEs [*Harrison*, 1994, and references therein].

6. Energetics: *Hundhausen* [1972] estimated that the specific energy release from the volume of a flare to generate a 10^{31} erg solar wind disturbance is about three orders of magnitude greater than the normal chromospheric thermal energy density. The average mechanical energy of a CME is 10^{31} erg [*Hundhausen*, 1997]. Also, there is a poor correlation between X-ray flare intensities and energies of associated CMEs [*Hundhausen*, 1997].

CMEs AND CORONAL SHOCKS

Type II bursts typically occur in the frequency range 100-20 MHz (i.e. wavelengths in the range 3-15 meters) [*Kundu*, 1964]. The radiation is plasma emission, and with a density model of the corona, the emission frequency can be related to a coronal height. Thus, meter wavelength type II bursts originate within about 3 solar radii of the Sun. The frequency drift rates can be used to determine a speed of the responsible agent, and it was this calculation that led to the realization of the presence of shocks. In the following text, I use the term coronal shock when discussing the agent responsible for meter wavelength type II bursts.

Since it was assumed that coronal shocks continued to propagate out from the Sun and become interplanetary shocks, it was thought that type II bursts would be well associated with CMEs, and their source locations would be in front of CMEs, where shocks should form. This was not found to be the case. The relevant observations are (see *Cane* [1984] for the specific references):

1. Type II bursts have been reported in the absence of CMEs despite good coronagraph coverage.
2. Radio imaging of type II bursts suggests that they are not located in front of CMEs.
3. Type II bursts are usually associated with flares (>70% [*Cane and Reames*, 1988]) with the radio event commencing almost simultaneously with the maximum in Hα intensity [*Bougeret*, 1985]. Furthermore type II bursts have never been

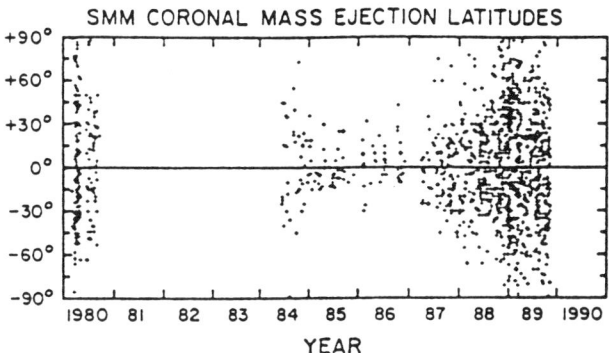

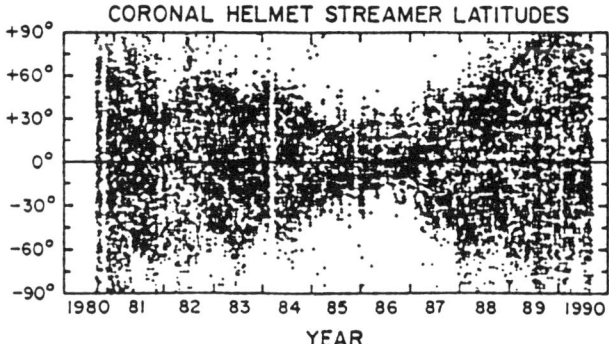

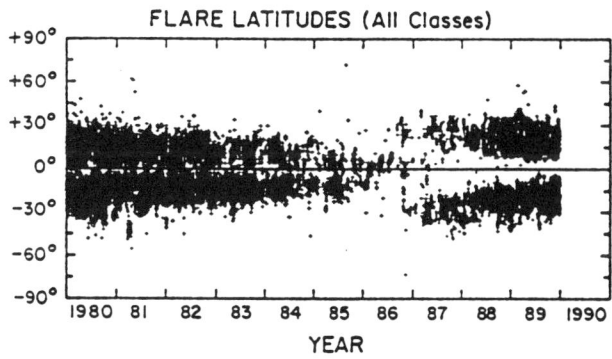

Figure 1. Plot of the central latitudes of CMEs, and latitudes of optical flares and bright coronal features from 1980-1991 (from *Hundhausen* [1993]).

seen to precede an associated flare. This implies a close coupling between coronal type II bursts and flares, in contrast to the situation for CMEs.

4. More than half of all type II bursts are associated with impulsive flares (durations less than 1 hour) [*Cane and Reames*, 1988]. In contrast, flares associated with CMEs generally have long durations [*Sheeley et al.*, 1983]. Figure 2

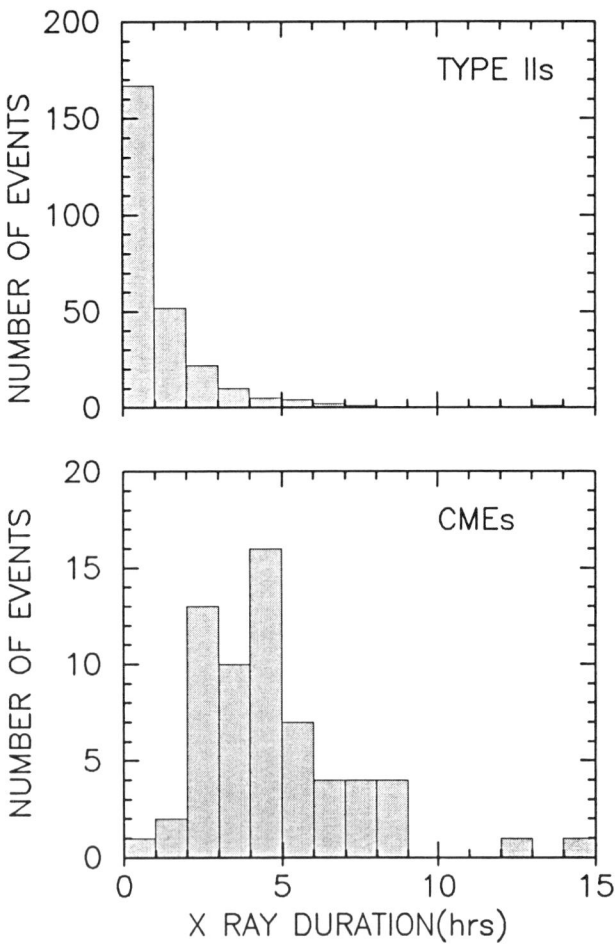

Figure 2. Histograms of soft X ray flare durations for flares associated with Solwind CMEs [*Sheeley et al.*, 1983] and with Culgoora type II bursts [*Cane and Reames*, 1988].

compares the distributions of soft X ray flares associated with CMEs and type II bursts.

5. Some type II bursts have very high starting frequencies implying the existence of coronal shocks in regions below the altitudes where CMEs form.

6. Shock speeds deduced from type II drift rates and CME speeds are inconsistent.

Thus, it has been suggested that coronal type II bursts are not directly related to CMEs and that there are two types of shocks. The shocks producing type II bursts would be blast waves related to the flare [*Wagner and MacQueen*, 1982; *Cane and Reames*, 1988; *Gopalswamy and Kundu*, 1995]. E. W. Cliver (private communication, 1996) notes, however, that there are many intense X-ray flares which are not accompanied by type II bursts. Furthermore, various arguments can be proposed which explain the observations in the context of a single shock model. For example, since the fastest CMEs are generally associated with type II bursts, this could account for the many slower CMEs seen without accompanying type II bursts. Also, since CMEs are difficult to observe when they originate far from the solar limbs, such selectivity could account for type II bursts without associated CMEs [cf. *Sawyer*, 1985]. Another possibility [e.g., *Robinson and Stewart*, 1985] is that some type II bursts originate from blast waves and some from CME-driven shocks. This seems unlikely given that there are no radio characteristics that distinguish two classes of type II bursts [*Cane and White*, 1989]. *Robinson et al.* [1986] reported that type II bursts associated with CMEs have low starting frequencies, but *Cane and Reames* [1988] showed that starting frequencies are related to the impulsiveness of the associated flare and that low starting frequencies are characteristic of gradual solar events, that are well associated with CMEs.

One study using particle data does support a close relationship between coronal shocks and CMEs. It has been found that a class of radio events at hectometer wavelengths, called SA (shock-associated) events, are well correlated with major proton events and with structure in coronal type II bursts indicative of electron streams (herringbone structure) [*Cane et al.*, 1981; *Kahler et al.*, 1986a]. In other words, the existence of SA events suggests a relationship between proton acceleration and coronal type II bursts. However, it has been established that there is a close link between major proton events and CMEs (see below) and so, taken together, the two associations (protons and coronal type II bursts and protons and CMEs) suggest that type II bursts and CMEs are related. It might be thought that SA events, herringbone structure, and proton events are a manifestation of big flares [*Kahler*, 1982] but it should be noted that at least one particle event preceded by an SA event (December 5, 1981) was not associated with a flare.

There is no doubt that radio emission at very long wavelengths (frequencies less than 1 MHz) does originate at interplanetary shocks. This emission, which must be observed from space, originates at heights extending from about 10 solar radii all the way to the Earth, where the radio emission coincides with the arrival of the shock. To avoid confusion with coronal type II bursts, these radio events (first well observed from ISEE-3 in the late 1970s) were called IP type II events [*Cane*, 1985]. *Cane et al.* [1987] found an excellent correlation between IP type II events and large, fast CMEs. In an analysis of the drift rates of IP type II events, there was difficulty relating emissions at frequencies below 1 MHz with those at higher frequencies; because the velocity curves were incompatible. Thus, *Cane* [1983] suggested a two shock model wherein the higher frequency emission came from a flare-initiated blast wave and the low-frequency emission originated in an interplanetary shock driven by a

CME. However, if the CME-driven shock only commences above about 10 solar radii, as it might if the radio emission below 1 MHz signifies the creation of the interplanetary shock, then this shock clearly cannot produce the highest-energy prompt particles, which commence before CMEs reach 5 solar radii [*Kahler*, 1994].

It is probable that the relationship between meter wavelength type IIs and CMEs/interplanetary shocks will only be resolved after radio events in the decameter and hectometer wavelength range have been observed by a single instrument. The WIND spacecraft radio astronomy experiment can do this, but the Sun has yet to produce some sufficiently energetic events.

ENERGETIC PARTICLES

The reason why flares have been related to energetic particle events is that most particle events are preceded by a flare. Assuming flares are the sources of energetic particles, one has then to explain why large energetic particle events last for days and can be observed on field lines which connect to the Sun more than 100 degrees away from the flare. Until the 1980s it was held that processes in the corona were responsible. The long durations could be accounted for by storage, and the huge traversals across the disk could be accounted for by a process called "coronal diffusion" [*Reid*, 1964] whose details were not identified. It is relevant to describe briefly the pre-1980s scenario for particle acceleration. It was thought that there were two acceleration phases. The first (impulsive phase) was directly related to flares and generated electrons up to an energy of a few 100 keV. These electrons also generated microwaves and hard X rays. The second phase consisted of the shock acceleration of protons and high energy electrons. The timing relationship was based on the occurrence of type II (shock generated) radio bursts several minutes after type III (electron stream generated) radio bursts. It was believed that the second phase required the first and that one could use radio emission associated with the impulsive phase to determine the proton intensities from the second phase. It was also known that interplanetary shocks accelerate particles [*Rao et al.*, 1967], but it was believed that the maximum energy achievable was about 10 MeV and that the whole effect was only seen locally as the shock passed. It was thought that the particles "from the flare" provided an energetic "seed" population on which the shock operated.

The following observations show this picture is not correct:

1. *Kahler et al.* [1978] found a correlation between CME speed and >4 MeV proton intensity. They also noted that the sizes of CMEs are comparable to the heliolongitude range over which particle events did not show delays to onset, the so-called "fast-propagation region" in which coronal diffusion was inferred to occur rapidly [*Reinhard and Wibberenz*, 1974].

2. *Domingo et al.* [1979] discussed an energetic proton event which was not associated with a flare, only an erupting prominence and an interplanetary shock. *Kahler et al.* [1986b] and *Cane et al.* [1986a] discussed other particle events of this type.

3. *Gosling et al.* [1981] showed that a particle enhancement associated with an interplanetary shock had a spectrum which continued smoothly to thermal energies, suggesting that these particles were drawn from the solar wind rather than accelerated flare particles. It is important to note, however, that this particle event was not detectable above 10 MeV.

4. *Cliver et al.* [1983] discussed a number of energetic proton events in which the impulsive phase was weak.

5. *Mason et al..* [1984] showed that the abundances of events did not vary as a function of the longitude of the event. This puts severe restraints on the "coronal" diffusion process, since different ions diffuse at different rates, but is reasonable if the particles are accelerated over a range of heliolongitudes at a large shock.

6. *Cane et al.* [1986b] found that interplanetary particle events could be divided into two classes. The large events were associated with the type of flare events associated with CMEs and interplanetary shocks. They could originate anywhere on the disk. The other particle events were associated with impulsive flares (without CMEs) and were only observed following flares in regions that were magnetically well connected to the observer, suggesting a much smaller source region.

7. *Luhn et al.* [1987] found that the ionization states of elements O-Fe in large events are roughly consistent with equilibrium temperatures of a few times 10^6 K. This means that the particles come from the corona and not flare-heated material.

8. *Cane et al.* [1988] showed that the intensity-time profiles of events were very well organized by the longitude of the event and could be explained in terms of the large-scale structure of interplanetary shocks (see Figure 3). The delays to maximum intensity of eastern events were a natural consequence of continued acceleration and improving connection to the stronger parts of the shock. The organization by observer location was substantiated by observations of the same event from separated spacecraft.

9. *Reames* [1988] obtained a bi-modal distribution (Figure 4) in the Fe/O abundances of all energetic particle events, substantiating the existence of two particle populations.

PRESENT STATUS

The relationship between flares and CMEs is not established. The most likely scenario is that flares occur under

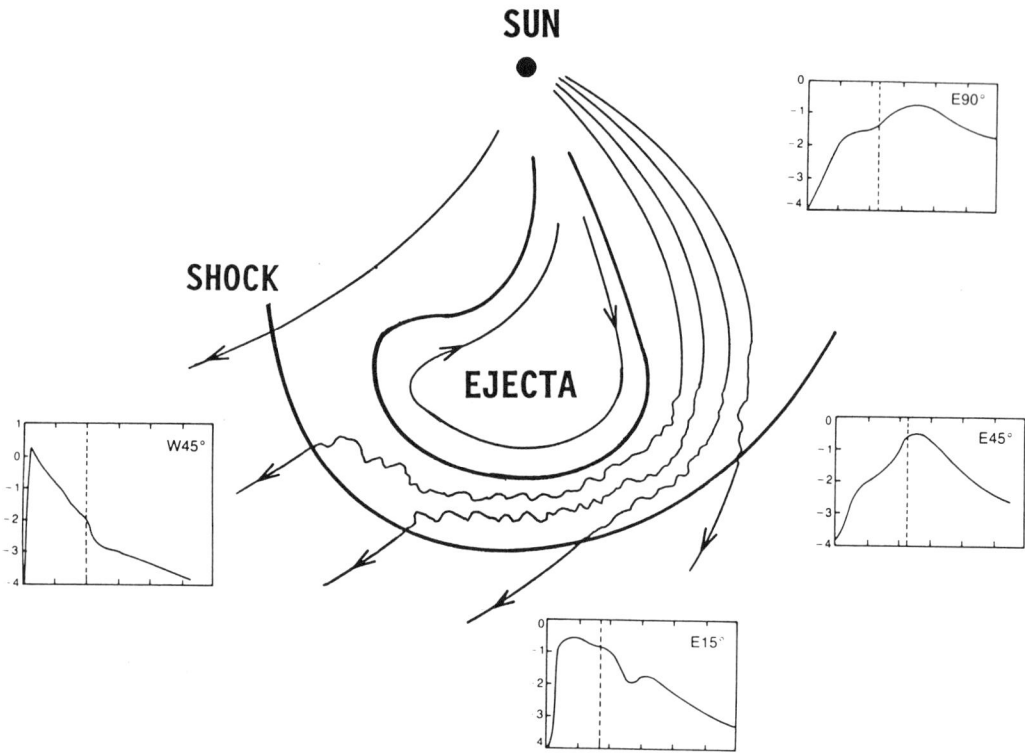

Figure 3. Representative profiles for actual events at 20 MeV for observers at different locations relatve to an interplanetary shock. The differing intensity profiles can be understood in terms of shock acceleration and the large scale structure of shocks. See *Cane et al.* [1988].

favorable conditions as a result of the same process that causes mass ejections. Although everyone agrees the ultimate energy source for CMEs is magnetic, the details have yet to be determined.

A full understanding of CMEs and how they propagate through, and affect, the interplanetary medium will be difficult to achieve until it is possible to relate the phenomena seen at the Sun with phenomena seen in space. There are a number of signatures that identify the interplanetary counterparts of CMEs [see, e.g., *Richardson and Cane*, 1993] but none appears to identify all events, and the timings of the different phenomena often disagree. It is probable that the most favored signature (that of counterstreaming suprathermal electrons [*Gosling et al.*, 1987]) does not uniquely identify CMEs [*Richardson and Cane*, 1993]. At present, interplanetary CMEs are divided into two classes: those with the char-acteristics of a magnetic flux rope (so-called magnetic clouds [*Zhang and Burlaga*, 1988]) and all others. It is probable that there is no fundamental difference between the classes and the observed structure is related to the history of the event and/or where it is intercepted [*Cane et al.*, 1997]. It is not clear that the magnetic cloud signature identifies the complete interplanetary CME.

It is now firmly established that solar energetic particles observed in the interplanetary medium are of two basic populations [*Cane et al.*, 1986b; *Lin*, 1987; *Reames*, 1993, 1995]. One population has its origins in flaring regions, and the events are Fe-rich, ^{3}He-rich, electron-rich, last for hours at most, extend over a limited region in longitude (<30°) and are not associated with CMEs. The associated electromagnetic flare emissions, e.g., Hα, are impulsive. In contrast, the so-called 'gradual' events are proton-rich, last for days, often spread over more than 180° in longitude and are well-associated with CMEs. The flares associated with these events are generally, but not always, gradual. On the other hand, these particle events certainly have more gradual intensity time profiles than do the impulsive particle events. Another indication for the presence of two particle populations is the inconsistency between the numbers of particles interacting at the Sun to produce gamma rays with those observed in space [*Cliver et al.*, 1989].

Recently, *Cliver* [1996] has cautioned researchers that the picture of two classes of particle events is overly simplistic. For example, the particle events associated with impulsive flares isolated by *Cane et al.* [1986b] were much more energetic than typical ^{3}He-rich events, and many were

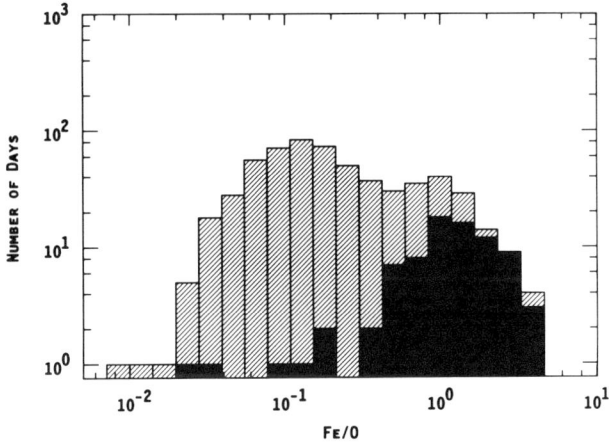

Figure 4. Histograms of the measured Fe/O values for ~2 MeV/amu solar ions. The dark shaded histogram is for days that were ^{3}He-rich (from *Reames et al.* [1988]).

associated with CMEs. *Cliver* [1996] suggests calling these events "mixed-impulsive". *Cane and Reames* [1990] suggested that in addition to "flare particles", protons accelerated at coronal shocks accompany many of the more energetic impulsive events. Also, *Van Hollebeke et al.* [1990] suggested that the high energy ^{3}He, Fe-rich interplanetary particles seen in the impulsive June 3, 1982, flare resulted from a second-phase shock acceleration of flare particles.

Cliver [1996] also discusses "mixed-gradual" events, in which there are flare particles along with shock-accelerated particles. Such mixed populations are only observed when there is good connection to the flare site. Mixed events were also mentioned by *Cane et al.* [1986b] and are most apparent when a large proton event has an Fe-rich component at onset [*Reames*, 1990]. It is possible that such a combination of acceleration processes could be responsible for the structure seen in some ground level enhancements [e.g., *Shea and Smart*, 1996].

Another modification to the "two class" picture is the existence of particle events with essentially no associated flare emissions (such as hard X rays and radio bursts), which are associated with interplanetary shocks caused by less energetic CMEs. These so-called "disappearing filament" events [*Cane et al.*, 1986a; *Kahler et al.*, 1986b; *Sanahuja et al.*, 1991] have steep energy spectra and typically do not extend above about 10 MeV. They consist of an intensity enhancement commenc-ing about a day before shock passage and peaking at the shock. Of course, there is a continuum of events, and the more energetic ones, that of December 5, 1981, [*Kahler et al.*, 1986] being the prime example, extend to higher energies and are almost indistinguishable from flare-associated particle events in the interplanetary medium. (It is possible that even the more energetic events have steeper spectra than flare-associated gradual events [*Kahler et al.*, 1986b].) It has been suggested that for the less energetic events, there is essentially no acceleration at or close to the Sun and that the relatively low maximum energy is a reflection of the limited extent to which shocks can accelerate particles beyond about 0.3 AU [*Kallenrode*, 1996]. Unfortunately, with few observations from widely spaced detectors, it is not possible to say whether a component of particles accelerated close to the Sun has been missed because of bad connection or because it was not accelerated. There is one event observed by multiple spacecraft where it would appear that the latter is the case [*Cane*, 1995]. Helios 1 was well connected to the source region of the shock of April 22-24, 1979, and yet did not detect any energetic particles.

Although it is now widely accepted that coronal/interplanetary shocks driven by CMEs are responsible for major proton events, there are many unanswered questions about the details. Probably the most important one is whether shocks can accelerate solar wind particles to energies of 10-100s of MeV locally in the interplanetary medium or whether the higher-energy ions are initially accelerated close to the Sun where shock acceleration is much more efficient (shocks are faster and the medium is presumably more turbulent) and then reaccelerated by the shock in the interplanetary medium. M.-B. Kallenrode and G. Wibberenz [Propagation of particles injected from interplanetary shocks: A black-box model and its consequences for acceleration theory and data interpretation, submitted to *J. Geophys. Res.*, 1996] argue that diffusive shock acceleration theory as applied to low energy (<~1 MeV) particle increases at interplanetary shocks [*Lee*, 1983] is unlikely to apply at high energies. In particular, a steady state is not reached. Furthermore, although upstream turbulence, which is critical for the process, is measured ahead of interplanetary shocks, there is no additional turbulence measured at wavelengths in resonance with 10s of MeV protons [*Wibberenz and Kallenrode*, 1995]. From their modelling work, B. Sanahuja and co-workers find that the shock acceleration efficiency drops off somewhere between 1 and 5 MeV (B. Sanahuja, private communication, 1996). *Kallenrode* [1996] suggests that particles which are accelerated close to the Sun undergo further interactions with the shock in the interplanetary medium (reacceleration). If the shock is sufficiently weak, then above some energy this population will have higher intensities than particles drawn out of the local solar wind. For such a model to work successfully, the interplanetary mean free path needs to be considerably less than the ~1 AU suggested by *Reames* [1993]. Otherwise particles accelerated close to the Sun will stream away and not scatter back to the shock.

There are also questions as to whether other processes in addition to shock acceleration (e.g., reconnection) contribute in large particle events. *Akimov et al.* [1993] suggest that energetic particles can be produced in post-flare loops, and *Chertok* [1995] suggests that these particles can escape to the interplanetary medium. The problem with these suggestions is that it is difficult to prove that post-flare-loop acceleration occurs because high-energy particles may be accelerated by a shock, and these would mask the particles originating lower in the corona. *Kahler* [1996] suggests that if this process does occur, it should be possible to identify a few ion events associated with slow CMEs that do not drive shocks. One important constraint when considering alternative acceleration processes is that charge-state measurements of elements O-Fe in large events are inconsistent with temperatures above a few times 10^6 K [e.g., *Mason et al.*, 1995]. Hence these particles cannot originate in high-temperature regions associated with flares or regions of reconnection. Another problem with any non-shock acceleration model is the difficulty in accounting for the wide-spread distribution of particles in the earliest stages of major events. This is especially true of any proposal related to small flare regions.

Kiplinger [1995] has found an association between interplanetary proton events and certain hard X-ray bursts and suggests that particles and a component of the hard X rays have the same source region. Although a physical mechanism is not specified by *Kiplinger* [1995], one assumes that post-flare loops would also be involved. As noted above, such a model is inconsistent with the widespread acceleration seen in major events. The algorithm that Kiplinger uses for predicting proton events incorporates the time scale of the X ray event, and it is not yet clear that this new algorithm has a better success rate than one using the time scale of the soft X rays [*Cliver and Cane*, 1990]. In either case, the most likely cause for the correlation is that gradual events are associated with CMEs. It should be noted that hard X ray signatures cannot be used to predict events like the one on December 5, 1981, which was associated with a CME/interplanetary shock but not a flare [*Kahler et al.*, 1986b].

Multispacecraft observations support the fact that in large events, particles are seen rather promptly from widespread regions (up to 300° in longitude) [*Cliver et al.*, 1995; *Cane*, 1996]. The wide-spread acceleration is difficult to reconcile in terms of the latitudinal extents of CMEs and the longitudinal extents of interplanetary shocks at 1 AU. It is possible that shocks rapidly decrease in longitudinal extent as they move away from the Sun [*Cane*, 1996]. Alternatively, the wide-spread acceleration of particles in the early stages of major events may be related to the presence of "global CMEs" revealed by the LASCO coronagraph [*Brueckner*, 1996]. Another possible scenario is that the earliest particles are accelerated at a shock not driven by a CME, i.e. that there are two shocks.

Acknowledgments. The author thanks JoAnn Joselyn and Nancy Crooker for organizing the conference and for the invitation to be a group discussion leader. This paper has benefitted from discussions with Ed Cliver, Steve Kahler, Don Reames, Ian Richardson and Gerd Wibberenz. Ian Richardson is thanked for the many suggestions which greatly improved the style of the manuscript. This work was supported at GSFC by a contract with Universities Space Research Association.

REFERENCES

Akimov, V. V., A. V. Belov, I. M. Chertok, V. G. Kurt, A. Magun, and V. F. Melnikov, High-energy gamma rays at the late stage of the large solar flare of June 15, 1991 and accompanying phenomena, *Proc. Int. Conf. Cosmic Rays 23rd, 3,* 111-114, 1993.

Barnden, L.F., The large-scale magnetic configuration associated with Forbush decreases, *Proc. Int. Conf. Cosmic Rays 13th, 2,* 1277-1282, 1973.

Bougeret, J.-L., Observations of shock formation and evolution in the solar atmosphere, in *Collisionless Shocks in the Heliosphere: Reviews of Current Research,* edited by B. T. Tsurutani and R. G. Stone, pp. 13-32, AGU, Washington, 1985.

Brueckner, G. E. and 14 others, The large angle spectroscopic coronagraph (LASCO), *Solar Phys., 162,* 357-402, 1995.

Brueckner, G. E., Dynamics of the solar corona as seen from the 3 LASCO coronagraphs on SOHO satellite (abstract), *Eos Trans. AGU, 77(17),* Spring Meet. Suppl., 204, 1996.

Burkepile, J.T. and O.C. St. Cyr, A revised and expanded catalogue of mass ejections observed by the Solar Maximum Mission coronagraph, NCAR/TN-369+STR, NCAR, Boulder, 1993.

Burlaga, L.F., Understanding the heliosphere and its energetic particles, *Proc. Int. Conf. Cosmic Rays 18th, 12,* 21-60, 1983.

Cane, H. V., Velocity profiles of interplanetary shocks, *Solar Wind Five,* NASA Conf. Publ., CP-2280, 703-709, 1983.

Cane, H. V., The evolution of interplanetary shocks, *J. Geophys. Res., 90,* 191-197, 1985.

Cane, H. V., The relationship between coronal transients, type II bursts and interplanetary shocks, *Astron. Astrophys., 140,* 205-209, 1984.

Cane, H. V., The structure and evolution of interplanetary shocks and the relevance for particle acceleration, *Nuclear Phys. B (Proc. Suppl), 39A,* 35-44, 1995.

Cane, H. V., Longitudinal extents of coronal/interplanetary shocks, in *High Energy Solar Physics,* edited by R. Ramaty, N. Mandzhavidze, and X.M. Hua, pp 124-130, AIP, Woodberry, N.Y., 1996.

Cane, H. V. and Reames, D. V., Soft X-ray emissions, meter-wavelength radio bursts, and particle acceleration in solar flares, *Astrophys. J., 325,* 895-900, 1988.

Cane, H. V. and White, S. M., On the source conditions for herringbone structure in type II solar radio bursts, *Solar Phys., 120,* 137-144, 1989.

Cane, H. V. and Reames, D. V., The relationship between energetic particles and flare properties for impulsive solar flares, *Astrophys. J. Suppl., 73,* 253-258, 1990.

Cane, H. V., R.G. Stone, J. Fainberg, R. T. Stewart, J.-L. Steinberg and S. Hoang, Radio evidence for shock acceleration of electrons in the solar corona, *Geophys. Res. Lett., 8,* 1285-1288, 1981

Cane, H. V., S. W. Kahler and N. R. Sheeley, Jr., Interplanetary shocks preceded by solar filament eruptions, *J. Geophys. Res., 91,* 13,321-13,329, 1986a.

Cane, H. V., R. E. McGuire and T. T. von Rosenvinge, Two classes of solar energetic particle events associated with impulsive and long duration soft X ray events, *Astrophys. J., 301,* 448-459, 1986b.

Cane, H. V., N. R. Sheeley, Jr., and R.A. Howard, Energetic interplanetary shocks, radio emission and coronal mass ejections, *J. Geophys. Res., 92,* 9869-9874, 1987.

Cane, H. V., D. V. Reames, and T. T. von Rosenvinge, The role of interplanetary shocks in the longitude distribution of solar energetic particle events, *J. Geophys. Res., 93,* 9555-9567, 1988.

Cane, H. V., I. G. Richardson, and G. Wibberenz, Helios 1 and 2 observations of particle decreases, ejecta, and magnetic clouds, in press, *J. Geophys. Res.,* 1997.

Carrington, R.C., Description of a singular appearance seen on the Sun on September 1, 1859, *Mon. Not. R. Astron. Soc., 20,* 13-15, 1860.

Chapman, S. and V. C. A. Ferraro, The electrical state of solar streams of corpuscles, *Mon. Not. R. Astron. Soc., 89,* 470-479, 1929.

Chertok, I. M., Post-eruption particle acceleration in the corona: A possible contribution to solar cosmic rays, *Proc. Int Conf. Cosmic Rays 24th, 4,* 78-81, 1995.

Cliver, E. W., Solar flare nomenclature, *Solar Phys., 157,* 285-293, 1995.

Cliver, E. W., Solar flare gamma-ray emission and energetic particles in space, in *High Energy Solar Physics,* edited by R. Ramaty, N. Mandzhavidze, and X.M. Hua, pp 45-60, AIP, Woodbury, N.Y., 1996.

Cliver, E. W. and H. V. Cane, X-class soft X-ray bursts and major proton events during solar cycle 21, *Proceedings of Solar Terrestrial Predictions Meeting,* Australia, 1989, Vol.1, pp 359-370, edited by R.J Thompson et al., NOAA/ERL, Boulder, CO, 1990.

Cliver, E.W., S. W. Kahler and P. S. McIntosh, Solar proton flares with weak impulsive phases, *Astrophys. J., 264,* 699-707, 1983.

Cliver, E. W., D. J. Forrest, H. V. Cane, D. V. Reames, R. E. McGuire, T. T. von Rosenvinge, S. R. Kane, and R. J. MacDowall, Solar flare nuclear gamma-rays and interplanetary proton events, *Astrophys. J., 343,* 953-970, 1989.

Cliver, E. W., S. W. Kahler, D. F. Neidig, H. V. Cane, I. G. Richardson, M.-B. Kallenrode and G. Wibberenz, Extreme "propagation" of solar energetic particles, *Proc. Int. Conf. Cosmic Rays 24th, 4,* 257-260, 1995.

Cocconi, G., K. Greisen, P. Morrison, T. Gold, and S. Hayakawa, The cosmic ray flare effect, *Nuovo Cimento Suppl. Ser. 10, 8,* No 2, 161-168, 1958.

Domingo, V. R., R.J. Hynds, and G. Stevens, A solar proton event of possible non-flare origin, *Proc. Int. Conf. Cosmic Rays 16th, 5,* 192-197, 1979.

Forbush, S.E., On the world-wide changes in cosmic ray intensity, *Phys. Rev., 54,* 975-988, 1938.

Gold, T., Discussion of shock waves and rarefied gases, in *Gas Dynamics of Cosmic Clouds,* edited by J. C. van de Hulst and J. M. Burgers, p 103, North-Holland, New York, 1955.

Gopalswamy, N. and M. R. Kundu, Coronal shocks, interplanetary shocks and coronal mass ejections, Proc. of the Second STIP Symp., *STEP GBRSC News, Vol. 5,* Special Issue, June 1, 1995.

Gosling, J.T., The solar flare myth, *J. Geophys. Res., 98,* 18,937-18,949, 1993.

Gosling, J.T., E. Hildner, R.M. MacQueen, R.H. Munro, A.I. Poland, and C.L. Ross, Mass ejections from the Sun: A view from Skylab, *J. Geophys. Res., 79,* 4581-4587, 1974.

Gosling, J.T., J. R. Ashbridge, S. W. Bame, W. C. Feldman, R. D. Zwickl, G. Paschmann, N. Sckopke and R. J. Hynds, Interplanetary ions during an energetic storm particle event: The distribution function from solar wind thermal energies to 1.6 MeV, *J. Geophys. Res., 86,* 547-554, 1981.

Gosling, J.T., D.N. Baker, S.J. Bame, W.C. Feldman, and R.D. Zwickl, Bidirectional solar wind heat flux events, *J. Geophys. Res., 92,* 8519-8535, 1987.

Harrison, R. A., E. Hildner, A.J. Hundhausen, D. G. Sime, and G. M. Simnett, The launch of coronal mass ejections: Results from the coronal mass ejection onset program, *J. Geophys. Res., 95,* 917-937, 1990.

Harrison, R. A., A statistical study of the coronal mass ejection phenomenon, *Adv. Space Res., 14(4),* 23-28, 1994.

Howard, R.A., N.R. Sheeley, Jr., M.J. Koomen and D.J. Michels, Coronal mass ejections: 1979-1981, *J. Geophys. Res., 90,* 8173-8191, 1985.

Howard, R. A., et al., Observations of CMEs from SOHO/LASCO, this volume.

Hudson, H. B. Haisch, and K. T. Strong, Comment on "The solar flare myth" by J. T. Gosling, *J. Geophys. Res., 100,* 3473-3477, 1995.

Hundhausen, A.J., Interplanetary shock waves and the structure of solar wind disturbances, in *Solar Wind,* edited by C. P. Sonett, P.J. Coleman, and J. M. Wilcox, NASA Spec.Publ. SP-308, 393-417, 1972.

Hundhausen, A.J., The origin and propagation of coronal mass ejections, *Solar Wind Six,* Tech. Note, edited by V. Pizzo, T. E. Holzer, and D. G. Sime, pp.181-214, Natl. Cent. for Atmos. Res., Boulder, Colorado, 1988.

Hundhausen, A. J., Sizes and locations of coronal mass ejections: SMM observations from 1980 and 1984-1989, *J. Geophys. Res., 98,* 13,177-13,200, 1993.

Hundhausen, A. J., Coronal mass ejections, in *Cosmic Winds and the Heliosphere,* edited by J. R. Jokipii, C. P. Sonett, and M. S. Giampapa, in press, University of Arizona Press, Tuscon, 1997.

Kahler, S. W., The role of the Big Flare Syndrome in correlations of solar energetic proton fluxes and associated microwave burst parameters, *J. Geophys. Res., 87,* 3439-3448, 1982.

Kahler, S. W., Solar flares and coronal mass ejections, *Ann. Rev. Astron. Astrophys., 30,* 113-141, 1992.

Kahler, S. W., Injection profiles of solar energetic particles as functions of coronal mass ejection heights, *Astrophys. J., 428,* 837-842, 1994.

Kahler, S. W., Coronal mass ejections and solar energetic particle events, in *High Energy Solar Physics,* edited by R. Ramaty, N.

Mandzhavidze, and X.M. Hua, pp 61-77, AIP, Woodberry, N.Y., 1996.

Kahler, S. W., E. Hildner, and M. A. I. van Hollebeke, Prompt solar proton events and coronal mass ejections, *Solar Phys., 57*, 429-443, 1978.

Kahler, S. W., E. W. Cliver, and H. V. Cane, The relationship of shock-associated kilometric radio emission with metric type II bursts and energetic particles, *Adv. Space Res., 6 (6)*, 319-322, 1986a.

Kahler, S. W., E. W. Cliver, H. V. Cane, R. E. McGuire, R. G. Stone and N. R. Sheeley, Jr., Solar filament eruptions and energetic particle events, *Astrophys. J., 302*, 504-510, 1986b.

Kahler, S. W., N. R. Sheeley Jr., and M. Liggett, Coronal mass ejections and associated X-ray flare durations, *Astrophys.J., 344*, 1026-1033, 1989.

Kallenrode, M.-B., A statistical survey of 5 MeV proton events at transient interplanetary shocks, *J. Geophys. Res., 101*, 24,393-24,409,1996.

Kiplinger, A., Comparative studies of hard X-ray spectral evolution in solar flares with high-energy proton events observed at earth, *Astrophys. J., 453*, 973-986, 1995.

Kundu, M. R., *Solar Radio Astronomy*, Interscience, New York, 1965.

Lee, M. A., Coupled hydromagnetic wave excitation and ion acceleration at interplanetary travelling shocks, *J. Geophys. Res., 88*, 6109-6119, 1983.

Lin, R. P., Solar particle acceleration and propagation, *Rev. Geophys., 25*, 676-684, 1987.

Lindemann, F. A., Note on the theory of magnetic storms, *Philos. Mag. 38*, 669-684, 1919.

Luhn, A., B. Klecker, D. Hovestadt, and E. Möbius, The mean ionic charge state of silicon in ^{3}He-rich solar flares, *Astrophys. J., 317*, 951-955, 1987.

MacQueen, R. M., Coronal transients: A summary, *Phil. Trans. Roy. Soc. Lond., A297*, 605-620, 1980.

Mason, G. M., G. Gloeckler, and D. Hovestadt, Temporal variations of nucleonic abundances in solar flare energetic particle events, II, Evidence for large-scale shock acceleration, *Astrophys. J., 280*, 902-916, 1984.

Miller, J. A., Much ado about nothing, *Eos Trans. AGU*, 76,401-407, 1995.

Mason, G. M., J. E. Mazur, M. D. Looper and R. A. Mewaldt, Charge state measurements of solar energetic particles observed with Sampex, *Astrophys. J., 452*, 901-911, 1995.

Morrison, P., Solar origin of cosmic ray time variations, *Phys. Rev., 101*, 1397-1404, 1956.

Munro, R. H., J. T. Gosling, E. Hildner, R. M. MacQueen, A. I. Poland, and C. L. Ross, The association of coronal mass ejections transients with other forms of solar activity, *Solar Phys., 61*, 201-215, 1979.

Newton, H. W., Solar flares and magnetic storms, *Mon. Not. R. Astron. Soc., 103*, 244-257, 1943.

Parker, E. N., Sudden expansion of the corona following a large solar flare and the attendant magnetic field and cosmic-ray effects, *Astrophys. J., 133*, 1014-1033, 1961.

Piddington, J. H., Interplanetary magnetic field and its control of cosmic ray variations, *Phys. Rev., 112*, 589-596, 1958.

Rao, U. R., K. G. McCracken, and R. P. Bukata, Cosmic ray propagation processes 2. The energetic storm particle event, *J. Geophys. Res., 72*, 4325-4341, 1967.

Reames, D. V., Bimodal abundances in the energetic particles of solar and interplanetary origin, *Astrophys. J. (Lett.), 330*, L71-L75, 1988.

Reames, D. V., Acceleration of energetic particles by shock waves from large solar flares, *Astrophys. J. (Lett.), 358*, L63-L67, 1990.

Reames, D. V., Non-thermal particles in the interplanetary medium, *Adv. Space Res., 13 (9)*, 331-339, 1993.

Reames, D. V., Solar energetic particles: A paradigm shift, *Rev. of Geophys., Supplement*, 585-589, 1995.

Reames, D. V., Ramaty, R, and T. T. von Rosenvinge, Solar neon abundances from gamma-ray spectroscopy and ^{3}He-rich particle events, *Astrophys. J. (Lett.), 332*, L87-L91, 1988.

Reid, G. C., A diffusive method for the initial phase of a solar proton event, *J. Geophys. Res., 69*, 2659-2667, 1964.

Reinhard, R. and G. Wibberenz, Propagation of flare protons in the solar atmosphere, *Solar Phys., 36*, 473-494, 1974.

Richardson, I. G. and H. V. Cane, Signatures of shock drivers in the solar wind and their dependence on the solar source location, *J. Geophys. Res., 98*, 15,295-15,304, 1993.

Richardson, I. G. and H. V. Cane, Particle flows observed in ejecta during solar event onsets and their implication for the magnetic field topology, *J. Geophys. Res., 101*, 27,521-27,532,1996.

Robinson, R. D. and R. T. Stewart, A positional comparison between coronal mass ejection events and solar type II bursts, *Solar Phys., 97*, 145-157, 1985.

Robinson, R. D., R. T. Stewart, N. R. Sheeley Jr, R. A. Howard, M. J. Koomen, and D. J. Michels, Properties of meter wavelength solar radio bursts associated with coronal mass ejections, *Solar Phys., 105*, 149-171, 1986.

Sabine, E. On periodical laws discoverable in the mean effects of the larger magnetic disturbances, No. 2, *Philos. Trans. R. Soc. London, 142*, 103, 1852.

Sanahuja, B., A. M. Heras, V. Domingo and J. A. Joselyn, Three solar filament disappearances associated with interplanetary low-energy particle events, *Solar Phys., 134*, 379-394, 1991.

Sawyer, C., Visibility and rate of coronal mass ejections, *Solar Phys., 98*, 369-378, 1985.

Shea, M. A. and Smart, D. F., in *High Energy Solar Physics*, edited by R. Ramaty, N. Mandzhavidze, and X.M. Hua, pp 131-139, AIP, Woodberry, N.Y., 1996.

Sheeley,N. R., Jr., R. A. Howard, M. J. Koomen, and D. J. Michels, Associations between coronal mass ejections and soft x ray events, *Astrophys. J., 272*, 349-354, 1983.

Sheeley,N. R., Jr., R. A. Howard, M. J. Koomen, D. J. Michels, R. Schwenn, K. H. Muhlhauser, and H. Rosenbauer, Coronal mass ejections and interplanetary shocks, *J. Geophys. Res., 90*, 163-175, 1985.

Sonnet, C. P., D. S. Colburn, L. Davis, E. J. Smith and P. J. Coleman, Evidence for a collision-free magnetohydrodynamic shock in interplanetary space, *Phys. Rev. Lett., 13*, 153-156, 1964.

St. Cyr, O. C. and D. F. Webb, Activity associated with coronal mass ejections at solar minimum: SMM observations from 1984-1986, *Solar Phys., 136*, 379-394, 1991.

Svestka, Z. and L. Fritzova-Svestkova, Type II bursts and particle acceleration, *Solar Phys., 36,* 417-431, 1974.

Tousey, R., The solar corona, *Space Res. XIII*, eds. M. J. Rycroft and S. K. Kuncorn, 713-730, Akademie-Verlag, Berlin, 1973.

Van Hollebeke, M. A. I., F. B. McDonald and J.-P. Meyer, Solar energetic particle observations of the 1982 June 3 and 1980 June 21 gamma-ray/neutron events, *Astrophys. J. Suppl., 73,* 285-296, 1990.

Wagner, W. J., SERF studies of mass motions arising in flares, *Adv. Space Res., 2 (11),* 203-219, 1982.

Wagner, W. J. and R. M. MacQueen, The excitation of type II radio bursts in the corona, *Astron. Astrophys., 120,* 136-138, 1983.

Webb, D. F., and A. J. Hundhausen, Activity associated with the solar origin of coronal mass ejections, *Solar Phys., 108,* 383-401, 1987.

Wibberenz, G. and M.-B. Kallenrode, Evolution of the particle injection at propagating interplanetary shocks, *Adv. Space Res., 15 (8/9),* 393-396, 1995.

Zhang, G. and L. F. Burlaga, Magnetic clouds, geomagnetic disturbances, and cosmic ray decreases, *J. Geophys. Res., 93,* 2511-2518, 1988.

H. V. Cane, Physics Department, University of Tasmania, GPO Box 252-21, Hobart 7001, Australia.

Energetic Particles and the Structure of Coronal Mass Ejections

Donald V. Reames

Laboratory for High Energy Astrophysics, NASA, Goddard Space Flight Center, Greenbelt, Maryland

The largest and most energetic solar-energetic-particle (SEP) events are associated with shock waves driven out from the Sun by coronal mass ejections (CMEs). The particles from these "gradual" events are clearly distinguished from flare-associated particles by their abundances, ionization states, associations, and distributions in space and time. Multi-spacecraft observations help us map the spatial distribution of the accelerated particles that flow out into the heliosphere from the evolving CME shock or those that remain trapped behind it.

1. INTRODUCTION

Evidence of high-energy particles from the Sun was first reported 50 years ago by *Forbush* [1946], long before the first observation of coronal mass ejections (CMEs) that we now know to be the dominant source of these particles. This unfortunate quirk of history led to many years of imaginative attempts to associate the large solar energetic particle (SEP) events with solar flares, the so-called "solar flare myth" [*Gosling*, 1993]. It is only in the last ~10 years that growing new evidence from the particles themselves has allowed us to distinguish those particles from impulsive and gradual events and to associate the most intense and energetic events with CMEs, *not* with flares [*Reames*, 1990a, 1993, 1995b; *Gosling* 1993].

This paper is a review of the recent evidence on the origin of the SEP events and of the complex spatial and temporal distribution of high energy particles that accompany an evolving CME.

2. IMPULSIVE AND GRADUAL EVENTS

Nearly all of the events seen by the early observers are what we now call "proton" events or large "gradual" events, a classification originally based on the duration of the associated soft X-ray event. The time profile of such an event is shown in Figure 1a. The source region for events like this can occur at any longitude on the visible disk, as shown in Figure 2, and even far behind the west limb. Notice that it is not easy to explain such SEP observation in terms of flares, since the particles are seen for much longer than flare photons and they have great difficulty crossing magnetic field lines to distant longitudes. This was "explained" by the *ad hoc* assumption of slow diffusion of particles around the corona and out into interplanetary space. It was known that SEPs often peak near interplanetary shock waves, even out near Earth, but these obviously shock-associated particles were considered to be a minor secondary process by most early workers.

In 1970 *Hsieh and Simpson* [1970] first reported ^{3}He-rich events. Such events were later found to have ^{3}He/^{4}He ratios in the range of 0.1 to 10, while the corresponding ratio in the corona and solar wind is 5×10^{-4}. Order of magnitude enhancements in heavy element abundances (e.g. Fe/O) were also seen [see *Reames, Meyer* and *von Rosenvinge*, 1994 and references therein], as compared with abundances in the solar corona or in large gradual events [see *e.g., Meyer* 1985; *Reames* 1995a]. In the 1980s we observed many of these events and found that they had time profiles like those in Figure 1b and were distributed in longitude as shown in Figure 2b. Near-relativistic electrons arrive only slightly after the photons in these impulsive events, and give good associations with

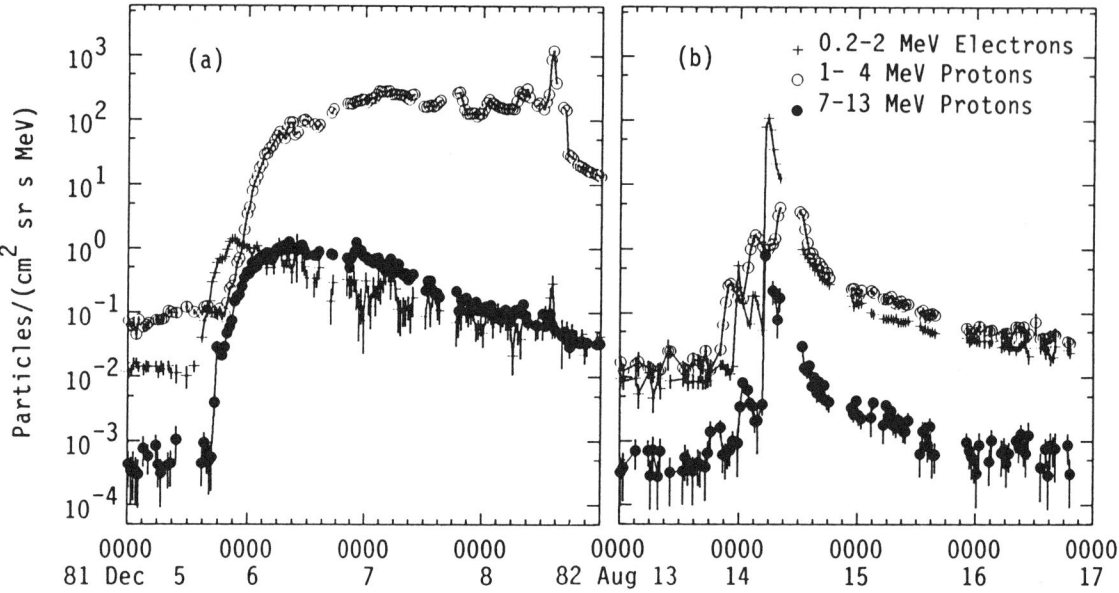

Figure 1. Time profiles of protons and electrons in *(a)* gradual (proton) event and in *(b)* impulsive (^{3}He-rich) events.

impulsive flares and type III radio bursts [*Reames et al.* 1988; *Reames, Cane,* and *von Rosenvinge,* 1990].

Thus, the particles that clearly came from impulsive flares showed only modest spread in longitude and time, contradicting the tremendous diffusion postulated for the large gradual events. The long time duration of the gradual events *must* be coming from continuing acceleration at an interplanetary shock; the longitude distribution screams "shock" to all who would listen. Shocks can easily cross field lines that particles cannot and they can accelerate the particles as they go. Since *Kahler et al.* [1984] showed a 96% correlation between proton events and CMEs, acceleration must occur at the CME-driven shock; blast-wave shocks, that are un-correlated with CMEs, must play little role.

The large gradual event shown in Figure 1a is one I show often since it comes from a well-known disappearing-filament event [*Kahler et al.* 1986] which has a CME but no associated flare. Such events show that no flare is required to produce a large particle event at 1 AU, nor is one required to provide "seed particles" for subsequent shock acceleration. Much of our progress in understanding the physical processes that occur in impulsive and gradual events has come from studying these smaller "pure" events with CMEs but no flares or with impulsive flares but no CMEs. The largest events are very complex and are plagued by "big flare syndrome" [*Kahler* 1982] so it is difficult to determine conclusively which particles are related to which phenomena at the Sun.

Even more compelling evidence for the origin of the particles in gradual events is found in the measurements of the ionization states of the elements C through Fe that are now measured by four different experiment groups on three different spacecraft over energies ranging from 0.3 to 600 MeV/amu. Table 1 summarizes the ionization states for Fe, the element most sensitive to the source plasma temperature. These ionization states are typical of plasma temperatures in the range 1 - 2 MK, like the ambient corona and solar wind. Even the C and O are not fully ionized in gradual events. Similar ionization states are measured for the ions of the solar wind itself. It is clear that the energetic ions do not come from hot plasma near a flare or reconnection region where temperatures exceed 10 MK and Fe XXV spectral lines are often seen. Even at the highest energies, the ions in gradual events are accelerated from the ambient corona and solar wind, far from a flare.

In contrast, a mean ionization state of Fe of 20.5±1.2 was found for impulsive-flare (^{3}He-rich) events [*Luhn et al.* 1987], indicating an electron temperature of >10 MK.

Not only do the energetic ions in gradual events originate at ambient coronal temperatures, but those above about ~10 MeV/amu would be further ionized by passing through dense coronal material. At the highest energies the Fe ions would be measurably stripped of additional

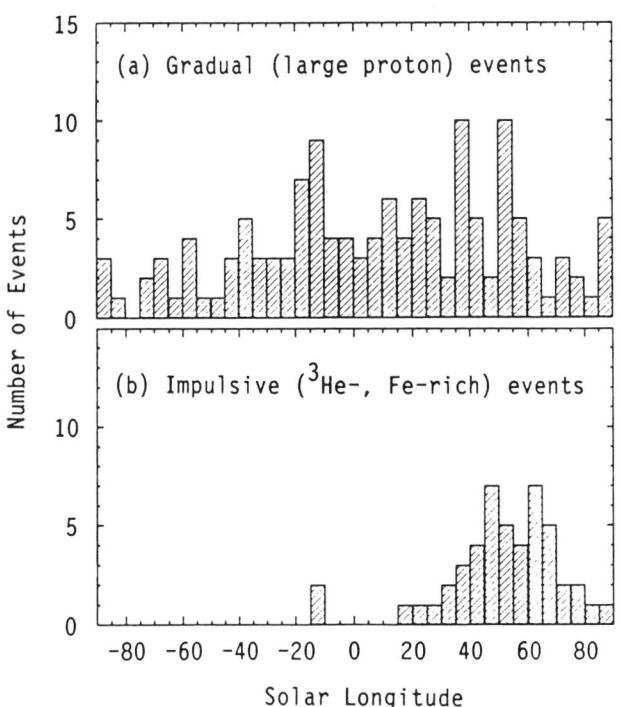

Figure 2. Longitude distribution of *(a)* gradual and *(b)* impulsive events [*Reames* 1995b].

Table 1. Mean Ionization States of Energetic Fe in Large Gradual Events

MeV/amu	Q_{fe}	Events	Reference
0.3 - 2	14.1±0.2	12	Luhn et al. 1987
0.5 - 5	11.0±0.2	2	Mason et al. 1995
15 - 70	15.2±0.7	2	Leske et al. 1995
200 - 600	14.1±1.4	3	Tylka et al. 1995

populations can occasionally be seen in a large event where both physical processes occur. In these cases, an Fe-rich population is seen early in the event from the impulsive phase, followed by Fe-poor material from the gradual phase [*Reames* 1990b]. *Cliver* [1996] has described these events as having a core and a halo, composed of particles from the impulsive and gradual populations, respectively. In some cases the Fe-rich flare component can be quite intense [*Van Hollebeke, McDonald and Meyer* 1990].

3. IMPULSIVE-FLARE EVENTS

It is presently thought that the characteristic abundance enhancements in impulsive events are produced by resonant wave-particle interactions in the flare plasma. Streaming 10-100 keV electrons, that are seen *via* the type III radio bursts and hard X-rays they produce, form velocity distributions that are unstable for production of hydrogen electromagnetic ion cyclotron (EMIC) waves [*Temerin and Roth* 1992]. These wave frequencies lie between the gyrofrequencies of H and of ^{4}He. The rare isotope ^{3}He is the only species whose gyrofrequency lies in this range to resonantly absorb these waves. EMIC waves are only produced in regions of high magnetic field near the base of the corona where the Alfvén speed is >2000 km/s (and in the Earth's aurorae where electrons, waves and ions are all measured). The acceleration is so efficient that the observed number of ^{3}He ions in space requires the acceleration of >10% of the ^{3}He in a typical flare volume near the base of the corona [*Reames* 1993]. The more modest enhancements of heavy elements are produced when these ions resonate with harmonics of these waves or with other wave modes [*Miller and Viñas* 1993].

In a few cases it has been possible to deduce ion abundances of the beam trapped on closed coronal loops from the broad γ-ray line intensities observed. These abundances show the same pattern of enhancements of ^{3}He, Ne, Mg, Si and Fe as are seen in the impulsive-flare particles we observe at 1 AU [*Murphy et al.* 1991].

electrons in less than 1 sec at a density of 10^{10} atoms cm^{-3} found in the low corona. The ionization time is much shorter than the acceleration time at this density. According to *Lee and Ryan* [1986] a time of tens of minutes is required to accelerate ions to GeV energies; this requires that the acceleration take place at a density of 10^7 atoms cm^{-3} or less to avoid stripping the energetic Fe.

Other evidence that high-energy particles are accelerated relatively far from the Sun has been presented by *Kahler* [1994]. He plots the solar injection intensity of particles as a function of the height of the leading edge of the CME, as shown in Figure 3. For most of the energies shown, the particle intensities approach a broad maximum beyond 12 $R_\odot$, while even at 21 GeV they peak between 5-10 $R_\odot$. In the 1989 September 29 event shown, the leading edge of the CME is observed to move at >1800 km/s.

Table 2 summarizes the properties of impulsive and gradual events. The two particle populations are clearly distinguished in their combined properties: events with the abundance signature of impulsive events have impulsive time profiles (like Figure 1b), come from well-connected flares, have type III radio bursts, impulsive X-ray profiles, rarely have associated CMEs, *etc*. However, both particle

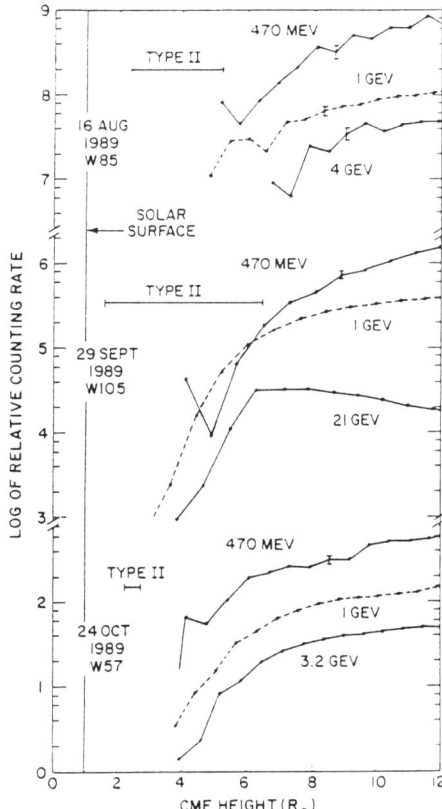

Figure 3. Source injection intensity of GeV particles *vs.* height of the leading edge of the CME [*Kahler* 1994].

Table 2. Properties of Impulsive and Gradual Events

	Impulsive	Gradual
Particles:	Electron-rich	Proton-rich
^{3}He/^{4}He	~ 1	~ 0.0005
Fe/O	~ 1	~ 0.1
H/He	~ 10	~ 100
Q_{fe}	~ 20	~ 14
Duration	Hours	Days
Longitude Cone	<30 deg	~180 deg
Radio Type	III, V (II)	II, IV
X-rays	Impulsive	Gradual
Coronagraph	-	CME (96%)
Solar Wind	-	IP Shock
Events/year	~1000	~10

4. PROTON EVENTS AND CME-DRIVEN SHOCKS

4.1 Shock Acceleration Profiles

To set a framework for understanding the time profiles of the shock-accelerated particles, it is useful to review briefly the theory of diffusive shock acceleration [see *Lee* 1983]. As accelerated particles stream away from a shock they generate Alfvén waves that resonantly scatter subsequent particles of that energy so as to reduce the streaming. These particles then scatter back and forth across the shock, gaining energy on each transit. As they stream outward at a higher energy they generate waves resonant at this new energy, *etc*. At a given energy an equilibrium is established so that the intensity of particles and resonant waves depends only on distance from the shock. This intensity pattern decreases with distance from the shock. Such patterns are observed near shock passage and are called "energetic storm particle (ESP)" events for historical reasons. Direct observations of both particles and waves have confirmed the predictions of the *Lee* [1983] theory at low energies where the measurements are easily made.

At some distance from the shock, particle intensities become too low to generate enough waves and they stream freely outward. This can lead to the flat profile shown in Figure 4a that is often seen in MeV particles. To understand this flat profile, suppose a shock wave accelerates some fraction of the local solar wind plasma to the energy of interest and the particles then expand like R^{-2} as they come out to the observer. If the source solar wind density also varies like R^{-2}, the intensity seen by an observer at distance R from the Sun will not vary with time. The intensity value of the flat profile is determined by saturation of the wave-particle equilibrium and is empirically found to be several x 100 protons (cm^2 sr s MeV)$^{-1}$ in the few MeV region [*Reames* 1990b]. This value was found consistent with theory by *Ng* and *Reames* [1994]. Their Table 2 shows that order-of-magnitude increases in particle intensity near the sun have little effect on intensities at 1 AU after this saturation value is reached.

At higher energies there may be too few particles to generate waves as the shock expands, so acceleration efficiency may decrease with time, leading to a profile more like the one shown in Figure 4b. At sufficiently high energy (or for a sufficiently weak shock) the ESP bump may not survive out to 1 AU, although cases of ESP events in ~500 MeV protons at 1 AU are known.

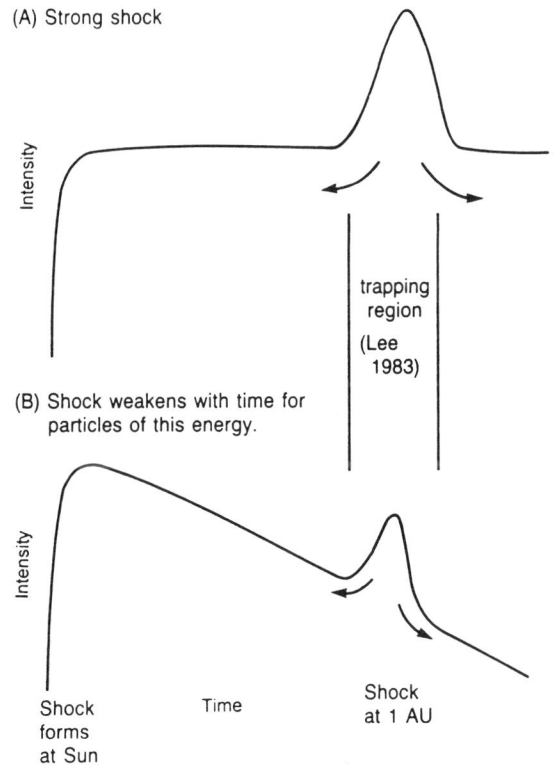

Figure 4. Intensity-time profiles for particles from shock acceleration that *(a)* remains constant out to the observer and *(b)* diminishes with time [*Reames, Barbier*, and *Ng* 1996].

4.2 Longitude Distributions

The foregoing discussion assumes that the observer remains magnetically connected to the same point on a shock as it propagates out to 1 AU. Unfortunately, this is never the case and the particle profiles are strongly affected because the observer's magnetic connection point sweeps across the face of the shock. This means that an observer viewing a CME from central meridian will nominally be connected 55° to the west of the source when it is near the Sun and the shock will first encounter his field line at right angles. As the CME moves outward, his connection point will swing across the face of the shock until he is connected to the nose of the shock when it arrives at 1 AU.

An observer viewing a CME at W55° will be directly connected to the nose of the shock when it leaves the Sun, but will be 55° around on its eastern flank when it arrives at 1 AU. Finally, the observer of an eastern event will see an improving connection with time, but he will only become connected to the nose of the shock after he crosses the local shock and arrives on field lines that connect to the nose of the shock from behind.

The pattern of typical particle intensity-time profiles resulting from these geometric effects is shown in Figure 5. The observer viewing a CME nearly head-on (E01°) sees flat time profiles after a slow rise, especially from a wide CME. The observer on the eastern flank sees a W53° event which peaks early, when he is connected to the nose of the shock, and long before shock arrival. For the observer on the western flank, viewing an E45° event, the particle intensities rise slowly as the magnetic connection improves, but peak intensity does not occur until the observer is behind the local shock on field lines that connect him to its nose.

4.3 Multi-Spacecraft Observations

The original study of particle profiles as a function of longitude leading to Figure 5 was done by *Cane, Reames,* and *von Rosenvinge* [1988] by examining profiles of 235 proton events (obtained over 20 years) in different source-longitude intervals. However, the sizes of CMEs and the strength of shocks differ greatly, so it is important to study the spatial distribution of particles around a single CME shock. A recent study, using spatially separated spacecraft, has been performed by *Reames, Barbier,* and *Ng* [1996].

Particle profiles in a fairly small gradual event are shown in Figure 6. The inset in the figure shows the spatial distribution of the *Helios 1, Helios 2,* and *IMP 8* spacecraft relative to the CME which is projecting downward, just as it was shown in the cartoon in Figure 5. In the upper panel of Figure 6, the profile seen at the centrally-located *Helios 1* shows a gradual rise to a plateau, followed by a pronounced ESP peak precisely at the time of shock passage. *Helios 2* and *IMP 8* are connected increasingly far around on the west flank of the shock. Their intensities slowly increase as their connection point swings eastward with time, but their intensities do not reach maximum until well after shock passage. This shock must have a steep gradient in intensity with longitude around from the nose.

The 30-45 MeV protons in the lower panel of Figure 6 have a similar pattern of rise until the shock reaches ~0.4 AU (for *Helios 1*) then the intensity begins to fall. This is understood as weakening in the acceleration efficiency with time as discussed in connection with Figure 4b.

Notice that long after shock passage the intensities at all spacecraft reach the same value that then decreases with time. This behavior is seen at all energies in all of the events where it could be studied. Behind the CME is a large, spatially uniform population of particles that is trapped in an expanding magnetic bottle [see *Reames, Barbier,* and *Ng* 1996].

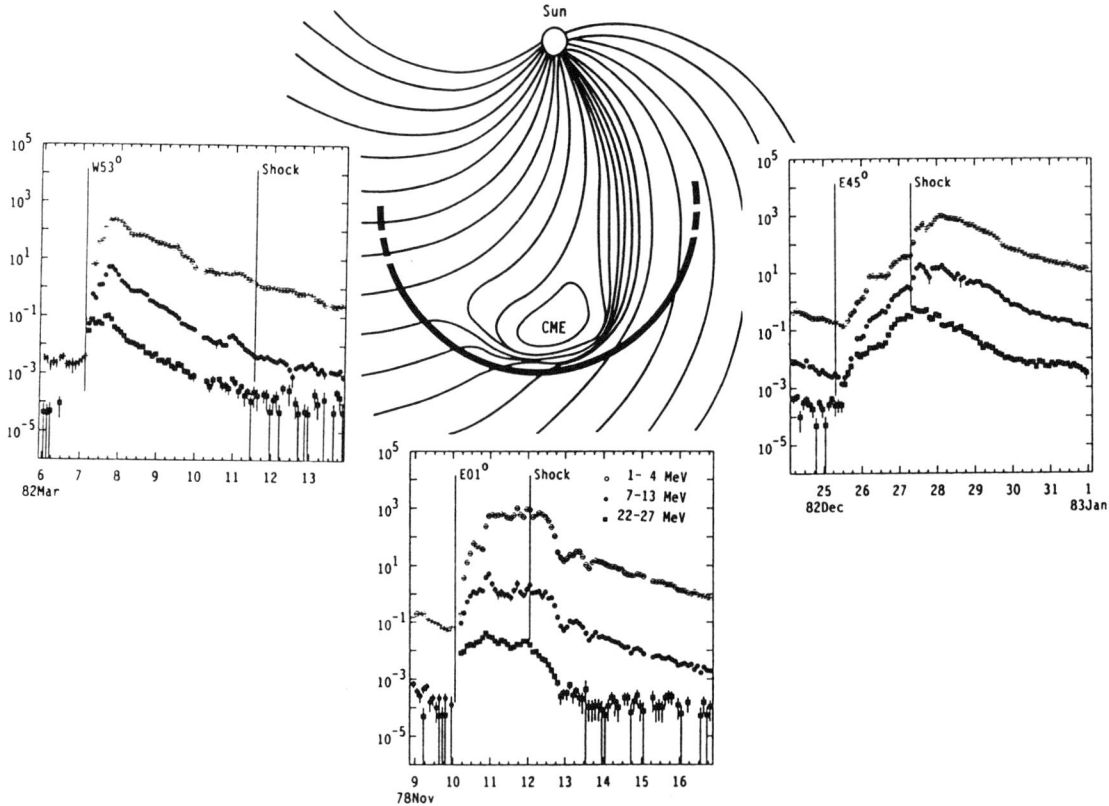

Figure 5. Intensity profiles for protons of different energy for observers viewing a CME from the three different longitudes indicated in the panels [*Reames, Barbier,* and *Ng* 1996].

A much larger event is shown in Figure 7. Despite a much wider separation of the spacecraft (*Helios 2* is 108° west of the source and its connection longitude is ~160° to the west), the intensities rise much more rapidly at all of the spacecraft and they remain high until after shock passage at all energies. This event has a broad shock front with little intensity variation along it. The shock continues to accelerate protons at least to 45 MeV out to 1 AU and beyond. Again, the intensities at all three spacecraft approach the same value late in the event when the shock and CME are beyond about 2 AU.

We can follow this same event further in time and space in Figure 8 where we see a second increase in intensity when the west flank of the shock re-encounters the field lines somewhere near *Voyager* causing a new increase there and sending a new burst of particles inward toward *IMP 8* [see *Reames, Barbier,* and *Ng* 1996] The intensities at the two spacecraft differ greatly during the first peak owing to an ~R^{-3} expansion of the peak intensity of particle distribution [*e.g. Parker* 1963]; the peak is delayed and broadened by particle scattering over the ~10 AU path length along the field. Intensities are similar on the second peak. The dip in the intensities at both spacecraft near October 1 tells us that the shock does not cover 360° with equal intensity; this decrease occurs after the east flank of the shock leaves the field line and before the west flank re-encounters it.

These enormous shocks propagate outward through the heliosphere, crossing and re-crossing each field line. Often, the strongest shocks overtake and merge with previous weaker ones, continuing to accelerate particles to the boundaries of the heliosphere.

Behind each large shock is a quasi-trapped population of particles that is spatially uniform in both longitude and radius over a large fraction of the inner heliosphere. As the bottle containing these particles expands, their intensity decreases adiabatically but their abundances and energy spectral shapes remain largely unaffected [*Reames, Barbier* and *Ng* 1996; *Reames et al.* 1996]. To some extent these particles also form a reservoir or seed-population that is available for re-acceleration and can continually feed the shock from behind.

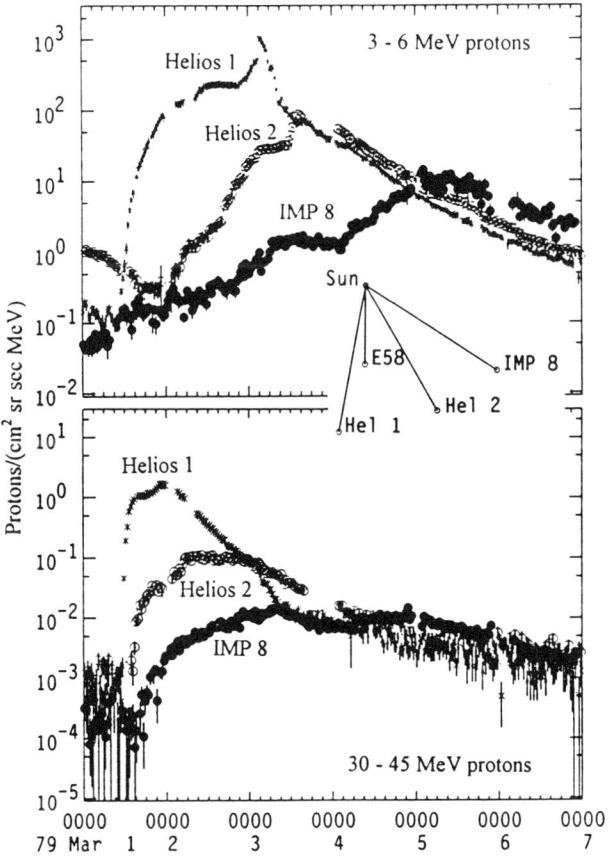

Figure 6. The spatial distribution of particles around a small gradual event as observed by three spacecraft [Reames, Barbier, and Ng 1996].

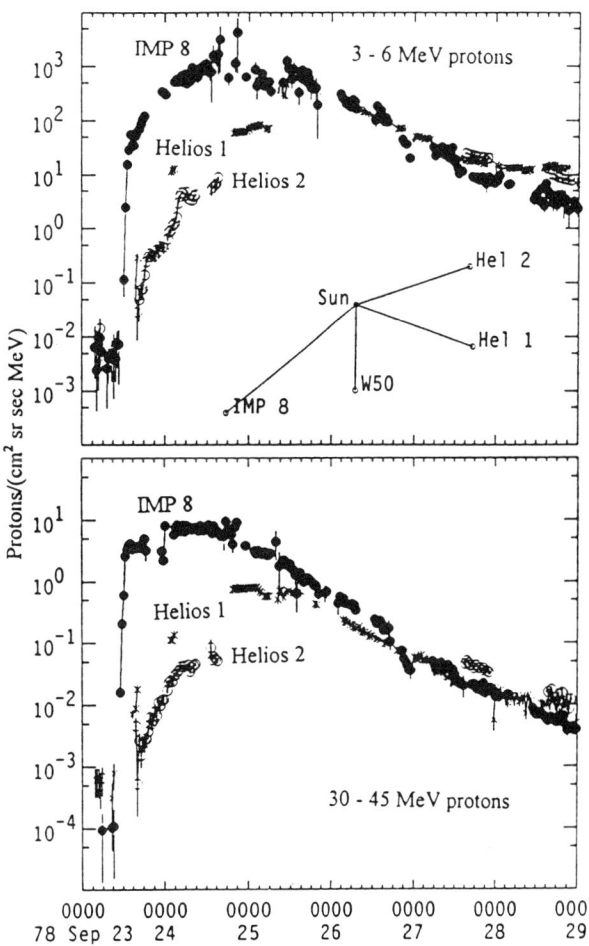

Figure 7. Intensities at widely separated spacecraft in the large event of 1978 September 23 show much less variation with longitude [Reames, Barbier, and Ng 1996].

4.4 Probing the Magnetic Cloud

Inside magnetic clouds in large events one finds a plateau in the proton intensity at a few percent of the intensity just ahead of the cloud (see 1978 Nov 13 in Figure 5). It is possible that the particles gain access to this closed field region by reconnection of a few percent of the internal field lines with external lines that thread the shock. This is also the region where bi-directional flows of ions and electrons are seen [*e. g. Richardson and Reames* 1993] as will be discussed elsewhere in this book.

However, new injections of particles from impulsive flares back at the Sun serve to probe the magnetic topology of this region [*Kahler and Reames* 1991]. The very existence of these particles tells us that at least one end of the field line is connected to the Sun, *i.e.* that *this* cloud is not a detached plasmoid. Very sensitive new measurements on the WIND spacecraft of the magnetic cloud of 1995 October 19 show a large number of tiny impulsive events that fill the cloud with Fe-rich material [*Reames et al.* 1996]. The electron observations in this cloud show a spatial patchwork of magnetic filaments that are alternately connected to and disconnect from the Sun [*Larson et al.* 1996].

When we examine the temporal distribution of impulsive events we find that they often occur in clusters from flares in a single active region; these clusters are frequently found in and behind a CME. We think this correlation is a consequence of the global geometry of the interplanetary magnetic field. Most of the field in interplanetary space has diverged out of coronal holes. However, flares do not occur in coronal holes, they occur in active regions. One of the best places to find field lines that connect to active regions where flares occur is in a magnetic cloud that has been ejected from an active region as part of a CME.

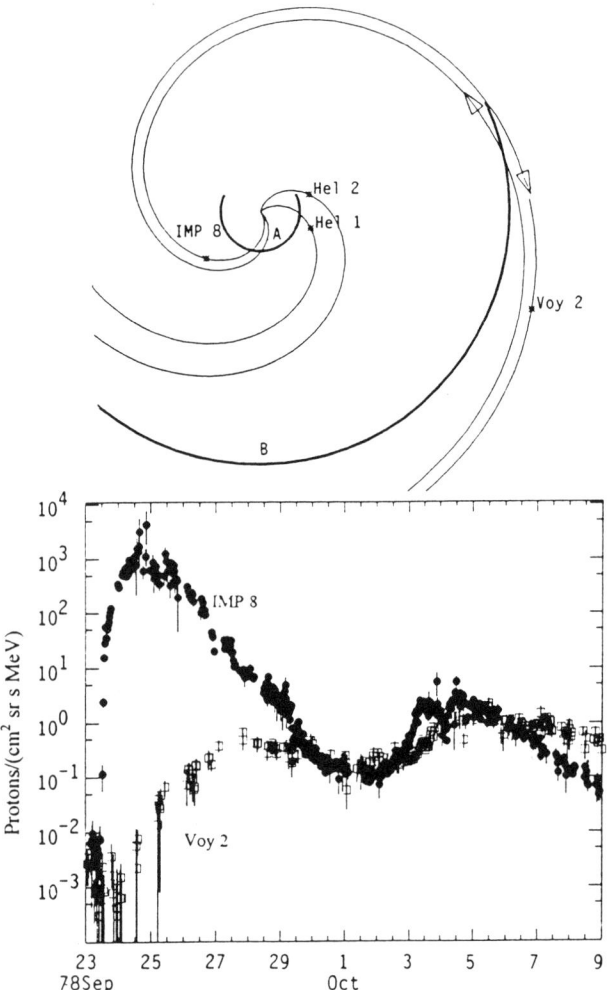

Figure 8. An expanded view of the 1979 September 23 event shows a second increase when the western flank of the shock re-encounters the field line connecting *IMP 8* and *Voyager 2* near 4 AU [adapted from *Reames, Barbier,* and *Ng* 1996].

It is important to note that the CME driver gas is *not* the source of the shock-accelerated ions. According to *Boberg, Tylka and Adams* [1996], the source gas for the accelerated ions is the *coronal* material overlying the CME rather than interplanetary material swept up at a later time. In the later phases, of course, an important source of high-energy particles may be those lower-energy particles accelerated earlier that have been trapped in and behind the shock.

5. CONCLUSIONS

Two distinct mechanisms are responsible for acceleration of the particles we observe in SEP events: 1) stochastic acceleration by resonant wave-particle interactions in the low corona in impulsive flares and 2) diffusive shock acceleration of the ambient plasma of the high corona and solar wind by the shock wave driven outward by a CME. The particle populations from each source can be distinguished, even when both mechanisms occur in a single large event.

The largest events, containing most of the particles seen at 1 AU, are gradual events that come from the CME shock source. Many particles are also accelerated in flares but most are trapped on flare loops and those that escape do so in a limited longitude region near the flare.

In large events many of the observed phenomena are, or appear to be, correlated. For example, X-rays are produced in the corona behind a CME, probably by electrons accelerated during magnetic reconnection. Meanwhile, protons and ions are accelerated at the shock in front of the CME. Thus protons are associated with long-duration X-ray events, which is the reason we call these events gradual. However, the fact that proton events are correlated with certain X-ray properties such as duration and, more recently, with a pattern of X-ray spectral hardening, does *not* imply that protons are accelerated at the same time and place as the electrons that produce X-rays. It only means that different phenomena occur in different parts of big fast CMEs.

Particle acceleration is strongest near the nose of the CME-driven shock. In most events it can decrease rapidly away from the nose, depending on the speed and longitude span of the CME. (In extremely large, powerful events the longitude gradients are more modest.) An observer's point of connection scans across the face of this shock with time leading to large intensity variations that are caused solely by this geometric effect. Western (eastern) events always have intensity maxima prior to (after) shock passage, because that is when the spacecraft is magnetically connected to the nose of the shock. Relatively few events, only those with the CME launched near central meridian, show the intensity peak exactly at the nose of the shock.

The energetic protons and other ions accelerated in gradual events are virtually invisible in photons. X-rays and radio emission are produced by electrons, not ions, and the acceleration occurs in low-density plasma where interactions are much too rare to produce either measurable X-rays or γ-rays. It is ironic that the most intense and pervasive particle events in the inner heliosphere can only be studied by direct *in situ* measurements of the particles themselves.

The flare myth dies hard. Most of us learned as students that SEP events came from flares and we learned our lessons well. Flares are familiar and easy to see. An army of physicists have dedicated careers to an intensive study of flares for over 100 years. However, the particles themselves have told a new and compelling story about the plasma from which they come. Most of them clearly do *not* come from flares, but from CME shocks; objections to this picture have withered as more measurements have been made. Once we accepted this origin we began to look for and find the rich spatial structure in the evolving SEP events that rivals that of the CMEs themselves. Even though we cannot image an SEP event, we can begin to appreciate the structure that such an image might reveal.

I would like to thank Chee K. Ng for numerous helpful discussions and Tycho von Rosenvinge for his comments on this manuscript.

REFERENCES

Boberg, P. R., A. J. Tylka, and J. H. Adams, Solar energetic Fe charge state measurements: Implications for acceleration by coronal mass ejection-driven shock, *Astrophys. J. (Letters)*, 471, L65, 1996.

Cane, H. V., D. V. Reames, and T. T. von Rosenvinge, The role of interplanetary shocks in the longitude distribution of solar energetic particles, *J. Geophys. Res.*, 93, 9555, 1988.

Cliver, E. W., Solar flare gamma-ray emission and energetic particles in space, in *High Energy Solar Physics*, edited by R. Ramaty, N, Mandzhavidze and X.-M. Hua, pp 45-60, AIP Conf. Proc 374, AIP press, Woodbury, NY, 1996

Forbush, S. E., Three unusual cosmic-ray increases possibly due to charged particles from the Sun, *Phys. Rev.*, 70, 771, 1946.

Gosling, J. T., The solar flare myth, *J. Geophys. Res.*, 98, 18949, 1993.

Hsieh, K. C., and J. A. Simpson, The relative abundances and energy spectra of ^{3}He and ^{4}He from solar flares, *Astrophys. J. (Letters)*, 162, L191, 1970.

Kahler, S. W., The role of the big flare syndrome in correlations of solar energetic proton fluxes and associated microwave burst parameters, *J. Geophys. Res.* 87, 3439, 1982.

Kahler, S. W., Injection profiles for solar energetic particles as functions of coronal mass ejection heights, *Astrophys. J.* 428, 837, 1994.

Kahler, S. W., E. W. Cliver, H. V. Cane, R. E. McGuire, R. G. Stone and N. R. Sheeley, Solar filament eruptions and energetic particle events, *Astrophys. J.* 302, 504, 1986.

Kahler, S. W., and D. V. Reames, Probing the magnetic topologies of magnetic clouds by means of solar energetic particles, *J. Geophys. Res.*, 96, 9419, 1991.

Kahler, S. W., N. R. Sheeley, Jr., R. A. Howard, M. J. Koomen, D. J. Michels, R. E. McGuire, T. T. von Rosenvinge, and D. V. Reames, Associations between coronal mass ejections and solar energetic proton events, *J. Geophys. Res.* 89, 9683, 1984.

Larson, D. E., R. E. Ergun, J. M. McTiernan, R. P. Lin, J. P. McFadden, C. W. Carlson, K. A. Anderson, M. McCarthy, G. Parks, H. Reme, J. M. Bosqued, K.-P, Wenzel, and T. R Sanderson, Wind spacecraft sbservations of energetic particles in the October 1995 magnetic cloud, to be published, 1996.

Lee, M. A., Coupled hydromagnetic wave excitation and ion acceleration at interplanetary traveling shocks, *J. Geophys. Res.*, 88, 6109, 1983

Lee, M. A. and J. M. Ryan, Time-dependent coronal shock acceleration of energetic solar flare particles, *Astrophys. J.* 303, 829, 1986.

Leske, R. A., J. R. Cummings, R. A. Mewaldt, E. C. Stone, and T. T. von Rosenvinge, Measurements of the ionic charge states of solar energetic particles using the geomagnetic field, *Astrophys J. (Letters)* 452, L149, 1995.

Luhn, A., B. Klecker, D. Hovestadt, and E. Mobius, The mean ionic charge of silicon in ^{3}He-rich solar flares, *Astrophys. J.*, 317, 951, 1987.

Mason, G. M., J. E. Mazur, M. D. Looper, and R. A. Mewaldt, Charge state measurements of solar energetic particles with SAMPEX, *Astrophys. J.* 452, 901 (1995)

Meyer, J. P., The baseline composition of solar energetic particles, *Astrophys. J.* Suppl., 57, 151, 1985.

Miller, J. A. and A. F. Viñas, Ion acceleration and abundance enhancements by electron beam instabilities in impulsive solar flares, *Astrophys. J.*, 412, 386, 1993.

Murphy, R. J., R. Ramaty, B. Kozlovsky, and D. V. Reames, Solar abundances from gamma-ray spectroscopy: comparisons with energetic particles and photospheric abundances *Astrophys. J.*, 371, 793, 1991.

Ng, C. K. and Reames, D. V., Focused interplanetary transport of ~1 MeV solar energetic protons through self-generated Alfvén waves, *Astrophys. J.* 424, 1032, 1994.

Parker, E. N., *Interplanetary Dynamical Processes*, (Interscience Publ., New York) 1963.

Reames, D. V., Energetic particles from impulsive solar flares, *Astrohys. J.* Suppl., 73, 235, 1990a.

Reames, D. V., Acceleration of energetic particles by shock waves from large solar flares, *Astrophys. J. (Letters)*, 358, L63, 1990b.

Reames, D. V., Non-thermal particles in the interplanetary medium, Adv. Space Res., 13 (No. 9), 331, 1993.

Reames, D. V., Coronal abundances determined from energetic particles, Adv. Space Res. 15, No.7, 41, 1995a.

Reames, D. V., Solar energetic particles: a paradigm shift, *Revs. Geophys Suppl.* 33, 585, 1995b.

Reames, D. V., L. M. Barbier, and C. K. Ng, The spatial distribution of particles accelerated by coronal mass ejection-driven shocks, *Astrophys. J.* 466, 473, 1996.

Reames, D. V., L. M. Barbier, T. T. von Rosenvinge, G. M. Mason, J. E. Mazur, and J. R. Dwyer, Energy spectra of ions accelerated in impulsive and gradual solar flares, *Astrophys. J.*, submitted 1996.

Reames, D. V., H. V. Cane, and T. T. von Rosenvinge, Energetic particle abundances in solar electron events, *Astrophys. J.*, 357, 259, 1990.

Reames, D. V., B. R. Dennis, R. G. Stone, and R. P. Lin, X-ray and radio properties of solar ^{3}He-rich events, *Astrophys. J.* 327, 998, 1988.

Reames, D. V., J. P. Meyer, and T. T. von Rosenvinge, Energetic-particle abundances in impulsive solar-flare events, *Astrophys. J. Suppl.*, 90, 649, 1994.

Richardson, I. G., and D. V. Reames, Bidirectional ~1 MeV/amu ion intervals in 1973-1991 observed by the Goddard Space Flight Center instruments on IMP-8 and ISEE-3/ICE, *Astrophys. J. Suppl.* 85, 411, 1993.

Temerin, M. and I. Roth, The production of ^{3}He and heavy ion enrichments in ^{3}He-rich flares by electromagnetic hydrogen ion cyclotron waves, *Astrophys. J. (Letters)* 391, L105, 1992.

Tylka, A. J., P. R. Boberg, J. H. Adams, Jr., L. P. Beahm, W. F. Dietrich, and T. Kleis, The mean ionic charge state of solar energetic Fe ions above 200 MeV/nucleon, *Astrophys. J. (Letters)* 444, L109 (1995).

Van Hollebeke, M. A. I.., F. B. McDonald, and J. P. Meyer, Solar energetic particle observations of the 1982 June 3 and 1980 June 21 gamma-ray/neutron events, *Astrophys. J. Suppl.*, 73, 285, 1990.

D. V. Reames, Code 661, NASA, Goddard Space Flight Center, Greenbelt, MD 20771 (email: reames@lheavx.gsfc.nasa.gov).

Particle Acceleration and Transport at CME-Driven Shocks

Martin A. Lee

Institute for the Study of Earth, Oceans and Space, University of New Hampshire, Durham, New Hampshire

The theory of shock acceleration is first reviewed briefly including a discussion of the basic diffusive transport equation and a list of its successes in accounting for many energetic particle populations throughout the heliosphere, in particular many energetic storm particle (ESP) events. The difficulties in applying the theory to the acceleration of solar energetic particles (SEPs) in "gradual" events at a CME-driven shock wave are then enumerated: complex temporal and spatial dependence, sensitivity of the predictions to the values of the transport coefficients, unknown injection rates, the importance of nearly scatter-free propagation in interplanetary space, and competing adiabatic deceleration in the solar wind. Nevertheless, a CME-driven shock is the most promising origin for the "gradual" ion events. It is shown that finite shock lifetime, while providing an energy cutoff of ~ 1 MeV/nucleon at the orbit of Earth (consistent with ESP events), may actually allow the acceleration of ions to ~ 100 MeV/nucleon when the shock is at ~ 20 solar radii. The shock origin also accounts for the observed elemental and charge-state composition of gradual events, and the global scale of their origin near the Sun.

1. INTRODUCTION TO SHOCK ACCELERATION

Solar energetic particles (SEP) may be divided into two classes of particle events: impulsive and gradual [*Reames*, 1988, this volume; *Klecker et al.*, 1990; *Reames et al.*, 1990]. Impulsive events last on the order of minutes, have associated neutron and γ-ray emission, are electron-rich, are enriched in heavy ions and in particular ^{3}He, and exhibit ion charge states indicative of source temperatures of ~ 2×10^7 °K. Gradual events are more energetic and less frequent, last on the order of hours or days, have associated type II radio bursts, are electron poor, have approximately solar wind ion composition, and have charge states indicative of a solar wind origin. Impulsive events are generally magnetically well-connected to a flare site, whereas gradual events are not. While these distinctions between the two classes are occasionally blurred, most events are in one class or the other. There appears to be no compelling reason to introduce a third class of events. Nevertheless I shall return to this possibility in Section 3, after discussing specific acceleration mechanisms.

Impulsive events exhibit no association with coronal mass ejections (CMEs), whereas gradual events are closely associated with CMEs. Therefore, in this paper I focus on gradual solar energetic particle events and their origins. Their observational characteristics are described in the companion paper by Reames [this volume].

The most likely acceleration mechanism for gradual ion events is diffusive shock acceleration at coronal/interplanetary shocks driven by CMEs. In a view prevalent at this time, a rapid CME provides the "piston" for a driven shock wave, which forms in the corona and precedes the CME driver gas into interplanetary space [e.g. *Kahler*, 1992]. The basic ion acceleration mechanism involves multiple traversals of the shock by the ions due to pitch-angle scattering of the ions on magnetic irregularities on both sides of the shock. A sample schematic ion trajectory is shown in Figure 1 at a shock viewed in the normal-incidence frame in which the shock is stationary and the upstream plasma velocity **V** is parallel to the shock normal **n**. Also shown is the irregular magnetic field, $\mathbf{B} = \mathbf{B}_0 +$

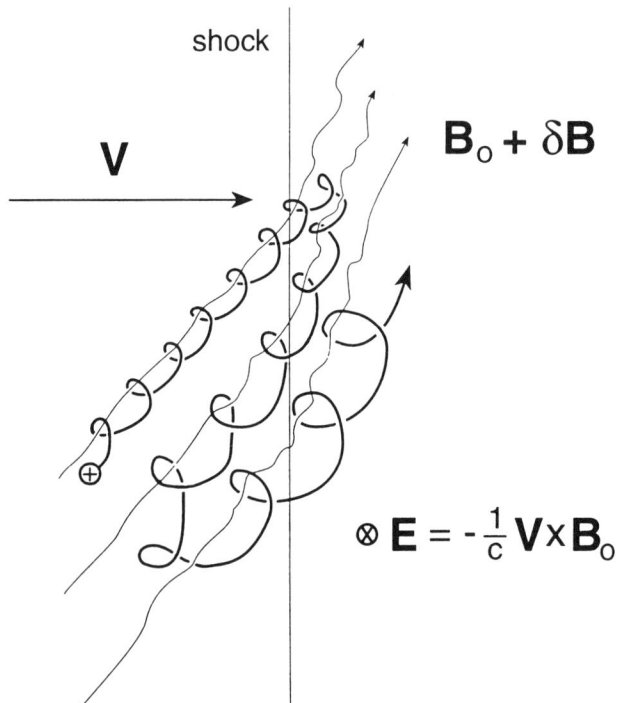

Figure 1. Schematic diagram of a planar "fast" shock in the normal-incidence frame with **V** the upstream plasma flow. Magnetic field lines are shown with both average, $\mathbf{B}_o$, and fluctuating, $\delta\mathbf{B}$, parts. The average "motional" electric field is indicated. A sample schematic ion trajectory is shown in which the ion traverses the shock and gains energy.

$\delta\mathbf{B}$, and the uniform motional electric field, $\mathbf{E} = -c^{-1}\mathbf{V} \times \mathbf{B}_o$. Not shown are the electric field fluctuations associated with $\delta\mathbf{B}$. Since $\mathbf{B}_o$ as shown increases downstream of the shock, this shock wave is a "fast" shock corresponding to the "fast" mode in magnetohydrodynamics. "Slow" shocks are rare in interplanetary space, if they exist at all, and are not thought to accelerate ions effectively [*Isenberg*, 1986].

The ions traversing the shock are coupled to the shock compression by the scattering and are thus accelerated in part by the process of first-order Fermi acceleration. In this process the ions find themselves effectively scattered between approaching massive walls, corresponding to the upstream and downstream plasma, and are accelerated, as a ball bouncing between paddles approaching one another. The ions also scatter within the upstream and downstream plasmas, but actual compressional energy gains only occur when the ions traverse the shock. The ions are also accelerated by the curvature and gradient of $\mathbf{B}_o$ at the shock, which causes them to drift parallel to **E**. This latter process exists in simplified form in the absence of irregularities, and was originally called "shock drift" acceleration [*Sarris and Van Allen*, 1974; *Pesses et al.*, 1979; *Decker*, 1988]. In fact, the division of shock acceleration into these two distinct processes is artificial [*Jokipii*, 1982]. The division is dependent on frame. Indeed, unless the shock is approximately perpendicular ($\mathbf{B}_o \cdot \mathbf{n} = 0$), the process may be viewed in the deHoffman-Teller frame in which $\mathbf{E} = 0$ and the "shock drift" contribution to acceleration vanishes. If the shock is perpendicular the "shock drift" contribution accounts for all of the acceleration.

The ions of primary interest are those which have been accelerated to high energy with speed $v \gg |\mathbf{V}|$. The very nature of the mechanism, which involves effective scattering adjacent to the shock, dictates that the velocity distribution of these ions is nearly isotropic, at least near the shock. Under these conditions the omnidirectional distribution function, $f(\mathbf{x}, p, t)$, of the ions, where p is ion momentum, satisfies the transport equation

$$\frac{\partial f}{\partial t} + \left(\mathbf{V} + \mathbf{V}_D\right) \cdot \nabla f - \nabla \cdot \mathbf{K} \cdot \nabla f - \frac{1}{3}\nabla \cdot \mathbf{V}\, p\, \frac{\partial f}{\partial p} = Q \quad (1)$$

originally proposed by Parker [1965] to describe the solar modulation of galactic cosmic rays. In equation (1), **V** is the plasma velocity (more specifically that of the scattering irregularities), **K** is the symmetric spatial diffusion tensor describing the spatial diffusion of the ions due in part to their pitch-angle scattering on magnetic irregularities, and Q is a source term. The drift velocity $\mathbf{V}_D$ was introduced by Jokipii et al. [1977] to include ion drift transport due to the inhomogeneous average magnetic field $\mathbf{B}_o$. Equation (1) is a statement of particle conservation while f varies due to convective, drift and diffusive transport, and to deceleration ($\nabla \cdot \mathbf{V} > 0$) or acceleration ($\nabla \cdot \mathbf{V} < 0$) of the particles.

All terms in equation (1) are essential to the process of diffusive shock acceleration. The term "diffusive shock acceleration" usually refers to shock acceleration in the limit of $v \gg |\mathbf{V}|$, effective scattering, and nearly isotropic distributions, as described by equation (1). The compression term describes acceleration at the shock, the convection term describes the advection of upstream ions to the shock and their loss downstream, the diffusion term describes the scattering required for multiple traversals of the shock, the drift term describes the non-diffusive transport of ions normal to the plane of Figure 1, and Q describes the injection rate of new particles into the process at low speeds $v \sim |\mathbf{V}|$ where the equation does not apply. Both compressive and shock drift energy gains are included in equation (1). If scattering at the shock is not effective, individual ions may still be accelerated, though by smaller factors, and the process is generally not termed diffusive shock acceleration and is not described by equation (1). In this case numerical or Monte Carlo methods are often employed [*Decker*, 1988; *Baring et al.*, 1993]. We shall return to this point in Section 2.

For an infinite planar shock at $x = 0$, at which spatial variation is only in the x-direction and $V_{x,u} = V_u$,

$V_{x,d} = V_d$, $K_{xx,u} = K_u$, and $K_{xx,d} = K_d$ are specified and independent of x and t ("u" refers to upstream quantities and "d" to downstream quantities), the following stationary distribution is derived from equation (1) for momenta $p > p_o$, where p_o is a characteristic injection momentum. The downstream distribution is independent of x and proportional to $Q(p/p_o)^{-\gamma}$, where $\gamma = 3 (1-X^{-1})$, and depends only on the shock compression ratio X. The upstream distribution, which also contains a ramp arising from a balance between convection toward the shock and diffusion away from it, is proportional to $Q(p/p_o)^{-\gamma} \exp(xV_u K_u^{-1})$. Since K_u generally increases with p, the upstream distribution "hardens" with distance upstream of the shock. These results have become synonymous with the process of diffusive shock acceleration, first developed by Axford et al. [1977], Krymsky [1977], Blandford and Ostriker [1978] and Bell [1978]. Extensive reviews of shock acceleration, and diffusive shock acceleration in particular, have been given for example by Blandford and Eichler [1987] and Jones and Ellison [1991].

If the particle configuration is not stationary, but rather arises from an injection Q which vanishes for $t < 0$ and is constant for $t > 0$, then the particle distribution both upstream and downstream of the shock contains the additional factor

$$\exp\{-t^{-1}[3\Delta V^{-1}\int_{p_o}^{p} dp'(p')^{-1}(K_u V_u^{-1} + K_d V_d^{-1})]\} \quad (2)$$

where $\Delta V = V_u - V_d$ [Axford, 1981]. The quantity in square brackets, $\tau(p)$, is the characteristic time required to accelerate particles to momentum p. Equation (2) describes the limitation to the efficacy of shock acceleration due to the finite lifetime of particle injection or of the shock. For lower energies such that $\tau(p) < t$ the stationary power-law spectrum results. At higher energies the spectrum exhibits an exponential cutoff which severely reduces particle intensities. In addition there may be other "loss" mechanisms which also contribute to reduced particle intensity at high energies: Particles may effectively escape the vicinity of the shock for shocks of limited spatial extent or at boundaries where scattering is reduced. Particles may also be decelerated in regions where $\nabla \cdot \mathbf{V} > 0$, which may in part counteract the shock acceleration.

The spatial diffusion tensor $\mathbf{K}$ is the most problematical ingredient in the transport equation (1), since it must describe particle transport resulting from stochastic trajectories in a turbulent magnetic field. The turbulent field $\delta \mathbf{B}$ may be decomposed according to wavevector $\mathbf{k}$ as $<|\delta \mathbf{B}|^2> = \Sigma_i \int d^3 k I_i(\mathbf{k})$, where $< >$ denotes an ensemble average and $I_i(\mathbf{k})$ is the intensity of the ith plasma mode. The largest element of $\mathbf{K}$, $K_{\parallel}$ corresponding to spatial diffusion along the average field $\mathbf{B}_o$ due to pitch-angle scattering, satisfies $K_{\parallel} \propto I^{-1}$. Here I is evaluated at wavevectors for which the fluctuations are resonant with the particle trajectory. The smaller element of $\mathbf{K}$, $K_{\perp}$ corresponding to diffusion across the average field, satisfies $K_{\perp} \propto I$ and contains resonant scattering of particles across $\mathbf{B}_o$, diffusion corresponding to the meandering of individual field lines, and drift due to the curvature and gradient of $\delta \mathbf{B}$ [Forman et al., 1974]. An additional complication may occur if the energetic particles have sufficient spatial or phase-space gradients to excite wave intensity enhancements [Bell, 1978; Lee, 1982, 1983]. Then $\mathbf{K}$ depends on f and the transport equation becomes nonlinear in f. Another possible nonlinearity is that in which the energetic particle pressure gradient can affect $\mathbf{V}$, or possibly $\mathbf{B}_o$ and $\mathbf{V}_D$, so that these quantities also depend on f. In most applications of diffusive shock acceleration in the heliosphere, however, the coefficients in equation (1) are simply specified.

The theory of diffusive shock acceleration has been successful in accounting for many of the energetic particle enhancements observed throughout the heliosphere. Whenever or wherever a shock is present, energetic particles are usually also present with characteristics which can generally be accounted for by the theory, with perhaps a different "twist" in the theory for each application. The diffuse ions sunward of Earth are accelerated at Earth's bow shock [Lee, 1982; Ellison et al., 1990]. Energetic storm particles (ESP) are accelerated at interplanetary traveling shocks [Scholer et al., 1983; Lee, 1983; van Nes et al., 1984]. The corotating ion events are accelerated at the forward and reverse shocks bounding corotating interaction regions in the solar wind [Palmer and Gosling, 1978; Fisk and Lee, 1980]. The cosmic ray anomalous component is accelerated at the termination shock of the solar wind [Pesses et al., 1981; Jokipii, 1986]. And finally, outside the heliosphere, the bulk of galactic cosmic rays are thought to be accelerated at supernovae shock waves [Axford, 1981; Blandford and Eichler, 1987].

In principle both electrons and ions may be accelerated at shocks. However, the high frequency turbulence required for the scattering of lower energy electrons is often not present, and is not readily excited by the electrons themselves. Presumably this is the reason that the heliospheric shock-associated particle enhancements are mostly ions. The gradual SEP events discussed below are electron-poor, which is consistent with a shock origin. In subsequent discussion I restrict myself to the dominant ions.

2. SHOCK ACCELERATION OF SOLAR ENERGETIC PARTICLES?

In view of the success of diffusive shock acceleration in accounting for energetic particle enhancements throughout the heliosphere, it would seem compelling to postulate a shock origin of the SEPs in gradual events. Indeed these events are strongly correlated with fast CMEs capable of producing shock waves, and often are associated with radio bursts produced by shock-energized electrons. Nevertheless, controversy over the origin of SEPs in gradual events has been substantial. Why?

The controversy surrounds the high-energy SEPs with energies greater than a few MeV/nucleon. The lower energy ions in gradual SEP events are dominated by the ESP events which clearly originate at the shock [*Bryant et al.*, 1962]. The maximum intensity of these ions generally occurs near the shock. In large events with substantial ion scattering adjacent to the shock the energy spectra are approximately the expected power law up to about 100 keV-1 MeV/nucleon [e.g. *Gosling et al.*, 1981], beyond which finite shock lifetime results in an energy cutoff according to equation (2). Smaller nearly scatter-free events, called "shock-spike" events, are due to shock drift acceleration, although these are generally not associated with the large gradual events. A survey of shock-associated low energy events observed by ISEE-3 is presented by van Nes et al. [1984]. They find that these events can be categorized as large diffusive events, events associated with quasi-perpendicular shocks which are irregular presumably due to large scattering mean free paths, shock-spike events, and weak shock events with little ion acceleration. Perhaps the single most intensively studied ESP event was that associated with the shock of 11,12 November 1978 [*Kennel et al.*, 1984a,b]. Kennel et al. [1986] compared observations of the event by ISEE 1,2 and 3 with the theory of Lee [1983], which includes wave excitation by the energetic ions. They concluded that the theory of diffusive shock acceleration accounted for ion energy spectra, gradients, and anisotropies, and the wave intensity spectra and polarization. Tan et al. [1989] compared the observed spatial gradients of several large ESP events upstream of the shock with the theoretical prediction of Lee [1983] and found excellent agreement. Altogether, ESP events up to hundreds of keV/nucleon are clearly shock accelerated although ineffective scattering at quasi-perpendicular shocks often results in large ion anisotropy and non-diffusive transport.

What is the origin of the high-energy ions, which generally precede the shock and are usually restricted to magnetic field lines which connect closer to the site of the CME eruption or of the associated flare if one is present? How can they be accelerated at a shock which even at the orbit of Earth is unable to accelerate ions to energies higher than ~ 1 MeV/nucleon? The difficulty in establishing a shock origin for these ions lies in part in the complex geometry and the intrinsic time dependence of the CME/shock/energetic ion configuration. Figure 2 shows a schematic spatial diagram at some fixed time of the Sun, a CME, the shock wave driven by the CME, and selected magnetic field lines. Considerable artistic license is involved in this rendition. The CME, shown here as a helical flux rope, moves out from the Sun as a volume of gas within the dot-dashed curve. The motion and expansion of this gas drives a shock wave shown by the smooth solid curve, which weakens along its flanks as indicated by the dashes.

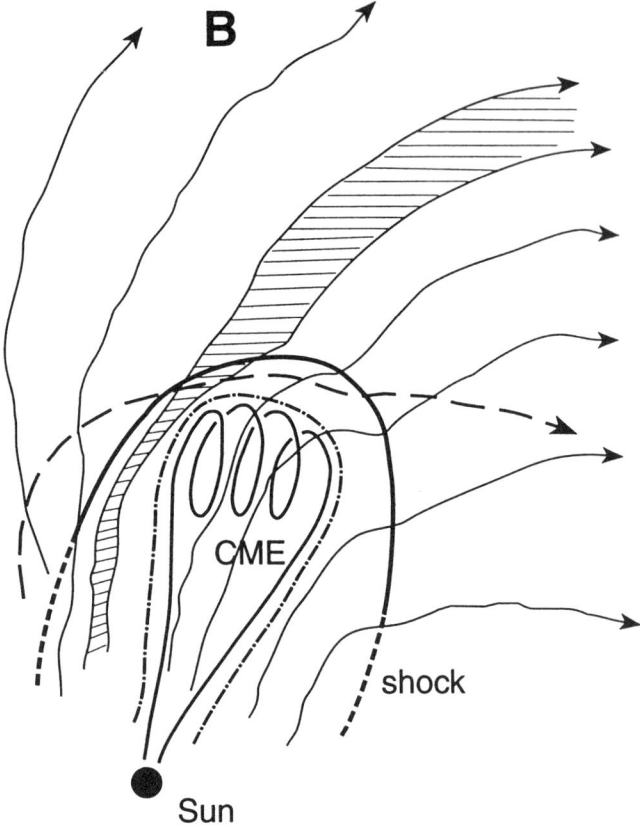

Figure 2. Schematic spatial configuration of a CME which has erupted from the Sun, its driven shock wave, and sample interplanetary magnetic field lines. The CME, shown with helical field structure, is bound by the dot-dashed curve. The shock is weaker at the flanks, as indicated by the dashed extensions of the shock curve. The dashed field line, at a lower level, may be wound tighter due to slower solar wind. The shaded flux tube is a typical flux tube, within which solar energetic particles are accelerated and transported, and escape to the orbit of Earth.

dashes. As the shock expands it intercepts more of the interplanetary magnetic field lines which extend on average in an Archimedes spiral pattern out to Earth and beyond. Occasionally, as indicated by the dashed underlying field line with perhaps a tighter winding due to slower solar wind, the shock may first encounter a given field line further from the Sun so that the field line passes through the shock twice. In any case particles are accelerated on any field line which passes through the shock surface. Many particles escape the vicinity of the shock and propagate into interplanetary space.

In the simplest case a given flux tube, indicated by shading in Figure 2, is intercepted once by the shock which moves along the tube as both are advected with the solar wind. If drift transport and diffusion normal to the field are neglected, usually a safe assumption, then the accelerated particles remain in the flux tube. Nevertheless the important transport parameters within the flux tube depend on heliocentric radial distance (r) and time (t). The Alfvén speed V_A (r) and the sound speed C_s(r) decrease substantially with distance from the Sun, as does the plasma density ρ(r). The shock speed relative to the solar wind V_s(t) decelerates with time, so that the Mach number M(t) and the compression ratio X(t) change. The overall geometry dictates that $θ_{BN}$(t), the angle between the shock normal and the upstream magnetic field, also changes, which in turn controls the rate at which particles are injected into the process of acceleration at the shock. The ion intensities at low energies in turn control the excitation of the scattering turbulence, which determines spatial diffusion along the flux tube and its energy dependence. Except right at the shock, $\nabla \cdot \mathbf{V} > 0$ in the expanding shocked and unshocked solar wind so that adiabatic deceleration tends to compete with acceleration at the shock. It is this adiabatic deceleration of ions trapped in the turbulence downstream of the shock that causes the particle intensity to decrease after shock passage [*Forman*, 1983; *Fisk and Lee*, 1980].

Thus, the acceleration of solar energetic particles at an evolving CME-driven shock wave is fundamentally time and space dependent with large variations in the key transport parameters. It is clear from equation (2) that parameters such as ΔV and **K** appear in the exponent so that their variation, or uncertainty in their values, stymies predictive theory. In addition the evolution of each flux tube is distinct. Some may connect to the stronger nose of the shock at early times, but to the flanks at later times. A given spacecraft measures particle intensities and anisotropies in a sequence of flux tubes during a SEP event. The history of each flux tube must be taken into account in order to interpret the observations.

A further difficulty in applying the theory of diffusive shock acceleration to SEP events is that Parker's transport equation (1) is not uniformly valid throughout the relevant domain of energy and space. The propagation of ions in interplanetary space, particularly at energies larger than ~ 10 MeV/nucleon, is often approximately scatter-free. Scattering mean free paths at Earth are inferred to be on the order of 1 AU [*Palmer*, 1982]. The SEPs which arrive first at Earth have high anisotropy consistent with scatter-free propagation from the vicinity of the Sun, or of the shock when it is close to the Sun. Clearly this mode of transport affects the time-intensity profiles of the observed particles, but it is not consistent with equation (1), which assumes effective scattering and near isotropy. Interestingly nearly scatter-free transport is inconsistent with the process of diffusive shock acceleration, which requires effective scattering at least in a sheath near the shock. This discrepancy poses a major theoretical challenge. Realistic treatments of interplanetary propagation are usually based on a transport equation describing evolution in pitch angle [*Roelof*, 1969; *Heras et al.*, 1991, 1995; *Hatzky et al.*, 1995; *Ruffolo and Khumlumlert*, 1995; *Ng and Reames*, 1994; *Kallenrode*, 1993], which accommodates large scattering mean free paths but does not readily describe the transport and acceleration adjacent to the shock.

Equation (1) is also not applicable at the low energies at which particles are extracted from the solar wind at the shock and "injected" into the process of diffusive shock acceleration. Thus the injection rate, Q, which depends on the highly variable quantities $θ_{BN}$, M, I (or **K**) and solar wind mass density ρ, is unknown quantitatively. The dependence of Q on M is particularly important. Acceleration may be difficult to initiate low in the corona. Even though the shock speed V_s may be as large as 2000 km/s for fast CMEs, the Mach number M can be small due to the large Alfvén speed. Thus the Mach number may not exceed a critical Mach number necessary for effective transmission of downstream shock-heated ions back upstream against the oblique magnetic field to initiate shock acceleration [*Edmiston et al.*, 1982]. In order to accelerate the high-energy ions early in a SEP event it is necessary to have injection close to the Sun.

Altogether, the space and time dependence of SEP acceleration at a CME-driven shock is forbiddingly complex. Important regimes of transport cannot be addressed with equation (1). There are no direct observations of shock acceleration close to the Sun where acceleration to high energies is expected to occur, although radio diagnostics of shock structure exist. Current models for the shock acceleration of SEPs are vastly oversimplified. The first to apply the theory of diffusive shock acceleration to SEP events was Achterberg and Norman [1980]. They started with the basic power law spectrum for stationary acceleration and incorporated in an ad hoc fashion modifications to account for shock lifetime and curvature, adiabatic deceleration, and Coulomb collisions of ions with ambient electrons. Lee and Fisk [1982] and Lee and Ryan [1986] calculated the time-dependent acceleration of particles at a spherically-symmetric blast wave propagating into a stationary surrounding medium. Although the energy spectra and time profiles they obtained were enticing, their neglect of the solar wind and their assumption that K_{rr} is independent of p forbid a quantitative comparison of the results with observations.

Although existing models are perhaps too simplistic to establish beyond doubt the shock origin of solar energetic particles in gradual events, I have no doubt that this origin is the correct one for most of the observed SEPs. One common criticism of shock acceleration is that it is too slow. ESP events at the orbit of Earth usually have energy cutoffs of a few hundred keV/nucleon due to finite shock lifetime as described by equation (2). How can such a shock accelerate ions up to 1 GeV/nucleon in less time earlier in its life? In fact this seemingly paradoxical result is not unreasonable. Neglecting constants and the diffusion coefficient downstream of the shock, K_d, the exponent in equation (2) is K_u $(t\Delta V V_u)^{-1}$. Using quasilinear theory restricted to wave propagation parallel to $\mathbf{B}_o$ [e.g. *Lee*, 1983] $K_{\parallel} \propto \gamma^2 v^3 B_o^2 \Omega^{-2} [I(\Omega \gamma^{-1} v^{-1})]^{-1}$, where $<|\delta B|^2> = \int dk\ I(k)$ and the cyclotron resonance condition for effective wave/particle interaction dictates that $k = \Omega \gamma^{-1} v^{-1}$. The particle gyrofrequency Ω is $eB_o\ m^{-1}\ c^{-1}$ and $\gamma = (1-v^2/c^2)^{-1/2}$. Neglecting variations in θ_{BN} so that $K_u \propto K_{\parallel}$, and setting $\Omega \gamma^{-1}\ v^{-1}\ I = <|\delta B|^2>_{res}$, the exponent in equation (2) is approximately proportional to ε $(\Delta V\ \Omega V t)^{-1}$ $(B_o^2/<|\delta B|^2>_{res})$. Here $<|\delta B|^2>_{res}$ is the variance of the magnetic field fluctuations in the wavenumber range cyclotron resonant with the particles and ε is ion kinetic energy. The radial dependence of the exponent may be approximated by noting that $\Omega \propto B_o \propto r^{-2}$ and by setting $Vt \sim r$. With substantial wave excitation by the ions, we take $(B_o^2/<|\delta B|^2>_{res})$, at least adjacent to the shock where most of the acceleration takes place, to be independent of r. ΔV certainly decreases dramatically with r; since $\Delta V \sim V_A$ and the Alfvén speed $V_A \propto B_o \rho^{-1/2} \propto r^{-1}$, we take $\Delta V \propto r^{-1}$. The result is an exponent proportional to $\varepsilon\ r^2$. Thus, if the cutoff energy due to finite shock lifetime is say 1 MeV/nucleon at the orbit of Earth, then it is 100 MeV/nucleon at 20 solar radii. Thus as long as injection is sufficient, such a shock should be able to accelerate ions to ~ 1 GeV/nucleon at ~ 20 solar radii. Closer to the Sun Q and ΔV will decrease and reduce the intensity of accelerated high-energy particles. The important points are simply that closer to the Sun (1) ΔV, and therefore the energy gain per shock traversal, is larger, (2) the gyroradius ($\propto r^2$) is much smaller, and therefore scattering mean free paths and acceleration times are correspondingly smaller, and (3) with larger densities and injection rates wave excitation could be much more rapid and contribute to even smaller acceleration times.

The most compelling evidence for the shock origin of gradual SEP events comes not from the theoretical models but from observations of SEP charge-state composition and the morphology of SEP intensity profiles as functions of time and heliolongitude. The charge state composition of gradual events indicates that the ions are accelerated out of typical coronal material at a temperature of ~2 x 10^6 °K [*Luhn et al.*, 1985; *Tylka et al.*, 1995]. The elemental composition is also indicative of a coronal origin [e.g. *Mason et al.*, 1984]. Acceleration at the site of the solar flare would yield charge-state distributions characteristic of plasma at temperatures of ~2 x 10^7 °K, as observed in impulsive events [*Luhn et al.*, 1987].

Furthermore, unlike impulsive events which are confined to narrow flux tubes magnetically connected to the flare which produced them, gradual events occur in flux tubes with inferred connection points in the corona up to ~ 60° in heliolongitude from the flare site (if an associated flare can be identified). The implication is that the particles are accelerated throughout a large volume of the corona, as expected at a shock wave. The time intensity profiles also support a shock origin [*Cane et al.*, 1988]: A spacecraft east of the location of the CME eruption (and possible flare) may be connected to the nose of the shock early in its life and observes high-energy particles arriving promptly. Later it is connected to the weak flank of the shock and observes a small ESP event. A spacecraft at the longitude of the CME is not connected to the shock initially but later connects to the nose of the shock. This spacecraft observes no prompt-arrival particles but a large ESP event. Finally a spacecraft west of the CME observes maximum particle intensity after the shock has passed when it is connected by magnetic field lines downstream of the shock back to the shock nose where acceleration is strong.

3. DISCUSSION

Is there an alternative to shock acceleration for the large gradual events? The only other mechanism which could occur globally throughout a large extent of the corona is stochastic or second-order Fermi acceleration. However, there would be no reason for the turbulence to be confined near the site of CME launch. Shocks can produce additional turbulence, but the shocks themselves then dominate the acceleration. Indeed generally the shock-accelerated ions are responsible for exciting the additional turbulence rather than the reverse. In any case stochastic acceleration is generally too slow, and the turbulence cannot provide the required power, particularly if most of the fluctuations propagate away from the Sun. There is a class of gradual or "long-duration" γ-ray flares, whose energetic ions are confined at the Sun and may be accelerated by stochastic acceleration in enhanced turbulence in a coronal loop [*Ryan and Lee*, 1991]. However, these events have nothing to do with the "gradual" events described in this paper.

There is one possible origin in addition to a CME-driven shock for a fraction of the particles observed in gradual events. In a few gradual events observed when the spacecraft is well-connected to the site of the flare and CME-eruption there is found to be a change in elemental composition from iron-rich early in the event (i.e. those ions arriving promptly) to normal coronal/gradual event composition later in the event [*Reames*, 1990]. It is possible that these iron-rich ions are accelerated directly at the flare site early in the event and are confined to the narrow flux tube which threads that site. Only spacecraft which are well-connected early in the event could observe them. These ions would then be analogous to those accelerated in impulsive flares, except that flares associated with CME eruption could be quite different from the compact flares low in the corona which produce impulsive events.

Litvinenko [1996] has calculated the acceleration of ions and electrons in the reconnecting current sheets thought to be the origin of flares. In small compact flares the sheets are not extensive enough to accelerate ions directly by the reconnection electric field. However, electrons are magnetized within the sheet and accelerated effectively. These electrons excite ion cyclotron waves which in turn accelerate ions stochastically, in particular ^{3}He, to produce the impulsive ion events [*Miller and Viñas*, 1993]. At flares higher in the corona the reconnecting current sheets can be more extensive and ions can return to the current sheet for additional direct acceleration by the electric field. These ions, accelerated at the more extended flare and current sheet associated with the eruption of a CME, could contribute to the gradual events as observed by spacecraft well-connected to the flare site. These ions could have somewhat different elemental and charge-state composition. This process could also be responsible for the occasional acceleration of ions to energies in excess of 1 GeV in "ground level events".

In conclusion, the challenge to theoreticians is to develop a model of shock acceleration which includes quantitatively the 3-D geometry and solar wind, the evolving shock wave, a realistic diffusion coefficient, possibly ion-excited waves, and a transition from scatter-dominated transport near the shock to nearly scatter-free transport in interplanetary space. Only then can the observed particle energy spectra and time profiles be used to learn about the origins of solar energetic particles close to the Sun.

Acknowledgments. I wish to thank H.V. Cane, E.W. Cliver, T. Forbes, A. Kiplinger, R.P. Lin, Y. Litvinenko, D.V. Reames, B. Sanahuja, and R. Turner, for stimulating discussions on the origins of solar energetic particles. The comments of two referees were very helpful. I also appreciate Nancy Crooker's and JoAnn Joselyn's patience in awaiting this paper under extenuating circumstances. This work was supported, in part, by NASA Space Physics Theory Program grant NAG5-1479 and by NSF grant ATM-9633366.

REFERENCES

Achterberg, A., and C.A. Norman, Particle acceleration by shock waves in solar flares, *Astron. Astrophys.*, 89, 353, 1980.

Axford, W.I., Acceleration of cosmic rays by shock waves, *Proc. Int. Conf. Cosmic Rays 17th*, 12, 155, 1981.

Axford, W.I., E. Leer, and G. Skadron, The acceleration of cosmic rays by shock waves, *Proc. Int. Conf. Cosmic Rays 15th*, 11, 132, 1977.

Baring, M.G., D.C. Ellison, and F.C. Jones, The injection and acceleration of particles in oblique shocks: A unified Monte Carlo description, *Astrophys. J.*, 409, 327, 1993.

Bell, A.R., The acceleration of cosmic rays in shock fronts, 1, *Mon. Not. R. Astron. Soc.*, 182, 147, 1978.

Blandford, R.D., and D. Eichler, Particle acceleration at astrophysical shocks: A theory of cosmic ray origin, *Phys. Reports*, 154, 1, 1987.

Blandford, R.D., and J.P. Ostriker, Particle acceleration by astrophysical shocks, *Astrophys. J.*, 221, L29, 1978.

Bryant, D.A., T.L. Cline, U. D. Desai, and F.B. McDonald, Explorer 12 observations of solar cosmic rays and energetic storm particles after the solar flare of September 28, 1961, *J. Geophys. Res.*, 67, 4983, 1962.

Cane, H.V., D.V. Reames, and T.T. von Rosenvinge, The role of Interplanetary shocks in the longitude distribution of solar energetic particles, *J. Geophys. Res.*, 93, 9555, 1988.

Decker, R.B., Computer modeling of test particle acceleration at oblique shocks, *Space Sci. Rev.*, 48, 195, 1988.

Edmiston, J.P., C.F. Kennel, and D. Eichler, Escape of heated ions upstream of quasi-parallel shocks, *Geophys. Res. Lett.*, 9, 531, 1982.

Ellison, D.C., E. Möbius, and G. Paschmann, Particle injection and acceleration at Earth's bow shock: Comparison of upstream and downstream events, *Astrophys. J.*, 352, 376, 1990.

Fisk, L.A., and M.A. Lee, Shock acceleration of energetic particles in corotating interaction regions in the solar wind, *Astrophys. J.*, 237, 620, 1980.

Forman, M.A., The effects of adiabatic deceleration and shock lifetime on energetic storm particle events, *Proc. Int. Conf. Cosmic Rays 18th*, 3, 153, 1983.

Forman, M.A., J.R. Jokipii, and A.J. Owens, Cosmic-ray streaming perpendicular to the mean magnetic field, *Astrophys. J.*, 192, 535, 1974.

Gosling, J.T., J.R. Asbridge, S.J. Bame, W.C. Feldman, R.D. Zwickl, G. Paschmann, N. Sckopke, and R.J. Hynds, Interplanetary ions during an energetic storm particle event: The distribution function from solar wind thermal energies to 1.6 MeV, *J. Geophys. Res.*, 86, 547, 1981.

Hatzky, R., G..Wibberenz, and J.W. Bieber, Pitch angle distribution of solar energetic particles (SEPs) and the transport parameters in interplanetary space, I: Properties of steady-state distributions, *Proc. Int. Conf. Cosmic Rays 24th*, 4, 261, 1995.

Heras, A.M., B. Sanahuja, Z.K. Smith, T. Detman, and M. Dryer, Large-scale effects of three interplanetary shocks on the associated particle events as a function of their solar origin longitude, *Proc. Int. Conf. Cosmic Rays 22nd*, 3, 284, 1991.

Heras, A.M., B. Sanahuja, D. Lario, Z.K. Smith, T. Detman, and M. Dryer, Three low-energy particle events: modeling the influence of the parent interplanetary shock, *Astrophys. J.*, 445, 497, 1995.

Isenberg, P.A., On a difficulty with accelerating particles at slow-mode shocks, *J. Geophys. Res.*, 91, 1699, 1986.

Jokipii, J.R., Particle drift, diffusion, and acceleration at shocks, *Astrophys. J.*, 255, 716, 1982.

Jokipii, J.R., Particle acceleration at a termination shock, 1, Applications to the solar wind and the anomalous component, *J. Geophys. Res.*, 91, 2929, 1986.

Jokipii, J.R., E.H. Levy, and W.B. Hubbard, Effects of particle drift on cosmic ray transport, l, General properties, applications to solar modulation, *Astrophys. J.*, 213, 861, 1977.

Jones, F.C., and D.C. Ellison, The plasma physics of shock acceleration, *Space Sci. Rev.*, 58, 259, 1991.

Kahler, S.W., Solar flares and coronal mass ejections, *Ann Rev. Astron. Astrophys.*, 30, 113, 1992.

Kallenrode, M.B., Particle propagation in the inner heliosphere, *J. Geophys. Res.*, 98, 19037, 1993.

Kennel, C.F., et al., Plasma and energetic particle structure upstream of a quasi-parallel interplanetary shock, *J. Geophys. Res., 89*, 5419, 1984a.

Kennel, C.F., et al., Structure of the November 12, 1978 quasi-parallel interplanetary shock, *J. Geophys. Res., 89*, 5436, 1984b.

Kennel, C.F., et al., A test of Lee's quasi-linear theory of ion acceleration by interplanetary traveling shocks, *J. Geophys. Res., 91*, 11917, 1986.

Klecker, B., E.W. Cliver, S.W. Kahler, and H.V. Cane, Particle acceleration in solar flares, *EOS, 71*, 1102, 1990.

Krymsky, G.F., A regular mechanism for the acceleration of charged particles on the front of a shock wave, *Dokl. Akad. Nauk SSSR, 234*, 1306, 1977.

Lee, M.A., Coupled hydromagnetic wave excitation and ion acceleration upstream of the earth's bow shock, *J. Geophys. Res., 87*, 5063, 1982.

Lee, M.A., Coupled hydromagnetic wave excitation and ion acceleration at interplanetary traveling shocks, *J. Geophys. Res., 88*, 6109, 1983.

Lee, M.A., and L.A. Fisk, Shock acceleration of energetic particles in the heliosphere, *Space Sci. Rev., 32*, 205, 1982.

Lee, M.A., and J.M. Ryan, Time-dependent coronal shock acceleration of energetic solar flare particles, *Astrophys. J., 303*, 829, 1986.

Litvinenko, Y.E., Particle acceleration in reconnecting current sheets with a nonzero magnetic field, *Astrophys. J., 462*, 997, 1996.

Luhn, A., D. Hovestadt, B. Klecker, M. Scholer, G. Gloeckler, F.M. Ipavich, A.B. Galvin, C.Y. Fan, and L.A. Fisk, The mean ionic charges of N, Ne, Mg, Si, and S in solar energetic particle events, *Proc. Int. Conf. Cosmic Rays 19th, 4*, 241, 1985.

Luhn, A., B. Klecker, D. Hovestadt, and E. Möbius, The mean ionic charge of silicon in ^{3}He-rich solar flares, *Astrophys. J., 317*, 951, 1987.

Mason, G.M., G. Gloeckler, and D. Hovestadt, Temporal variations of nucleonic abundances in solar flare energetic particle events. II. Evidence for large-scale shock acceleration, *Astrophys. J., 280*, 902, 1984.

Miller, J.A., and A.F. Viñas, Ion acceleration and abundance enhancements by electron beam instabilities in impulsive solar flares, *Astrophys. J., 412*, 386, 1993.

Ng, C.K., and D.V. Reames, Focused interplanetary transport of ~1 MeV solar energetic protons through self-generated Alfvén waves, *Astrophys. J., 424*, 1032, 1994.

Palmer, I.D., Transport coefficients of low-energy cosmic rays in interplanetary space, *Rev. Geophys. Space Phys., 20*, 335, 1982.

Palmer, I.D., and J.T. Gosling, Shock-associated energetic proton events at large heliocentric distances, *J. Geophys. Res., 83*, 2037, 1978.

Parker, E.N., The passage of energetic charged particles through interplanetary space, *Planet. Space Sci., 13*, 9, 1965.

Pesses, M.E., B.T. Tsurutani, J.A. Van Allen, and E.J. Smith, Acceleration of energetic protons by interplanetary shocks, *J. Geophys. Res., 84*, 7297, 1979.

Pesses, M.E., J.R. Jokipii, and D. Eichler, Cosmic ray drift, shock wave acceleration, and the anomalous component of cosmic rays, *Astrophys. J., 246*, L85, 1981.

Reames, D.V., Bimodal abundances in the energetic particles of solar and interplanetary origin, *Astrophys. J., 330*, L71, 1988.

Reames, D.V., Acceleration of energetic particles by shock waves from large solar flares, *Astrophys. J., 358*, L63, 1990.

Reames, D.V., Energetic particles and the structure of coronal mass ejections, this volume.

Reames, D.V., H.V. Cane, and T.T. von Rosenvinge, Energetic particle abundances in solar electron events, *Astrophys. J., 357*, 259, 1990.

Roelof, E.C., Propagation of solar cosmic rays in the interplanetary magnetic field, in Lectures in High Energy Astrophysics, edited by H. Ögelman and J.R. Wayland, *NASA Spec. Publ., SP-199*, 111, 1969.

Ruffolo, D., and T. Khumlumlert, Propagation of coherent pulses of solar cosmic rays, *Proc. Int. Conf. Cosmic Rays 24th, 4*, 277, 1995.

Ryan, J.M., and M.A. Lee, On the transport and acceleration of solar flare particles in a coronal loop, *Astrophys. J., 368*, 316, 1991.

Sarris, E.T., and J.A. Van Allen, Effects of interplanetary shock waves on energetic charged particles, *J. Geophys. Res., 79*, 4157, 1974.

Scholer, M., F.M. Ipavich, G. Gloeckler, and D. Hovestadt, Acceleration of low-energy protons and α particles at interplanetary shock waves, *J. Geophys. Res., 88*, 1977, 1983.

Tan, L.C., G.M. Mason, G. Gloeckler, and F.M. Ipavich, Energetic particle diffusion coefficients upstream of quasi-parallel interplanetary shocks, *J. Geophys. Res., 94*, 6552, 1989.

Tylka, A.J., P.R. Boberg, J. H. Adams, Jr., L.P. Beahm, W.F. Dietrich, and T. Kleis, The mean ionic charge state of solar energetic Fe ions above 200 MeV per nucleon, *Astrophys. J., 444*, L109, 1995.

van Nes, P., R. Reinhard, T.R. Sanderson, and K.-P. Wenzel, The energy spectrum of 35- to 1600-keV protons associated with interplanetary shocks, *J. Geophys. Res., 89*, 2122, 1984.

M. A. Lee, Space Science Center, Morse Hall, University of New Hampshire, Durham, NH 03824.

Mass Ejections Observed in Radio Propagation Measurements Through the Solar Corona

Richard Woo

Jet Propulsion Laboratory, California Institute of Technology, Pasadena, California

A wide variety of radio propagation and scattering phenomena observed when a radio source is occulted by the solar corona, often referred to as radio occultation measurements, has formed the basis for probing the corona for over four decades. These measurements serve as an important bridge between white-light coronagraph and *in situ* plasma measurements beyond 0.3 AU. In the past, temporal variations that seemed different from the background solar wind were usually identified as transients and thought to represent propagating interplanetary disturbances such as coronal mass ejections (CMEs) observed in white-light coronagraphs. Recent progress has surprisingly shown that these temporal variations can also represent the rotation of quasi-stationary spatial structures such as coronal streamers across the radio path. This paper summarizes our current understanding of temporal variations, including CMEs, observed in radio occultation measurements.

Two regions of enhanced density and density fluctuations appear to be associated with CMEs. The first region represents the compressed plasma ahead of the CME, while the second region appears to be associated with the main body of the CME observed in white-light coronagraph measurements and the CME identified at 1 AU by *in situ* plasma measurements based on counterstreaming suprathermal electrons. Single magnetic field polarity reversals are generally found ahead of the second region and most likely represent the deflection of the ambient magnetic field ahead of the advancing CME, or the draping of the ambient magnetic field around the front of the CME. Multiple magnetic field polarity reversals, which are sometimes observed within the second region of the CME, are suggestive of large internal field rotation, magnetic ropes, and magnetic clouds.

1. INTRODUCTION

Imagine a white-light coronagraph whose field of view extends to 1 AU, yielding precise, high-sensitivity and high-time-resolution measurements of polarized brightness, but only for one point in the plane of the sky. Or imagine a spacecraft passing through the corona and instrumented to make solar wind measurements integrated along a linear path between the spacecraft and Earth. Measured (not necessarily all the time, and certainly not simultaneously) are: (1) density with high precision, high sensitivity, and high-time resolution, (2) solar wind velocity with lower precision and higher uncertainty, and (3) the product of density and the component of magnetic field along the path. Then ask the question: What can these measurements tell us about coronal mass ejections (CMEs) observed in white-light measurements?

The preceding circumstances describe some of the information provided by radio occultation measurements of the solar corona, measurements that reflect a variety of radio propagation and scattering phenomena when a radio source is occulted by the corona. Both natural radio sources and spacecraft radio signals have been used in these experiments, which started in the late 1950s with angular broadening measurements [*Hewish*, 1958; *Vitkevitch*, 1961; *Erickson*, 1964], before the existence of the solar wind had even been established. Information on density fluctuations is available from angular broadening, intensity scintillation, phase/Doppler scintillation, and spectral broadening measurements; on density from ranging and phase measurements; on magnetic field from Faraday rotation measurements; and on solar wind velocity from single- and multiple-station intensity scintillation measurements. Recent reviews of these measurements and their results have been given by *Watanabe and Schwenn* [1989], *Bird and Edenhofer* [1990], *Coles* [1993], *Hewish* [1993], and *Woo* [1993a].

Since few studies of the interplanetary manifestation of CMEs based on *in situ* solar wind measurements at Earth orbit have been based on signatures of either plasma density or solar wind velocity [e.g., Gosling, 1990; Neugebauer and Goldstein, this volume], the value of radio occultation measurements for investigating CMEs might be questioned. However, the strength in radio occultation measurements lies in the fact that they remotely probe the same solar wind where white-light CMEs are observed, with ranging actually observing the same plasma parameter as polarized brightness, while at the same time providing plasma measurements that are closely related to the solar wind measurements made directly by spacecraft beyond 0.3 AU. Radio occultation measurements, therefore, serve as an important and natural bridge between solar and *in situ* spacecraft measurements.

In spite of the often fragmented and disparate results produced by decades of radio occultation measurements, our knowledge of the morphology of the corona based on these unique but limited measurements has been advancing rapidly [*Woo et al.*, 1995; *Woo*, 1996a, b]. In the past, temporal variations observed in radio occultation measurements that seemed different from the background solar wind were usually thought to represent propagating interplanetary disturbances such as CMEs. Recent progress, however, has surprisingly shown that these events can also represent the rotation of quasi-stationary spatial structures such as coronal streamers across the radio path.

The purpose of this paper is to summarize our current understanding of temporal events observed in radio occultation measurements of the solar corona. After reviewing the early history of temporal events and recent results on coronal streamers, features of CMEs observed in radio occultation measurements are described and synthesized.

2. TEMPORAL EVENTS IN RADIO OCCULTATION MEASUREMENTS

2.1. Early History

Conspicuous temporal events were detected as soon as space exploration began and spacecraft radio signals were occulted by the solar corona. Not surprisingly, these early events, observed in 1968 by Pioneer 6 spectral broadening [*Goldstein*, 1969] and Faraday rotation measurements [*Levy et al.*, 1969], indicated a corona that was complex and dynamic. Additional observations of temporal events followed [*Cannon*, 1976]. The Pioneer 6 Faraday rotation events were interpreted as magnetic bottles [*Schatten*, 1970], while a major event, observed in Voyager 1 measurements of spectral broadening, intensity scintillation, and phase on August 18, 1979 at 13 R_0, was identified as an interplanetary shock associated with a solar flare [*Woo and Armstrong*, 1981; *Cane et al.*, 1982]. The Voyager 1 measurements are notable not only for observing a very fast CME and its associated interplanetary shock but also for providing rare separate profiles of density, density fluctuations, and solar wind speed, thus showing the relationship between these solar wind parameters. In general, whenever it was possible to compare spectral broadening and Faraday rotation temporal events with simultaneous white-light CMEs, a one-to-one correspondence was found between them [*Woo et al.*, 1982; *Bird et al.*, 1985].

Routine Doppler measurements taken in the course of tracking interplanetary spacecraft provide an extensive data base for investigating Doppler scintillation temporal events [*Woo et al.*, 1985; *Woo*, 1993b]. When measurements of the time of arrival of Doppler scintillation events representing interplanetary shocks were combined with white-light measurements of CMEs, shock velocity profiles as a function of heliocentric distance were obtained [*Woo et al.*, 1985]. An important result was that the range of shock velocities near the Sun was found to be significantly greater than that at 1 AU deduced from *in situ* plasma measurements. Although the number of cases was not large, the more rapidly decelerating interplanetary shocks appeared to be associated with flares, while the slowly decelerating ones seemed to be associated with eruptive prominences.

The frequency of occurrence of Doppler scintillation events and its variation over a solar cycle was investigated and found to be similar to that of CMEs observed in white-light measurements [*Woo*, 1993b], reinforcing the hypothesis based on individual comparisons between CMEs and spectral broadening transients that CMEs and Doppler scintillation events are different manifestations of the same physical phenomenon. A comparison with *in situ* plasma measurements over a limited 3-month period in 1981–1982 showed that there was a near one-to-one correspondence between Doppler scintillation events and interplanetary shocks [*Woo and Schwenn*, 1991]. Since fast CMEs produce interplanetary shocks, this result is consistent with the close association between CMEs and shocks found when white-light and *in situ* plasma measurements were correlated [*Sheeley et al.*, 1985].

2.2. Coronal Streamers

A striking pattern was revealed when 1984 Pioneer Venus Orbiter (PVO) Doppler scintillation temporal events were compared with synoptic maps of solar wind parameters based on 6-month averages of direct observations by IMP 8 at 1 AU and PVO at the orbit of Venus [*Woo and Gazis*, 1993]. Although temporal events were present in the high-density, slow solar wind associated with the streamer belt, some of which represented CMEs observed by the Solwind white-light coronagraph, they were conspicuously absent in the low-density, fast wind associated with coronal holes. Subsequent studies based on Doppler scintillation measurements of the heliospheric current sheet conducted when CMEs were not present [*Woo et al.*, 1995] surprisingly revealed that temporal events also represent the rotation of coronal streamer stalks observed in white-light coronagraph measurements [*Koutchmy*, 1977] across the radio path. These streamer stalks, which measure 1–2° in angular size, rotate across the radio path in about two hours. The enhanced filamentary structures that comprise the streamer stalk give rise to the observed scintillation event. It is now clear that those 1984 PVO events associated with the peaks in density observed in the synoptic maps were most likely coronal streamers and that the striking PVO scintillation pattern was the manifestation of the close association of CMEs and coronal streamers observed in white-light measurements [*Hundhausen*, 1993].

As in the case of a single spacecraft making *in situ* measurements, occultation measurements cannot always unambiguously distinguish corotating from outward propagating features. Still, the interpretation of prominent 2-hour long Doppler scintillation events in terms of coronal streamers rotating across the radio path was not anticipated. Streamer passage has also been identified in ranging measurements of path-integrated density [*Woo et al.*, 1995] as well as the electron density spectra inferred from phase scintillation and spectral broadening measurements [*Woo and Habbal*, 1997], the latter result demonstrating that the small-scale filamentary structures within the streamer stalks are not only stronger but also finer than those in the fast wind from coronal holes. Most surprising, however, has been the tracing of the three 1968 Pioneer 6 Faraday rotation and spectral broadening events to streamer stalk passages [*Woo*, 1997]. This long delayed streamer identification of the Pioneer 6 events raises confidence in the general interpretation and usefulness of Faraday rotation events, while the reversal of magnetic field polarity deduced from them provides observational evidence confirming what has previously only been inferred from modeling [*Pneuman and Kopp*, 1971], that streamers observed in white-light measurements are the manifestation of the heliospheric current sheet.

3. CORONAL MASS EJECTIONS

Simultaneous radio occultation and white-light coronagraph measurements of CMEs are crucial not only for establishing the association between them but also for understanding features observed by both techniques. There have been only a few cases of simultaneous measurements, but features that seem common to CMEs are emerging. We will discuss these first in terms of density and density fluctuations and then magnetic field.

3.1. Density and Density Fluctuations

Figure 1, reproduced from *Bird and Edenhofer* [1990], shows the Solwind difference images and the corresponding Helios 2 time series of spectral broadening bandwidth and Faraday rotation (which will be discussed in section 3.2) for a CME observed on the west limb on October 24, 1979. A finer time resolution (2-min integration time) version of the spectral broadening time series

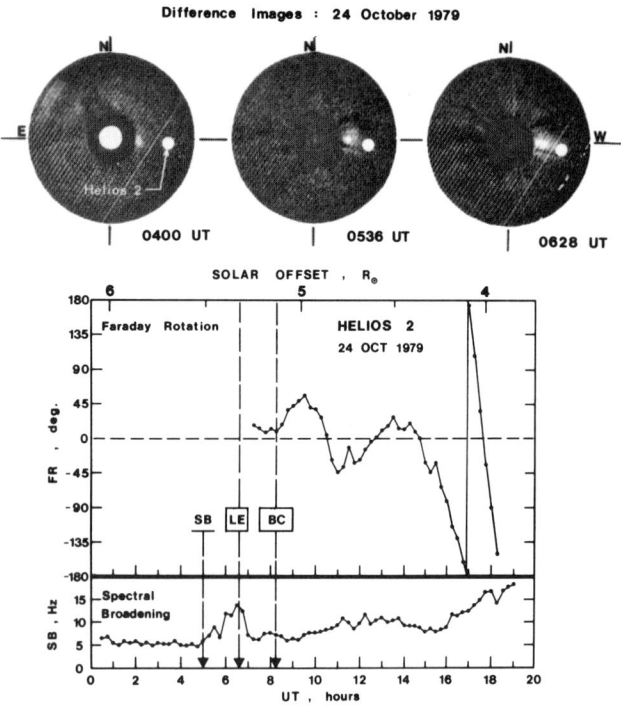

Figure 1. Time histories of Faraday rotation (FR) and spectral broadening (SB), and Solwind white-light coronagraph images of the October 24, 1979 CME (reproduced from Bird and Edenhofer [1990]). The three Solwind difference images at the top show the geometrical relationship between the CME and the apparent position of Helios 2 off the west limb of the Sun indicated by the white dots.

is given in Figure 3 of *Woo et al.* [1982], and additional Solwind images for later times are shown in *Bird et al.* [1985]. The leading edge (LE) and a following bright core (BC) (i.e., CME body) were two features that were tracked by Solwind during their radial expansion, giving estimated projected velocities of 160 ± 50 km/s and 120 ± 40 km/s for the leading edge and bright core, respectively. The start of the spectral broadening event is marked SB, while the estimated arrival times at the Helios radio path of the leading edge and CME body are indicated by the vertical dashed lines. The widths of the boxes LE and BC indicate the estimated errors of these arrival times. As first pointed out by *Woo et al.* [1982], the start of the spectral broadening event (SB) precedes the leading edge of the CME (LE). Another example of the detection of spectral broadening enhancement ahead of the CME boundary is provided by Helios 2 and Solwind coronagraph measurements of a northern CME observed on October 23, 1979, near 8 R_o (see Figure 2 of *Bird et al.* [1985]).

Another feature that appears common to CMEs are the two regions of enhanced density (enhanced phase) and enhanced density fluctuations (enhanced spectral broadening) observed in the interplanetary shock event of August 18, 1979 [*Woo and Armstrong*, 1981]. These are also evident in the spectral broadening measurements of the CMEs shown in Figures 1–3. Figures 2 and 3 correspond to CMEs on October 23, 1979, and November 16, 1979, respectively, and are reproduced from *Bird et al.* [1985]. Phase measurements of density were not available for these cases, but evidence that density is also enhanced is provided by the fact that the leading edges of the CME bodies of the October 23 and October 24 CMEs seen in the Solwind white-light measurements, and, hence, indicating the start of enhanced density, occur near the start of the second region of enhanced density fluctuations. Also consistent is the fact that a northern CME observed on November 15, 1979, by Helios 1 exhibited only one region of enhanced spectral broadening, but a comparison with the Solwind white-light measurements showed that the CME body missed the Helios 1 probing point (see Figures 10 and 11 of *Bird et al.* [1985]).

These results suggest that the first region represents the compressed plasma ahead of the CME while the second region of enhanced density and density fluctuations is associated with the CME bodies observed in white-light measurements. The first region may or may not be preceded by an interplanetary shock and is probably related to the forerunners observed in white-light measurements [*Jackson and Hildner*, 1973]. Detection by white-light observations is dependent on adequate spatial resolution and sensitivity of the measurements; but another significant factor, illustrated by *in situ* measurements of density and density fluctuations and discussed below, accounts for the difference between radio and white-light measurements.

Shown in Figure 4 are the levels of mean density n, density fluctuations Δn and fractional density fluctuations $\Delta n/n$ for various types of solar wind flow based on 5-min ISEE 3 solar wind measurements at 1 AU averaged over one hour [*Huddleston et al.*, 1995]. Transient flows from CMEs were identified on the basis of counterstreaming suprathermal electrons by *Gosling et al.* [1987] and/or the detection of enhanced helium abundance. In the remainder of this paper, apostrophes will be used to distinguish CMEs identified in this fashion from those observed in the white-light measurements. Interaction regions (i.e., sheaths) are the regions between interplanetary shocks and the leading edges of the 'CME' driver gas that follow them. The

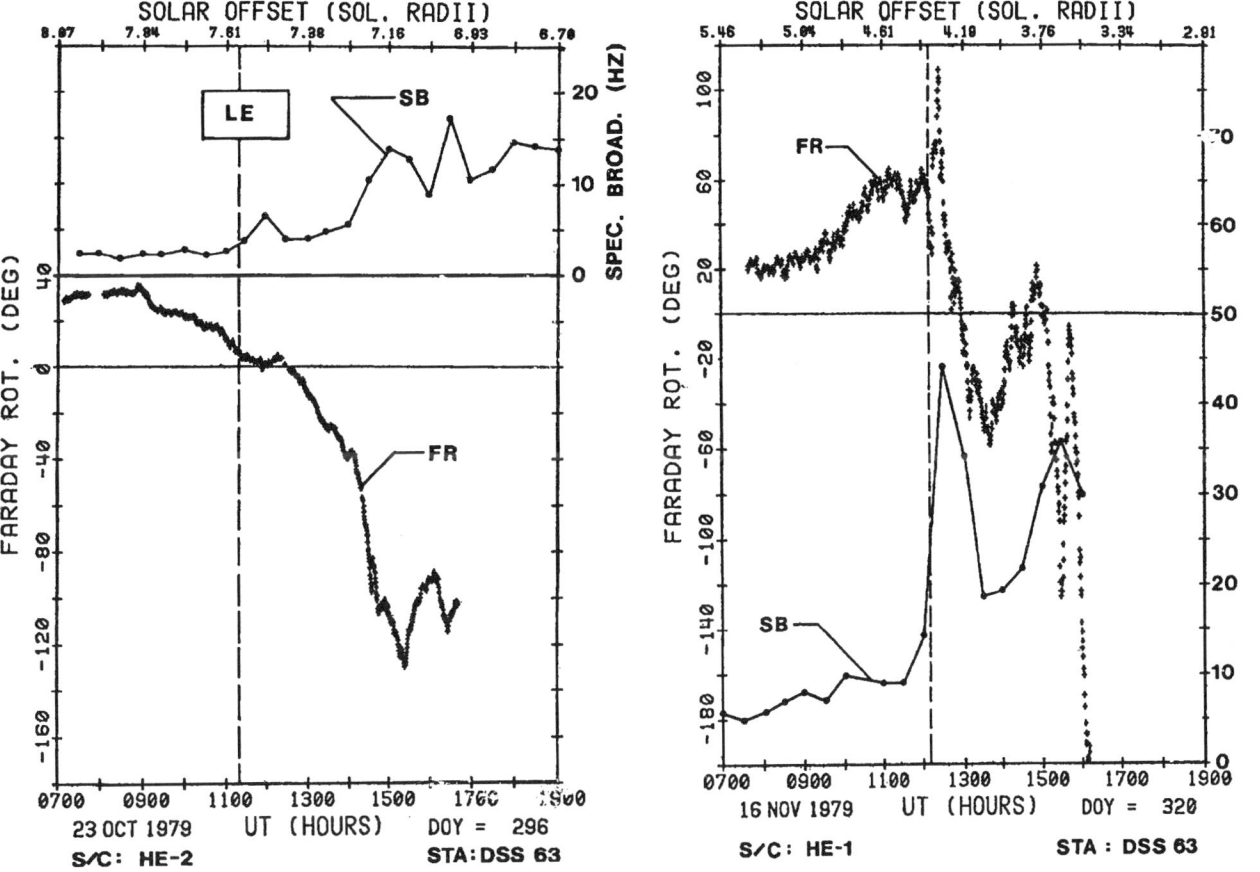

Figure 2. Time histories of Faraday rotation (FR) and spectral broadening (SB) for the October 23, 1979 CME observed by the Solwind white-light coronagraph (reproduced from Bird et al. [1985]).

Figure 3. Time histories of Faraday rotation (FR) and spectral broadening (SB) for the November 16, 1979 CME observed by the Solwind white-light coronagraph (reproduced from Bird et al. [1985]).

heliospheric plasma sheet encompasses the heliospheric current sheet [*Winterhalter et al.*, 1994] and appears to represent the interplanetary counterpart of the streamer stalks detected in Doppler scintillation measurements, as discussed earlier [*Bavassano et al.*, 1996].

The results in Figure 4 show that density and density fluctuations are enhanced in the compressed plasma ahead of the 'CME' (interaction region) and within the 'CME' but that the enhancement factor of the density fluctuations is significantly higher than that of density, demonstrating why radio occultation measurements that sense density fluctuations (angular broadening, spectral broadening, intensity and phase/Doppler scintillation) have been especially effective in detecting CMEs. While these measurements may sense density fluctuations in different frequency ranges, because the density fluctuations associated with CMEs appear to be broadband, all are more sensitive to CMEs than those sensing density.

Some individual time series of the ISEE 3 solar wind measurements across 'CMEs' show enhancements in density and density fluctuations that coincide with the 'CME' boundaries (see, e.g., Figure 4 of *Gosling* [1990]), suggesting that the boundary of the second region of enhanced density and density fluctuations of CMEs corresponds to the boundary of the 'CME' driver gas.

3.2. *Magnetic Field*

As seen in the case of coronal streamers, Faraday rotation observations are unique among radio occultation measurements because they respond to magnetic field in addition to density, making it possible to probe the magnetic field. A linearly polarized radio wave propagating

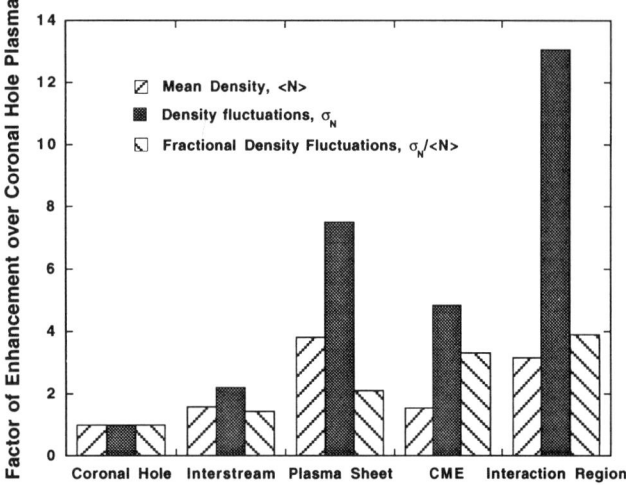

Figure 4. Ratios of mean density n, density fluctuations Δn, and fractional density fluctuations Δn/n for various solar wind flow to the values in coronal hole flow (reproduced from Huddleston et al. [1995]).

through a magnetized plasma such as the solar corona will display a rotation in its plane of polarization by an angle $\Delta\psi$:

$$\Delta\psi = \frac{e^3}{2\pi m_e^2 c^2 f^2} \int n_e \mathbf{B} \cdot d\mathbf{s} \quad (1)$$

where n_e is electron density, **B** the coronal magnetic field, ds the incremental path length, f radio frequency, e electron charge, m_e electron mass, and c the speed of light. The angle of rotation is positive when the magnetic field has a component in the direction of radio propagation.

Unfortunately, unraveling Faraday rotation measurements, especially when separate measurements of density are not available, can be difficult. The interpretation is not unique because different combinations of magnetic field structure and electron density can produce the same net polarization rotation. For instance, in the case of the Pioneer 6 Faraday rotation measurements of a coronal streamer, the observed Faraday rotation throughout the event was of one sign, suggesting that the magnetic field was unipolar [Bird et al., 1985], when in fact there was a polarity reversal in magnetic field [Woo, 1997]. The situation with corotating streamers is, of course, unusual. In the case of interplanetary disturbances propagating outward from the Sun, field reversals are more likely to be manifested as sign changes in Faraday rotation.

As seen for the October 23 and November 16 CMEs shown in Figures 2 and 3, simultaneous measurements of spectral broadening and Faraday rotation show a polarity reversal of magnetic field taking place ahead of the second region of enhanced spectral broadening. Also, several polarity reversals are observed inside the second region of enhanced spectral broadening in the October 24 CME of Figure 1, but not in the October 23 and November 16 CMEs of Figures 2 and 3. Strictly speaking, the sign change in Faraday rotation applies to the combination of background solar wind and CME, but we assume that the CME contribution is dominant. Moreover, when the background is removed, the locations of the polarity reversals do not appear to change significantly [Bird et al., 1985].

Reversals in magnetic field polarity are also a common feature ahead of and inside 'CMEs' observed at 1 AU in solar wind measurements [Gosling, 1990]. The reversals ahead of the 'CME' can be interpreted as representing the deflection of the ambient plasma or the draping of the ambient magnetic field around the front of the 'CME' [Gosling and McComas, 1987], while the reversals inside the 'CME' are characteristic of magnetic flux ropes and magnetic clouds [Burlaga et al., 1987].

Finally, the time histories of spectral broadening and Faraday rotation of the October 23 CME in Figure 2 reinforce the notion that density is enhanced in the second region of enhanced density fluctuations. Density rather than magnetic field variation appears to be dominating the Faraday rotation event after the polarity reversal. The apparent anticorrelation of spectral broadening and Faraday rotation after 1330 UT suggests that density fluctuations are varying in step with density, as would be expected if density fluctuations were proportional to density.

4. MEASUREMENTS BEYOND 0.5 AU

Extensive meter-wavelength intensity scintillation observations, often referred to as IPS for interplanetary scintillation, have also been used to investigate density fluctuation and solar wind velocity in the heliocentric distance range of 0.3–1.0 AU. The emphasis of this paper is on the solar corona, but it interesting to make a few comments about these measurements and the evolution of the solar wind.

Farther from the Sun, large-scale features characterized by enhanced density fluctuations in the meter-wavelength IPS measurements include the compressed plasma at the leading edges of high speed streams (corotating interaction regions) in addition to streamers and CMEs [*Houminer*, 1977; *Kakinuma*, 1977; *Watanabe and Schwenn*, 1989; *Hewish*, 1993; *Hick et al.*, 1995; *Woan*, 1995]. Furthermore, evolution and path-integration effects make it increasingly difficult to distinguish these large-scale features, as apparent in attempts to search for coronal streamers [*Houminer and Gallagher*, 1993]. There is a tendency for corotating interaction regions and coronal streamers to become confused because of close proximity to each other as a result of evolution, as observed in coronal streamers tracked by Doppler scintillation measurements from near the Sun to 1 AU [*Woo et al.*, 1995], *in situ* measurements of density fluctuations at 1 AU [*Huddleston et al.*, 1995], and meter-wavelength IPS measurements beyond 0.5 AU [*Ananthakrishnan et al.* 1980]. Finally, the difference in enhancement between density and density fluctuations of these large-scale solar wind features appears to be one of the reasons why meter-wavelength IPS measurements have been more effective in detecting propagating interplanetary disturbances and corotating features beyond 0.5 AU than meter wavelength ranging (group delay) measurements [*Croft*, 1979].

5. SUMMARY AND DISCUSSION

The ability of radio occultation measurements to probe remotely the enigmatic solar corona and bridge the gap between white-light coronagraph and *in situ* plasma measurements has been the main reason for the appeal and long history of these unique measurements. Realizing the full potential of these measurements, however, has not been easy. Since they are conducted during solar conjunctions and often sporadically, radio occultation measurements have generally been constrained to brief discontinuous periods. Compared with *in situ* plasma measurements, the solar wind parameters sensed by radio occultation measurements are limited; density and density fluctuations are most directly observed, but estimates of solar wind velocity and some information on magnetic field can also be deduced. Furthermore, since it is commonplace for radio occultation measurements reflecting only one radio scattering or propagation phenomenon to be conducted at a time, in many instances, only one solar wind parameter is measured. Added to these measurement limitations are the facts that the nature of the density fluctuations (that they represent raylike structures as well as turbulence [*Woo and Habbal*, 1997]) is only now becoming understood and that the complex and dynamic nature of the solar corona makes it difficult to distinguish temporal and spatial variations. Still, in spite of these difficulties, significant advances in establishing the morphology of the solar corona and understanding CMEs have been made. Although earlier investigations were carried out before it was realized that temporal events could also represent streamers, the short durations of the streamer events usually precluded them from these studies [*Woo*, 1993b].

One of the reasons for the success of radio occultation measurements in probing the corona is that although small-scale density fluctuations represent a relatively small fraction of the mean density, their variation between coronal holes, streamers and CMEs near the Sun is both striking and abrupt, with vestiges still evident in the *in situ* density measurements at 1 AU. Thus, measurements such as spectral broadening and Doppler scintillation, which sense only small-scale density fluctuations (a parameter that has not usually received much attention in solar wind studies based on *in situ* measurements), readily detect large-scale structures such as CMEs and coronal streamers. A consequence of this result is that measurements that sense density fluctuations define the context of those that detect density. Only when placed into context by spectral broadening measurements did Faraday rotation measurements yield significant information on magnetic field, leading to the demonstration that the heliospheric current sheet coincides with streamer stalks, and providing hints of the magnetic topology of CMEs.

Radio occultation measurements complement white-light measurements of CMEs in ways other than providing important information on solar wind velocity and magnetic field. Partly because density fluctuations are more enhanced than density, and partly because radio occultation measurements are much more sensitive than white-light measurements, features such as the compressed plasma ahead of CMEs and interplanetary shocks that may not appear in white-light measurements are readily detected in radio occultation measurements.

There have been only a few cases for which simultaneous radio occultation and white-light measurements of CMEs could be compared, but the following picture based on a synthesis of common features and hints from the *in situ* plasma measurements of CMEs at 1 AU seems to be emerging. Two regions of enhanced density and density fluctuations appear to be associated with CMEs. The first region represents the compressed plasma ahead of the CME, while the second region appears to be associated with the CME body observed in white-light, and the CME identified at 1 AU by *in situ* plasma measurements based on counterstreaming suprathermal electrons. An interplanetary shock may or not precede the compressed plasma. The compressed plasma seems likely to be related to the forerunners observed in white-light measurements, and its detection in white-light depends on the sensitivity and spatial resolution of the white-light measurements.

Single magnetic field polarity reversals are generally found ahead of the second region. Most likely they represent the deflection of the ambient magnetic field ahead of the advancing CME or the draping of the ambient magnetic field around the front of the CME. Multiple magnetic field polarity reversals that have sometimes been observed within the second region of the CME (observed in the October 24, 1979 but not in the October 23, 1979 and November 16, 1979 CMEs) are suggestive of large internal field rotation, magnetic ropes, and magnetic clouds. This occasional presence of multiple magnetic field polarity reversals is consistent with the result based on *in situ* measurements at 1 AU showing that coherent internal field rotations, which would lead to polarity reversals, are only observed within 30% of the 'CMEs' [*Gosling*, 1990].

Although angular scattering or angular broadening was the first radio propagation phenomenon to be observed through the solar corona, measurements have not been as extensive as those of intensity or Doppler scintillation. Some transients or temporal events, however, have been reported [*Erickson et al.*, 1981]. Angular scattering is anisotropic near the Sun due to organization of the density fluctuations by the magnetic field. The unique measurements of anisotropy provided by angular scattering offers possibly an additional means for distinguishing CMEs and coronal streamers.

In the temporal events, the enhancement in scattering is sometimes accompanied by a rotation of the major axis of the structure function by as much as 45° from the radial direction. In other cases, the enhancements are manifested by an increase in axial ratio of the anisotropic structure function but no rotation of its major axis. The results of this paper suggest that the rotation events reflect the deflection of the ambient magnetic field ahead of a CME and hence represent the passage of CMEs. Those events that show an increase in axial ratio but no rotation probably correspond to the passage of a coronal streamer. Since the magnetic field is always oriented in the radial direction, no rotation in anisotropic scattering is observed. The axial ratio increases because of the significant decrease in the size of filamentary structures within a streamer stalk [*Woo and Habbal*, 1997].

It is important to further exploit radio occultation measurements to study CMEs near the Sun. Future measurements would be most useful if they yielded simultaneous profiles of density, density fluctuations, solar wind velocity and magnetic field, which means observing different radio phenomena at the same time. Improved understanding of the signatures of coronal mass ejections, streamers, and plumes increases the chances that other coronal features that would also appear as temporal events in radio occultation measurements could be identified, e.g., soft X-ray jets [*Shimojo et al.*, 1996] or the interplanetary manifestation of soft X-ray bright points. Current and future NASA missions with which radio occultation measurements of the corona can be conducted include Galileo, Mars Global Survey, Cassini, and DS1; simultaneous observations with SOHO, Yohkoh and direct spacecraft measurements would be especially valuable.

Acknowledgments. It is a pleasure to thank J. Armstrong for many useful discussions, M. Bird for making available the originals of his figures, W. Erickson for discussion about the angular scattering transients, and C. Copeland for producing this paper. This paper describes research carried out at the Jet Propulsion Laboratory, California Institute of Technology, under a contract with the National Aeronautics and Space Administration.

REFERENCES

Ananthakrishnan, S., W.A. Coles, and J.J. Kaufman, Microturbulence in solar wind streams, *J. Geophys. Res.*, 85, 6025–6030, 1980.

Bavassano, B., R. Woo, R. Bruno, and H. Rosenbauer, The heliospheric plasma sheet in near-the-Sun solar wind, paper presented at the *Third Symposium on Solar and Interplanetary Transient Phenomena*, Beijing, China, October 14–18, 1996.

Bird, M.K., H. Volland, R.A. Howard, M.J. Koomen, D.J. Michels, N.R. Sheeley, Jr., J.W. Armstrong, B.L. Seidel, C.T. Stelzried, and R. Woo, White-light and radio sounding observations of coronal transients, *Solar Phys.*, 98, 341–368, 1985.

Bird, M.K., and P. Edenhofer, 1990, Remote sensing observations of the solar corona, in *Physics of the Inner Heliosphere, I*, edited by R. Schwenn and E. Marsch, pp. 13–08, Springer-Verlag, Berlin, 1990.

Burlaga, L.F., L. Klein, N.R. Sheeley, Jr., D.J. Michels, R.A. Howard, M.J. Koomen, R. Schwenn, and H. Rosenbauer, A magnetic cloud and a coronal mass ejection, *Geophys. Res. Lett.*, 9, 1317–1320, 1982.

Cane, H.V., R.G. Stone, and R. Woo, Velocity of the shock generated by a large east limb flare on August 18, 1979, *Geophys. Res. Lett., 9*, 897–900, 1982.

Cannon, A.R., Radio-frequency probing of the solar corona, PhD thesis, 426 pp., University of California, Berkeley, 1976.

Coles, W.A., Scintillation in the solar wind (IPS), in *Wave propagation in random media (Scintillation)*, edited by V. Tatarskii, A. Ishimaru, and V. Zavorotny, pp. 156–168, SPIE, Bellingham, Washington, 1993.

Croft, T.A., A graphical summary of solar wind electron content observations by Pioneer 6, 8, and 9, *J. Geophys. Res., 84*, 439–449, 1979.

Erickson, W.C., The radio-wave scattering properties of the solar corona, *Astrophys. J., 139*, 1290–1311, 1964.

Erickson, W.C., M.J. Mahoney, and W.M. Cronyn, Radio wave scattering in the outer solar corona, *BAAS, 13*, 841, 1981.

Goldstein, R.M., Superior conjunction of Pioneer 6, *Science, 166*, 598–601, 1969.

Gosling, J.T., Coronal mass ejections and magnetic flux ropes in interplanetary space in *Physics of Magnetic Flux Ropes, Geophys. Monogr. 58*, edited by C. T. Russell, E. R. Priest, and L. C. Lee, pp. 343–364, AGU, Washington D.C., 1990.

Gosling, J.T., and McComas, D.J., Field line draping about fast coronal mass ejecta: A source of strong out-of-the-ecliptic interplanetary magnetic fields, *Geophys. Res. Lett., 14*, 355–358, 1987.

Gosling, J.T., D.N. Baker, S.J. Bame, W.C. Feldman, R.D. Zwickl, and E.J. Smith, Bidirectional solar wind electron heat flux events, *J. Geophys. Res., 92*, 8519–8535, 1987.

Hewish, A., The scattering of radio waves in the solar corona, *Mon. Not. Roy. Astr. Soc., 118*, 534–546, 1958.

Hewish, A., Interplanetary scintillation imaging of disturbances in the solar wind, in *Wave propagation in random media (Scintillation)*, edited by V. Tatarskii, A. Ishimaru, and V. Zavorotny, pp. 261–270, SPIE, Bellingham, Washington, 1993.

Hick, P., B.V. Jackson, S. Rappoport, G. Woan, G. Slater, K. Strong, and Y. Uchida, Synoptic IPS and Yohkoh soft X-ray observations, *Geophys. Res. Lett., 22*, 643–646, 1995.

Houminer, Z., Scintillation observations of the interplanetary plasma, in *Study of Travelling Interplanetary Phenomena 1977*, edited by M.A. Shea, D.F. Smart, and S.T. Wu, pp. 119–141, Reidel, Dordrech, Holland, 1977.

Huddleston, D.E., R. Woo, and M. Neugebauer, Density fluctuations in different types of solar wind flow at 1 AU and comparison with results from Doppler scintillation measurements near the Sun, *J. Geophys. Res., 100*, 19951–19956, 1995.

Hundhausen, A., Sizes and locations of coronal mass ejections: SMM observations from 1980 and 1984–1989, *J. Geophys. Res., 98*, 13177–13200, 1993.

Jackson, B.V., and E. Hildner, Forerunners: Outer rims of solar coronal transients, *Solar Phys., 60*, 155, 1973.

Kakinuma, T., Observations of interplanetary scintillations: Solar wind velocity, in *Study of Travelling Interplanetary Phenomena 1977*, edited by M.A. Shea, D.F. Smart, and S.T. Wu, pp. 101–118, Reidel, Dordrech, Holland, 1977.

Koutchmy, S., Study of the June 30, 1973 trans-polar coronal hole, *Solar Phys., 51*, 399–407, 1977.

Levy, G.S., T. Sato, B.L. Seidel, C.T. Stelzried, J.E. Ohlson, and W.V.T. Rusch, Pioneer 6: Measurement of transient Faraday rotation phenomena observed during solar occultation, *Science, 166*, 596–598, 1969.

Neugebauer, M., and R. Goldstein, Particle and field signatures of coronal mass ejections in the solar wind, this volume.

Pneuman, G., and R.A. Kopp, Gas-magnetic field interactions in the solar corona, *Solar Phys., 18*, 258–270, 1971.

Schatten, K.H., Evidence for a coronal magnetic bottle at 10 solar radii, *Solar Phys., 12*, 484–491, 1970.

Sheeley, N.R., Jr., R.A. Howard, M.J. Koomen, D.J. Michels, R. Schwenn, K.H. Mülhäuser, and H. Rosenbauer, Coronal mass ejections and interplanetary shocks, *J. Geophys. Res., 90*, 163–175, 1985.

Shimojo, M., S. Hashimoto, K. Shibata, T. Hirayama, H.H. Hudson, and L.W. Acton, Statistical study of solar X-ray jets observed with the Yohkoh soft X-ray telescope, *Publ. Astron. Soc. Japan, 48*, 123–136, 1996.

Vitkevitch, V.V., Radio astronomical observations of moving plasma clouds in the solar supercorona, *Sov. Astron., 4*, 897–903, 1961.

Watanabe, T., and R. Schwenn, Large-scale propagation properties of interplanetary disturbances revealed from IPS and spacecraft observations, *Space Sci. Rev., 51*, 147–173, 1989.

Winterhalter, D., E.J. Smith, M.E. Burton, N. Murphy, and D.J. McComas, The heliospheric plasma sheet, *J. Geophys. Res., 99*, 6667–6680, 1994.

Woan, G., Observations of long-lived solar wind streams during 1990–1993, *Ann. Geophys., 13*, 227–236, 1995.

Woo, R., Spacecraft radio scintillation and solar system exploration, in *Wave propagation in random media (Scintillation)*, edited by V. Tatarskii, A. Ishimaru, and V. Zavorotny, pp. 50–83, SPIE, Bellingham, Washington, 1993a.

Woo, R., Solar cycle variation of interplanetary disturbances observed as Doppler scintillation transients, *J. Geophys. Res., 98*, 18999–19004, 1993b.

Woo, R., Coronal structures observed by radio propagation measurements, in *Proc. Solar Wind Eight*, edited by D. Winterhalter, J. Gosling, S.R. Habbal, W. Kurth, and M. Neugebauer, pp. 38–43, AIP, New York, 1996a.

Woo, R., Kilometre-scale structures in the Sun's corona, *Nature, 379*, 321–322, 1996b.

Woo, R., Evidence for the reversal of magnetic field polarity in coronal streamers, *Geophys. Res. Lett., 24*, 97–100, 1997.

Woo, R., and J.W. Armstrong, Measurements of a solar flare-generated shock wave at 13.1 R_0, *Nature, 292*, 608–610, 1981.

Woo, R., and P. Gazis, Large-scale solar-wind structure near the Sun detected by Doppler scintillation, *Nature, 366*, 543–545, 1993.

Woo, R., and S.R. Habbal, Finest filamentary structures of the inner corona in the slow and fast solar wind, *Astrophys. J., 474*, L139–142, 1997.

Woo, R., and R. Schwenn, Comparison of Doppler scintillation and *in situ* spacecraft plasma measurements of interplanetary disturbances, *J. Geophys. Res., 96*, 21227–21244, 1991.

Woo, R., J. W. Armstrong, N.R. Sheeley, Jr., R.A. Howard, D.J. Michels, and M.J. Koomen, Simultaneous radio scattering and white light observations of a coronal transient, *Nature, 300*, 157–159, 1982.

Woo, R., J.W. Armstrong, N.R. Sheeley, Jr., R.A. Howard, M.J. Koomen, and D.J. Michels, Doppler scintillation observations of interplanetary shocks within 0.3 AU, *J. Geophys. Res., 90*, 154–162, 1985.

Woo, R., J.W. Armstrong, M.K. Bird, and M. Pätzold, Fine-scale filamentary structure in coronal streamers, *Astrophys. J., 449*, L91–L94, 1995.

R. Woo, Jet Propulsion Laboratory, California Institute of Technology, Pasadena, CA 91109.

Particle and Field Signatures of Coronal Mass Ejections in the Solar Wind

Marcia Neugebauer and Raymond Goldstein

Jet Propulsion Laboratory, California Institute of Technology, Pasadena, CA

Abstract. When CME plasma passes over a spacecraft in interplanetary space, it can often be recognized by a number of characteristic signatures in the properties of the plasma and magnetic field. Those signatures are briefly summarized. When two or more signatures are present, they are often not synchronized with each other. As an example, the low-temperature signature is often encountered ahead of the bidirectional streaming of suprathermal electrons. Periods of quiet, nearly radial fields are found in the trailing portions of approximately one quarter of the fast or energetic CME events. It is suggested that the radial fields may be manifestations of the legs of magnetic loops carried into space by the CME. Another feature of the trailing portions of some CME events is a strong flux of outward propagating Alfvén waves. In some events these waves probably represent a return to the ambient solar wind through which the CME is propagating, but we suggest that in other events the waves may be a signature of a transient coronal hole at the footpoints of the CME.

There is no single distinctive feature exhibited by all the plasma clouds resulting from CMEs. Rather, the identification of CME plasma clouds in the interplanetary medium must rely on several features that may appear singly or jointly with other features in any particular event. What are those features or CME signatures?

(1) Unusually low kinetic temperatures of ions and/or electrons for a given solar-wind speed [*Gosling et al.*, 1973; *Montgomery et al.*, 1974]. In the quasi-stationary solar wind, ion temperatures are positively correlated with flow speed. Figure 1 shows a scatter plot of proton temperature versus speed for intervals of known flow type — open symbols (squares, circles, or triangles) for quasi-stationary flows and line symbols (x, + or -) for CME flows observed by ISEE 3. The method of event selection and identification is explained in a paper by *Neugebauer and Alexander* [1991]. A diagonal line has been drawn, by eye, to separate a region containing only CME points from a region with a mixture of types of flow. We have defined a "thermal index" I_{th}, by

$$I_{th} = (500v_p + 1.75 \times 10^5)/T_p \qquad (1)$$

which is >1 for points below the line and <1 for points above it. If $I_{th} > 1$, the plasma is likely to be associated with a CME, but if $I_{th} < 1$, the flow might or might not be CME-associated. This method of identifying CME plasma is qualitatively similar to the method developed and used by *Richardson and Cane* [1995] and others. The reason that the CME plasma has a lower temperature, or a higher Mach number, than the ambient, quasi-stationary wind is probably that it pushes aside the ambient wind to expand into a larger volume than it would occupy if it were channeled flow like the quasi-stationary wind. This effect is illustrated by the cartoon in Figure 2.

(2) An unusually pronounced anisotropy of the proton distribution with $T_\parallel > T_\perp$ [*Zwickl et al.*, 1983; *Gosling et al.*, 1987] caused by the conservation of the magnetic moment of the ions as the plasma expands.

246 INTERPLANETARY SIGNATURES

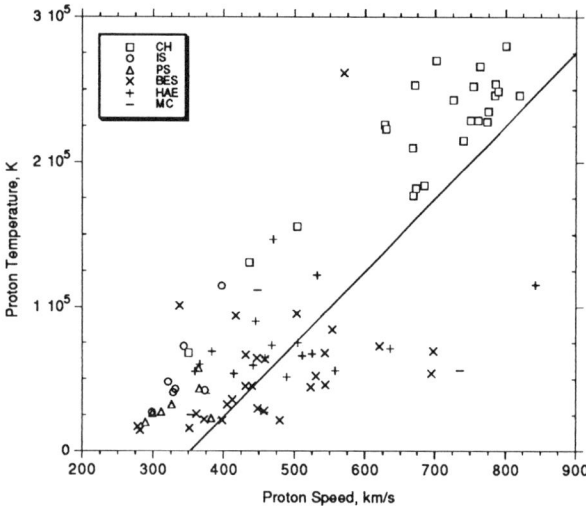

Figure 1. Scatter plot of proton temperature versus proton speed for intervals clearly identified as flow associated with either coronal holes (CH), interstream (IS), heliospheric plasma sheet (PS), bidirectional electron streaming (BES), helium abundance enhancements (HAE), or magnetic clouds (MC). The line is the trace of equation (1) for $I_{th} = 1$.

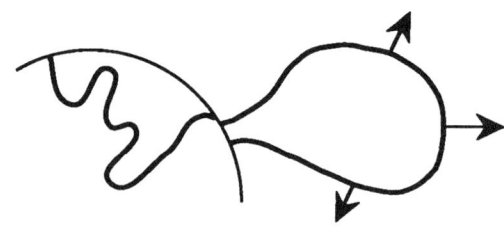

Figure 2. A cartoon depicting the difference in the modes of expansion of the channeled flow in the quasi-stationary wind and the more explosive expansion of CMEs.

(3) Unusually high helium abundance [*Hirshberg et al.*, 1972; *Borrini et al.*, 1982]. The ratio of the number density of helium n_α to that of protons n_p hardly ever exceeds ~0.08 except within or in the vicinity of an interplanetary cloud generated by a CME. The occurrence of high helium abundance is very patchy, and sometimes appears outside of what would otherwise be considered to be the CME cloud proper; see, for example, the appearance of $n_\alpha/n_p > 0.08$ just prior to the interplanetary shock in Figure 3. The cause of the helium enhancements has not yet been established theoretically. One possibility is that it is a sludge removal phenomenon wherein helium left behind at the base of the flow tubes, especially those in the slow wind near the heliospheric current sheet, is cleaned out by the explosive event.

(4) Anomalies in the abundances of other ion species [*Bame et al.*, 1979; *von Steiger et al.*, 1992; *Galvin et al.*, 1987]. This topic is reviewed in the paper by *Galvin* [this volume].

(5) Bidirectional streaming of suprathermal electrons [*Montgomery et al.*, 1974; *Temnyi and Vaisberg*, 1979; *Bame et al.*, 1981; *Gosling et al.*, 1987; *Pilipp et al.*, 1987] and energetic ions [*Rao et al.*, 1967; *Palmer et al.*, 1978; *Kutchko et al.*, 1982; *Sarris and Krimigis*, 1982; *Sanderson et al.*, 1983; *Marsden et al.*, 1987; *Tranquille et al.*, 1987; *Richardson and Reames*, 1991; 1993; *Richardson*, this volume]. This feature is considered to be indicative of a closed magnetic configuration with both footpoints of the field lines rooted in the Sun.

(6) Quiet, strong magnetic fields which, when combined with the low temperatures, leads to low β [*Hirshberg and Colburn*, 1969; *Burlaga and King*, 1979; *Burlaga et al.*, 1981; *Neugebauer*, 1983; *Gosling et al.*, 1987]. In CME plasma, the proton β is often less than 0.1.

(7) Rotations of the magnetic field that can be modeled as flux ropes [*Burlaga et al.*, 1981; *Klein and Burlaga*, 1982; *Lepping et al.*, 1990] Some of these configurations qualify to be called "magnetic clouds" if the field strength increases by a factor > 2, if at least one component of the field has a large, smooth rotation, and if the ion temperature is low. Magnetic clouds and flux ropes are discussed further in other papers [*Osherovich; Farrugia; Marubashi; Bothmer and Rust*, all in this volume].

(8) Decreased fluxes of low energy cosmic rays [*Barnden*, 1973; *Marsden et al.*, 1987; *Cane*, 1988; *Sanderson et al.*, 1990; *Cane*, 1993]. This topic is also covered by *Richardson* [this volume].

(9) Unusual ionization states of heavy ions [*Bame et al.*, 1979; *Fenimore*, 1980; *Gosling et al.*, 1980; *Schwenn et al.*, 1980; *Zwickl et al.*, 1982; *Bochsler*, 1983; *Ipavich et al.*, 1992; *Galvin et al.*, 1993], indicative of a plasma source in either hot coronal loops or (very occasionally) in relatively cold prominence material. See the paper by *Galvin* [this volume] for further discussion.

To this list of CME signatures, we can add other features that are associated with the more energetic events which result in plasma flows significantly faster than the ambient, quasi-stationary solar wind. These features include a forward shock ahead of the plasma cloud, a sheath of compressed, noisy plasma between the shock and the cloud, the draping of the interplanetary magnetic field around the cloud, and local maxima (sometimes spikes) in the pressure and density at the cloud's leading edge.

Figure 3 illustrates some of the signatures and features discussed above for a CME event observed by ISEE 3 on days 232-234 (Aug 20-22), 1979. The event started with an interplanetary shock denoted by the vertical line labeled S. The sheath, with its increased temperature, density, pressure, and field strength, is shown between the shock and the discontinuity D1, where the spacecraft entered the CME cloud. Note the spike in proton density at D1, the start of an interval of high helium abundance (n_α/n_p), and a high value of the thermal index I_{th}. The horizontal bar in the top panel denotes the interval of bidirectional streaming of suprathermal electrons (J. T. Gosling, personal communication). This event did not contain a magnetic cloud. If the CME plasma is assumed to extend from D1 to D3, where I_{th} is high, the patchiness of the helium abundance enhancements is quite evident.

Since not all the CME signatures are synchronized with each other, some appearing at different times than others or not at all, it is of interest to determine if there are any significant temporal patterns. We are currently studying this question and can show only a few preliminary results here. Figure 4A shows a superposed epoch histogram of the fraction of each hour that bidirectional streaming of suprathermal electrons was observed by ISEE 3 (based on the data in the list given by *Gosling et al.* [1987]), where the zero epoch is the hour in which the bidirectional streaming started. It is seen that the duration of bidirectional electron streaming is typically 8 to 10 hours. Figure 4B shows the fraction of hours with $I_{th} > 1$ as a function of time before and after the same zero-epoch times as in Figure 4A. It is clearly seen that the $I_{th} > 1$ indicator extends for ~2 days after the zero epoch, thus lasting much longer than the bidirectional electron streaming. This pattern is consistent with the field lines in the CME plasma reconnecting with the ambient field in the manner postulated by *Gosling et al.* [1995a].

The horizontal line in figure 4B represents the fraction of all hours for which ISEE-3 velocity and temperature were available and for which $I_{th} > 1$; i.e., on average, I_{th} exceeds unity 6.4% of the time. It is probably significant that I_{th} is greater than average for the day preceding the zero epoch time at which the bidirectional streaming starts. In many individual events, intervals of $I_{th} > 1$ are seen before the appearance of bidirectional streaming. Figure 4C shows a histogram similar to those in Figures 4A and 4B, except that the zero epoch time is chosen as

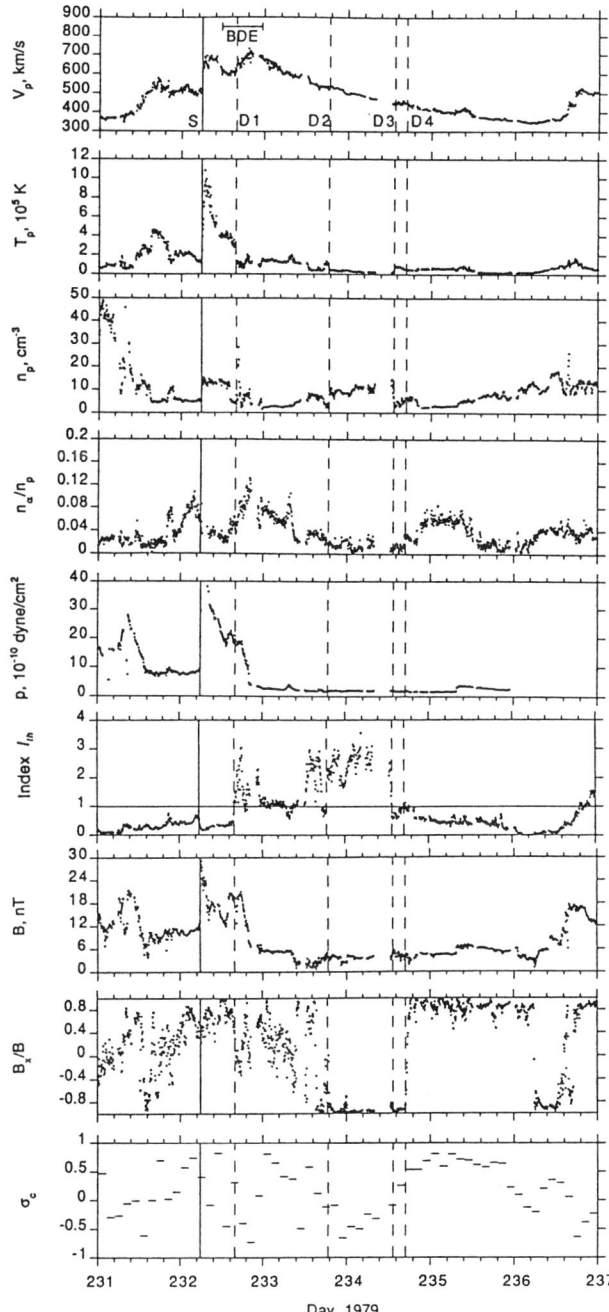

Figure 3. From top to bottom, the parameters plotted are proton speed, proton temperature, proton density, the ratio of the alpha-particle to proton densities, the total gas plus magnetic pressure in the plasma frame, an index I_{th} defining the relation between proton temperature and speed as defined in Equation (1), the magnetic field strength, the ratio of the radial component of the field divided by field strength, and the normalized helicity as defined by Equation (2). Each plasma data point is a 5-minute sample and each field data point is a 5-minute average.

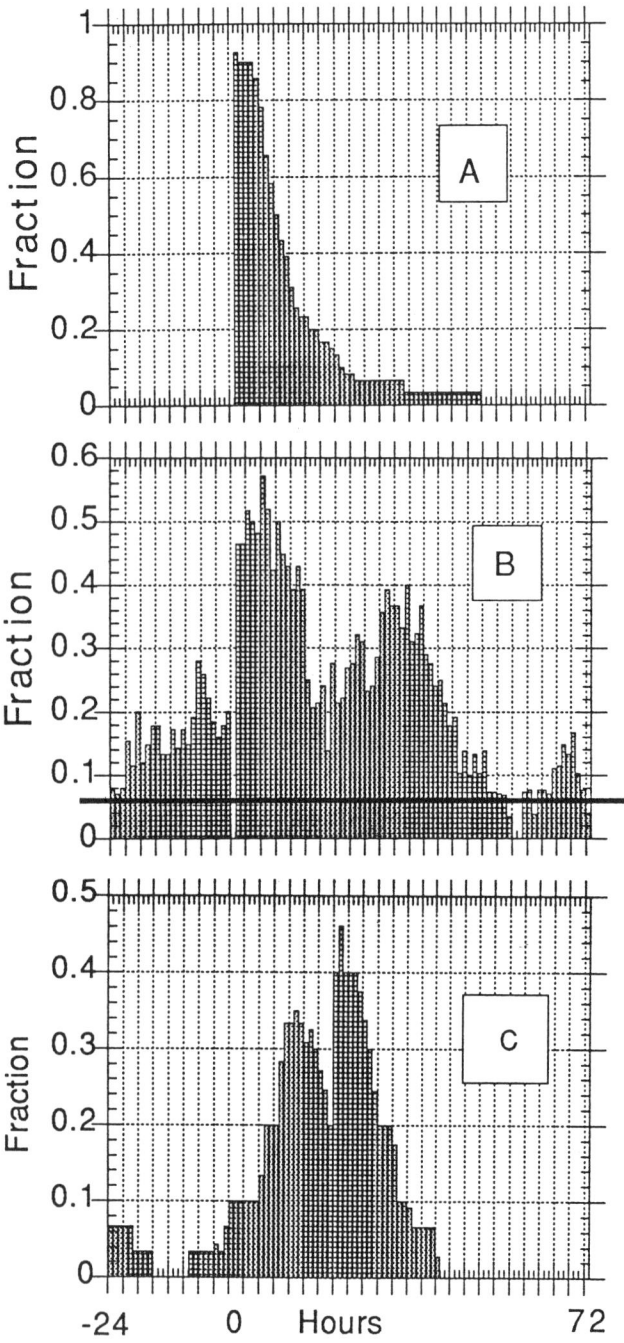

Figure 4. Histograms of the frequency of detection of bidirectional streaming of suprathermal electrons (BDES) and of the occurrence thermal index $I_{th} > 1$ in a superposed epoch format. (A) Frequency of detection of BDES as a function of time from a zero epoch defined by the first detection of BDES. (B) Frequency of $I_{th} > 1$ with zero epoch defined by start of BDES. (C) Frequency of detection of BDES with zero epoch defined by start of $I_{th} > 1$.

the start of $I_{th} > 1$. In panel 4C, it can be seen that the bidirectional electron streaming generally starts ≥ 10 hours into the CME as identified by $I_{th} > 1$. This result suggests that the fields at the leading edges or the noses of the CME clouds tend to be reconnected to the interplanetary field. This is not surprising because it is at the nose where the CME and interplanetary fields are most strongly pressed together, at least for that part of its trajectory where the CME is faster than the ambient wind.

One feature of fast CMEs that has not been previously reported is the appearance of extended intervals of quiet radial fields in their trailing edges. Figure 3 contains an example of this phenomenon; the next to bottom panel shows that B_x/B, where B_x is the radial component of the field in heliographic solar ecliptic coordinates, was close to -1 between D2 and D4 (days 233.8-234.7). We have found intervals with $|B_x|/B > 0.9$ for ≥6 consecutive hours in the trailing portions of about one quarter of the major CMEs observed by ISEE 3. About half the time, the quiet, radial fields overlapped other CME signatures such as high n_α/n_p, high I_{th}, or bidirectional streaming (as in Figure 3), and the rest of the time they immediately followed those signatures. In every case, the CME speed was greater than the ambient wind speed and the quiet radial field was observed during a period of strongly decreasing solar wind speed. None of the radial-field events occurred in a CME that had been designated as a magnetic cloud by *Zhang and Burlaga* [1988]. We suggest that a CME-associated quiet radial field interval may mean the spacecraft is situated in a leg of a magnetic loop (or perhaps even a flux rope) whose leading edge has been stretched out into space by the fast plasma at the front of the cloud. A paper presenting further information on the relation of quiet radial fields to CMEs is in preparation.

Next, consider the bottom panel of Figure 3 which shows a plot of the normalized cross helicity σ_c

$$\sigma_c = \frac{2 < \delta v \cdot \delta B >}{< \delta v^2 + \delta B^2 >} \qquad (2)$$

where the sign of σ_c has been "corrected" for the sector structure such that $\sigma_c = +1$ indicates Alfvén waves propagating outward from the Sun. The values of σ_c were calculated from the variations of 5-minute values of velocity v and magnetic field B over periods of 2.4 hours. Figure 3 shows an extended period of high σ_c starting at D4 (where the quiet radial field ended) and persisting for over a day. Outward propagating Alfvén waves are prevalent in high-speed flow from coronal holes, and may be remnants of the wave field responsible for the acceleration of the wind in an open-field geometry. We therefore speculate that the waves seen on day 235 may have originated in a transient coronal hole created by the eruption of the CME. Examples of such transient coronal holes were shown in the poster paper presented by *Webb et al.* at the Chapman Conference on CMEs. They are interpreted as the foot-points of flux ropes evacuated by the CME. Similar intervals of intense wave activity can be found behind many CMEs, but there is a question whether they indicate a CME-associated feature or re-entry into the ambient solar wind. For the particular event shown in Figure 3, we believe the Alfvénic wave flux is part of the CME because (1) it occurs on the velocity gradient caused by the CME and (2) in the quasi-stationary wind, Alfvénic fluctuations are not commonly found at solar wind speeds as low as the ~400 km/s average speed observed during this particular Alfvénic event.

Almost everything discussed above has been based on the properties of CMEs observed near the ecliptic plane. There are both a number of similarities and some important differences between the CMEs described above and those studied at high latitudes by Ulysses. The similarities include the decreased ion temperatures, bidirectional streaming, and occasional flux ropes, magnetic clouds, and quiet radial field intervals. It is the differences that are more interesting. First, each of the high-latitude CME plasma clouds had roughly the same speed (>700 km/s) as the ambient quasi-stationary solar wind from the polar coronal holes [*Gosling et al.*, 1994a]. This finding may mean that, except for the most energetic CME events, once the CME plasma is released from the Sun, the same processes accelerate the transient and quasi-stationary winds. Alternatively, *Gosling and Riley* [1996] suggest that the acceleration of the CME plasma may be caused by the dynamic interaction of the slower plasma cloud with the higher-speed ambient wind ahead of and behind it. At low latitudes, where the CME plasma is embedded in slow solar wind, the magnetic field in the plasma cloud is strong and steady, showing little evidence of Alfvénic fluctuations, but the CME clouds in the fast, high-latitude wind are just as Alfvénic as the ambient plasma, suggesting an important role for wave acceleration of the high-latitude CME plasma

If the high-latitude CMEs have nearly the same speed as the ambient high-latitude wind, one might not expect them to be preceded by an interplanetary shock. In fact, however, many of the high-latitude CMEs are observed to be preceded by a forward shock and followed by a reverse shock [*Gosling et al.*, 1994b]. The driver for these shocks (or rather for a single 3-D shock which crosses the spacecraft twice) is the rapid expansion of the CME into the surrounding plasma rather than its overtaking slower plasma in its path.

Another departure of high-latitude CMEs from the typical behavior near the ecliptic is the absence of helium abundance enhancements. As reported by *Galvin* (this volume), the differences in ionization charge states between CMEs and quasi-stationary plasma is also smaller at high latitudes than near the ecliptic.

Gosling et al. [1995b] have compared the properties of a single CME observed by both IMP 8 near Earth and Ulysses at 54°S and a heliocentric distance of 3.5 AU. In addition to the differences discussed just above, this comparison shows that because of the higher speed at high latitude, the instantaneous shape of the plasma cloud must be very different from the familiar sketches or cartoons of CMEs.

SUMMARY

We have reviewed some of the features used to identify the interplanetary plasma associated with CMEs. These include low temperatures, strong fields and magnetic flux rope structures, unusual elemental abundances and charge states, bidirectional streaming of suprathermal electrons and energetic ions, and reduction of cosmic ray fluxes. The presence or absence of each of the signatures varies from one CME event to the next. We have found a tendency for the low-temperature signature to precede the bidirectional streaming signature, and to last longer. Two additional signatures, reported here for the first time, are periods of nearly radial magnetic fields and Alfvénic fluctuations in the trailing regions of many CMEs. Finally, observations by Ulysses revealed that CMEs within the flow from the polar coronal holes have higher speeds than their low-latitude counterparts and do not show the variations in ion abundances or charge states seen in low-latitude events.

Acknowledgments. We thank Doug Clay for his contribution to the compilation of the relative timing of different CME features. This research was performed at the Jet Propulsion Laboratory under a contract between the California Institute of Technology and the National Aeronautics and Space Administration.

REFERENCES

Bame, S. J., J. R. Asbridge, W. C. Feldman, E. E. Fenimore, and J. T. Gosling, Solar wind heavy ions from flare-heated coronal plasma, *Sol. Phys., 62*, 179, 1979.

Bame, S. J., J. R. Asbridge, W. C. Feldman, J. T. Gosling, and R. D. Zwickl, Bi-directional streaming of solar wind electrons >80 eV: ISEE evidence for a closed-field structure within the driver gas of an interplanetary shock, *Geophys. Res. Lett., 8*, 173, 1981.

Barnden, L. R., Forbush decreases 1966-1972: Their solar and interplanetary associations and their anisotropies, *Conf. Pap. Int. Cosmic Ray Conf. 13th, 2*, 1271, 1973.

Bochsler, P., Mixed solar wind originating from coronal regions of different temperatures, in *Solar Wind Five, NASA Conference Publication 2280*, edited by M. Neugebauer, pp. 613, National Aeronautics and Space Administration, Washington, DC, 1983.

Borrini, G., J. T. Gosling, S. J. Bame, and W. C. Feldman, Helium abundance enhancements in the solar wind, *J. Geophys. Res.*, *87*, 7370, 1982.

Bothmer, V., and D. M. Rust, The field configuration of magnetic clouds and the solar cycle, this volume.

Burlaga, L. F., and J. H. King, Intense interplanetary magnetic fields observed by geocentric spacecraft during 1963-1975, *J. Geophys. Res.*, *84*, 6633, 1979.

Burlaga, L., E. Sittler, F. Mariani, and R. Schwenn, Magnetic loop behind an interplanetary shock: Voyager, Helios, and IMP 8 observations, *J. Geophys. Res.*, *86*, 6673, 1981.

Cane, H. V., The large-scale structure of flare-associated interplanetary shocks, *J. Geophys. Res.*, *93*, 1, 1988.

Cane, H. V., Cosmic ray decreases and magnetic clouds, *J. Geophys. Res.*, *98*, 3509, 1993.

Farrugia, C. J., Recent work on modelling the global field line topology of interplanetary magnetic clouds, this volume.

Fenimore, E. E., Solar wind flows associated with hot heavy ions, *Astrophys. J.*, *235*, 245, 1980.

Galvin, A. B., Ion composition and charge states in solar transient related solar wind, this volume.

Galvin, A. B., G. Gloeckler, F. M. Ipavich, C. M. Shafer, J. Geiss, and K. Ogilvie, Solar wind composition measurements by the Ulysses SWICS experiment during transient solar wind flows, *Adv. Space Res.*, *13*, (6) 75, 1993.

Galvin, A. B., F. M. Ipavich, G. Gloeckler, D. Hovestadt, S. J. Bame, B. Klecker, M. Scholer, and B. T. Tsurutani, Solar wind iron charge states preceding a driver plasma, *J. Geophys. Res.*, *92*, 12069, 1987.

Gosling, J. T., J. R. Asbridge, S. J. Bame, W. C. Feldman, and R. D. Zwickl, Observations of large fluxes of He^+ in the solar wind following an interplanetary shock, *J. Geophys. Res.*, *85*, 3431, 1980.

Gosling, J. T., D. N. Baker, S. J. Bame, W. C. Feldman, R. D. Zwickl, and E. J. Smith, Bidirectional solar wind electron heat flux events, *J. Geophys. Res.*, *92*, 8519, 1987.

Gosling, J. T., S. J. Bame, D. J. McComas, J. L. Phillips, B. E. Goldstein, and M. Neugebauer, The speeds of coronal mass ejections in the solar wind at mid heliographic latitudes: Ulysses, *Geophys. Res. Lett.*, *21*, 1109, 1994a.

Gosling, J. T., J. Birn, and M. Hesse, Three-dimensional magnetic reconnection and the magnetic topology of coronal mass ejection events, *Geophys. Res. Lett.*, *22*, 869, 1995a.

Gosling, J. T., D. J. McComas, J. E. Phillips, L. A. Weiss, V. J. Pizzo, B. E. Goldstein, and R. J. Forsyth, A new class of forward-reverse shock pairs in the solar wind, *Geophys. Res. Lett.*, *21*, 2271, 1994b.

Gosling, J. T., D. J. McComas, J. L. Phillips, V. J. Pizzo, B. E. Goldstein, R. J. Forsyth, and R. P. Lepping, A CME-driven solar wind disturbance observed at both low and high heliographic latitudes, *Geophys. Res. Lett.*, *22*, 1753, 1995b.

Gosling, J. T., V. Pizzo, and S. J. Bame, Anomalously low proton temperatures in the solar wind following interplanetary shock waves -- Evidence for magnetic bottles?, *J. Geophys. Res.*, *78*, 2001, 1973.

Gosling, J. T., and P. Riley, The acceleration of slow coronal mass ejections in the high-speed solar wind, *Geophys. Res. Lett.*, *23*, 2867, 1996.

Hirshberg, J., S. J. Bame, and D. E. Robbins, Solar flares and solar wind helium enrichments: July 1965-July 1967, *Sol. Phys.*, *23*, 467, 1972.

Hirshberg, J., and D. S. Colburn, Interplanetary field and geomagnetic variations: A unified view, *Planet. Space Sci.*, *17*, 1183, 1969.

Ipavich, F. M., A. B. Galvin, J. Geiss, K. W. Ogilvie, and F. Gliem, Solar wind iron and oxygen charge states and relative abundances measured by SWICS on Ulysses, in *Solar Wind Seven*, edited by E. Marsch and R. Schwenn, pp. 369, Pergamon, Oxford, 1992.

Klein, L. W., and L. F. Burlaga, Interplanetary magnetic clouds at 1 AU, *J. Geophys. Res.*, *87*, 613, 1982.

Kutchko, F. J., P. R. Briggs, and T. P. Armstrong, The bi-directional particle event of October 12, 1977, possibly associated with a magnetic loop, *J. Geophys. Res.*, *87*, 1419, 1982.

Lepping, R. P., J. A. Jones, and L. F. Burlaga, Magnetic field structure of interplanetary magnetic clouds at 1 AU, *J. Geophys. Res.*, *95*, 11957, 1990.

Marsden, R. G., T. R. Sanderson, C. Tranquille, K.-P. Wenzel, and E. J. Smith, ISEE 3 observations of low-energy proton bidirectional events and their relation to isolated interplanetary magnetic structures, *J. Geophys. Res.*, *92*, 11009, 1987.

Marubashi, K., Interplanetary flux ropes and solar filaments, this volume.

Montgomery, M. D., J. R. Asbridge, S. J. Bame, and W. C. Feldman, Solar wind electron temperature depressions following some interplanetary shock waves: Evidence for magnetic merging?, *J. Geophys. Res.*, *79*, 3103, 1974.

Neugebauer, M., Observational constraints on solar wind acceleration mechanisms, in *Solar Wind Five*, edited by M. Neugebauer, pp. 135, NASA, Washington DC, 1983.

Neugebauer, M., and C. J. Alexander, Shuffling footpoints and magnetohydrodynamic discontinuities in the solar wind, *J. Geophys. Res.*, *96*, 9409, 1991.

Osherovich, V. A., and L. F. Burlaga, Magnetic clouds, this volume.

Palmer, I. D., F. R. Allum, and S. Singer, Bidirectional anisotropies in solar cosmic ray events: evidence for magnetic bottles, *J. Geophys. Res.*, *83*, 75, 1978.

Pilipp, W. G., H. Miggenrieder, M. D. Montgomery, K.-H. Mühlhäuser, H. Rosenbauer, and R. Schwenn, Characteristics of electron velocity distribution functions in the solar wind derived from the Helios plasma experiment, *J. Geophys. Res.*, *92*, 1075, 1987.

Rao, U. R., K. G. McCracken, and R. P. Bukata, Cosmic ray propagation processes, 2, The energetic storm particle event, *J. Geophys. Res.*, *72*, 4325, 1967.

Richardson, I. G., Using energetic particles to probe the

magnetic topology of ejecta, this volume.

Richardson, I. G., and H. V. Cane, Regions of abnormally low proton temperature in the solar wind (1965-1991) and their association with ejecta, *J. Geophys. Res., 100*, 23397, 1995.

Richardson, I. G., and D. V. Reames, Bidirectional MeV/n ion intervals: Observations from the Goddard Space Flight Center instruments on the ISEE 3/ICE, IMP 8, Helios 1 and Helios 2 spacecraft, *Conf. Pap. Int. Cosmic Ray Conf. 22nd, 3*, 292, 1991.

Richardson, I. G., and D. V. Reames, Bidirectional ~1 MeV/amu ion intervals in 1973-1991 observed by the Goddard Space Flight Center instruments on IMP 8 and ISEE 3/ICE, *Astrophys. J. Suppl. Ser., 85*, 411, 1993.

Sanderson, T. R., J. Beeck, R. G. Marsden, C. Tranquille, K.-P. Wenzel, R. B. McKibben, and E. J. Smith, Cosmic ray, energetic ion and magnetic field characteristics of a magnetic cloud, *Conf. Pap. Int. Cosmic Ray Conf. 21st, 6*, 255, 1990.

Sanderson, T. R., R. G. Marsden, R. Reinhard, K.-P. Wenzel, and E. J. Smith, Correlated particle and magnetic field observations of a large-scale magnetic loop structure behind an interplanetary shock, *Geophys. Res. Lett., 10*, 916, 1983.

Sarris, E. T., and S. M. Krimigis, Evidence for magnetic loops beyond 1 AU, *Geophys. Res. Lett., 9*, 167, 1982.

Schwenn, R., H. Rosenbauer, and K.-H. Mühlhäuser, Singly-ionized helium in the driver gas of an interplanetary shock wave, *Geophys. Res. Lett., 7*, 201, 1980.

Steiger, R. v., J. Geiss, G. Gloeckler, H. Balsiger, A. B. Galvin, U. Mall, and B. Wilken, Magnesium, carbon, and oxygen abundances in different solar wind flow types, as measured by SWICS on Ulysses, in *Solar Wind Seven*, edited by E. Marsch and R. Schwenn, pp. 399, Pergamon Press, Oxford, 1992.

Temnyi, V. V., and O. L. Vaisberg, A dumbbell distribution of epithermal electrons in the solar wind based on observations on the Prognoz 7 satellite, *Cosmic Res., 17*, 476, 1979.

Tranquille, C., T. R. Sanderson, R. G. Marsden, K.-P. Wenzel, and E. J. Smith, Properties of a Large-Scale Interplanetary Loop Structure as Deduced from Low-Energy Proton Anisotropy and Magnetic Field Measurements, *J. Geophys. Res., 92*, 6, 1987.

Zhang, G., and L. F. Burlaga, Magnetic clouds, geomagnetic disturbances, and cosmic ray decreases, *J. Geophys. Res., 93*, 2511, 1988.

Zwickl, R. D., J. R. Asbridge, S. J. Bame, W. C. Feldman, and J. T. Gosling, He$^+$ and other unusual ions in the solar wind: A systematic search covering 1972-1980, *J. Geophys. Res., 87*, 7379, 1982.

Zwickl, R. D., J. R. Asbridge, S. J. Bame, W. C. Feldman, J. T. Gosling, and E. J. Smith, Plasma properties of driver gas following interplanetary shocks observed by ISEE-3, in *Solar Wind Five; NASA Conference Proceedings 2280*, edited by M. Neugebauer, pp. 711, NASA, Washington, DC, 1983.

M. Neugebauer and R. Goldstein, Jet Propulsion Laboratory, 4800 Oak Grove Dr., Mail Stop 169-506, Pasadena, CA 91109

Minor Ion Composition in CME-Related Solar Wind

A. B. Galvin

Department of Physics, Univ. of Maryland, College Park, Maryland

Elemental and ionic charge state composition measurements play an important role in the identification of different structures in the solar wind. Solar wind associated with coronal mass ejections (CMEs) frequently exhibit helium and minor ion (Z > 2) elemental abundances (relative to protons) that are enhanced compared to nominal solar wind values. Another aspect of CME-related composition is an apparent enrichment of elemental abundances relative to photospheric values for elements with low (< 10 eV) first ionization potentials, the so-called "FIP effect". Charge states of CME-related solar wind usually indicate hotter than normal coronal conditions at the solar wind freezing-in site. However, the charge state signature varies among different events, and a high charge state signature within an event does not preclude the simultaneous presence of lower charge states that paradoxically suggest lower coronal temperatures. The compositional distinctions are not as prevalent in CME-related solar wind observed at higher heliographic latitudes, as measured by the Ulysses spacecraft. At these higher latitudes, the observed CME-related ionization temperatures and the He/H abundance ratios are closer to the prevailing polar coronal hole solar wind values.

1. GENERAL CHARACTERISTICS OF CMES IN THE SOLAR WIND

The solar transient phenomenon known as a coronal mass ejection (CME) is a special source of solar wind whose *in situ* manifestation is identified by the presence of one or more distinguishing signatures. The magnetic field within the interplanetary CME structure typically has a high and steady field strength [*Burlaga et al.*, 1981; *Pudovkin et al.*, 1977; *Zwickl et al.*, 1983]. The structure may be a "magnetic cloud", showing a smooth magnetic field rotation through a large angle [*Burlaga et al.*, 1981; *Klein and Burlaga*, 1982; *Smith*, 1983]. The solar wind usually exhibits abnormally depressed proton and electron kinetic temperatures [*Gosling et al.*, 1973; *Montgomery et al.*, 1974] and an increase in the ratio of the proton kinetic temperature's parallel to perpendicular components relative to the magnetic field direction [*Zwickl et al.*, 1983]. Plasma beta is typically low [*Neugebauer*, 1983].

The bidirectional electron (BDE) heat flux is an important signature [*Bame et al.*, 1981; *Gosling et al.*, 1987], although the BDE event may occur only during a portion of the entire CME-related interval. Bidirectional energetic proton events [*Marsden et al.*, 1987] and bidirectional solar energetic electrons [*Palmer et al.*, 1978] may be present, but are not always coincident with the BDE.

Solar wind speeds may be high, low, or nominal, but will show a small decrease in fluctuations within the event proper [*Zwickl et al.*, 1983]. Solar wind minor ions in CME-related solar wind move at similar velocities to the protons [*Ipavich et al.*, 1986]. When the speed of the expelled CME is sufficiently high (supersonic) compared to that of the overtaken ambient solar wind, a shock forms in front of the CME (hence the historical terms "post shock flow" and "driver plasma" for this type of CME-related solar wind).

It is rare to have all these features present within a given solar wind interval. (The CME event described by *Galvin et al.* [1987] is one of the notable exceptions.) Nor are

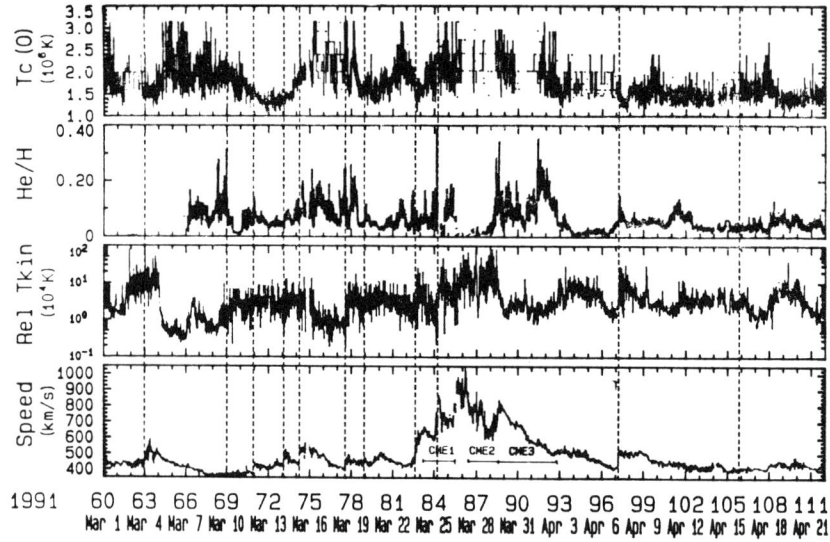

Figure 1. Solar wind plasma parameters (speed, relative kinetic temperature, He/H abundance ratios, and oxygen ionization temperatures) for the March - April 1991 time interval (taken from *Galvin et al.* [1993]). Several shocks were observed [*Burton et al.*, 1992], shown here as dashed lines. The identification of three BDEs, as marked in the bottom panel, were reported by *Phillips et al.* [1992]. The term "relative" kinetic temperature refers to the fact that SWICS does not have full directional information and therefore does not provide the standard defined value for kinetic temperature.

these features when taken individually necessarily unique to CME-related solar wind. Hence, it is useful to have additional tools available in the "art" of CME identification. Elemental and charge state compositions of minor ions in CME-related solar wind provide significant telltales. Even more fundamentally, the solar wind compositions (both charge state and elemental) yield information on the solar processes involved at the coronal site where the solar wind is generated and accelerated, and hence may provide clues regarding the solar CME process itself.

2. ELEMENTAL COMPOSITION SIGNATURES

2.1 Elemental Composition Relative to Protons

Before it became established that coronal mass ejections are the responsible solar source for this type of solar wind, one early solar wind classification scheme keyed in on unusual aspects of the *in situ* compositional signature: hence the categorization as "HAEs", Helium Abundance Enhancements [e.g., *Hirshberg et al.*, 1972]. Helium to proton density ratios in CME-related flows are frequently enhanced over nominal solar wind values. For coronal hole solar wind, the He/H density ratio is ~ 4 to 5%. Interstream solar wind is much less constant, but has typical values that are low, dipping to ~3% at sector boundaries [e.g., *Wimmer*, 1994; *von Steiger et al.*, 1995]. The He/H signature is highly variable among CME-related events, and is frequently non-uniformly distributed within a given event, but He/H values greater than 8% are often observed [*Hirshberg et al.*, 1970, 1972; *Bame et al.*, 1979; *Borrini et al.*, 1982; *Zwickl et al.*, 1983]. Figure 1 shows several intervals occurring during March - April 1991 in which the He/H density ratio frequently exceeds 10% [*Galvin et al.*, 1993]. An entire series of CME-related interplanetary shocks are present [*Burton et al.*, 1992]. Higher than average charge states are observed for the minor ions during the CME-related periods, illustrated in this figure by the oxygen ionization temperatures. (These ionization temperatures are derived from the relative abundance of oxygen ions with charge states +6 and +7 [*Arnaud and Rothenflug*, 1985].) It is well established that periods of enhanced He/H exhibit heavy ion charge states indicative of hotter than normal conditions in the corona [*Fenimore*, 1980].

Minor ions (Z>2) in CME-related solar wind also exhibit enrichment relative to protons, an effect that has been most extensively reported for iron. *Ipavich et al.* [1986] observe Fe/H density ratios of 135×10^{-6} and 100×10^{-6} for two CME-related events (compared to values of $32-33 \times 10^{-6}$ for two coronal hole periods). *Mitchell et al.* [1983] report Fe/H flux ratios enhanced by a factor of 4 to 5 over coronal hole solar wind. *Bame et al.* [1979] report an average Fe/H ratio that is ~2 times higher than interstream solar wind measurements.

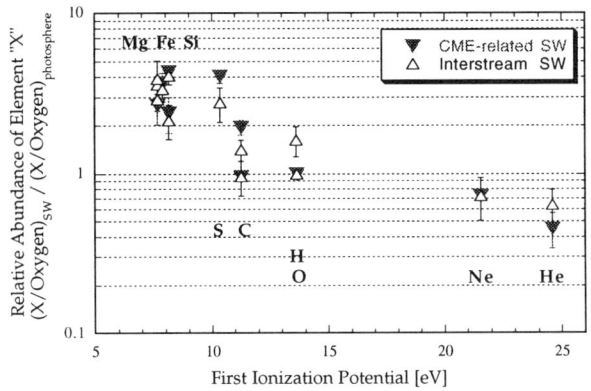

Figure 2. Interstream and CME-related solar wind abundances, normalized to oxygen and photospheric values, as a function of first ionization potential (FIP). (Data in the figure are compiled from *Ogilvie et al.* [1992]; *Galvin et al.* [1992, 1993]; *Ipavich et al.* [1992]; *von Steiger et al.* [1992, 1995]; *Cohen* [1995]; and references therein.)

The reason for the helium and minor ion enrichments, and the local variability within and among events, is not completely understood, but has generally been interpreted as evidence that the plasma originates in the low corona, where local, non-homogeneous enrichments could be expected to occur as a result of dynamic fractionation or static stratification [e.g., *Geiss et al.*, 1995a].

2.2 Elemental Composition Relative to Photospheric Abundances

First reported as a compositional trend observed in solar energetic particles [*Hovestadt*, 1974], it is now well established [cf., *Meyer*, 1993] that elemental abundances in interstream solar wind are subject to the so-called "FIP-effect", in which elements with a first ionization potential less than ~10 eV are enriched by a factor of ~3 to 4 compared to photospheric values. Elements with FIP above ~10 eV are close to photospheric values (helium is often a notable exception). Elements such as sulfur and carbon, which have FIPs near 10 eV, typically show in-between "transitional" values. The 10 eV "breakpoint" is likely related to the Lyman-alpha energy of hydrogen (10.2 eV). The FIP effect is not a universal feature of all solar wind. The effect is much reduced or perhaps absent in coronal hole solar wind [*Geiss et al.*, 1995a].

Abundance ratios for CME-related solar wind have been reported for selected elements and are shown in Figure 2 along with interstream values. The pattern indicates the FIP effect is operating within CME-related solar wind, as it is for interstream. However, as discussed in a later section, there are occasional observations in CME-related solar wind of such low charge states that an influx of prominence material is suspected. If this is the case, there may be

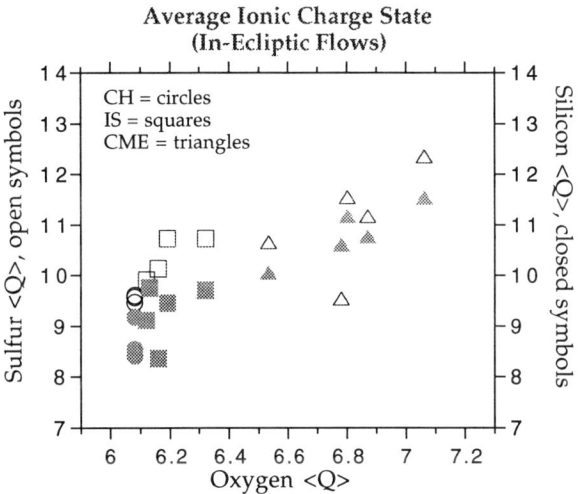

Figure 3. Average charge states of sulfur and silicon plotted against the corresponding oxygen average charge state, for three categories of solar wind flow. (Derived from *Cohen* [1995].)

instances in CME-related solar wind of "FIP-free" composition. This possibility is under investigation (R. Neukomm, private communication, 1995).

Several theories have been proposed to explain the FIP effect. Essentially the various models invoke ways to affect the atom-ion separation process in the chromosphere before the chromospheric material is fed into the corona. Whether the relevant physical or atomic characteristic is actually the first ionization potential, or the first ionization time, or the ionization diffusion length, or some other parameter [*e.g., Peter*, 1996] is an active area of modeling efforts. (See discussions by *Geiss et al.* [1994]; *von Steiger* [1996], and references therein.)

3. CHARGE STATE COMPOSITION SIGNATURES

The relative ionization states of ions in the corona depend on the local electron temperature and density, the ions' collisional ionization and radiative and dielectronic recombination rates, and the ions' outflow velocities [*Hundhausen et al.*, 1968]. The local coronal density and temperature change with altitude. As the solar wind expands outward, the coronal electron density decreases to the extent that the solar wind ion expansion time scale is short compared to the ionization and recombination time scales. The relative ionization states become constant, forever reflective of the conditions at the freezing-in altitude. The solar wind ions maintain their chemical and charge state identity as they continue to propagate through the outer solar atmosphere and into interplanetary space. Hence the charge state composition of *in situ* solar wind is used to infer coronal electron temperatures.

256 MINOR ION COMPOSITION IN CMES

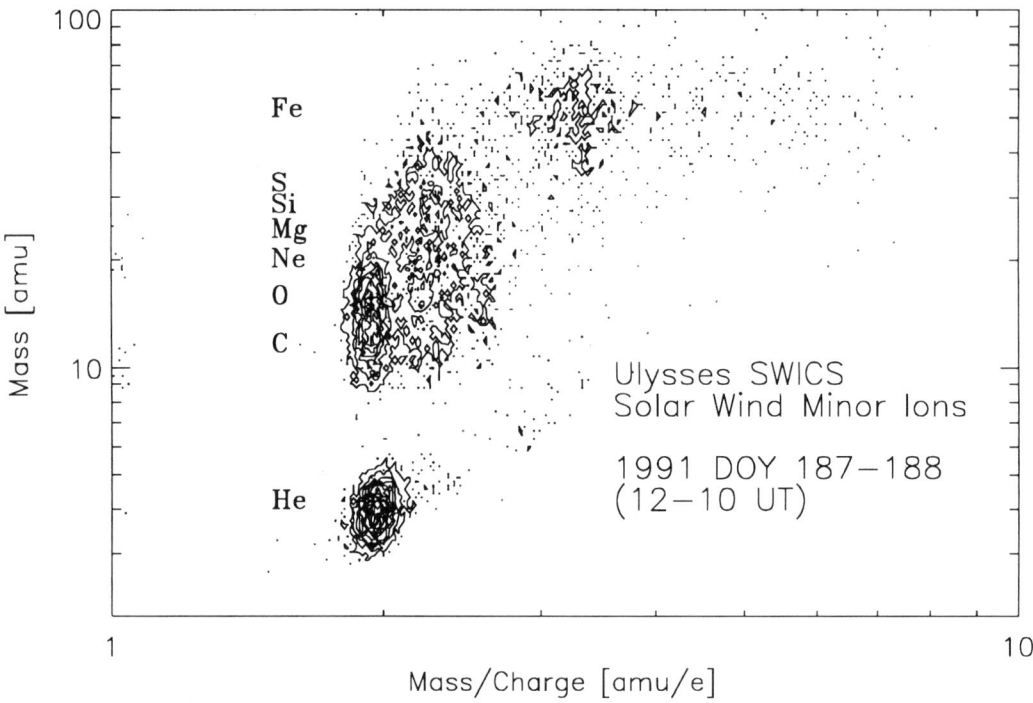

Figure 4. Minor ion composition during a particularly "hot" CME-related solar wind. The data are accumulated from 1200 UT Doy 187 until 1000 UT Doy 188, 1991. Mass is plotted against Mass per Charge, with the relative intensity indicated by contours. Protons are not shown. Oxygen and carbon are mainly fully stripped (+8, +6, respectively). Silicon is principally +12. Iron peaks near +17.

This is overly simplistic, perhaps especially so in the case of coronal mass ejections. One should consider whether the CME-related solar wind expansion is too rapid for local thermodynamic equilibrium to be attainable; whether the expansion time profile differs for CMEs [*Fenimore*, 1980]; whether the local electron velocity distribution is non-Maxwellian [*Owocki and Scudder*, 1983]; whether coronal shock effects are involved [*Owocki and Hundhausen*, 1983]; whether in some cases photo-ionization occurs as a result of nearby flare activity [*Mullan and Waldron*, 1986]. These scenarios require a significant modeling effort. When this paper refers to ionization temperatures, the standard tables are used [*Arnaud and Rothenflug*, 1985; *Arnaud and Raymond*, 1992]. The reader is warned that this is just a convenience for illustrating differences in charge state composition.

The charge state composition is often useful in distinguishing the type of solar wind flow. For example, although both coronal hole and CME-related solar winds have been observed with high speeds, the coronal hole oxygen charge states indicate temperatures of about 1.3 MK, while the CME-related solar wind is typically above 2 MK. The charge states of other minor ions typically track the oxygen temperatures, as seen in Figure 3. The average sulfur and silicon charge states are plotted against the average oxygen charge state for three types of solar wind (coronal hole, interstream, and CME-related). The data are taken during the in-ecliptic phase of the Ulysses mission [*Cohen*, 1995]. The coronal hole solar wind has the lowest observed charge states. There is some scatter among the interstream flows and the CME-related solar wind, however on average the CME-related events are "hotter" than the interstream events, which in turn are "warmer" than coronal hole events.

High ionization states are indicative of CME origins. (However, the inverse of that statement is not always true. Some CME-related solar wind exhibit nominal charge states, or occasionally even abnormally low charge states.) Iron charge states of +15, +16 have been reported [*Bame et al.*, 1979; *Ipavich et al.*, 1986], and even a case with Fe +17, +18 [*Fenimore*, 1980]. Shown in Figure 4 is a CME-related case with exceptionally high charge states. The oxygen ionization state is predominately charge state +8 (the ionization temperature from the +7 and +6 charge state ratio is above 3 MK); carbon is mostly +6. This is an instance where a traditional solar wind energy-per-charge analyzer instrument would see most of the oxygen and carbon "disappear" under the mass-per-charge peak at M/q=2, which is dominated by doubly-charged helium. As seen in Figure 5, the iron peaks at charge state +17, but the spectrum extends to +19 [*Liu*, 1994]. Notice that lower iron charge states (+8, +9, +10) are present. CME-related

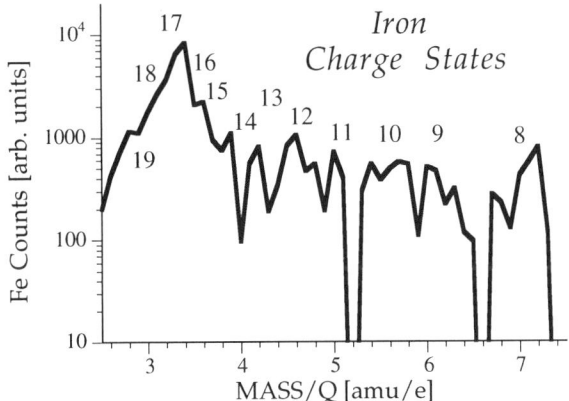

Figure 5. The iron charge state distribution for the time period shown in Figure 4. Iron peaks at about +16, +17, +18, but exhibits a broad range of charge states, including charge states +8, +9, +10.

solar wind frequently exhibits a range of charge states within the same event [*Fenimore*, 1980]. *Galvin et al.* [1993] report a CME-related solar wind in which high charge states are observed for silicon (e.g., +12) along with lower charge states (e.g., +7, +8, +9), and a similar trend is seen for iron in the same period. *Ipavich et al.* [1986] report events where the iron shows a mixed contribution by both high and lower charge states. *Fenimore* [1980] has suggested a CME expansion model which assumes an expansion time increasing with height (opposite to the normal assumption), resulting in a wide spread of freezing-in altitudes, and hence different representative temperatures. More work needs to be done in this area.

4. THE SHEATH REGION

If the CME is traveling fast enough relative to the ambient solar wind, a shock forms. The shock defines the leading edge of a "sheath region" of accelerated, heated and compressed ambient solar wind. This region is characterized by turbulent and increased plasma flow speed and magnetic field strength [e.g., *Pudovkin et al.*, 1977; *Burlaga et al.*, 1981; *Borrini et al.*, 1982]. When observed at the earth's orbit (1 AU), the sheath region typically endures for 6-18 hours after the shock passage, with an average thickness of 0.14 AU [*Borrini et al., 1982*]. The sheath region is followed by a contact surface (a tangential discontinuity) which may mark the boundary with the CME-related solar wind. Sheath He/H abundances are usually nominal solar wind values [*Hirshberg et al.*, 1970; *Borrini et al.*, 1982]. However, compositional signatures can be mixed for both elemental abundances and charge states [*Zwickl et al.*, 1983; *Galvin et al.*, 1987]. There is no definitive interpretation of this phenomenon, but it may indicate precursor activity, extended photo-ionization at the acceleration site, mixed solar origins, or some other scenario entirely.

5. RARE CHARGE STATES

The expected He^+/He^{++} solar wind abundance ratio for a million-degree corona is on the order of 10^{-6} [*Feldman et al.*, 1974; *Schwenn*, 1986]. Upper limits of 5×10^{-4} have been reported for selected time intervals in low speed solar wind [*Geiss et al.*, 1992]. It is important to distinguish solar He^+ from other possible origins [*Feldman et al.*, 1974], as interstellar pickup He^+ ions have been observed [*Möbius et al.*, 1985]. Nonetheless, there have been reported cases, albeit rare, of CME-related solar wind containing singly charged He, with observed He^+/He^{++} ranging from less than 1% to as much as 30% [*Bame et al.*, 1968; *Gosling et al.*, 1980; *Schwenn et al.*, 1980; *Zwickl et al.*, 1982; and review by *Bame*, 1983].

Schwenn [1983] reports CME-related solar wind associated with an eruptive prominence in which not only He^+ was observed, but also O^{+2}. An extensive survey conducted by *Zwickl et al.* [1982] incorporating nearly eight years of IMP-7, IMP-8, ISEE-1 and ISEE-3 observations found only three distinct He^+ events. This study observed other minor ions with unusually low charge states (for any type of solar wind) in the same events, such as Fe charge states extending as low as +5, O charge states +2 and +3, and C charge state +4. In a three year (1991-1993) study with Ulysses, no discernible solar wind He^+ events have been observed, and a resultant upper limit of 2.4×10^{-4} has been estimated [Mall, Gloeckler, and Geiss, private communication, 1995]. Some interesting and suggestive mass-per-charge peaks have been seen in a recent CME-related solar wind event observed by WIND instruments on January 11, 1997, however the analysis is still preliminary.

It has been suggested that the low charge states may be prominence material, and therefore chromospheric in origin. If so, this would be an area to look for the "FIP-free" elemental composition.

6. LATITUDINAL VARIATIONS

With the unique solar polar orbit of Ulysses, it has been possible to make the first latitudinal measurements of the composition of the solar wind. The results have been surprising. *Gosling et al.* [1994] determined that the high latitude CME-related solar wind always has high speeds comparable to that of the polar hole solar wind in which it is embedded (typically between 700-800 km/s). *Phillips et al.* [1995] report that the CME-related He/H abundance ratios become close to the ambient polar hole values (i.e., close to 5%) at high latitudes. Similar results are observed by the solar wind composition experiment: the CME-related solar wind minor ion charge states at the higher

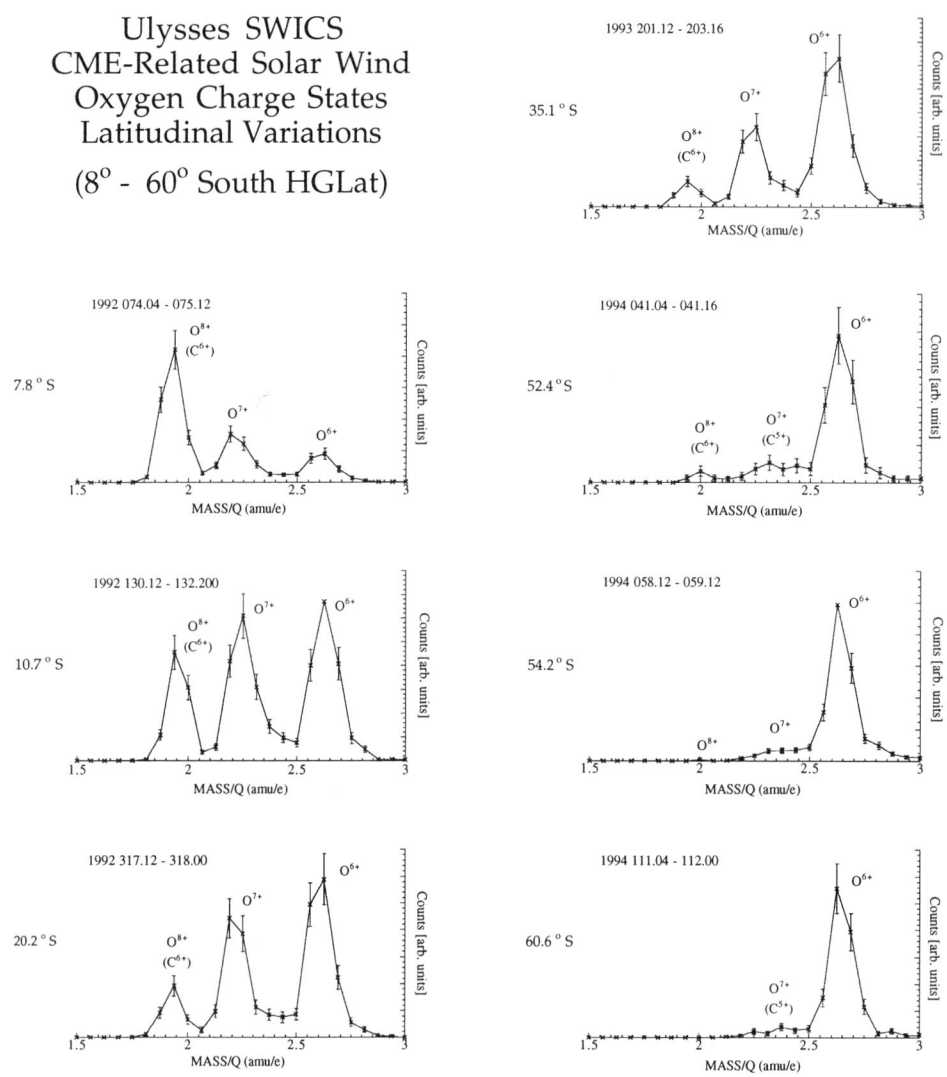

Figure 6. Oxygen charge states observed for individual CME-related solar wind time periods. The panels correspond (from top left to bottom right) to higher heliographic latitudes, as the Ulysses spacecraft headed from the ecliptic plane towards the south pole of the sun. The CME oxygen is predominantly +8 near the ecliptic, eventually being dominated by the +6 charge state at higher (i.e., poleward) latitudes.

latitudes, that is, poleward of 38° S, are very similar to the polar coronal hole solar wind charge state values. The oxygen ionization temperatures are ~ 1.2-1.4 MK in the CME-related events, compared to 1.1-1.2 MK for the ambient polar coronal hole wind [*Galvin et al.*, 1995; *Geiss et al.*, 1995b]. As seen in Figure 6, the charge state transition occurs systematically as more poleward heliographic latitudes are reached by the spacecraft. Silicon and iron behave similarly to oxygen. CME-related silicon typically peaks near +12 for in-ecliptic events, while the Si charge state +8 is more dominant for high latitude CME-related events. Iron shows a transition from in-ecliptic events normally peaking around charge state +16, while at the high latitudes the dominant charge states are +10, +11.

The underlying cause of the high latitude composition changes are not yet understood. Possibilities include a cooler solar origin or the involvement of a different kind of CME (e.g., polar arcades *vs.* erupting prominences). Perhaps high latitude observations are more likely to cover only the outer edge of the CMEs, as the solar wind from the polar coronal hole may push the main body of the structures towards the ecliptic, away from the spacecraft. It will be interesting to pursue the high latitude CME studies during the Ulysses' second set of polar passes in 2000 and 2001, which occur close to solar maximum conditions.

Acknowledgments. I thank the SWICS Principal Investigators, G. Gloeckler and J. Geiss, and other members of the SWICS team, and C.M.S. Cohen, who left a legacy of plotting programs as well as her thesis behind. I thank the organizers and other participants of the Chapman Conference on CMEs, in particular I acknowledge useful discussions with E. Hildner. This work was supported under NASA/JPL Contract 955460.

REFERENCES

Arnaud, M., and R. Rothenflug, An updated evaluation of recombination and ionization rates, *Astron. Astrophys. Suppl. Ser., 60*, 425, 1985.

Arnaud, M., and J. Raymond, Iron ionization and recombination rates and ionization equilibria, *Astrophys. J., 398*, 394, 1992.

Bame, S.J., A.J. Hundhausen, J.R. Asbridge, and I.B. Strong, Solar wind ion composition, *Phys. Rev. Lett., 20*, 393, 1968.

Bame, S.J., Solar wind minor ions - recent observations, in *Solar Wind Five, NASA Conf. Publ. CP-2280*, ed. M. Neugebauer, NASA Scientific and Technical Information Branch, 573 pp., 1983.

Bame, S. J., J. R. Asbridge, W.C. Feldman, E. E. Fenimore, and J. T. Gosling, Solar wind heavy ions from flare-heated coronal plasma, *Solar Phys., 62*, 179, 1979.

Bame, S. J., J. R. Asbridge, W. C. Feldman, J. T. Gosling, and R. D. Zwickl, Bidirectional streaming of solar wind electron > 80 eV; ISEE evidence of a closed-field structure within the driver gas of an interplanetary shock, *Geophys. Res. Lett., 8*, 173, 1981.

Borrini, G., J. T. Gosling, S. J. Bame, and W. C. Feldman, An analysis of shock wave disturbances observed at 1 AU from 1971 through 1978, *J. Geophys. Res., 87*, 4365, 1982.

Burlaga, L., E. Sittler, F. Mariani, and R. Schwenn, Magnetic loop behind an interplanetary shock: Voyager, Helios, and IMP 8 observations, *J. Geophys. Res., 86*, 6673, 1981.

Burton, M. E., E. J. Smith, B. E. Goldstein, A. Balogh, R. J. Forsyth, and S. J. Bame, Ulysses: interplanetary shocks between 1 and 4 AU, *Geophys. Res. Lett., 19*, 1287, 1992.

Cohen, C.M.S., Measurements of solar wind sulfur abundance and charge states, Ph. D. dissertation, University of Maryland, 1995.

Feldman, W. C., J. R. Asbridge, S. J. Bame, and P.D. Kearney, Upper limits for solar wind He+ content at 1 AU, *J. Geophys. Res., 79*, 1808, 1974.

Fenimore, E. E., Solar wind flows associated with hot heavy ions, *Astrophys. J., 235*, 245, 1980.

Galvin, A.B., F.M. Ipavich, G. Gloeckler, D. Hovestadt, S. J. Bame, B. Klecker, M. Scholer, and B. T. Tsurutani, Solar wind iron charge states preceding a driver plasma, *J. Geophys. Res., 92*, 12069, 1987.

Galvin, A.B., F. M. Ipavich, G. Gloeckler, R. von Steiger, and B. Wilken, Silicon and oxygen charge state distributions and relative abundances in the solar wind measured by SWICS on Ulysses, in *Solar Wind Seven, Cospar Colloquia Series, 3*, ed. E. Marsch and R. Schwenn, 337 pp., 1992.

Galvin, A. B., G. Gloeckler, F. M. Ipavich, C. M. Shafer, J. Geiss, and K. Ogilvie, Solar wind composition measurements by the Ulysses SWICS experiment during transient solar wind flows, *Adv. Space Res., 13*, 675, 1993.

Galvin, A. B., F. M. Ipavich, C.M.S. Cohen, G. Gloeckler, R. von Steiger, Solar wind charge states measured by Ulysses/SWICS in the south polar hole, *Space Sci. Rev., 72*, 65, 1995.

Geiss, J., K. W. Ogilvie, R. von Steiger, U. Mall, G. Gloeckler, A. B. Galvin, F. Ipavich, B. Wilken, and F. Gliem, Ions with low charge states in the solar wind as measured by SWICS on board Ulysses, in *Solar Wind Seven, Cospar Colloquia Series, 3*, ed. E. Marsch and R. Schwenn, 341 pp., 1992.

Geiss, J., G. Gloeckler, and R. von Steiger, Solar and heliospheric processes from solar wind composition measurements, *Phil. Trans. R. Soc. Lond. A, 349*, 213, 1994.

Geiss, J., G. Gloeckler, and R. von Steiger, Origin of the solar wind from composition data, *Space Sci. Rev., 72*, 49, 1995a.

Geiss, J., et al., The southern high speed stream: Results from SWICS/Ulysses, *Science, 268*, 1995b.

Gosling, J. T., V. Pizzo, and S. J. Bame, Anomalously low proton temperatures in the solar wind following interplanetary shock waves: Evidence for magnetic bottles?, *J. Geophys. Res., 78*, 2001, 1973.

Gosling, J. T., J. R. Asbridge, S. J. Bame, W. C. Feldman, and R. D. Zwickl, Observations of large fluxes of He+ in the solar wind following interplanetary shock waves: Evidence for magnetic bottles, *J. Geophys. Res., 85*, 3431, 1980.

Gosling, J. T., D. N. Baker, S. J. Bame, W. C. Feldman, R. D. Zwickl, and E. J. Smith, Bidirectional solar electron heat flux events, *J. Geophys. Res., 92*, 8519, 1987.

Gosling, J. T., S.J. Bame, D.J. McComas, J.L. Phillips, B.E. Goldstein, and M. Neugebauer, The speeds of coronal mass ejections in the solar wind at mid-heliographic latitudes: Ulysses, *Geophys. Res. Lett., 21*, 1109, 1994.

Hirshberg, J., A. Alksne, D. S. Colburn, S. J. Bame, and A. J. Hundhausen, Observation of a solar flare-induced interplanetary shock and helium-enriched driver gas, *J. Geophys. Res., 75*, 1, 1970.

Hirshberg, J., S. J. Bame and D. E. Robbins, Solar flares and solar wind helium enrichments: July 1965 - July 1967, *Solar Phys., 23*, 467, 1972.

Hovestadt, D., Nuclear composition of solar cosmic rays, in *Solar Wind Three*, ed. C. T. Russell, University of California, 2. pp., 1974.

Hundhausen, A. J., H. E. Gilbert, and S. J. Bame, Ionization state of the interplanetary plasma, *J. Geophys. Res., 73*, 5485, 1968.

Ipavich, F. M., A. B. Galvin, G. Gloeckler, D. Hovestadt, S. J. Bame, B. Klecker, M. Scholer, L. A. Fisk, and C. Y. Fan, Solar wind Fe and CNO measurements in high-speed flows, *J. Geophys. Res., 91*, 4133, 1986.

Ipavich, F. M., A. B. Galvin, J. Geiss, K. W. Ogilvie, and F. Gliem, Solar wind iron and oxygen charge states and relative abundances measured by SWICS on Ulysses, in *Solar Wind Seven, Cospar Colloquia Series, 3*, ed. E. Marsch and R. Schwenn, 369 pp., 1992.

Klein, L. W., and L. F. Burlaga, Interplanetary magnetic clouds at 1 AU, *J. Geophys. Res., 87*, 613, 1982.

Liu, S., Beobachtungen und analyse plasmaphysikalischer prozesse im Sonnenwind unter spezieller berücksichtigung von α-teilchen und schweren ionen, Ph. D. dissertation, George-August-Universität zu Göttingen, 1994.

Marsden, R. G., T. R. Sanderson, C. Tranquille, K.-P. Wenzel, and E. J. Smith, ISEE 3 observations of low-energy proton bidirectional events and their relation to isolated interplanetary magnetic structures, *J. Geophys. Res., 92*, 11009, 1987.

Meyer, J.-P., Elemental Abundances in Active Regions, Flares, and Interplanetary Medium, *Adv. Space Res.*, *13*, 377, 1993.

Mitchell, D.G., E.C. Roelof, and S. J. Bame, Solar wind iron abundance variations at speeds >600 km sec^{-1}, 1972-1976, *J. Geophys. Res.*, *88*, 9059, 1983.

Möbius, E., D. Hovestadt, B. Klecker, M. Scholer, G. Gloeckler, and F.M. Ipavich, Direct observation of He$^+$ pick-up ions of interstellar origin in the solar wind, *Nature*, *318*, 426, 1985.

Montgomery, M. D., J. R. Asbridge, S. J. Bame, and W. C. Feldman, Solar wind electron temperature depressions following some interplanetary shock waves: Evidence for magnetic merging?, *J. Geophys. Res.*, *79*, 3103, 1974.

Mullan, D. J., and W. L. Waldron, Ionic charge states of solar energetic particles: Effects of flare x-rays, *Astrophys. J.*, *308*, L21, 1986.

Neugebauer, M., Observational constraints on solar wind acceleration mechanisms, in *Solar Wind Five, NASA Conf. Publ. CP-2280*, ed. M. Neugebauer, NASA Scientific and Technical Information Branch, 135 pp., 1983.

Ogilvie, K.W., M. A. Coplan, and J. Geiss, Solar wind composition from sector boundary crossings and coronal mass ejections, in *Solar Wind Seven, Cospar Colloquia Series, 3*, ed. E. Marsch and R. Schwenn, 379 pp., 1992.

Owocki, S. P., and A. J. Hundhausen, The effect of a coronal shock wave on the solar wind ionization state, *Astrophys. J.*, *274*, 414, 1983.

Owocki, S. P., and J. D. Scudder, The effect of a non-Maxwellian electron distribution on oxygen and iron ionization balances in the solar corona, *Astrophys. J.*, *270*, 758, 1983.

Palmer, I. D., F. R. Allum, and S. Singer, Bidirectional anisotropies in solar cosmic ray events: Evidence for magnetic bottles, *J. Geophys. Res.*, *83*, 75, 1978.

Peter, H., Velocity-dependent fractionation in the solar chromosphere, *Astron. Astrophys.*, *312*, L37, 1996.

Phillips, J. L., S. J. Bame, J. T. Gosling, D. J. McComas, B. E. Goldstein, E. J. Smith, A. Balogh, and R. J. Forsyth, Ulysses plasma observations of coronal mass ejections near 2.5 AU, *Geophys. Res. Lett.*, *19*, 1239, 1992.

Phillips, J. L., *et al.*, Ulysses solar wind plasma observations at high southernly latitudes, *Science*, *268*, 1030, 1995.

Pudovkin, M.I., S. A. Zaitseva, L. P. Oleferenko, and A. D. Chertkov, The structure of the solar flare stream magnetic field, *Solar Phys.*, *54*, 155, 1977.

Schwenn, R., Direct correlations between coronal transients and interplanetary disturbances, *Space Sci. Rev.*, *34*, 85, 1983.

Schwenn, R. Relationship of coronal transients to interplanetary shocks: 3D aspects, *Space Sci. Rev.*, *44*, 139, 1986.

Schwenn, R., H. Rosenbauer, and K.H. Mühlhäuser, Singly-ionized helium in the driver gas of an interplanetary shock wave, *Geophys. Res. Lett.*, *7*, 201, 1980.

Smith, E. J., Observations of interplanetary shocks: Recent progress, *Space Sci. Rev.*, *34*, 101, 1983.

von Steiger, R., J. Geiss, G. Gloeckler, H. Balsiger, A. B. Galvin, U. Mall, and B. Wilken, Magnesium, carbon, and oxygen abundances in different solar wind flow types, as measured by SWICS on Ulysses, in *Solar Wind Seven, Cospar Colloquia Series, 3*, ed. E. Marsch and R. Schwenn, 399 pp., 1992.

von Steiger, R., R.F. Wimmer Schweingruber, J. Geiss, and G. Gloeckler, Abundance variations in the solar wind, *Adv. Space Res.*, *15*, 73, 1995.

von Steiger, R., Solar wind composition and charge states, in *Solar Wind Eight, Proceedings of the Eighth International Solar Wind Conference, AIP Conference Proceedings 382*, ed. D. Winterhalter, J.T. Gosling, S.R. Habbal, W.S. Kurth, M. Neugebauer, 193 pp., 1996.

Wimmer Schweingruber, R. F., Oxygen, helium, and hydrogen in the solar wind: SWICS/Ulysses results, Ph.D. dissertation, University of Bern, 1994.

Zwickl, R. D., J. R. Asbridge, S. J. Bame, W. C. Feldman, and J. T. Gosling, He$^+$ and other unusual ions in the solar wind: A systematic search covering 1972-1980, *J. Geophys. Res.*, *87*, 7379, 1982.

Zwickl, R. D., J. R. Asbridge, S. J. Bame, W. C. Feldman, J. T. Gosling, and E. J. Smith, Plasma properties of driver gas following interplanetary shocks observed by ISEE-3, in *Solar Wind Five, NASA Conf. Publ. CP-2280*, ed. M. Neugebauer, NASA Scientific and Technical Information Branch, 711 pp., 1983.

A. B. Galvin, Department of Physics, University of Maryland, College Park, MD 20742.

Global Modeling of CME Propagation in the Solar Wind

V. J. Pizzo

Space Environment Center, NOAA, Boulder, Colorado

Begging the question of build-up, initiation, and launch, our ability to model the propagation of CMEs out into the interplanetary medium depends upon (1) how well we can characterize the timing, location, geometry, and dynamical content of the ejecta in the lower corona; (2) how well we know the background flow into which the CME propagates; and (3) how well we can describe the subsequent interaction between CME and the structured, background, quasi-steady wind. Simulations of CMEs are are needed to help interpret observations of complicated transient structures and are essential to space weather applications. We assess where we stand in terms of observational and simulation capabilities for CME propagation models and offer suggestions for further development.

PURPOSE

While some conventional review material is included in this paper, it is intended more as a broad statement of the physical considerations involved in successfully modeling the propagation of CMEs from the corona out to Earth and beyond; it also offers an assessment of where things stand in various aspects of the overall problem. This approach is motivated partly to spur space weather applications developments and partly in hopes of channeling basic research efforts in directions deemed most fruitful.

We will begin with a brief overview of essential CME phenomena in interplanetary space, as regards both observations and concepts. Although no explicit consideration is given to the fascinating solar processes leading to and responsible for the initiation of CMEs, the starting point for discussion of CME propagation logically extends well down into the corona, since that is where many of the characteristic properties of interplanetary CMEs are impressed upon the flow. We will also focus on CMEs in the inner heliosphere (out to $\sim$ 5AU), for we wish to highlight opportunities for advances in the global understanding of these phenomena arising from the collective trove of data issuing from new and archived spacecraft observations, as well as from ever-improving ground-based contributions. The evolution of CMEs in the distant heliosphere is a weighty topic on its own, meriting a separate review.

OVERVIEW OF THE PHYSICAL PHENOMENON

Whatever nomenclature we attach to them (we will use the term "CME" throughout, for simplicity), CMEs may be defined as large-scale disturbances which originate in the lower solar atmosphere and which eject in the process substantial amounts of material from the Sun; these ejecta subsequently propagate out through interplanetary space to Earth and beyond. CME disturbances have long been viewed remotely at the solar limb from Earth by white-light coronagraphs, and various manifestations of these events have been detected on disk and off (both directly and indirectly) in selective bands across virtually the entire observable spectrum, including H-α, coronal line, X-ray, EUV, and metric and shorter wavelength radio emissions.

However, it was not until the first combined space-based coronagraph, X-ray, and interplanetary in-situ observations became available some two decades ago that the modern picture of the true nature of CMEs and their relation to the global solar magnetic structure and its evolution began to emerge. Most recently this endeavor has been boosted by the striking long-term X-ray record provided by Yohkoh observations in concert with the global, interplanetary perspective afforded by Ulysses. Now, with the advent of SOHO and WIND observations and ongoing improvements in ground-based capabilities, we are at a juncture where we may expect exciting advances in our overall understanding of the CME phenomenon.

CMEs have come to be recognized as inherently magnetic phenomena [*B. C. Low*, this volume], with their ultimate origins lying in the large aggregations of magnetic flux which erupt from the convective zone in organized, albeit complex patterns. While the basic bipolar connectivities with which active region fields emerge may be preserved over long periods, they are inexorably modified and rearranged as random surface motions, differential rotation, emerging flux from adjacent regions, and reconnection on small scales – all parts of the global solar dynamo process – gradually meld them into the global-scale magnetic distribution which forms the basis for the coronal streamer belt structure.

Although the exact processes spawning CMEs at any given location and time remain unclear, we can say for sure that vast amounts of mass and energy become stored in the evolving large scale structures suspended above the solar surface, and that such arrangements can be stable for some time. While it can happen that CMEs may be spawned from isolated, rapidly evolving active region complexes, it is much more common for CMEs to be initiated from in and around the coronal streamer belt [*Hundhausen*, 1993]. In either case, some instability in the overall force balance of a segment (sometimes quite large, as in the case of polar crown events observed by Yohkoh [e.g., *McAllister et al.*, 1996]) of a coronal magnetic arcade leads to its impulsive expulsion.

One of the most troublesome hurdles for interplanetary propagation models is the extreme variability in CME properties at the solar source, as inferred from white light and other observations. The range of velocities, masses, spatial extents, etc., is quite substantial, and the variance from case to case cannot be reconciled and organized with respect to the observed parent structure. That is, there is no discernible pattern relating, say, fast CMEs with any characteristic feature in the coronal or underlying surface magnetic structures from which it originates. This probably stems from the aforementioned fact that CMEs are inherently magnetic in nature. White light and X-ray observations, the two best-developed diagnostics of coronal structures, provide direct information only on densities and temperatures. Visible structures are thought to trace the magnetic field configuration, but even that is compromised by line-of-sight ambiguities and emission threshold effects (intense fields with too cool or insufficient material entrained will be invisible) accruing from the optically thin nature of the medium. The lack of quantitative field measurements in the corona makes it difficult to identify with certitude the characteristic topologies involved, or, indeed, whether one or several fundamental mechanisms are at play. From the modeling standpoint, all this means that our knowledge of what is being injected into the solar wind (and why) is decidedly limited.

Nevertheless, it is clear that substantial amounts of mass, momentum, and energy (as compared to normal ambient conditions) are impulsively injected into the interplanetary medium in CME events. Much of this material is drawn from the overlying corona, but some of it may also make its way up from deeper layers, all the way down to the chromosphere and perhaps below. Moreover, embedded in the outflow are discrete magnetic structures, which probably become detached from the Sun in the latter stages of the event. These possibly undergo further 3-D reconnection and other structural evolution [*Gosling*, 1996] as they push into and interact with the overlying corona and interplanetary medium beyond.

CME structure has been probed *in situ* by interplanetary spacecraft over a broad range of heliocentric distances and latitudes for over three decades. While a fairly representative sampling of interplanetary CME events has been obtained, the sparse distribution of spacecraft has stymied the development of a comprehensive view of CMEs as a class or individually. Thus, our knowledge of the structural and dynamical elements of interplanetary disturbances remains statistical and composite in nature. In light of the intrinsic variation in CME structures input into the interplanetary medium, this poses a formidable challenge to our ability to model, and, more importantly, to test in detail the propagation of CMEs to Earth.

The lack of detailed knowledge obscures our understanding of the mechanism by which particular CMEs activate the Earth's magnetosphere. Even with the advent of L1 monitors like the WIND spacecraft, we are privy to but a very selective slice through an extended structure whose global properties vary substantially over its bulk, and which likely contains considerable substructure. It is natural to expect that the distribution of field and material across a CME, and perhaps lateral expansion effects as well, (all of which are poorly known) may play a pivotal role in the interaction with the magnetosphere. Thus, the fine-grained sample afforded by upstream monitors generally provides only a limited view of the oncoming disturbance.

COMPONENTS OF A CME PROPAGATION MODEL

In this section, we review systematically the essential components of any model intended to simulate the propagation of a CME all the way from the Sun to Earth.

1. *Input at the Sun*

At a bare minimum, the following information concerning the launch of the CME from the corona is needed:
- start time and duration
- location (centroid) and spatial extent
- speed and direction
- physical content of the ejecta

The latter includes gross estimates of mass, momentum, and energy input to the solar wind; also, it implies some knowledge of the internal structure of the CME, such as the distribution of material and the magnetic configuration of any embedded plasmoid.

CMEs at the limb provide the best opportunity to obtain the above information; where an interplanetary spacecraft is favorably situated to sample the resultant solar wind flow, the dynamics of propagation can studied most advantageously. However, limb events are poorly placed for collecting important ancillary information, such as the surface magnetic field distribution, loop topology as seen in X-rays, etc, which are best observed on the disk. Conversely, disk center CMEs appear in coronagraphs only as "halo" events, which provide at best timing and a general sense of direction.

In short, the physical properties of the input ejecta cannot be adequately inferred from present observations, and, by virtue of the known intrinsic variability in CME structure, this remains by far the weakest link in the entire chain of modeling needs.

2. *Background into which CME flows*

Real CMEs do not propagate into the featureless, spherically symmetric background flow commonly presumed in theoretical studies. Rather, they tend to be launched in close proximity to the complicated structures of the streamer belt, so that they are often forced into direct, immediate interaction with relatively dense overlying structures. Moreover, as they move out from the Sun, they experience differing conditions depending upon which regime of the background global solar wind structure (fast flow from holes, slow flow from streamer belt, or compound flow from previous, slower transients) they intersect along the way.

Few, if any, transients are so overpowering that they simply blow aside all intervening material. To the contrary, CMEs interact quite significantly with the pre-existing flow enroute to Earth, so that good knowledge of the solar wind structure into which the CME propagates is required for realistic modeling. In this regard, it is useful to bear in mind that shocks in the interplanetary medium (most expressly including CME-driven shocks) are generally of moderate strength and that shocks are a form of wave; thus their evolution is intimately tied to the background upon which they propagate.

This view has recently been reinforced by Yohkoh and Ulysses observations of CMEs spanning a broad range of heliographic latitudes and flow states. These observations strongly suggest that the appearance and speed of a given CME in the interplanetary medium is critically linked to the background flow into which it is ejected [*Gosling et al.*, 1995]. Moreover, observations and simulations provide support for the idea that a single, initially homogeneous CME can be distorted into two or more seemingly unrelated, morphologically discordant fragments by interplanetary dynamical processes [*Riley et al.*, 1997]. That is, for a large CME whose bulk extends across fast and slow background flow regimes, the two parts may evolve in dramatically different ways in interplanetary space.

Modeling CME propagation requires accurate description of the global, quasi-steady structure of the corona prior to CME eruption and from this structure determination of the global solar wind flow pattern extending out to several AU.

Since direct measurement or reliable remote sensing of the medium over most of this spatial volume is not feasible, heavy reliance on modeling is required to flesh out the picture of physical conditions in the background coronal structure and solar wind outflow. 3-D MHD simulations are now able to replicate with some accuracy the gross coronal structure, for select Carrington rotations, out to several solar radii above the surface [e.g., *Linker et al.*, 1996]. This kind of model, like earlier source-surface models, uses direct observations (to this point, ground-based) of the large-scale surface magnetic field distribution as the essential input. The fidelity of the modeling is tested by comparing actual white light observations with simulated coronagraph images generated from the model; thus the criterion of comparison is the large-scale density distribution. (The topology of field lines as traced out by polar plumes and other fine structure may additionally be compared with projections of field lines from the model.) Although the evaluation remains somewhat subjective, the general impression gained from such comparisons is quite favorable.

Encouraging though all this may be, global, quasi-steady solar wind models face two major obstacles:

1. The surface magnetic field distributions, which constitute the prime input to the model, impose certain limitations. First, they are usually assembled synoptically, which introduces temporal inconsistency into the boundary conditions. The non-stationarity of the input is perhaps tolerable in the late declining and minimum phases of the solar cycle, but rapid changes in the global surface field distribution around solar maximum compromises the viability of such models during that epoch. Second, projection effects preclude obtaining any useful information on fields much above 60° heliographic latitude. Since the large scale magnetic field may be relatively concentrated toward the poles and would therefore contribute significantly to the spreading of the polar holes with height in the corona, this lack of information is a serious shortcoming.

2. The physics of the coronal acceleration region remains a mystery. Without a realistic description of the underlying mechanism, the velocity profile in the fast outflow from holes cannot be adequately reproduced, nor can the velocity contrast between hole and slow, streamer flow be replicated except through ad-hoc parameterizations. Statistical profiles of CME height-velocity profiles provide a measure of the acceleration in the lower corona, but this is well below the region where conventional wisdom indicates the preponderance of the overall acceleration takes place. Coronagraph observations from SOHO may be expected to shed some light on this phenomenon, hopefully significantly, in the near future.

Difficulties in modeling the background flow are compounded if the adjacent corona has recently undergone a transient disruption and a stable, new ambient flow has not had time to be established. It should also be recognized that the physics of slow, streamer-associated flows is none too well understood, either. Recent observations [*Breuckner*, 1996] and analyses [*Crooker et al.*, 1993] suggest that such flows may be episodic in nature, as opposed to truly steady, and may be more complicated morphologically than previously thought.

3. Physical description of CME interaction with ambient medium

This section is included to draw a fine, but significant, distinction. And that is that for simulation purposes it is insufficient to simply characterize (through observations, say) the gross properties of a CME upon launch from the lower corona and to map out the background structure into which it propagates. Rather, modeling additionally requires an accurate physical description of the processes by which the CME couples with the background coronal acceleration, detaches itself from the corona, and interacts dynamically with the ambient, structured interplanetary solar wind flows in its path.

Fortunately, it appears that an inviscid (except at shocks), ideal MHD description is adequate for modeling the large scale propagation of CMEs, at least in interplanetary space. This is because the high mach number of the flow outside the Alfvenic point ($\gtrsim 20R_\odot$) assures that the dynamics there are highly momentum dominated. Thus, such considerations as thermal disparities between protons and electrons, the contributions of minor ions, dissipation and reconnection at current sheets, energetic particle generation and acceleration, and most wave processes as well – can all be either parameterized or neglected entirely without impacting the modeling of the interplanetary dynamics too severely. Moreover, the momentum content of the interplanetary magnetic field falls off sufficiently rapidly with heliocentric distance that (with the possible exception of intense magnetic clouds) the effects of the field upon the dynamics are generally minor and are readily treated.

In short, modeling large scale aspects of CMEs in interplanetary space is fundamentally more straightforward than simulating the propagation in the corona because the underlying physics is simpler.

The dynamical simplicity afforded by the momentum-dominated character of the flow does come at a price, however. It has the consequence that secondary processes and attributes, such as ionic composition and the details of the magnetic and thermal structure, will be poorly modeled. Even when dynamically unimportant, these and related quantities (e.g., the southward component of the magnetic field, energetic particle acceleration, cosmic ray modulation effects) can be extremely significant for space weather applications; for basic research purposes, they are valuable as diagnostics and indicators of the flow state. Therefore, improving the treatment of these secondary phenomena should be a priority for future simulations.

4. Geomagnetic Considerations

Although the interaction of a CME with the terrestrial environment depends, at least in part, upon the initial state of the magnetosphere itself, three primary interplanetary factors contribute to the response:

1. Interplanetary disturbances sweep up and compress the ambient magnetic field and solar wind flows in their path. The interplanetary material thus accumulated and intensified at the leading of a CME is thrust up against Earth's magnetosphere during the initial stages of the interaction. In particular, the draping of the background interplanetary magnetic field (including any intervening current sheets) about an oncoming CME [*McComas and Gosling*, 1988] is known to contribute to the geomagnetic response. Fortunately, since the physical processes involved are relatively clear, this aspect of CME propagation appears quite amenable to modeling [e.g., *Detman et al.*, 1991].

2. Even more germane to the terrestrial response is the speed and internal structure of the CME. It is well established that the duration and intensity of southward-pointing magnetic field in an interplanetary disturbance is the prime factor in driving large magnetic storms; fast CMEs containing a well-defined, favorably oriented magnetic cloud configuration (preferably, southward leading) are known to be the most geoeffective [*Gosling*, 1993]. For those CMEs not including a magnetic cloud (perhaps a sizable fraction of all CMEs), the speed with which draped southward interplanetary fields are pushed into the Earth's magnetosphere is the chief agent for driving geomagnetic disturbances. Where a magnetic cloud is contained within the CME ejecta, the relaxation of magnetic and gas pressure forces within the cloud may play a significant role in the evolution of the internal structure of the CME with heliocentric distance. Certainly, Ulysses observations of high latitude transients support this view, at least where speed differences between the CME and the background medium are not too large. Indeed, there may be instances where nonradial expansion effects driven by the internal relaxation may significantly affect the overall propagation of the CME.

Models for evaluating the nature and extent of dynamical processes in magnetic clouds embedded within transient flows are only now coming online and need to be further developed [*V. Osherovich and L. F. Burlaga*, this volume; *Cargill et al.*, 1996; *Vandas et al.*, 1996].

3. Finally, knowledge of the centroid of the CME relative to Earth is an important (if not widely appreciated) aspect of the solar-terrestrial interaction. It has been known for decades that the disturbance front driven by a CME is of considerably greater spatial extent than the ejecta per se. Recent observations by Ulysses and other spacecraft underscore the intrinsic spatial inhomogeneity of such disturbances. Depending upon whether a given CME strikes the Earth a direct or glancing blow, it is reasonable to expect differences in response on this basis alone, particularly when the orientation of the Earth's magnetic field is taken into account. At certain times of the year, CMEs passing to the north of the Earth, say, may interact with the polar regions of the geomagnetic environment more directly than would the same disturbance six months later, particularly if the CME is undergoing strong lateral expansion due to the internal forces mentioned above. Hence whatever simulations may tell us concerning the nonradial motion and expansion of CMEs between solar source and 1 AU is of importance to space weather applications.

5. Verification

The final link in the simulation chain is rigorous comparison of simulated properties against observations. Aside from the obvious fact that observations near the Sun are few and quite limited in scope and that more detailed *in situ* interplanetary observations are likewise sparse, the major impediment to proper evaluation of large scale CME simulations is the extreme variation in CME properties as observed at the coronal source and as perpetuated out into the solar wind. Simulations have thus far been most successful when applied to understanding patterns of behavior common to CMEs, as opposed to attempts aimed at modeling any given CME. We cannot expect any drastic change in this regard until more comprehensive observations characterizing the full, extended, 3-D nature of the disturbances, become available.

FUTURE DIRECTIONS

It should be clear that the overriding bottleneck to accurate, realistic simulation of CMEs in the interplanetary medium is observational. Specifically, knowledge of the requisite inputs – the global, physical properties of CME structures injected into the solar wind – is sorely lacking and is likely to remain so for the near future, at least. Likewise, the comprehensive global data coverage necessary to compare against the simulations to improve their performance dramatically does not now exist. And the outstanding physical question to be resolved, the nature and form of the coronal acceleration mechanism operating in the mid to upper corona, will probably only be divined through observations.

The technical ability to conduct 3-D MHD simulations of CME structures in the solar wind has undergone rapid development over the last decade. Despite the limitations of present-day models, their capabilities far outstrip the available data and, oftentimes, the physical insight needed to guide their application. It can be predicted with great certitude that expertise in simulation techniques will continue to expand dramatically in scope and in the ability to treat physical complexities. If and when a breakthrough is made concerning our understanding of the coronal acceleration mechanism, it will be just a matter of time – and not much of it, at that – before an adequate accounting of the process is successfully incorporated into a global MHD model.

For very practical reasons, we cannot be so sanguine about overcoming the observational hurdles with the same alacrity. Let us therefore consider the kinds of observations, feasible over the next decade or so, that would go furthest in alleviating the empirical roadblocks outlined above.

1. *Key New Observations*

• Observations of the Sun-Earth system taken from multiple viewing perspectives would be most informative. Something like the proposed solar STEREO concept, which envisions one or more spacecraft observing from large angles to the Sun-Earth line, would afford for the first time truly global imaging of CME structures all the way from the Sun to Earth and beyond. Depending on the instrumentation, one would obtain at the very least 3-D morphological information on a variety of coronal and interplanetary structures, and some limited quantitative data as well. Moreover, it should be possible to establish a direct, causal link between specific events on the Sun and specific events at Earth for the very first time. (Recent SOHO and WIND observations provide tantalizing evidence for this link [*Carlowicz*, 1977], but the single view precludes the kind of concise, concrete analysis required for space weather applications.) Multiperspective measurement of the surface magnetic field of the Sun (such as proposed in the MAGSONAS concept [*J. Feynman*, this volume]) would offer a major boost to global modeling efforts by eliminating many of the problems associated with the use of synoptic data. Finally, if an imaging spacecraft can be lofted to high heliocentric latitudes, measurement of the large scale magnetic field distribution near the poles of the Sun (crucial to accurate modeling of the global background corona and solar wind flow and to certain high-latitude CME processes) will at last be enabled.

• Improved on-disk monitoring capabilities in the soft X-ray regime would go far toward establishing the timing and morphology of CME events, especially those most important from a geospace perspective. The upcoming Solar X-ray Imager to be launched aboard GOES platforms beginning in the year 2000 will provide a frequent, regular observing cadence in X-ray bands more conducive to observing CMEs than previous spacecraft.

• As stated at the outset, the global structure of the Sun and CMEs in particular are inherently magnetic phenomena. Hence measurement of magnetic field strengths in the solar corona are a high priority. Recent interest has turned to detection of these fields via infrared emissions, but Faraday rotation measurements, utilizing spacecraft radio transmissions from the far side of the Sun [e.g., *Hollweg et al.*, 1982] could also be exploited for this purpose. The potential rewards here are so great that any promising avenue should be pursued.

2. *Revitalized Observational Analysis*

While awaiting all these new observations, there is much to be done in the realm of extracting greater insight from the inputs currently available. Taking the view that development of ever more sophisticated numerical models will take place in any case, it is to be stressed that greater efforts should be expended toward integrating the many long-known signatures and indicators of solar and coronal activity into a more coherent and comprehensive picture than now exists. It should be possible to garner significant advances in on-disk CME detection and characterization, for example, through the application of what may generally be called pattern recognition techniques. Modern computer resources should make it possible to splice together the available observables (e.g., loop and arcade formation indicative of major reconnection events; transient regional dimming in X-ray emissions; long-duration, hard X-ray signatures; type II radio bursts, helicity conservation, etc.) in innovative ways to probe such events. Similarly, we now have in store white light synoptic maps of the corona, solar surface magnetic maps, and the large-scale distribution of filaments, prominence chains, and other features in the lower atmosphere spanning several solar cycles. Surely there are characteristic patterns of behavior here, which, when coupled with modern X-ray and other observations, should help illuminate the processes leading to CME production, suggest some organization to their properties, and offer some of the inputs we so desperately need for modeling.

3. *Simulation Needs*

The most pressing modeling needs include:

• Improved models of the transonic region for both background and CME-related flows depend upon knowledge of the coronal acceleration mechanism. Pending resolution of that problem, however, incorporation of an ad-hoc heating term or some similar parameterization would still mark a major improvement over the polytropic or classic conduction treatments usually employed. Tentative steps in this direction have already appeared in the literature [*Steinolfson*, 1988; *Suess et al.*, 1996], and with guidance from SOHO and other observations it should be possible to cobble a much improved approximation to the actual behavior.

- Treatment of the internal dynamics of CMEs in general and magnetic clouds in particular stands much improvement. In addition to more realistic thermodynamics, inclusion of 3-D reconnection processes to better model the evolving magnetic topology and to provide a basis for interpreting bi-directional streaming of electrons associated with CME structures is a major goal [see also *Chen*, 1996, and *V. Osherovich and L. F. Burlaga*, this volume].
- Self-consistent global modeling of energetic particle acceleration associated with CMEs has not yet been attempted. All that is currently available are detailed local models shock front or current sheet processes and empirical large scale models derived from statistical analyses of CMEs. Considering the importance of particle generation for space weather applications, it is somewhat surprising that so little effort has thus far been devoted to the development of such capabilities.
- Finally, the most outstanding technical problem for simulation of CMEs is the problem of scales – such structures are marked by large areas of relatively smooth, dynamically benign flow which is interspersed with tiny regions where much of the important dynamics takes place. Shock fronts, current sheets, and reconnection processes are all examples of small scale phenomena whose dynamical impact and observational value is significant. Adaptive-grid models [e.g., *Gombosi et al.*, 1996; *Zelasak et al.*, 1995] seem the most promising way to approach these features, and they merit heavy development in the near future. In particular, the ability to model accurately over a large range of spatial scales should prove indispensable for tracking the propagation of CME structures for large heliocentric distances.

REFERENCES

Brueckner, G., Global coronal disturbances as the source for the low latitude solar wind, *EOS*, abstract only, 1996.

Cargill, P. J., J. Chen, D. S. Spicer, and S. T. Zelasak, Magnetohydrodynamic simulations of the motion of magnetic flux tubes through a magnetized plasma, *J. Geophys. Res.*, 101, 4855, 1996.

Carlowicz, M., Satellites, scientists track storm from Sun to surface, *EOS*, 78, #5, pg. 48, 1997.

Chen, J., Theory of prominence eruption and propagation: Interplanetary consequences, *J. Geophys. Res.*, 101, 27499, 1996.

Crooker, N. U., G. L. Siscoe, S. Shodhan, D. F. Webb, J. T. Gosling, and E. J. Smith, Multiple heliospheric current sheets and coronal streamer belt dynamics, *J. Geophys. Res.*, 98, 9371, 1993.

Detman, T. R., M. Dryer, T. Yeh, S. M. Han, S. T. Wu, and D. J. McComas, A time-dependent 3-D MHD numerical study of interplanetary magnetic draping around plasmoids in the solar wind, *J. Geophys. Res.*, 96, 9531, 1991.

Feynman, J., Evolving magnetic structures and their relation to coronal mass ejections, this volume.

Gombosi, T.I., D.L. De Zeeuw, R.M. Haberli, and K.G. Powell, A 3D multiscale MHD model of cometary plasma environments, *J. Geophys. Res.*, submitted, 1996.

Gosling, J. T., Coronal mass ejections: The link between solar and geomagnetic activity, *Phys. Fluids, B*, 5(7), 2638-2645, 1993.

Gosling, J. T., Magnetic topologies of coronal mass ejection events: Effects of 3-dimensional reconnection, *Solar Wind Eight*, ed. D. Winterhalter, J. T. Gosling, S. R. Habbal, W. S. Kurth, and M. Neugebauer, American Inst. of Physics, Conf Proc 383, NY, pp. 438-441, 1996.

Gosling, J. T., D. J. McComas, J. L. Phillips, V. J. Pizzo, B. E. Goldstein, R. J. Forsyth, and R. P. Lepping, A CME-driven solar wind disturbance observed at both low and high heliographic latitudes, *Geophys. Res. Lett.*, 22, 1753, 1995.

Hollweg, J. V., M. Bird, H. Volland, P. Edenhofer, C. Stelzried, and B. Seidel, Possible evidence for coronal Alfven waves, *J. Geophys. Res.*, 87, 1, 1982.

Hundhausen, A. J., Sizes and locations of coronal mass ejections: SMM observations from 1980 and 1984-1989, *J. Geophys. Res.*, 98, 13177, 1993.

Linker, J. A., Z. Mikic, and D. D. Schnack, Global coronal modeling and space weather prediction, in "Solar drivers of interplanetary and terrestrial disturbances," *Astron. Soc. Pac. Conf.*, 95, 208, 1996.

Low, B.-C., The role of coronal mass ejections in solar activity, this volume.

McAllister, A. H., M. Dryer, P. McIntosh, H. Singer, and L. Weiss, A large polar crown coronal mass ejection and a "problem" geomagnetic storm: April 14-23, 1994, *J. Geophys. Res.* 101, 13,497, 1996.

McComas, D. J., and J. T. Gosling, Magnetic field draping about coronal mass ejections, in Proc. of the Sixth Intl. Solar Wind Conf., eds V. J. Pizzo, T. E. Holzer, and D. G. Sime, NCAR TN-306, vol. 1, p291, 1988.

Osherovich, V., and L. F. Burlaga, Magnetic clouds, this volume.

Riley, P., J. T. Gosling, and V. J. Pizzo, A Two-dimensional simulation of the radial and latitudinal evolution of a solar wind disturbance driven by a fast, high-pressure coronal mass ejection, *J. Geophys. Res.*, submitted, 1997.

Steinolfson, R. S., Density and white light brightness in loop-like coronal mass ejections: Importance of the pre-event atmosphere, *J. Geophys. Res.*, 93, 14261, 1988.

Suess, S. T., A.-H. Wang, and S. T. Wu, Volumetric heating in coronal streamers, *J. Geophys. Res.*, 101, 19957, 1996.

Vandas, M., S. Fischer, M. Dryer, Z. Smith, and T. Detman, Simulation of magnetic cloud propagation in the inner heliosphere in two-dimensions, 2, loop parallel to the ecliptic plane and the role of helicity, *J. Geophys. Res.*, 101, 2505, 1996.

Zelasak, S. T., D. S. Spicer, R. Lohner, and S. A. Curtis, Three-dimensional unstructured modeling of the terrestrial magnetosphere and coupled ionosphere: results using adaptive mesh refinement, *EOS*, 76, #46, pg. 498, 1995.

V. J. Pizzo, Mail Code /R/E/SE, NOAA/SEC, 325 South Broadway, Boulder, CO 80303. (e-mail: vpizzo@sec.noaa.gov)

Extending Coronal Models to Earth Orbit

Jon A. Linker and Zoran Mikić

Science Applications International Corporation, San Diego, California

Solar wind conditions at Earth play a primary role in the initiation of geomagnetic activity. The forecasting of solar wind conditions at Earth based on remote observations of the Sun is thus a key element of space weather prediction. We describe how observations of the photospheric magnetic field can be incorporated into three-dimensional MHD computations of the solar corona and inner heliosphere. We show that the resulting solutions compare favorably with observations and that this same capability can be used to model the initiation of coronal mass ejections and their propagation out into the heliosphere. These encouraging results suggest that an operational computational solar wind model can eventually be developed, suitable for forecasting solar wind properties at Earth.

1. INTRODUCTION

Coronal mass ejections (CMEs) are exceedingly complex phenomena. From their initiation on the Sun to their propagation through the heliosphere, CMEs span a large range of both distance and physical parameter space. Understanding how CMEs, typically observed as loop-like structures in white-light coronagraphs [e.g., *Hundhausen*, 1993], are initiated and how they ultimately manifest themselves in interplanetary space is a fundamental challenge for solar and heliospheric science. Apart from the intellectual attraction of such a challenge, solution of this problem has significant practical applications. It is well known that solar wind conditions upstream of the Earth's magnetosphere play an important role in geomagnetic activity, and that CMEs in particular are associated with the largest geomagnetic storms [*Gosling*, 1993]. Geomagnetic storms can cause disruption of satellite operations, communications, navigation, and electric power distribution grids, and create a hazardous environment for astronauts engaged in extra-vehicular activities. The prediction of such "space weather" phenomena has thus become recognized as an important problem for the space science community, as evidenced by the National Space Weather Program Strategic Plan [*Wright et al.*, 1995].

Remote observations of the magnetic and plasma environment of the Sun have been made routinely for some time. With the SOHO spacecraft now operational, the amount and quality of such measurements, including the detection of CMEs, has greatly increased. In the context of space-weather forecasting, using remote solar observations to accurately predict the characteristics of the solar wind at Earth orbit, especially the arrival time and properties of CMEs, is one of the primary services to be provided by solar and heliospheric science.

To predict effects at Earth from events occurring on the Sun, solar observations must be incorporated into a physical model. The magnetohydrodynamic (MHD) fluid description is an appropriate starting point for modeling the solar wind. Even with multi-fluid and kinetic effects neglected, consideration of the important physical processes in multi-dimensional geometry renders the MHD equations intractable to an analytic approach, and a computational solar wind model is necessary if we are to forecast solar wind conditions at Earth orbit. This is not surprising, as computational models of the atmosphere have long played an important role in terrestrial weather prediction [*Houghton*, 1977]. Our computational solar wind model must fulfill two requirements: (1) computation of the "background" solar corona and solar wind, and (2) calculation of the initiation and propagation of CMEs. Requirement (1) arises because the geoeffectiveness of CMEs is related to the structure of the interplanetary magnetic field (IMF) [*Gosling et al.*, 1990]; [*Crooker et al.*, 1992]. Also, apart from the effect of CMEs, the background solar wind plays an important role in geomagnetic activity. It has long been known that 27 day recurrences in geomagnetic activity are directly linked to the solar rotation period [e.g., *Maunder*, 1905]. Fast streams in the solar wind, which originate in coronal holes [e.g., *Altschuler et al.*, 1972], are generally believed to be the cause of recurrent geomagnetic storms [*Hundhausen*, 1979; *Zirin*, 1988; *Foukal*, 1990]. Recently the role of the streamer belt and corotating interaction regions in producing recurrent geomagnetic activity has also been recognized [*Crooker and Cliver*, 1994].

At the outset, we must recognize the formidable nature of the goal we have described and how distant we are from achieving it. While many obstacles contribute to the difficulty of solar wind forecasting, perhaps the most obvious problem is our lack of understanding of how CMEs are initiated. Nevertheless, to make progress on this task, we must outline a strategy for model development. With this purpose in mind, we demonstrate in this paper how MHD models of the solar wind can supply the two key requirements for solar wind forecasting. Section 2 briefly discusses the methodology of our computations. In section 3 we describe realistic computations of the solar corona, and we show that our results compare favorably with coronal and heliospheric observations. In sections 4 and 5 we demonstrate a model computation of the solar corona and inner heliosphere at solar minimum, and we show how one candidate process for CME initiation (shearing of the magnetic footpoints by differential rotation) can be modeled and the heliospheric effects of the resulting disturbance studied. Section 6 summarizes our present capabilities and indicates future directions.

2. METHODOLOGY

To compute MHD solutions for the large-scale corona, we solve the following form of the equations in spherical coordinates:

$$\nabla \times \mathbf{B} = \frac{4\pi}{c}\mathbf{J} \qquad (1)$$

$$\frac{1}{c}\frac{\partial \mathbf{B}}{\partial t} = -\nabla \times \mathbf{E} \qquad (2)$$

$$\mathbf{E} + \frac{\mathbf{v} \times \mathbf{B}}{c} = \eta \mathbf{J} \qquad (3)$$

$$\frac{\partial \rho}{\partial t} + \nabla \cdot (\rho \mathbf{v}) = 0 \qquad (4)$$

$$\rho\left(\frac{\partial \mathbf{v}}{\partial t} + \mathbf{v} \cdot \nabla \mathbf{v}\right) = \frac{1}{c}\mathbf{J} \times \mathbf{B} - \nabla p - \nabla p_w$$
$$+ \rho \mathbf{g} + \nabla \cdot (\nu \rho \nabla \mathbf{v}) \qquad (5)$$

$$\frac{\partial p}{\partial t} + \nabla \cdot (p\mathbf{v}) = (\gamma - 1)(-p\nabla \cdot \mathbf{v} + S) \qquad (6)$$

where $\mathbf{B}$ is the magnetic field intensity, $\mathbf{J}$ is the electric current density, $\mathbf{E}$ is the electric field, $\mathbf{v}$, ρ, and p are the plasma velocity, mass density, and pressure. The gravitational acceleration is $\mathbf{g}$, γ is the ratio of specific heats, η is the resistivity, ν is the viscosity, S represents energy source terms, and the wave pressure p_w represents the acceleration due to Alfvén waves [*Jacques*, 1977; *Holweg*, 1978].

The term S in equation (6) includes the effects of coronal heating, thermal conduction parallel to $\mathbf{B}$, radiative losses, and Alfvén wave dissipation (viscous and resistive dissipation can also be included). A simplified model of the corona, known as the "polytropic model", is obtained when an adiabatic energy equation with a reduced γ is used [*Parker*, 1963]. This is a crude way of modeling the complicated thermodynamics in the corona with a simple energy equation. This choice results by setting $S = 0$ in Eq. (6) and $p_w = 0$ in Eq. (5). With this model, values of γ close to 1 ($\gamma = 1.05$ for the results shown in this paper) are necessary to produce density and temperature profiles that are similar to coronal observations; this indicates that the terms included in S are in fact important for describing the energy balance of the corona. In this paper we describe computations using the polytropic model. Computations using the full equations (1–6) have been performed [*Mikić et al.*, 1996ab] and will be described in a future paper.

Mikić and Linker [1994] describe the method used to solve equations (1–6). To compute coronal and heliospheric solutions, the equations must be supplemented with appropriate boundary and initial conditions. In spherical geometry, two boundaries appear in the simulation: the physical inner radial boundary at r = R_s (the solar radius) and an artificial outer radial boundary at r = R_1, which we typically place in the range 20–215 R_s.

Specification of the appropriate boundary conditions is facilitated by examining the characteristic form of equations (1–6) [*Courant and Friedrichs*, 1948]. (See *Hu and Wu*, [1984] for an example using the MHD equations.) Characteristics traveling into the domain require that physical information be provided. At the inner boundary, four characteristics point into the domain, and require that four quantities be specified. We specify the distributions of ρ and p at r = R_s. When no surface motions are included in the calculation, we also specify that E_θ and $E_\phi = 0$ at r = R_s; this condition fixes the radial magnetic field at the inner boundary (B_{r0}) at its initial value and is equivalent to setting $\mathbf{v}_\perp$ (velocity perpendicular to $\mathbf{B}$) = 0 there. When surface motions such as the solar rotation are included, E_θ and E_ϕ are specified to be consistent with this motion (from equation 3); note that in this case the distribution of B_{r0} can be modified by the surface motions. The outer boundary is typically placed well beyond the critical points, so all characteristics are outgoing and no physical boundary conditions are required. To advance the solution at the outer boundary, we use the characteristic equations to compute $\mathbf{v}$. The plasma β ($= 8\pi p/B^2$, the ratio of plasma pressure to magnetic pressure) is typically 1 or greater at the outer boundary, and we find that characteristics based on the gas equations are sufficient. The staggering of the mesh then allows all other quantities to be computed in the same manner as the interior points [*Mikić and Linker*, 1994]. Characteristic equations are also used to compute $\mathbf{v}_\parallel$ (velocity parallel to $\mathbf{B}$) at the lower boundary.

For the initial condition, a potential magnetic field consistent with the specified distribution of B_r at the lower boundary, and a wind solution [*Parker*, 1963] consistent with the specified ρ and p are used. Equations (1–6) are then integrated forward in time until a steady state is reached. The computations are performed on a mesh that is nonuniform in the r and θ directions: $\Delta r \approx .01 \ R_s$ near the inner boundary and $\Delta r \approx 10 \ R_s$ at r = 215 R_s; $\Delta\theta$ varied between .03 and .06 radians. The longitudinal (ϕ) coordinate is treated using a pseudospectral method (this requires a uniform distribution of points in ϕ). Our higher resolution cases used 101 × 101 × 64 (r,θ,ϕ) points; cases extending out to approximately 1 A.U. (1 astronomical unit = 1.49 × 10^6 km = 214 solar radii) used 111 × 51 × 32 points.

Previous coronal and solar wind solutions of (1–6) have typically been performed with idealized magnetic fields [*Endler*, 1971; *Pneuman and Kopp*, 1971; *Steinolfson et al.*, 1982; *Washimi et al.*, 1987; *Linker et al.*, 1990; *Wang et al.*, 1993; *Linker and Mikić*, 1995], or with an inner boundary beyond the critical points [*Smith and Dryer* 1990; *Detman et al.*, 1991; *Pizzo*, 1991; *Odstrcil*, 1994]. To perform a realistic 3-D MHD computation of the corona that can be compared with specific observations, it is necessary to incorporate solar observations into the boundary conditions [*Usmanov*, 1993; *Mikić and Linker*, 1996; *Linker et al.*, 1996]. One of the most readily available observational data sets is the magnetic field at the photosphere. This is also the most important observation to address for coronal and heliospheric modeling. We have used Wilcox Solar Observatory synoptic maps (collected during a solar rotation by daily measurements of the line-of-sight magnetic field at central meridian) to specify the radial magnetic field at the photosphere (in the manner described by *Wang and Sheeley* [1992]).

3. COMPARISONS WITH CORONAL AND HELIOSPHERIC DATA

Solutions obtained in the manner described in section 2 can in principle provide a 3D description of the corona and inner heliosphere, including the detailed distribution of magnetic fields, currents, plasma density, and temperature. However, the validity of this approach can only be verified through comparison with observations. As a test of our coronal modeling capability, we used our computations to predict the large-scale structure of the solar corona during the October 24, 1995 eclipse (occurring during Carrington rotation (CR) 1901), visible in a number of locations in the eastern hemisphere. We carried out a simulation using photospheric magnetic field data from the previous rotation (CR1900; September 2 – September 29, 1995) on October 5, 1995, and put the results on the World Wide Web (more detailed comparisons and new results can be viewed at http://iris023.saic.com:8000/corona/modeling.html). We also presented the results at the Sacramento Peak workshop on October 18, 1995 [*Linker et al.*, 1996]. Figure 1 (leftmost frame) shows the magnetic field lines from our calculation. The view angle corresponds to the approximate time of the eclipse. The solution shows the formation of helmet streamers; these are regions with closed magnetic fields that trap coronal plasma flowing out of the Sun. Along open magnetic field lines, the solar wind streams freely, reaching supersonic speeds.

To directly compare our results with observations, we develop images of the polarization brightness (pB; pro-

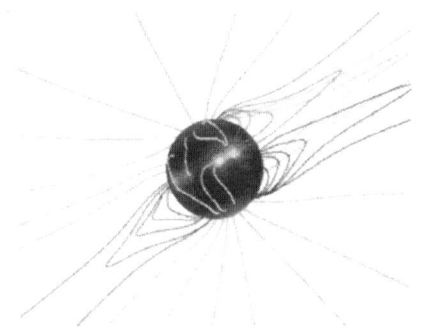

Figure 1. A prediction of the structure of the solar corona during the October 24, 1995 solar eclipse. The MHD simulation was carried out on October 5, 1995, using Wilcox synoptic magnetic data for the previous rotation. From left to right, the frames show: Field lines, polarization brightness computed from the simulation, and an eclipse photograph taken by F. Diego (UCL) in white light with F=910 mm and a two-second exposure time.

portional to the line-of-sight integral of the product of the electron density and a scattering function that varies along the line of sight). This quantity is frequently observed with coronagraphs. Using the plasma density from our coronal model, we can compute pB to simulate an eclipse or coronagraph image and compare it with the actual data. Radially graded filters are applied to eclipse images to compensate for the rapid fall-off of coronal density with radial distance; we detrend our computed pB in a similar manner. The polarization brightness of the corona predicted by our simulation, as it would be seen on October 24, 1995, at 05:00UT is shown in Figure 1 (middle frame), along with an image of the eclipse taken by F. Diego of University College, London (rightmost frame). The helmet streamers and open field regions predicted by the computation agree reasonably well with the eclipse observations. We have performed a similar comparison for the November 3, 1994 eclipse and CR1888 [*Mikić and Linker*, 1996; *Linker et al.*, 1996]. These computations support the long-held belief that the magnetic field distribution on the Sun controls the position and shape of the streamer belt.

We have also compared the results of our calculations with interplanetary observations. As a first test, we computed an MHD model of the solar corona for CR1869 (May–June 1993). This rotation was of particular interest for Ulysses observations, as the Ulysses spacecraft ceased to observe sector-boundary crossings during that time period [*Smith et al.*, 1993]. Figure 2 shows a comparison of the heliospheric current sheet (HCS) predicted by our MHD computation with that of the source-surface model [e.g., *Schatten et al.*, 1969; *Altschuler and Newkirk*, 1969; *Hoeksema*, 1991; *Wang and Sheeley* 1988, 1992], a frequently used tool for approximating heliospheric structure. Ulysses' latitude position for this time period (near 30° S latitude) is also shown. The source-surface model predicts crossings for this time period, whereas the MHD simulation correctly predicts no HCS crossings.

During February–April of 1995 (before and after the spacecraft approached perihelion), the Ulysses spacecraft sampled a wide range of heliographic latitude in a short period of time. Figure 3 shows the HCS predicted by our MHD computation for CR1892, the start of this fast latitude scan. Also shown is the Ulysses trajectory projected in solar latitude and Carrington longitude (back at the Sun) and published Ulysses HCS crossings indicated by crosses [*Smith et al.*, 1995]. The different line styles on the trajectory plots indicate the Carrington rotation at that time. During CR1892 (the time period for which the calculation is most valid), the two Ulysses crossings occur almost exactly where predicted by the MHD computation. Later in time (CR1893 and CR1894), the overall shape of the MHD HCS agrees well with Smith et al.'s empirically derived HCS, but the Ulysses crossings occur above the MHD HCS. The reason for this can be seen in Figure 4, which shows the predicted source-surface model for the 3 rotations. The source-surface model suggests that the solar magnetic field is changing during this time period, as evidenced by the changing HCS. Therefore, MHD computations of CR1893 and CR1894 are required for a complete comparison; this work is presently underway.

While the favorable comparisons between our computational results and coronal and heliospheric observations are encouraging, it should be noted that there

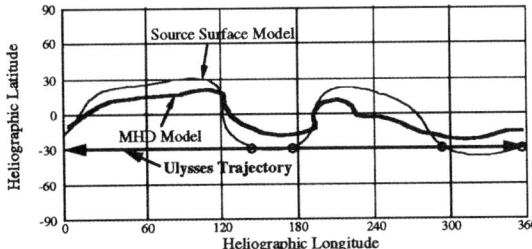

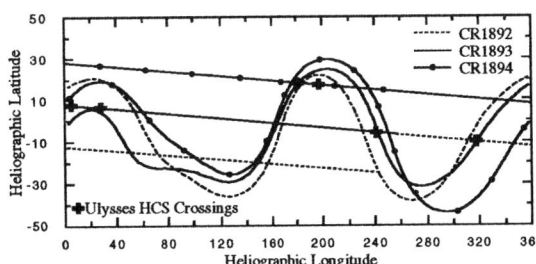

Figure 2. A comparison of the heliospheric current sheet predicted by the source-surface model and an MHD calculation for Carrington rotation (CR) 1869 (May 10–June 6, 1993). The Ulysses spacecraft, which did not observe current sheet crossings during this rotation, was situated at 30° latitude. The circles indicate the crossings predicted by the source-surface model.

Figure 4. Variation of the heliospheric current sheet predicted by the source-surface model for the rotations occuring during the fast-latitude scan. The extent of the HCS varies during this time period.

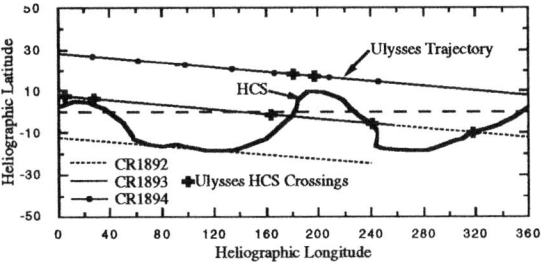

Figure 3. The heliospheric current sheet (HCS) predicted by the MHD model for CR1892, with the Ulysses trajectory for the fast-latitude scan superimposed. HCS crossings identified by *Smith et al.* [1995] are indicated by black crosses. The times of the different rotations (CR1892, CR1893, and CR1894) are coded by line style on the trajectory plot.

are also some differences between our simulations and observations. Fine-scale details of the corona do not appear in our computations. Higher resolution magnetograms (such as those from the National Solar Observatory at Kitt Peak or the SOI/MDI instrument aboard SOHO), coupled with higher resolution computations, may help to capture some of these fine-scale features. Streamers in eclipse images typically show a stronger nonradial tendency than in our the computations. This may be related to the poor estimation of polar fields in the Wilcox data, due to projection effects, and may also be improved by better magnetograms. Most important, our computations (using a polytropic model) fail to reproduce the fast (800km/s) solar wind observed by Ulysses at high latitude. Improvement of this aspect of the calculation requires consideration of the momentum and energy source terms discussed in section 2. Our preliminary 1D and 2D calculations including these terms

show promising results [*Mikić et al.*, 1996ab], and further investigation of these solutions is ongoing.

4. A MODEL OF THE CORONA AND INNER HELIOSPHERE AT SOLAR MINIMUM

To demonstrate how CMEs may be initiated in the corona and propagate out into the heliosphere, we developed a model configuration of the solar corona (and its extension to 1 A.U.) at solar minimum. Guided by our experience with the Wilcox photospheric magnetic field data and our comparisons with eclipse images, we specified an initial magnetic flux distribution of the form:

$$B_r = A_0 \cos^3\theta + A_1 \sin^2\theta \cos 2\phi + A_2 \sin^3\theta \sin 3\phi \quad (7)$$

with $A_0 = 13.3$ G (Gauss), $A_1 = 1.3$ G, $A_2 = 0.33$ G, and the distribution rotated by -20° around the y axis. We then computed an equilibrium configuration by integrating the MHD equations to steady state, as described in section 2, with the additional constraint that the Sun's rigid rotation rate was imposed (corresponding to a sidereal period of 26 days). The resulting configuration is shown in Figure 5. The magnetic field lines and polarization brightness near the Sun (Fig. 5a and 5b) show a configuration similar to that often seen at solar minimum. As we move farther from the Sun, the magnetic field lines show the expected spiral behavior (Figure 5d shows the field lines out to 1 A.U.). With this configuration, we can investigate how different processes can affect coronal evolution, how CMEs might be initiated, and the heliospheric consequences of these events.

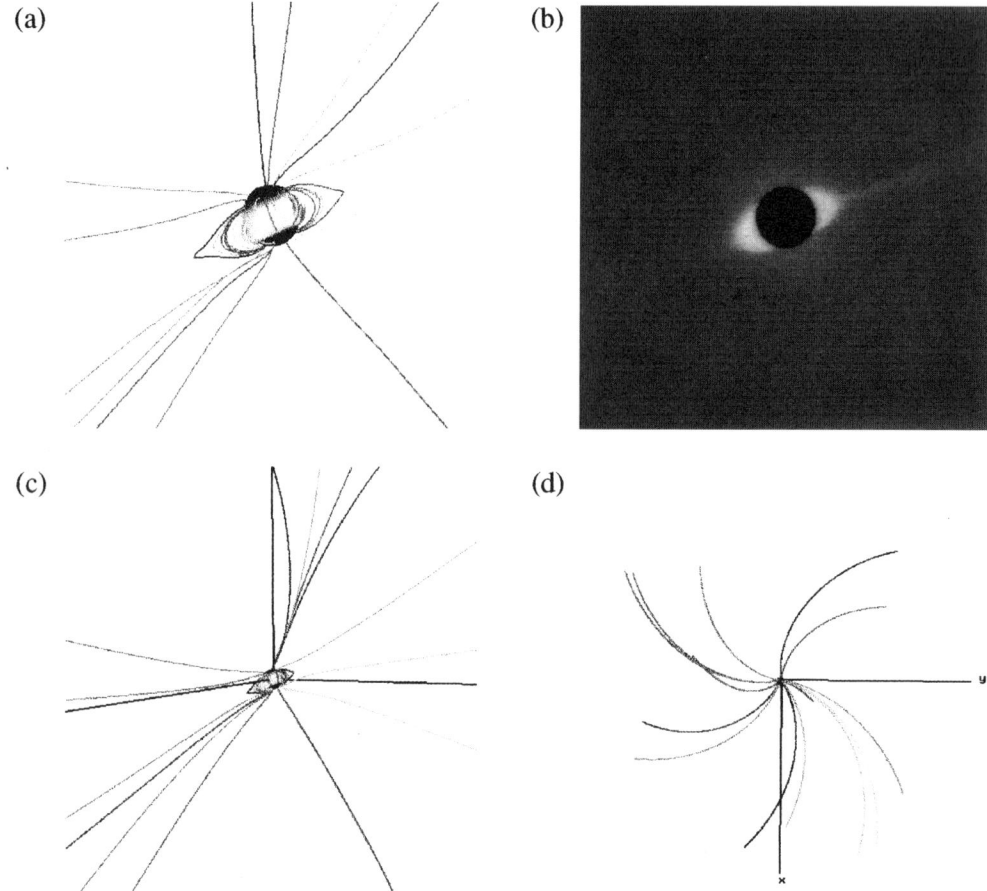

Figure 5. An MHD simulation of the solar corona and inner heliosphere for a solar-minimum type configuration. The computation is performed in the inertial frame, so the magnetic flux distribution on the Sun rotates rigidly. (a) Field lines viewed close to the Sun, showing a helmet streamer configuration. (b) polarization brightness from the same view as (a). (c) Field lines from the same view angle as (a) and (b), but farther from the Sun. (d) Field lines from 1 A.U. above the Sun's North pole. The spiral structure is apparent. Field lines that appear "shorter" actually are receding from or approaching the viewpoint.

5. MODELING CME INITIATION AND PROPAGATION

While the exact cause of CMEs is unknown, it is generally agreed that the amount of energy in the coronal magnetic field (above the energy of the corresponding potential field) is an important factor in determining if and when the coronal magnetic field erupts. Unfortunately, line-of-sight magnetograms do not contain information about the amount of parallel current that is flowing in the coronal magnetic field, so we do not know the magnetic energy state of the corona from these measurements. As is well known from studies of active regions, a vector magnetogram is required to uniquely specify the field. Full-disk vector magnetograms, if available with sufficient accuracy, could allow us to compute coronal configurations that match the measured twist (or shear) in the photospheric magnetic field and determine the magnetic energy state of the corona. Lacking this information, the coronal equilibria we compute (such as those shown in Figure 1 and 5) correspond to helmet streamers with a minimum amount of twist in the magnetic field and probably have an unrealistically low magnetic energy.

One process that might initiate CMEs is shearing of the magnetic field footpoints by photospheric motions [*Mikić and Linker*, 1994; *Linker et al.*, 1994; *Romeliotis*, 1994; *Linker and Mikić*, 1995]. *Mikić and Linker* [this volume] describe how an idealized photospheric shear profile causes the coronal configuration of Figure

5 to evolve, resulting in eruption of a portion of the closed field lines. Because the calculation is starting with a magnetic field that is in a much lower energy state than the actual corona, an artificially long shearing time is necessary to initially energize the field. As we have shown previously [*Linker et al.* 1994; *Linker and Mikić*, 1995], after a sheared helmet streamer erupts, re-formation of the helmet streamer by magnetic reconnection releases only a portion of the available magnetic energy and does not return the configuration to the pre-sheared state. Subsequent eruptions thus require less shear.

Here we discuss a computation where a differential rotation profile, rather than the idealized shear profile described by *Mikić and Linker* [this volume], was introduced. To accomplish the initial energization of the field more rapidly (and reduce computing time), the rotation rate of the Sun was increased by a factor of 10 when we introduced differential rotation. The resulting disruption of the helmet streamer configuration was similar to that described by *Mikić and Linker* [this volume]. The introduction of shear initially causes the magnetic field to expand slowly. When a critical shear is reached, the magnetic field erupts rapidly outward, followed by reconnection of magnetic field lines. In this respect our 3D results are similar to previous 2D results [*Linker et al.*, 1994; *Linker and Mikić*, 1995], but the evolution of the magnetic field in 3D, particularly the magnetic reconnection, is more complicated. As an example of how we can investigate the heliospheric consequences of such eruptions, Figure 6 shows the magnetic field line evolution out to 0.5 A.U. (a portion of the total simulation domain). The black field lines extend out to 1 A.U. in the equilibrium shown in Figure 5; note that the artificially increased rotation rate of the Sun increases the spiral angle for these field lines. A magnetic eruption at $t = 44$ hours results in a portion of the helmet streamer of Figure 5 expanding rapidly outward into the heliosphere (the gray field lines). A subsequent eruption beginning at $t = 76$ hours results in more magnetic flux being carried out into the heliosphere.

In the simpler geometry of the previously mentioned 2D studies, a completely detached plasmoid (a torus surrounding the Sun) propagated outward. In the more complicated 3D case shown here, no magnetic field lines that are completely detached from the Sun are apparent, but their presence has not yet been ruled out. These calculations represent our first efforts to investigate coronal mass ejections in three dimensions and should only be

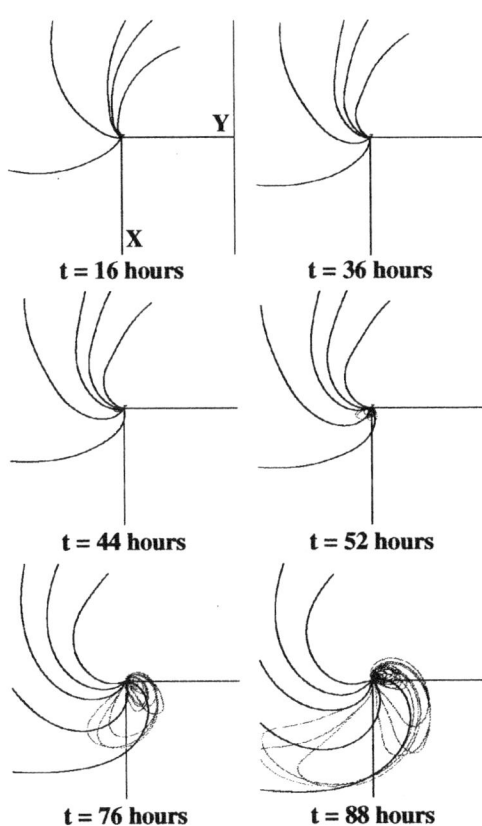

Figure 6. Evolution of magnetic field lines viewed from above the Sun's north pole at 0.5 A.U., after the solar-minimum configuration of Figure 5 is subjected to an enhanced differential rotation profile. Black field lines were open in the initial configuration; gray field lines were initially part of the closed-field helmet streamers. Magnetic eruptions at 44 hours and again at 76 hours cause magnetic flux that previously closed near the Sun to be carried out into the heliosphere.

regarded as a first step. We plan in further studies to examine the details of the magnetic topology both in the corona and far from the Sun, as well as studying other possible initiation mechanisms.

6. SUMMARY

An important element of predicting geomagnetic activity is forecasting solar wind conditions at Earth orbit. Essential to this effort is a computational model of the solar wind capable of describing (1) the structure of the solar corona and inner heliosphere, and (2) the initiation and propagation of coronal mass ejections.

We have described an MHD model that can in principle provide these capabilities. The favorable comparisons of our model with coronal and heliospheric data indicate that the first of these goals is the most feasible, although a more sophisticated treatment of energy transport in the solar wind is necessary to accurately compute solar wind velocities. The second goal is the more difficult task, since our present understanding of CMEs and their initiation is limited. Simulations like those we have described are an attempt to understand the basic phenomena of CMEs. With the SOHO mission now operational and with continued Yohkoh and ground-based observations, the next few years should see a rapid growth in our knowledge of CMEs. The confluence of improved observations and more sophisticated theory and modeling may lead to an improved understanding of CMEs and, eventually, to the capability to forecast CME effects at the Earth.

Acknowledgments. We would like to thank the Wilcox Solar observatory for the use of their synoptic charts, and Serge Koutchmy for providing us with the October 24, 1995 eclipse photograph shown in Figure 1. We also thank Daniel Winterhalter of JPL for providing us with the Ulysses trajectory. This research was supported by NASA, NSF, and AFPL; computations were performed at the San Diego Supercomputer Center and at the National Energy Research Supercomputer Center.

REFERENCES

Altschuler, M. D., D. E. Trotter and F. Q. Orral, *Sol. Phys.*, *26*, 354, 1972.

Altschuler, M. D., and G. Newkirk, *Sol. Phys.*, *9*, 131, 1969.

Courant, R., and K. O. Friedrichs, *Supersonic Flow and Shock Waves*, Interscience Publishers, New York, 1948.

Crooker, N. U., E. W. Cliver and B. T. Tsurutani, *Geophys. Res. Lett.*, *19*, 429, 1992.

Crooker, N. U. and E. W. Cliver, *J. Geophys. Res.*, *99*, 23383, 1994.

Detman, T. R., M. Dryer, T. Yeh, S. M. Han, S. T. Wu and D. J. McComas, *J. Geophys. Res.*, *96*, 9531, 1991.

Endler, F., *Interaction between the Solar Wind and Coronal Magnetic Fields*, Ph.D. Thesis, Gottingen Univ., 1971.

Foukal, P., *Solar Astrophysics*, Wiley-Interscience, New York, 1990.

Gosling, J. T., S. J. Bame, D. J. McComas and J. L. Philips, *Geophys. Res. Lett.*, *17*, 901, 1990.

Gosling, J. T., *J. Geophys. Res.*, *98*, 18937, 1993.

Hoeksema, J. T., Tech. Rep. CSSA-ASTRO-91-01, Center for Space Science and Astronomy, Stanford University, California, 1991.

Holweg, J. V., *Rev. Geophys. Space Phys.*, *16*, 689, 1978.

Houghton, J. T., *The Physics of Atmospheres*, Cambridge University Press, Cambridge, 1977.

Hu, Y. Q. and S. T. Wu, *J. Comp. Phys.*, *55*, 33, 1984.

Hundhausen, A. J., *Rev. Geophys.*, *17*, 2034, 1979.

Hundhausen, A. J., *J. Geophys. Res.*, *98*, 13177, 1993.

Jacques, S. A., *Astrophys. J.*, *215*, 942, 1977.

Linker, J. A., G. Van Hoven and D. D. Schnack, *Geophys. Res. Lett.*, *17*, 2281, 1990.

Linker, J. A. and Z. Mikić, *Astrophys. J.*, *438*, L-45, 1995.

Linker, J. A., Z. Mikić and D. D. Schnack, in *Solar Drivers of Interplanetary and Terrestrial Disturbances*, (K. S. Balasubramaniam, S. L. Keil, and R. N. Smartt, eds.), Astron. Soc. Pac. Conf., *95*, 208, 1996.

Linker, J. A., Z. Mikić and D. D. Schnack, in *Solar Dynamic Phenomena and Solar Wind Consequences*, Proc. Third SOHO Workshop, Estes Park, Colorado, (ESA SP-373), 249, 1994.

Maunder, E. W., *Mon. Not. R. Astron. Soc.*, *65*, 2, 1905.

Mikić, Z. and J. A. Linker, *Astrophys. J.*, *430*, 898, 1994.

Mikić, Z., and J. A. Linker, *Solar Wind 8*, AIP Conf. Proceedings 382, 104, 1996.

Mikić, Z., and J. A. Linker, this volume.

Mikić, Z., J. A. Linker and J. A. Colborn, *EOS Trans. AGU* (abstract), 77, 1996a.

Mikić, Z., J. A. Linker and J. A. Colborn, AAS/SPD Meeting (abstract), Madison, Wisconsin, 1996b.

Odstrcil, D., *J. Geophys. Res.*, *99*, 17653, 1994.

Parker, E. N., *Interplanetary Dynamical Processes*, Interscience Publishers, New York, 1963.

Pizzo, V., *J. Geophys. Res.*, *96*, 5405, 1991.

Pneuman, G. W. and R. A. Kopp, *Solar Phys.*, *18*, 258, 1971.

Roumeliotis, G., P. A. Sturrock, and S. K. Antiochos, *Astrophys. J.*, *432*, 847, 1994.

Schatten, K. H., J. M. Wilcox and N. Ness, *Solar Phys.*, *6*, 442, 1969.

Smith, Z., and M. Dryer, *Solar Phys.*, *129*, 1990.

Smith, E. J., M. Neugebauer, A. Balogh, S. J. Bame, G. Erdos, R. J. Forsyth, B. E. Goldstein, J. L. Phillips and B. Tsurutani, *Geophys. Res. Lett.*, *20*, 2327, 1993.

Smith, E. J., A. Balogh, M. E. Burton, G. Erdos and R. J. Forsyth, *Geophys. Res. Lett.*, *22*, 3325, 1995.

Steinolfson, R. S., S. T. Suess and S. T. Wu, *Astrophys. J.*, *255*, 730, 1982.

Usmanov, A. V., *Solar Phys.*, *146*, 207, 1993.

Wang, A. H., S. T. Wu, S. T. Suess and G. Poletto, *Sol. Phys.*, *147*, 55, 1993.

Wang, Y. M. and N. R. Sheeley, *J. Geophys. Res.*, *93*, 11, 227, 1988.

Wang, Y. M. and N. R. Sheeley, Jr., *Astrophys. J.*, *392*, 310, 1992.

Washimi, H., Y. Yoshino and T. Ogino, *Geophys. Res. Lett.*, *14*, 487, 1987.

Wright, J. M., Jr., et al., *National Space Weather Progrom Strategic Plan*, FCM-P30-1995, 1995.

Zirin, H., *Astrophysics of the Sun*, Cambridge University Press, New York, 1988.

Jon A. Linker and Zoran Mikić, Science Applications International Corporation, San Diego, CA 92121.

Coronal Mass Ejections, Corotating Interaction Regions, and Geomagnetic Storms

A. H. McAllister[1]

High Altitude Observatory, PO Box 3000, Boulder, Colorado

N. U. Crooker

Center for Space Physics, Boston University, Boston, Massachusetts

The geomagnetic activity index, *Dst*, solar wind data, and solar coronal images have been analyzed and compared for 10 solar rotations from August 1993 to May 1994. A description of the background solar wind patterns is given for the period, and the modification of the geoeffectiveness of this pattern by seasonal effects is reviewed. Most of the geomagnetic storms are found to be associated with passage of the sector boundaries in the interplanetary magnetic field. The geoeffectiveness of these storms is determined by three components. One is the corotating interaction region (CIR) formed at the leading edge of high-speed streams from major coronal holes, a traditional source for recurrent geomagnetic storms. All of the storms near sector boundaries were followed by high-speed streams, and by assumption, were associated with CIRs. In the study period, the modification of the geoeffectiveness of these CIRs by the second component, seasonal effects, was quite strong. The final component is due to coronal mass ejection (CME) driven solar wind transients. Association of solar signatures with each storm suggests that all major storms, and probably most moderate storms, have a transient component. We conclude that major geomagnetic storms, both recurrent and nonrecurrent, are the result of the combined effects of CMEs and CIRs.

1. INTRODUCTION

The large eruptions of solar coronal magnetic fields and plasma, called coronal mass ejections (CMEs) [*Hundhausen*, 1996], are believed to be the main source of the strong interplanetary disturbances and shocks that cause many nonrecurrent geomagnetic storms [*Sheeley et al.*, 1985; *Gosling et al.*, 1991] and may play a role in the largest recurrent storms as well [*Crooker and Cliver*, 1994; *Crooker and McAllister*, 1997]. The association of CMEs with geomagnetic disturbances is covered in reviews by Gosling [1993], Hundhausen et al. [1984], Kahler [1992], and Webb [1993]. Direct observations of CMEs, made primarily with white-light coronagraphs (e.g., NASA's SMM, HAO's Mauna Loa, and NRL's Solwind) are limited to events within about $35°$ of the solar limb [*Hundhausen*, 1993]. The ability to effectively predict geomagnetic storms requires information about CMEs that occur on the solar disk and

[1] Now at Helio Research, La Crescenta, California

propagate toward Earth. Past attempts to use proxies (e.g., filament eruptions [*Joselyn and McIntosh*, 1981]), have proved useful, but with limitations (see review by Neugebauer, [1988]).

One of the discoveries made by Skylab was that filament eruptions and CMEs are often followed by coronal x-ray brightenings of loop or arcade structures [*Sheeley et al.*, 1975; *Webb, Krieger, and Rust*, 1976]. These events are often referred to as 'long duration events' (LDEs) and are marked by a characteristic signal in the integrated x-ray flux (when visible above background). Similar dynamic arcade structures associated with filament eruptions are a common feature of the Yohkoh Soft X-ray Telescope (SXT) observations [*Hanaoka et al.*, 1994; *McAllister et al.*, 1992, 1996a]. The Yohkoh SXT data also confirm that coronal arcades form in the aftermath of some CMEs on the quiet Sun [*Hiei, Hundhausen, and Sime*, 1993; *Sime, Hiei, and Hundhausen*, 1994; *Hundhausen*, 1996]. Statistical work by McAllister and Hundhausen [1996] has shown that x-ray arcades, like CMEs [*Hundhausen*, 1993], form mainly at the base of the coronal streamer belts. Based on their survey of 240 quiet Sun events over 18 months during the declining phase of solar cycle 22, and the ongoing accumulation of direct and indirect association of arcades with CMEs, they conclude that medium and large scale arcades are a consistent signature of eruptive events associated with CMEs. It is also generally accepted that large active region LDEs with an arcade appearance are reliable signatures of CMEs, although no systematic study has yet been done.

The dynamic coronal arcades allow the accurate mapping of the partial photospheric footprints of a large fraction of all CMEs [*McAllister and Hundhausen*, 1996] and give us a list of potentially geoeffective interplanetary events. The timing of the launching of a CME and the appearance of an x-ray arcade are similar to that between a filament, or prominence, eruption and the appearance of the following two ribbon flare, ranging from 10s of minutes to perhaps several hours. The arcades thus also give us a launch time to within an hour or two, being more accurate for fast events and less so for slow events. Not all interplanetary disturbances created by CMEs, however, are equally geoeffective [*Gosling et al.*, 1991]. Only those that have a suitable orientation, structure, and velocity relative to the background solar wind, such that a sustained large southward magnetic field (Bz) is driven hard into the nose of the magnetosphere, can generate large geomagnetic storms [*Tsurutani and Gonzalez*, 1987; *Tsurutani et al.*, 1988; *Gosling et al.*, 1991; and references therein]. A comparison of the number of on-disk arcade events (≈ 100) [*McAllister and Hundhausen*, 1996], during the period of the present study with the number of storms (21, though some probably do not involve transients) gives a rough estimate of the odds.

Besides the location and timing of a CME, the coronal arcades can provide the size and orientation of the arcade structure (and other associated signatures, [e.g. *Hudson, Acton, and Freeland*, 1996]), and some sense of the geometry of the CME as it leaves the Sun [*Martin and McAllister*, 1997, also this volume]. These factors are not sufficient, however, to predict the characteristics of the downstream disturbances at Earth, as no reliable information about the speed and direction of the CME is available. Study of the SMM data set has shown there is no strong correlation between the magnitude of an associated x-ray event and the size or speed of a CME (A.J. Hundhausen, keynote address, Chapman Conference on CMEs, Bozeman, 1996). Moreover, it is almost certain that once a CME is launched, it begins to interact with the background solar wind. The location of a CME relative to major coronal structures (e.g., coronal holes and the heliospheric current sheet) provides clues to the interplanetary environment through which it propagates and, therefore, to its evolution en route to Earth [*Crooker and Cliver*, 1994; *Crooker and McAllister*, 1997].

To be able to improve predictions of geomagnetic activity, we need to answer two questions: (1) What does a CME look like on the disk? (2) Which solar events (CMEs) are geoeffective? The first question has only been partially answered by the recent work on coronal arcades. The use of arcades as CME proxies in conjunction with measurements of the solar wind near Earth [e.g. *Crooker and McAllister*, 1997; *McAllister, Knipp, and Crooker*, 1997] offers the opportunity to clarify the associations of solar signatures and geomagnetic events. Worked in reverse, from the solar signatures to the near-Earth solar wind, these data sets provide a tool for exploring the extent to which CMEs interact with, and are modified by, the background solar wind.

In this paper, we present results of a comprehensive analysis of a 10 solar rotation period from August 12, 1993, to May 8, 1994, (partially covered by Crooker and McAllister [1997]) to show how geomagnetic, solar wind, and coronal data can be applied to these questions. We start with a description of the background solar wind pattern for the period and the modification of the geoeffectiveness of this pattern by seasonal effects (Section 2). We then describe the role of corotating interaction regions (CIRs) that occur at the leading edge of the high-speed streams from the major coronal holes (Section 3) and the role of CMEs (Section 4) in recurrent geomagnetic activity, and we conclude with a summary and discussion (Section 5).

2. BACKGROUND SOLAR WIND STRUCTURE

The geomagnetic index Dst from August 1993 to May 1994 is shown in Figure 1, where the data are plotted in 27 day strips corresponding to the solar rotation period. Intervals of large negative Dst indicate significant geomagnetic activity. During this period, there were 21 intervals with $D_{st} < -50$ nT, defined as storms in this paper, and numbered in Figure 1. These storms are due to a combination of quasi-steady state solar wind structures, giving rise to 'recurrent' behavior, and solar wind transients from CMEs.

The period of our study covers the last half of the declining phase of solar cycle 22. The background solar wind structure reflected the slowly evolving global structure of the solar corona [*Hundhausen*, 1977]. In the ecliptic plane, this resulted in a sequence of slow- and high-speed flows, the former associated with the magnetically closed streamer belts and the latter with the open coronal holes. The structure of the solar wind speed near Earth during this period is shown in Figure 2.

The shading in Figure 1 indicates solar wind sectors in which the interplanetary magnetic field points predominantly away from or toward the Sun. Each sector is associated with an equatorward extension of one of the polar coronal holes. The heliospheric current sheet (HCS) forms the boundary between the sectors and maps back to the streamer belt on the Sun. As the streamer belt warps around the coronal hole extensions, the HCS is similarly warped, and the rotation of the Sun consequently brings first one polarity and then the other to bear on Earth. Figure 3 shows an example of the solar configuration that corresponds to the toward-away transition of early November 1993, in which the ecliptic fields were connected first to the south polar coronal hole and later to the north polar coronal hole, separated by a passage of the HCS.

As the Sun rotates and the high-speed stream from the coronal hole catches up with the slow flow associated with the closed streamer belt region, a corotating interaction region (CIR) forms on the leading edge of the high-speed stream [e.g., *Pizzo*, 1978]. The resulting compression peak closely follows HCS passage at 1 AU [*Gosling et al.*, 1978]. Since the compression can produce an enhanced southward Bz component from the magnetic field fluctuations CIRs, are usually associated with peak recurrent storm strength [*Crooker and Cliver*, 1994; *Tsurutani et al.*, 1995; and references therein]. The following high-speed flow often produces a long period of enhanced activity that can last for more than a week [*Tsurutani and Gonzales*, 1987]. This pattern can be seen in Figure 2 in the main away sector in the

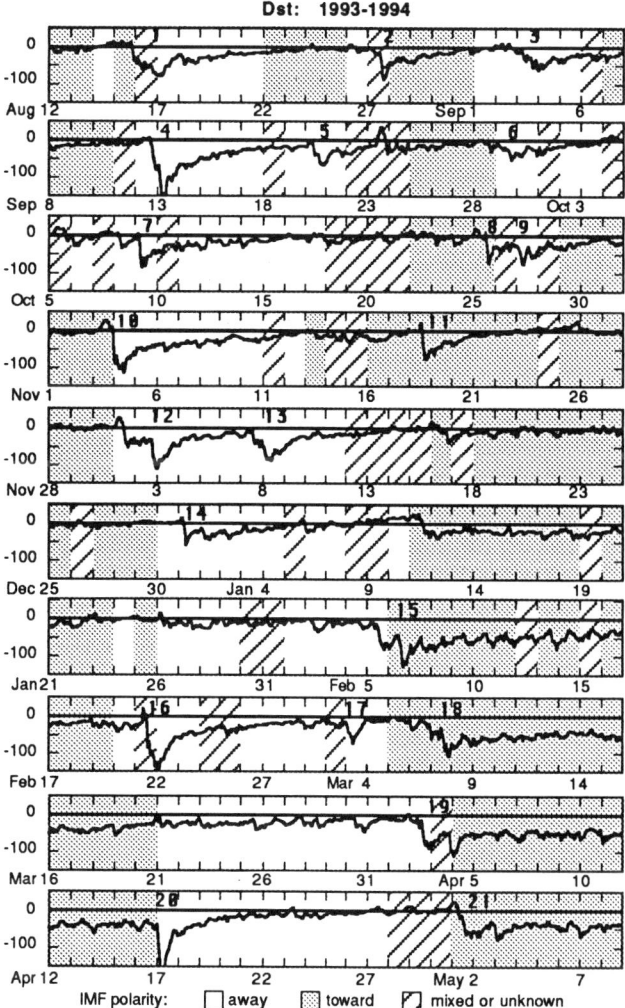

Figure 1. Dst for 10 solar rotations (Bartels, shifted by 5 days) from August 1993 - April 1994. The data was obtained from the National Space Science Data Center's OMNI data set. The numbers at the top indicate the 21 storms discussed in the text (adapted from Figure 1 in *Crooker and McAllister* [1997]).

fall of 1993 and especially clearly in the toward sector (shaded in the spring of 1994; see, also, below).

During the early fall months of 1993 (rotations 1-3), a four-sector structure was giving way to a two-sector structure that was maintained for the rest of the Figure 1 period. The initial configuration was driven by 3 significant polar coronal hole extensions at lower latitudes, one in the south and two in the north. The sector structure change reflected the disappearance of one of the northern extensions and the ongoing evolution of the other two. The resulting two-sector structure

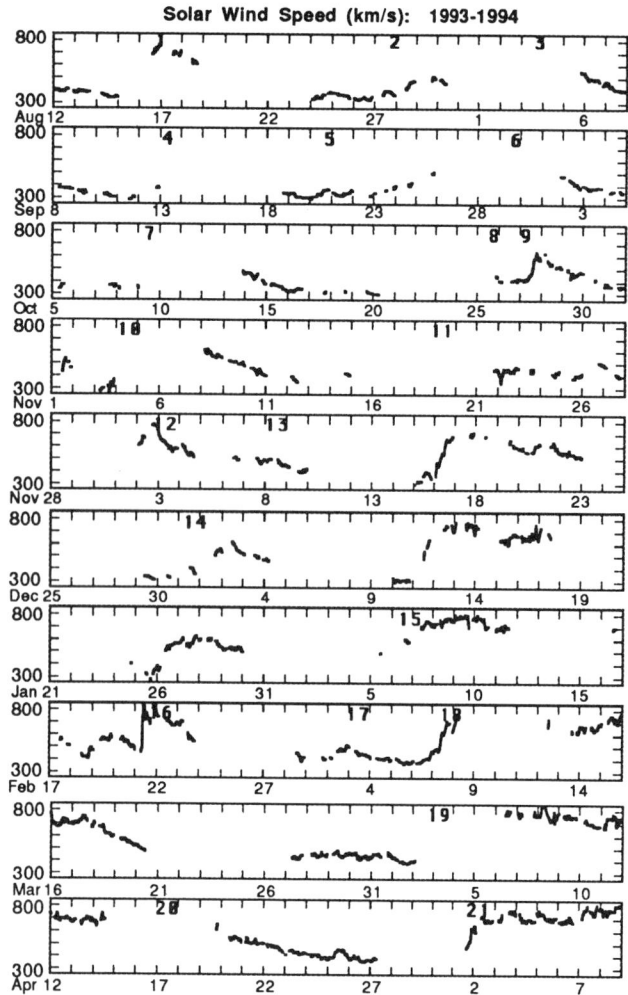

Figure 2. The solar wind speed measured by the IMP-8 satellite for 10 solar rotations from August 1993 to April 1994. The data were obtained from the National Space Science Data Center's OMNI data set, (adapted from Figure 2 in Crooker and McAllister [1997]).

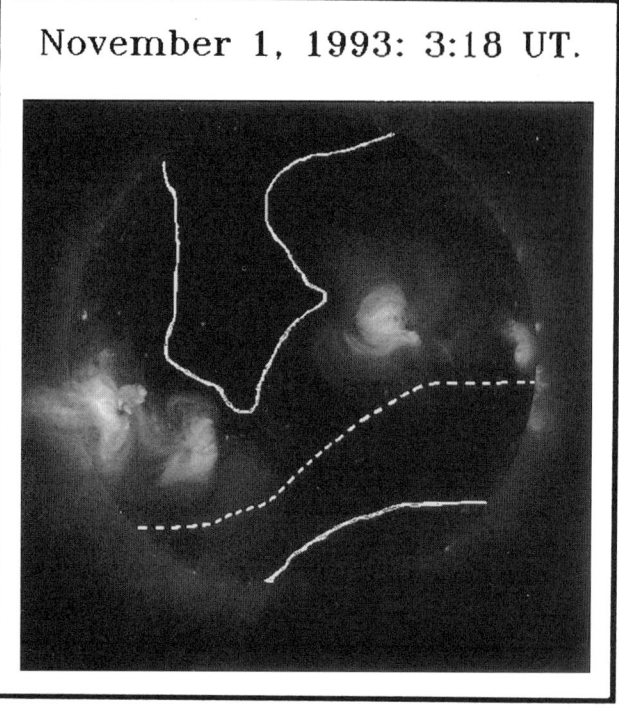

Figure 3. A Soft X-ray Telecope image from 3:18 UT November 1, 1993. The contours highlight the large extension of the northern coronal hole (corresponding to an away sector), the obscured south polar coronal hole (corresponding to a toward sector), and the projection of the heliospheric current sheet (dashed), which also indicates the base of the main coronal streamer belt.

is typical of the late declining phase of the solar cycle [Hundhausen, 1977], when the corona can be roughly described as a tilted dipole.

There are, unfortunately, significant data gaps in Figure 2 due to the passage of the IMP-8 satellite through Earth's magnetotail roughly twice during each solar rotation. The dependence of the main solar wind high-speed streams on the solar coronal hole locations, coupled with the general recurrent nature of these structures and the trends visible in the available velocity data, allow us to qualitatively fill in many of these gaps. The Dst plot serves as a check for the continuity and timing of the main high-speed streams and the existence of geoeffective transients during this period. In a few instances (e.g., November 3-8, 1993), we can explicitly fill in the gap using GEOTAIL solar wind data. These various data streams combine to provide a consistent estimate of the large scale solar wind structures across all but a few of the gaps.

The geomagnetic impact of the CIRs, which we assume exist on the leading edge of each observed and deduced high-speed stream, is conditioned by three main factors, two of which are seasonal [Crooker et al., 1996; and the references therein]. The best known is the Russell-McPherron effect [Russell and McPherron, 1973], due to the changing relative orientation of the solar and terrestrial axes as Earth orbits the Sun. This results in periods when the magnetic field along the Parker spiral project a southward Bz component in Earth's tilted dipole system, thus enhancing the geoeffectiveness of ecliptic fields during these periods. The effect is strongest in the fall, when it enhances activity in the away sector, and in the spring, when it enhances activity in the toward sector, and the effect reverses its hemispheric (northern/southern) association with each 11 year solar cycle. A second seasonal effect, also due to the changing axial relationship, places Earth more deeply in the high-speed streams associated with the southern polar coronal hole extensions in spring and the northern extensions in the fall [*Murayama*, 1974; *Bohlin*, 1977; *Crooker et al.*, 1996].

The third factor is the evolution of the specific coronal holes that give rise to the high-speed solar wind. In the period covered by this study, the main extension of the north polar coronal hole (see Figure 3) was reduced by newly emerging flux after December 1993, while the equatorward extension of the south polar coronal hole grew significantly after January 1994. The coronal hole changes produced solar wind high-speed stream changes that served to enhance the seasonal emphasis given to the main away-toward boundary in the fall of 1993 and the toward-away boundary in the spring of 1994. The cumulative nature of these three effects thus 'favored' certain sector passages, while others remained 'unfavored'. In particular, there is pronounced sustained geomagnetic activity associated with the favored spring toward sectors (see Figure 1).

3. THE GEOEFFECTIVENESS OF CIRS

During the declining phase of the solar cycle, geomagnetic activity for most of each rotation is described (to first order) by the broad streams of the the steady solar wind, the evolution of these streams as the global coronal configuration changes, and the modifying seasonal factors. In tabulating geomagnetic activity, the following activity levels have been used: $D_{st} < -25$ nT indicates disturbed conditions and $D_{st} < -50$ nT indicates storm conditions, with $D_{st} < -75$ nT for moderate and $D_{st} < -100$ nT for major storms. While the extended high-speed streams can cause significant sustained geomagnetic activity (e.g., the spring toward sectors), the peaks of the recurrent storms occur in association with the CIRs at the leading edge of the high-speed streams.

During the 10 Bartels rotations of our study, there were 21 storms ($D_{st} < -50$ nT), of which 15 were associated with deduced CIRs, 3 [S3, S9, S11] were possibly associated (but are not counted as such in this paper), and 3 [S5, S13, S17] were definitely unassociated. Figure 4 shows histograms of storm strength, separated into storms associated with CIRs and those that were not. The CIR-associated storms are ones that have traditionally been classed as recurrent storms, and the remainder as nonrecurrent storms [*Legrand and Simon*, 1989]. During this period, the majority (79%) of the peak geomagnetic activity occurred in association with CIRs. Moreover, the average strength of the

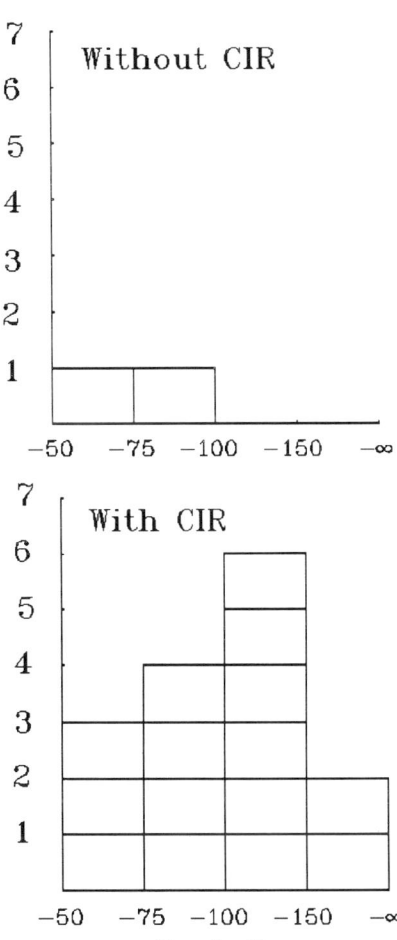

Figure 4. The distribution in peak *Dst* of storms associated, and not associated with CIRs. The tabulation covers the same period as the data in Figures 1 & 2.

mid-sector storms (-78 nT) is significantly lower than that for those associated with the CIRs near the sector boundaries (-105 nT), and all 8 of the major storms are CIR-associated.

There were 28 sector boundary crossings in the period of the study: 6 in rotation 1; 4 in rotations 2 and 3; and 2 in the remaining rotations. Of these, we deduce that at least 19 had associated CIRs, based on the following high-speed streams. (The August 27, 1993 storm [S2], even though it has not been counted as having an associated CIR, was followed by a very moderate stream 200 km/s faster than the surrounding slow flow).

In Section 2, we introduced the concept of 'favored' sectors, those which are favored by the seasonal effects and are thus more likely to generate geomagnetic activity. In the period of this study, the favored sectors are the away sectors in August-December 1993 and the toward sectors in February-May 1994. Combined with the above result, this suggests that the leading edges of 'favored' sectors may be particularly geoeffective.

Figure 5 shows histograms of the peak Dst associated with the leading edges of the seasonally favored and unfavored sectors. The difference in activity is clear. Only 4 of the unfavored sectors are associated with leading edge activity at storm level [S2, S14, S16, S20], of which 3 had probable transient components [S2, S16, S20]. On the other hand, only 1 of the favored sectors had leading edge activity which peaked below the storm level, and that sector lacked high-speed flow (September 23, 1993). Although half of the unfavored sectors had no leading CIR (all of these fall passages), there were 4 that contained significant high-speed streams with presumably strong CIRs (from late December 1993 on, see also the discussion of Figure 6). The clearest case for the ability of seasonal factors to suppress, as well as to enhance, the effect of a high-speed stream is seen by comparing Figure 1 and Figure 2 for January 26 through 30, 1994. This major high-speed stream is associated with very quite geomagnetic conditions.

The contribution of the high-speed streams (and assumed CIRs) in this period is highlighted in Figure 6, in which the magnitude of peak Dst following each sector boundary crossing (3 cases were not plotted due to uncertainties in the solar wind speed estimation) is tabulated against the qualitative significance of any associated high-speed streams (the top two categories were assumed to have generated CIRs). The lack of cases in the lower right corner of this table strongly indicates that a large storm will not occur in the absence of a high-speed stream. The 2 cases which came closest to being exceptions are associated with storms S2 and S8, while the 4 cases to the upper left are the 4 mentioned above as being apparently suppressed by the unfavorable seasonal effects. These results show a nearly bimodal correlation between stream strength and the magnitude of peak Dst following each crossing. However, the range of peak Dst associated with the leading edges of moderate and large streams is considerable. The presence of a CIR near the leading edge of a favored sector nearly guarantees a storm-level magnetospheric reaction, but its magnitude cannot be predicted with any precision.

4. THE ROLE OF CMES.

That the presence of a CIR, modified by seasonal effects, does not explain all the variation in geomagnetic activity suggests that transients as well as steady

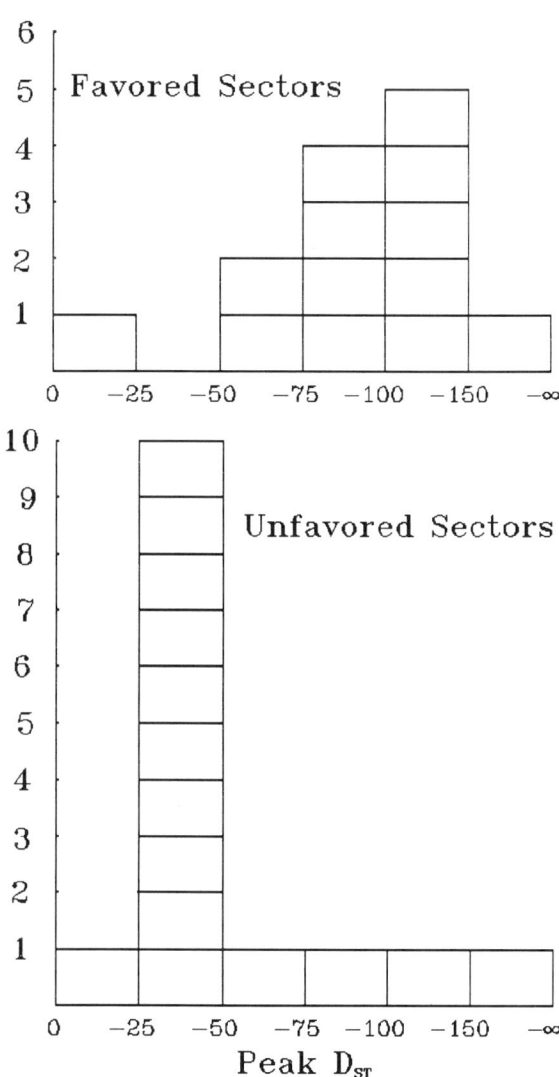

Figure 5. The distribution of peak Dst following the sector boundaries during the period covered in Figures 1 & 2. The crossings into seasonally favored and unfavored sectors are shown separately.

streams play an important role. The wide range of storm strength associated with CIRS at the leading edges of both favored (peak Dst = -53 to -169 nT) and unfavored (peak Dst = -52 to -201 nT) sectors (only the borderline August 27 storm [S2] not being counted as associated with a CIR) suggests another variable is involved (see Figure 5). The separation of the CIRs into these two classes should remove most of the purely seasonal effects, and the general evolution of the coronal

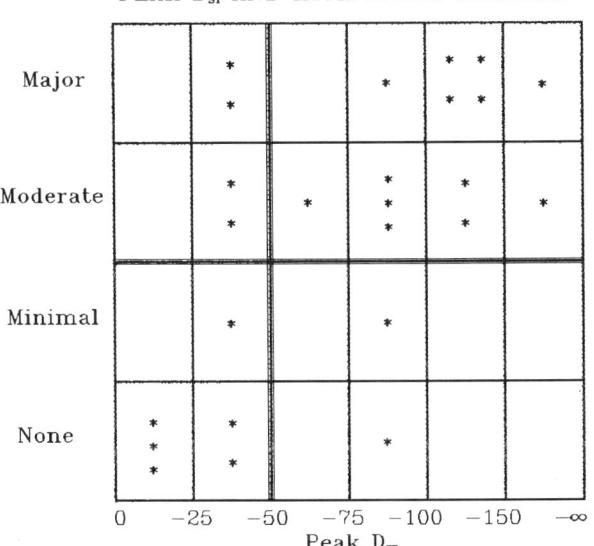

Figure 6. The estimated qualitative high-speed stream significance following the sector boundary crossings during the period covered in Figures 1 & 2. The four categories are: none, flows below 400 km/s; minimal, flows up to about 500 km/s; moderate and major, roughly 600 km/s-800 km/s, depending on a combination of peak speed and duration. The stream significance is plotted against the peak Dst in tabular form.

holes is not rapid enough to explain all the variation within each class. Secondly, there are several storms that show a 'double dip' characteristic (e.g. S12, S14, S18). While it is reasonable for one of these dips to be due to a CIR, it seems likely that this is not true for both of them (e.g. December 3, 1993, see below). Finally, it is generally accepted that major storms cannot be generated by CIRs alone [*Gosling*, 1993; *Tsurutani et al.*, 1995]. Therefore the presence of major storms (peak D_{st} < -100 nT) at the leading edges of both favored and unfavored sectors also suggests the presence of transients.

In order to test for the contribution of CMEs, we have worked backward from Earth to the Sun to find associated solar events, i.e., coronal arcades and other eruptive signatures. This process involves estimating the Sun-Earth transit time, defining a temporal window, and searching the Yohkoh SXT data for the coronal CME signatures described in Section 1 (for more detail, see *McAllister, Knipp and Crooker* [1997]; *Weiss et al.* [1996]; *Crooker and McAllister* [1997]). Preference was given to events within 45° of disk center, and note was taken of the relationship of the signature to the HCS at the Sun compared to the relationship of the storm to the sector boundary crossing at Earth. We have carried out this process for all the events used above (both storms and unactive sector leading edges) and have tabulated the results in Figure 7. The associations are categorized as: *strong*, when there are clear eruptive signatures and we have a high confidence that we can select the specific event; *moderate*, when there were probable eruptive signatures, or several eruptives, none of which had better timing and location; *weak*, when there were only possible signatures, or probable signatures with poor timing or location; and *none*, when there were no clear eruptive signatures in the temporal and spatial window.

Figure 7 shows a general trend for the stronger associations to correspond to the larger storms. Overall, 18 of the 20 storms (S13 was omitted due to a one-day gap in the SXT data) had moderate to strongly associated signatures of CMEs at the Sun. On the other hand, 11 of the 13 sector leading edges without storms had no or weak associations. For storm [S10] we have listed two associations [*McAllister, Knipp, and Crooker*, 1997], as there were two separate transients. These statistics speak strongly in favor of the importance of CME driven disturbances contributing to the 'recurrent' events as well as the 'nonrecurrent' events during this period. The expanded sample analyzed here continues to support the conclusion of Crooker and McAllister [1997] that while CIRs are usually associated with storms near sector boundaries, the variation in storm size is consistent with variations in a CME driven transient component, as suggested by the coronal signatures (see also Crooker and Cliver, [1994]).

This relationship can be seen best in some cases where the transient and CIR components of the solar wind were separated. The toward-away passage of early December 1993 [S12] is a good illustration. In Figure 1 we can see the effects of CIR passage on December 1 pushing Dst down to $\approx$ -50 nT. On December 3, there is another sharp drop to -111 nT, which is well associated with a large south polar crown dynamic arcade (an excellent CME signature). We suggest that the occasional separation between a CME and an associated CIR may account for other 'double dip' events, as well.

The two major storms [S16, S20] in the unfavored category are those associated with the solar events of February 20 and April 14, 1994. The first one was associated with an x-ray class M 4.0 active region LDE and a very fast CME, the second with a large polar crown arcade and an above average speed CME. These associations are well studied and very strong [*Lemen et al.*, 1996; *McAllister et al.* 1996b; *Weiss et al.*, 1996], They show how strong transients, possibly in combination with CIRs, have been able to produce storms at the leading edge of unfavored sectors, when similar CIRs

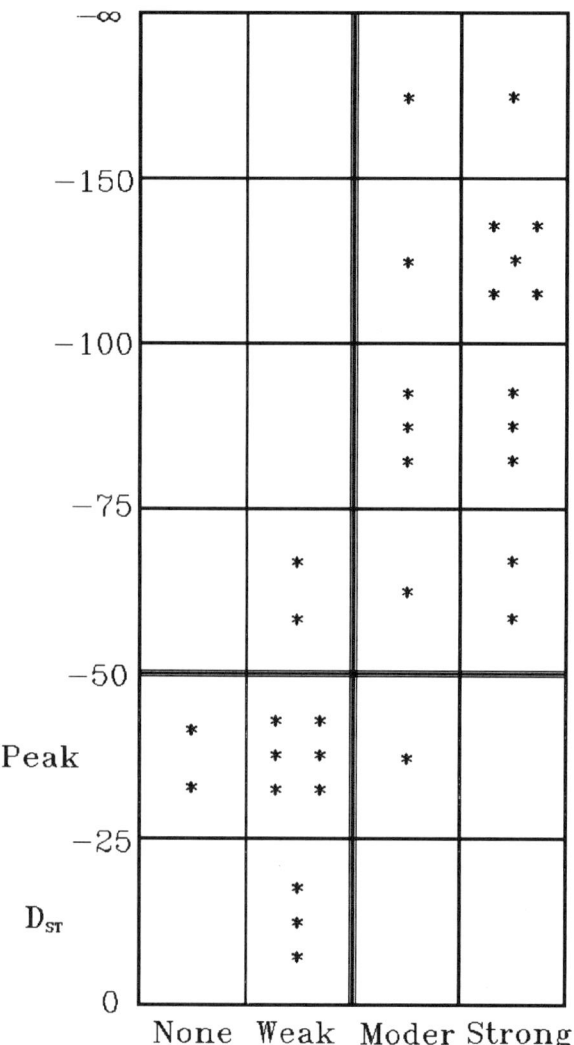

Figure 7. The strength of the solar associations for each storm and non-storm sector leading edge included in the statistics. The strength of the associations depends both on the probability that a solar signature indicates an eruptive event and on the timing relative to the period of interest at Earth. The solar associations are plotted against the peak Dst in tabular form.

(high-speed streams) were not able to do so (January 26 and March 21, 1994). A third case, at the leading edge of a favored sector, has also been analyzed in detail. The November 3, 1993 storm (S10) was caused by transients in combination with a CIR [*McAllister, Knipp, and Crooker*, 1997]. The S10 analysis benefited from excellent solar wind coverage by the GEOTAIL satellite. Likewise storm A of Tsurutani et al. [1995] had indications of 3 transient streams (see, also, Cliver [1997]).

Overall, the associated solar events were divided between 14 medium and large quiet Sun arcades (6 along the polar crown; 2 on switchbacks, diagonal neutral lines connecting to the polar crown; and 6 on midlatitude neutral lines) and 5 active region LDEs (plus one possible), two of which had multiple likely candidates. Thus there was a predominance of quiet Sun arcades over active region LDEs during this declining phase of the solar cycle.

5. SUMMARY AND DISCUSSION.

The comparison of the geomagnetic activity index, Dst, with near-Earth solar wind data and solar coronal imagery has been undertaken for the period of August 1993 to May 1994. Comparison of the location of coronal holes with the terrestrial solar wind data indicates that the broad strokes of geomagnetic activity are painted by the background solar wind: the interplay of the slow and high-speed streams which produce CIRs. Sustained geomagnetic activity correlates well with the seasonally favored larger high-speed streams, and most of the geomagnetic storms are associated with CIRs. All of the large storms were near sector boundaries and only a few small and moderate storms occurred in midsector.

The geoeffectiveness of the CIRs is strongly modified by the long-term evolution of the coronal holes and by the seasonal effects due to the changing relative orientation of Earth and the Sun. When these effects are used to define favored sectors, we have found that 12 of 16 CIR-associated storms occurred in favored sectors. Moreover, there are no favored sectors with peak $D_{st} < -50$ nT, and only one favored sector with $D_{st} > -50$ nT. It is, therefore, clear that during the declining phase of the solar cycle, the CIRs at the leading edge of these favored sectors are geomagnetically important.

These results confirm the well-known dominance of recurrent storms in the declining phase of the solar cycle and appear to agree with the hypothesis that these storms are driven by the CIRs at the leading edge of high-speed streams in the solar wind. The wide range in peak storm strength (even after factoring out the seasonal effects), the presence of major storms, and the double dip character of others, all suggest, however, that we must look for a transient component as well. To do this, we searched for signatures, or absence of associated solar events. The results strongly indicate that 1) the range of CIR-associated storm strengths, after factoring for seasonal effects, is due to the frequent presence of CME driven transients, and that 2) major storms, $D_{st} < -100$ nT, are associated with the leading edge of a high-speed stream, where there would normally be a CIR and one or more CME driven transients. Our survey suggests that a transient component is also often present in moderate storms (-75 nT$>D_{st} > -100$ nT).

Past efforts to classify storms as being either recurrent (i.e. related to a coronal hole stream) or nonrecurrent (i.e., related to a CME driven disturbance) seem to have been too restrictive. In 1974, during the declining phase of solar cycle 20, Tsurutani et al. [1995] found only 3 major storms ($D_{st} < -100$ nT). These were associated with sector boundary passages and determined to be induced by nonrecurring streams. However, there were associated recurring coronal hole streams, as well, and the data appear to be entirely consistent with CME driven transients arriving along the HCS, followed by recurring high-speed streams. A recurring stream onset is listed on day 186, one day after the start of the first major storm, and the solar wind data clearly show a strong extended high-speed stream following this event. The high-speed stream associated with the second and third storms was weaker than on previous passages, and arrived about a day after the storms; however, it fits well in a recurring sequence. The general solar wind configuration is similar to the one we have found two solar cycles later, and our results suggest that the associated high-speed streams are not irrelevant.

But why should this configuration be so common? As mention in Section 1, the majority of CMEs occur in the main streamer belts at the base of the HCS [*McAllister and Hundhausen*, 1996]. Figure 3 illustrates the fact that in the declining (tilted dipole) phase of the solar cycle, the streamer belt crosses the ecliptic in conjunction with lower-latitude extensions of the polar coronal holes. Assuming radial propagation, the CMEs that are most likely to impact Earth will, therefore, occur along those sections of the streamer belt that are associated with the sector boundary crossings in the ecliptic and are usually followed by high-speed streams and CIRs [see, also, *Crooker and Cliver*, 1994]. This relationship is tightened further by the clustering of CMEs along certain sections of the main magnetic neutral line under the base of the heliospheric current sheet [*McAllister et al.*, 1996c]. These clusters are directly related to coronal hole locations and the global coronal dynamics that lead to changes in coronal hole size and location. While these clusters are not always near-equatorial crossings, many of them are, and many others are so large that their ends may extend to the crossing (e.g., April 14, 1994, the solar event associated with storm [S20] [*McAllister et al.*, 1996b]. We therefore find a natural association of CMEs, ecliptic crossings, and coronal holes.

The association of coronal holes and CMEs has been noted previously [*Crooker and Cliver*, 1994; *Bravo*, 1995; *Bhatnager*, 1996]. Bravo and coworkers have moved further to suggest that CMEs which are associated with either permanent or temporary enlargements of coronal holes are followed by new high-speed streams and that it is these CMEs which are responsible for the largest geomagnetic storms. Our survey confirms that the association of CMEs with high-speed streams is indicated for major storms. It is also well known from the Yohkoh SXT data that arcade events are sometimes accompanied by transient coronal holes [e.g., McAllister et al., Hiei et al., and Webb and Hudson, posters at the Chapman Conference on CMEs]. However, there are also large numbers of events that appear to lack this signature, and in those that do have it, the size of the new holes is often fairly small. At least some major storms (e.g., [S10, S16]) do not show changes in the associated coronal holes, and some (e.g., [S20]) are associated with shrinking coronal holes. Our results suggest that the presence of a significant high-speed stream is one key to major geomagnetic activity. While this stream may, in some cases be supplied, or supplemented, by a new coronal hole associated with the CME itself, there is in most cases in our survey an existing high-speed stream associated with storms on the sector boundary. Major geomagnetic storms in the declining phase are apparently dual-cause phenomena.

In terms of the questions posed in the introduction, our results indicate the general direction of the answer to question (1). The coronal signatures associated with enhanced geomagnetic activity (at least in the period of this study) are dominated by the dynamic coronal arcade events that have become so well known due to the Yohkoh SXT images. These include both large scale quiet Sun events (often of low overall intensity) and compact active region LDEs. There are, however, likely to be other, mostly fainter signatures that are not yet well understood. The ability to work from Earth back to the Sun is a powerful tool for furthering our knowledge of CME signatures. That there are many more CMEs than geomagnetic storms leads to question (2). The tendency for geoeffective CME driven disturbances to arrive at Earth in conjunction with CIRs strongly supports the suggestion that the geoeffectiveness of CMEs is related to their coronal context. Although this is not the only factor involved, in the absence of direct observations of CME velocities and energies, coronal context can be an important tool for predictive purposes, and the continued study of solar wind and coronal data sets can contribute significantly to informing us when CMEs are likely to be geoeffective.

Acknowledgments. This work was supported by the National Science Foundation through HAO/NCAR and grant #ATM94-21814. We thank the Yohkoh and SXT teams for the efforts expended in producing the SXT images. The authors would like to acknowledge the contributions to this work by A.J. Hundhausen and D. Knipp.

REFERENCES

Bhatnager, A., Solar mass ejections and coronal holes, in *Proceedings of IAU Colloquium No. 154, Pune India*, in press, Astrophys. Space Sci., 1996.

Bohlin, J. D., XUV observations of coronal holes: I. Locations, sizes and evolution of coronal holes, June 1973 - January 1974, *Sol. Phys.*, 51, 377–398, 1977.

Bravo, S., A solar scenario for the associated occurrence of flares, eruptive prominences, coronal mass ejections, coronal holes, and interplanetary shocks, *Sol. Phys.*, 161, 57–65, 1995.

Cliver, E. W., Comment on "Interplanetary origin of geomagnetic activity in the declining phase of the solar cycle" by B.T. Tsurutani et al., *J. Geophys. Res.*, 101(A12), 27,625–27,629, 1997.

Crooker, N. U., and E. W. Cliver, Postmodern view of m-regions, *J. Geophys. Res.*, 99(A12), 23,383–23,390, 1994.

Crooker, N. U., and A. H. McAllister, Transients associated with recurrent storms, *J. Geophys. Res.*, in press, 1997.

Crooker, N. U., A. J. Lazerus, R. P. Lepping, K. W. Ogolvie, J. T. Steinberg, A. Szabo, and T. G. Onsager, A two-stream, four-sector, recurrence pattern: Implications from Wind for the 22-year geomagnetic activity cycle, *Geophys. Res. Lett.*, 23(10), 1275–1278, 1996.

Gosling, J. T., The solar flare myth, *J. Geophys. Res.*, 98(A11), 18,937–18,949, 1993.

Gosling, J. T., J. R. Asbridge, S. J. Bame, and W. C. Feldman, Solar wind stream interfaces, *J. Geophys. Res.*, 83, 1401–1412, 1978.

Gosling, J. T., D. J. McComas, J. L. Phillips, and S. J. Bame, Geomagnetic activity associated with earth passage of interplanetary shock disturbances and coronal mass ejections, *J. Geophys. Res.*, 96(A5), 7831–7839, 1991.

Hanaoka, Y., et al., Simultaneous observations of a prominence eruption followed by a coronal arcade formation in radio, soft X-ray and Hα, *Publ. Astron. Soc. Japan*, 46, 205–216, 1994.

Hiei, E., A. J. Hundhausen, and D. G. Sime, Reformation of a coronal helmet streamer by magnetic reconnection after a coronal mass ejection, *Geophys. Res. Lett.*, 20(24), 2785–2788, 1993.

Hudson, H. S., L. W. Acton, and S. Freeland, A long-duration solar flare with mass ejection and global consequences, *Astrophys. J.*, 470, 629 1996.

Hundhausen, A. J., An interplanetary view of coronal holes, in *Coronal Holes and High Speed Streams*, edited by J. B. Zirker, pp. 225–329, Colo. Assoc. Univ. Press, Boulder, 1977.

Hundhausen, A. J., The size and locations of coronal mass ejections: SMM observations from 1980 and 1984-1989, *J. Geophys. Res.*, 98(A8), 13,177–13,200, 1993.

Hundhausen, A. J., Coronal mass ejections, in *Cosmic Winds and The Heliosphere*, edited by J. R. Jokipii, C. P. Sonett, and M. S. Giampapa, in press, University of Arizon Press, 1997.

Hundhausen, A. J., et al., Coronal transients and their interplanetary effects, in *Solar-Terrestrial Physics: Present and Future*, edited by D. M. Butler and K. Papadopoulos, pp. 6-3-6-32, NASA Reference Publication 1120, 1984.

Joselyn, J. A., and P. S. McIntosh, Disappearing solar filaments: A useful predictor of geomagentic activity, *J. Geophys. Res.*, 86(A6), 4555–4564, 1981.

Kahler, S. W., Solar flares and coronal mass ejections, *Annu. Rev. Astron. Astrophys.*, 30, 113–141, 1992.

Legrand, J.-P., and P. A. Simon, Solar cycle and geomagnetic activity: A review for geophysicists, part I, The contributions to geomagnetic activity of shock waves and of the solar wind, *Ann. Geophys.*, 7, 565–578, 1989.

Lemen, J. R., L. W. Acton, D. Alexander, A. B. Galvin, K. L. Harvey, J. T. Hoeksema, X. Zhao, and H. S. Hudson, Solar identification of solar-wind disturbances observed at Ulysses, in *Solar Wind Eight*, edited by D. Winterhalter, J. Gosling, S. R. Habbal, W. Kurth, and M. Neugebauer, p. 92–95, New York, AIP, 1996.

Martin, S. F., and A. H. McAllister, Predicting the Sign of Helicity in Erupting Filaments and Coronal Mass Ejections, in this volume, 1997.

McAllister, A., Y. Uchida, S. Tsuneta, K. Strong, L. Acton, E. Hiei, M. Brunner, T. Watanabe, and K. Shibata, The structure of the coronal soft x-ray source associated with the dark filament disappearence of September 28, 1991 using the Yohkoh Soft X-ray Telescope, *Publ. Astron. Soc. Japan*, 44, L205–L210, 1992.

McAllister, A. H., and A. J. Hundhausen, The relation of Yohkoh coronal arcades to coronal streamers and CMEs, in *Solar Drivers of Interplanetary and Terrestrial Disturbances*, edited by K. S. Balasubramaniam, S. L. Keil, and R. N. Smartt, vol. ASP Conference Series, vol. 95, pp. 171–179, San Francisco, Ca. Astro. Soc. Pac., 1996.

McAllister, A. H., A. J. Hundhausen, J. T. Burkepile, P. McIntosh, and E. Hiei, Declining phase coronal evolution: The statistics of X-ray arcades, in *Magnetohydrodymanic Phenomena in the Solar Atmosphere, Prototypes of Stellar Magnetic Activity, Proceedings of the 153rd IAU Colloquium*, edited by Y. Uchida, H. Hudson, and T. Kosugi, pp. 123–124, Kluwer Acad., Norwell, Mass., 1996a.

McAllister, A. H., H. Kurokawa, K. Shibata, and N. Nitta, A filament eruption and accompanying coronal field changes on November 5, 1992, *Sol. Phys.*, 169, 123–149, 1996b.

McAllister, A. H., M. Dryer, P. McIntosh, H. Singer, and L. A. Weiss, A large polar crown coronal mass ejection and a "problem" geomagnetic storm: April 14-23, 1994, *J. Geophys. Res.*, 101(A6), 13,497–13,515, 1996c.

McAllister, A. H., D. Knipp, and N. U. Crooker, Identification of the solar drivers of the 3-4 November 1993 geomagnetic storm, *J. Geophys. Res.*, submitted, 1997.

Murayama, T., Origin of semi-annual variation of geomagnetic K_p indices, *J. Geophys. Res.*, 79(3), 297–300, 1974.

N. R. Sheeley, J., R. A. Howard, M. J. Koomen, D. J. Michels, R. Schwenn, K.-H. Muhlhauser, and H. Rosenbauer, Coronal mass ejections and interplanetary shocks, *J. Geophys. Res.*, 90(A1), 163–175, 1985.

Neugebauer, M., The problem of associating solar and interplanetary events, in *Proceedings of the Sixth International Solar Wind Conference*, edited by V. J. Pizzo, T. Holzer, and D. G. Sime, pp. 243–259, NCAR/TN-306+PROC, Boulder Colo., 1988.

Pizzo, V., A three-dimensional model of corotating streams in the solar wind, 1. Theoretical foundations, *J. Geophys. Res.*, *83*(A12), 5563, 1978.

Russell, C. T., and R. L. McPherron, Semiannual variation of geomagnetic activity, *J. Geophys. Res.*, *78*(1), 92–108, 1973.

Sheeley, N. R., et al., Coronal changes associated with a disappearing filament, *Sol. Phys.*, *45*, 377–392, 1975.

Sime, D. G., E. Hiei, and A. J. Hundhausen, Coronal eruptive events on April 4 and May 4, 1992, in *X-ray Solar Physics from Yohkoh*, edited by Y. Uchida, T. Watanabe, K. Shibata, and H. S. Hudson, pp. 197–200, Universal Academy Press, Tokyo, 1994.

Tsurutani, B. T., and W. D. Gonzalez, The cause of high intensity long-duration continuous AE activity (HILDCAAS): Interplanetary Alfven wave trains, *Planet. Space Sci.*, *35*, 405–412, 1987.

Tsurutani, B. T., W. D. Gonzalez, F. Tang, S. I. Akasofu, and E. J. Smith, Origin of interplanetary southward magnetic fields responsible for major magnetic storms near solar maximum (1978-1979), *J. Geophys. Res.*, *93*(A8), 8519–8531, 1988.

Tsurutani, B. T., W. D. Gonzalez, A. L. C. Gonzalez, and F. Tang, Interplanetary origin of geomagnetic activty in the declining phase of the solar cycle, *J. Geophys. Res.*, *100*(A11), 21,717–21,733, 1995.

Webb, D. F., The heliospheric manifestations and geoeffectiveness of solar mass ejections, in *Solar-Terrestrial Predictions IV, Vol. II, Proceedings of a Workshop at Ottawa, Canada, May 18-22, 1992.*, edited by J. Hruska, M. A. Shea, D. F. Smart, and G. Heckman, pp. 71–89, NOAA/SEL, Boulder, Colo., 1993.

Webb, D. F., A. S. Krieger, and D. M. Rust, Coronal x-ray enhancements associated with Hα filament disappearances, *Sol. Phys.*, *48*, 159, 1976.

Weiss, L. A., J. T. Gosling, A. H. McAllister, J. L. Phillips, D. J. McComas, A. J. Hundhausen, J. T. Burkepile, K. T. Strong, and G. L. Slater, A comparison of Yohkoh soft X-ray coronal events with observations of coronal mass ejections at Ulysses, *Astron. Astrophys.*, *316*, 384–395, 1996.

N. U. Crooker, Center for Space Physics, Boston University, Boston, Massachusetts, 02215.

A. H. McAllister, High Altitude Observatory, National Center for Atmospheric Research, P.O. Box 3000, Boulder, CO 80307-3000. (e-mail: ahm@ncar.ucar.edu)

CMEs and Space Weather

J. G. Luhmann

Space Sciences Laboratory, University of California, Berkeley, California

Although it has been appreciated for some time that solar wind disturbances drive major geomagnetic activity, there is now a better understanding of the solar causes of these disturbances. In particular, the identification of CMEs at the Sun as the primary source of the most "geoeffective" disturbances has given new impetus and inspiration to the space weather forecasting enterprise. Basic research on the CME initiation and propagation processes is expected to lead to improved physical understanding that translates to predictive schemes. The National Space Weather Program Implementation Plan briefly outlines what needs to be done in a set of "bullets" that are repeated here for perspective. These plans aside, we can already see new ideas and data sets in this volume that have the potential to be incorporated into space weather applications, in many ways the ultimate test of our understanding.

1. INTRODUCTION

In the past, solar flares held center stage in descriptions of solar activity effects on the Earth and its technological systems. Without question, the generation of bursts of energetic photons at UV, X-ray and sometimes gamma ray wavelengths produce ionospheric effects within the light travel time of the flares' occurrence on the Sun. Similarly, flare-associated radio bursts interfere with communications around the time of the flare, while probable flare site-accelerated energetic particles produce anomalous ionization in the atmosphere at high latitudes and sometimes add to the satellite radiation environment. Nevertheless, especially since the publication of the consciousness-raising *Journal of Geophysical Research* article by Gosling [*Gosling*, 1993], new appreciation has been gained for the importance of the less visible coronal eruptions called CMEs in producing "space weather" events. Indeed, as several papers in this volume imply, the CME phenomenon can be regarded as perhaps the greatest challenge of space weather forecasting. Some of the reasons are briefly reviewed here.

2. GEOMAGNETIC DISTURBANCES: SPACE WEATHER'S STORMS

One has only to look at extended geomagnetic records, like the Dst index interval reproduced in Figure 1 [from *Tsurutani et al.*, 1995], to confirm the episodic nature of space weather. Each major (>100 nT) reduction in this index reflects a decrease in the Earth's surface magnetic field at mid and low latitudes caused by increased currents in the magnetosphere. Experience has taught us that such episodes are generally accompanied by enhanced auroral activity and all of the associated atmospheric, ionospheric, and induced ground-current effects that collectively make a magnetic storm.

It has been understood for some time that solar wind disturbances lead to magnetic storms, and that they have their greatest effects when the disturbance has the combination of large plasma velocities V, and large southward components of the interplanetary magnetic field (-Bz). In fact, most physically-inspired solar wind energy

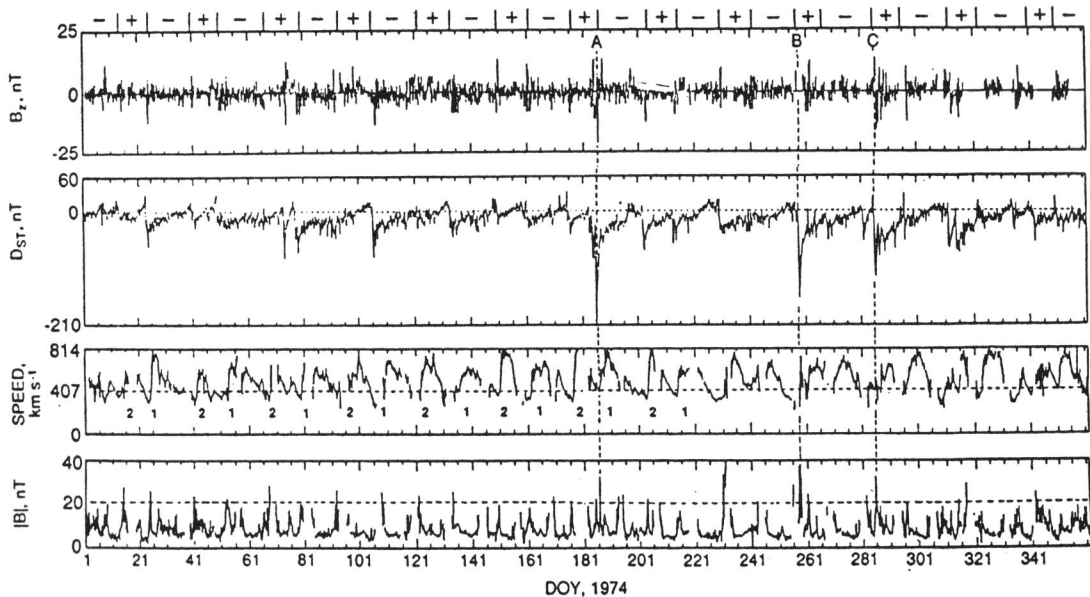

Figure 1. Record of the geomagnetic activity index Dst for the year 1974 [from *Tsurutani et al.*, 1995]. The Dst index is a particularly good measure of major geomagnetic activity, or magnetic storms, because it is heavily based on mid to low latitude ground based magnetic measurements. Key solar wind parameters (velocity and north-south interplanetary field Bz) are also shown. Several large storms are indicated by the dashed vertical lines. Each major Dst disturbance is likely to be the response to arrival of a CME-caused interplanetary disturbance.

or momentum "coupling functions" involve some combination of these two parameters [e.g., *Gonzalez et al.*, 1994]. One particularly popular combination is the product VBz, which is the component of the solar wind convection electric field (E=-VXB) related to -Bz. The physical basis for this choice is illustrated by Figure 2 [from *Hughes*, 1995]. A southward interplanetary field efficiently interconnects with the Earth's magnetic field, thereby mapping the solar wind electric field into the magnetosphere and ionosphere along the approximately "equipotential" interconnected field lines.

3. INTERPLANETARY DISTURBANCES

The two predominant causes of solar wind disturbances producing intervals of enhanced interplanetary VBz are CMEs and solar wind stream interaction regions [e.g., *Lindsay et al.*, 1995, and references therein]. The effects associated with a CME in the solar wind near 1 AU and a stream interaction region near 1 AU are illustrated by the examples in Figure 3. Each type of disturbance has a distinctive character corresponding to its physical nature. The CMEs (e.g. Figure 3a), if moving supermagnetosonically with respect to the ambient solar wind, are preceded by an interplanetary shock that is

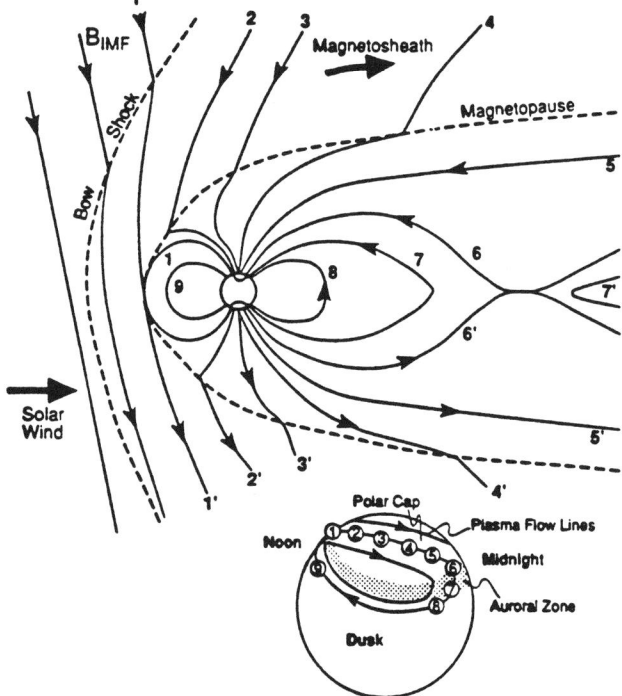

Figure 2. Illustration of how southward interplanetary field interconnects with Earth's field to allow efficient transfer of solar wind energy and momentum to the magnetosphere [from *Hughes*, 1995].

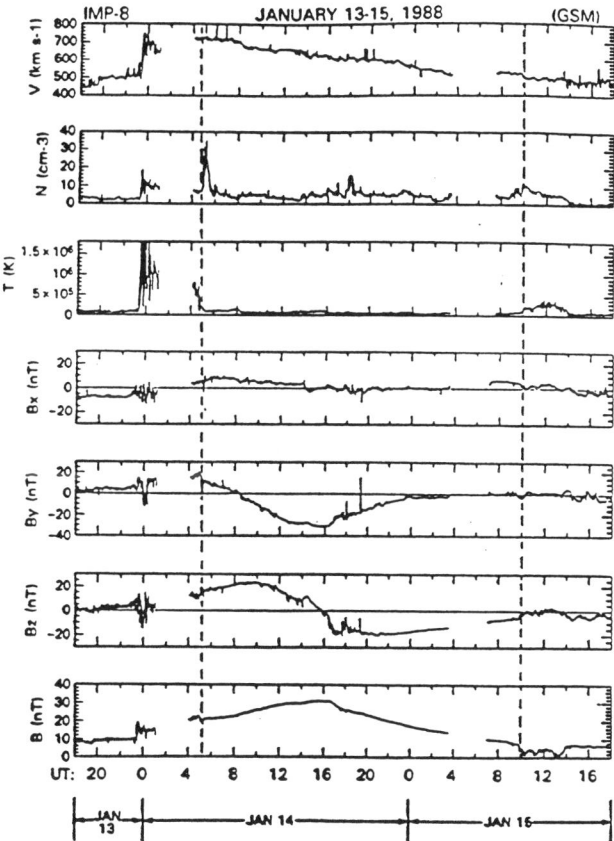

Figure 3a. Example showing the disturbed interplanetary conditions produced by the passage of coronal mass ejecta [from *Farrugia et al.*, 1993].

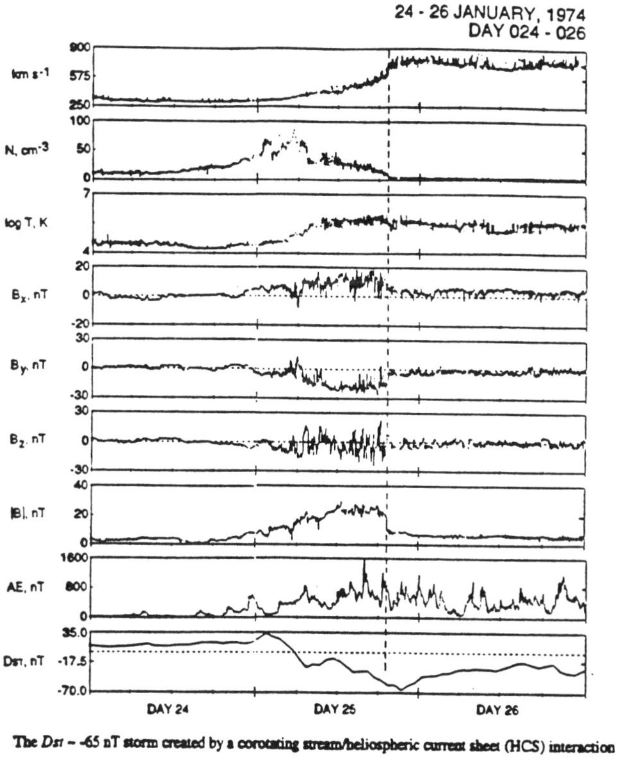

Figure 3b. Example of disturbed interplanetary conditions associated with a particularly strong stream interaction region at 1 AU [from *Tsurutani et al.*, 1995].

followed by enhanced density, velocity and magnetic fields reflecting the pile-up of ambient solar wind ahead of the CME as it plows outward. The magnetic field in this "sheath" region is often deflected out of the ecliptic, thereby enhancing its potential "-Bz" contribution [*McComas et al.*, 1989]. The sheath region passage is typically followed by what appears to be the evolved coronal ejecta, another source of enhanced VBz that sometimes resembles a huge flux-rope [see *Marubashi*, this volume]. It should also be appreciated that in addition to the VBz effects caused by fast CMEs, the preceding shocks can act as broad sources of interplanetary energetic particle events.

The stream interaction regions (e.g., Figure 3b), in contrast, are rarely accompanied by shocks at 1 AU. They further tend to have more rapidly fluctuating and smaller Bz enhancements, and the higher velocities, which appear near the end of the interaction region passage, do not necessarily coincide with the largest Bz enhancements.

This behavior is understood from the different physical nature of the stream interaction, where compression of the slow stream is caused by a highly oblique interaction with the fast stream, and the fast stream is not a discrete entity but rather a part of the general solar wind flow [e.g., *Pizzo*, 1991].

While stream structure disturbances like that in Figure 3b can cause geomagnetic disturbances that affect indices like Dst, and in fact are the likely cause of "recurrent" storms that reappear with the 27 day rotation period of the Sun, they are generally smaller than fast CME disturbances. As seen in the results reproduced in Figure 4, statistical studies of the "geoeffectiveness" parameter VBz in stream interaction regions and CME disturbances show clearly the greater impact of CMEs. Of course *McAllister et al.* [this volume] and others have shown that CME disturbances can be reinforced by interaction with stream structure that magnifies the already enhanced solar wind parameters. These complex or compound disturbances may in fact be responsible for the very largest storms. It is also worthwhile in this context to mention the role of solar

294 CMES AND SPACE WEATHER

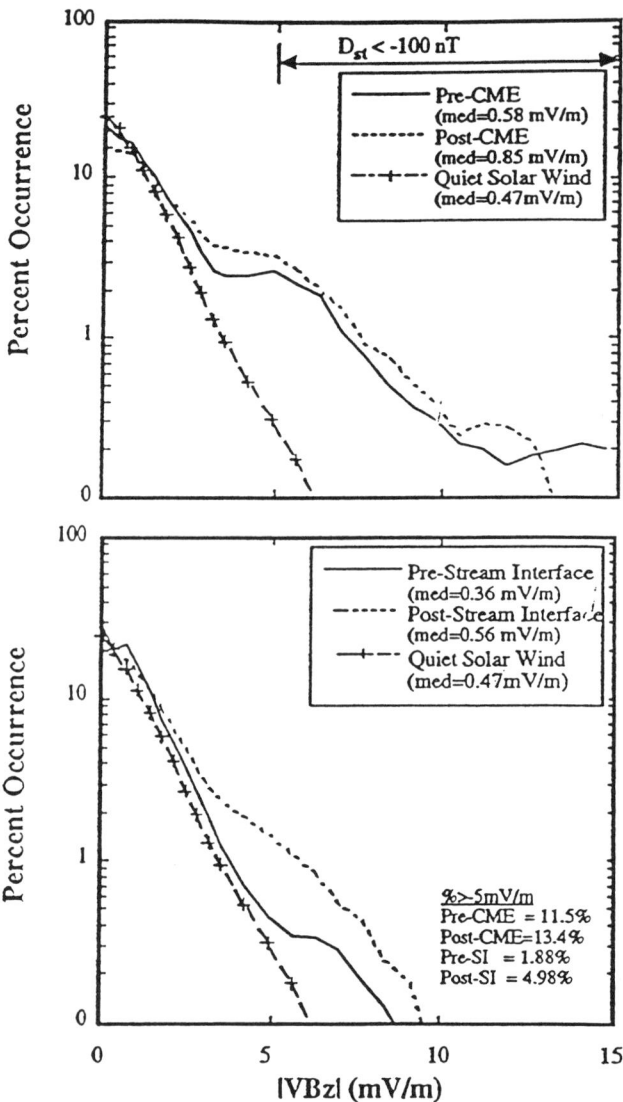

Figure 4. Statistical analysis of the interplanetary parameter VBz associated with passing CMEs and stream interaction regions [from *Lindsay et al.*, 1995]. Note the larger tail on the distribution for CMEs, which make the strongest disturbances.

wind dynamic pressure as it is currently understood. Studies that relate geomagnetic activity indices such as Dst to solar wind parameters find that increases in dynamic pressure can produce disturbances, and can significantly enhance the effects of increased VBz [e.g., *Scurry and Russell*, 1991]. It has recently been found that in cases where the compression of the magnetosphere occurs suddenly in response to the arrival of an interplanetary shock, transient new radiation belt populations may appear [*Blake et al.*, 1992]. However, large VBz appears to be the primary factor in producing the wide range of space weather effects associated with major storms.

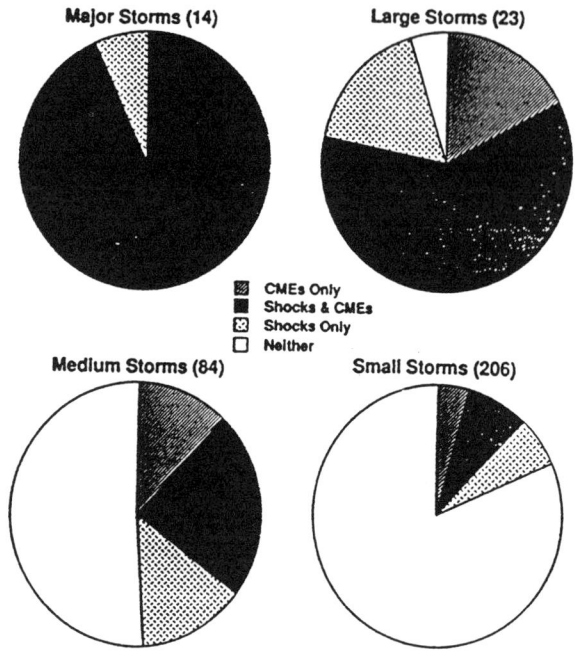

Figure 5. Pie charts from *Gosling et al.* [1991] showing the causes of magnetic storms of various sizes based on analysis of the prevailing interplanetary conditions. (It should be noted that Gosling used a different definition of "major" and "large" storms than is used by the NOAA Space Environmental Center.)

4. THE INTERPLANETARY/GEOMAGNETIC DISTURBANCE CONNECTION

The goal of predicting major geomagnetic activity is of course closely tied to the determination of the cause(s) of large interplanetary VBz. Substantial progress was made when the results contained in the pie charts in Figure 5 [from *Gosling et al.*, 1991] made the CME-magnetic storm connection explicit. Of 14 major and 23 large storms, ~90% and ~65% were associated with some interplanetary signature of a CME disturbance. In many of these cases, a leading interplanetary shock, signalling the passage of a particularly fast cloud of ejecta, provided another measure of geoeffectiveness. Over long time scales, the appearance of a sunspot cycle-phased modulation of many geomagnetic indices, as shown by the example in Figure 6 [from *McPherron*, 1995], implies the dominance of CMEs as geomagnetic activity drivers. Figure 7 [from *Webb and Howard*, 1994], and Figure 8 [from *Lindsay et al.*, 1994] in combination show that virtually all signatures of CME occurrence at the Sun and in interplanetary space increase and decrease with the sunspot number. In contrast, the stream interaction-related disturbances occur with a frequency almost in antiphase with the sunspot cycle.

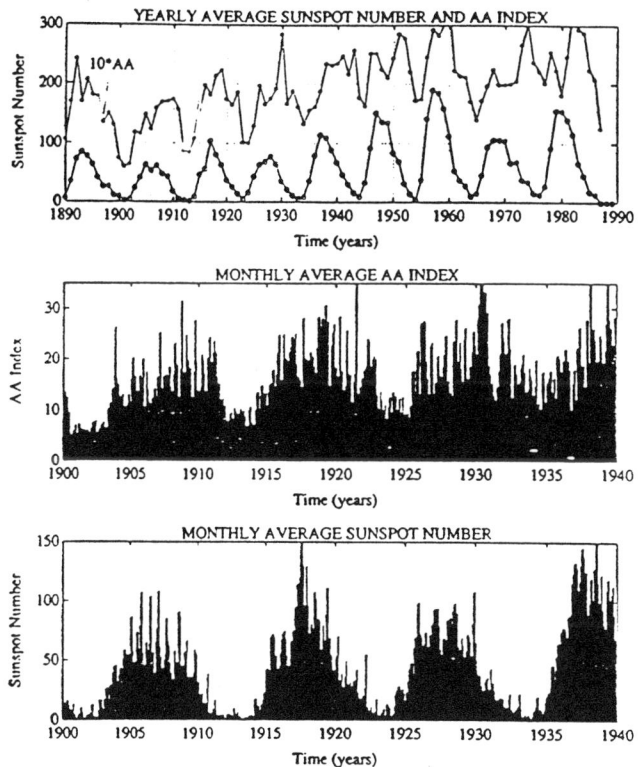

Figure 6. Record showing the solar cycle dependence of the geomagnetic AA index (also based on ground magnetic field deviations) compared to the sunspot number [from *McPherron*, 1995].

This conclusion regarding CMEs' primary contribution to geomagnetic activity explains the poor record for magnetic storm forecasts mentioned by Hildner and illustrated in Figure 9 [from *Joselyn*, 1995]. While the preponderance of geomagnetically quiet days can generally be predicted, the chance of accurately predicting the occurrence of a magnetic storm is well-exceeded by the chance of giving a "false alarm" and the chance of missing a storm that occurs. An ability to detect the departure of a potentially geoeffective (e.g., fast), Earth-bound CME at the Sun would probably provide a more noticeable improvement in our forecasting capabilities than any other single accomplishment. This is the essence of why the understanding of CMEs figures so prominently in the plans of the National Space Weather Program.

5. CME GOALS FOR THE SPACE WEATHER PROGRAM

The philosophy behind the National Space Weather Program initiative is to use and improve our physical

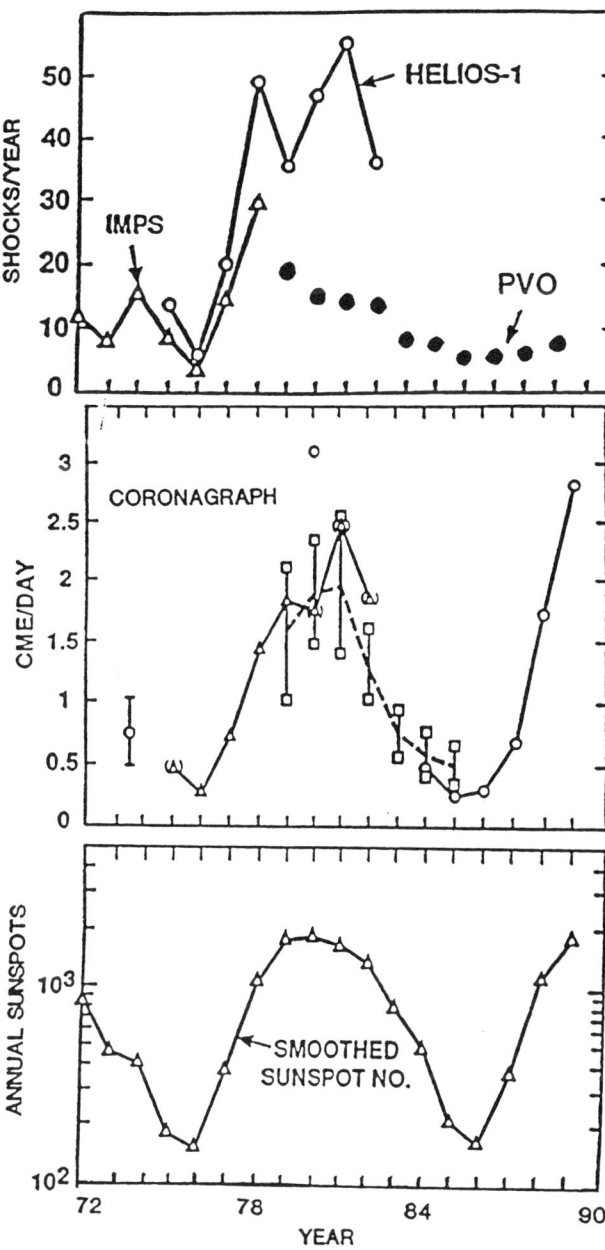

Figure 7. Comparison of CME frequency derived from coronagraph records and interplanetary observations with sunspot number [from *Webb and Howard*, 1994].

understanding of "space weather" to provide accurate predictions of space environment conditions. The manner in which this challenge can be addressed has been given some attention in the National Space Weather Program Implementation Plan (http://www.geo.nsf.gov/atm/nswp/nswp.htm). For CMEs this entails understanding:

296 CMES AND SPACE WEATHER

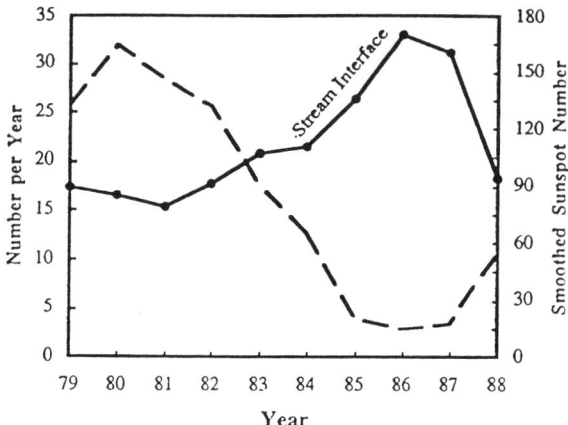

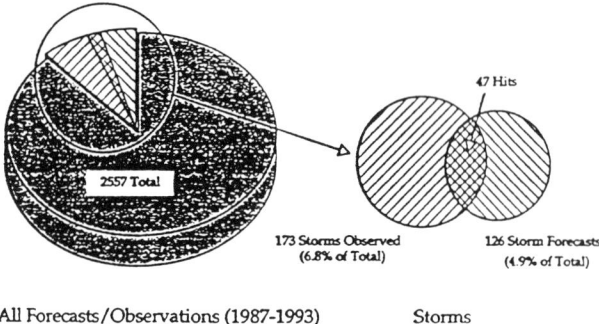

Figure 9. Illustration of the degree of success of magnetic storm forecasting using present techniques [from *Joselyn*, 1995].

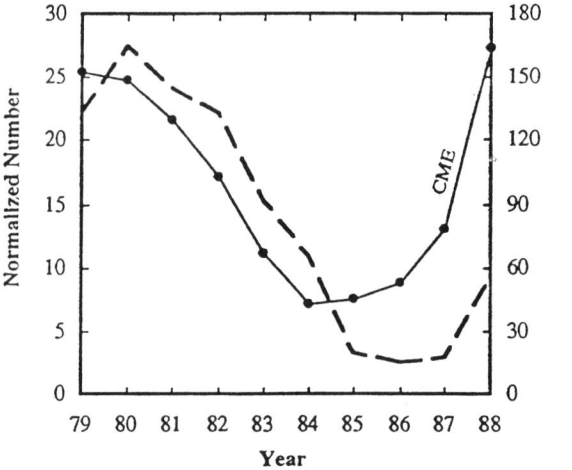

Figure 8. Comparisons of CME occurrence and stream interaction region occurrence as observed on the Pioneer Venus Orbiter with the sunspot number [from *Lindsay et al.*, 1994].

- The physics of the CME initiation process, the factors which determine their sizes, shapes, masses, speeds, and internal field strengths and topologies.
- How to predict the above on the basis of planned observing systems.
- How to predict CME-caused solar wind disturbances and solar energetic particle events near the Earth.

A combination of modeling and observational developments is advocated in the Implementation Plan, the list of which is essentially repeated here:

Models

- Models of the CME initiation process that use realistic observable boundary conditions at the Sun to predict "injection" speed, mass, and intrinsic magnetic field attributes of the ejecta.
- 3D MHD models of the ambient solar wind.
- 3D MHD models of CME-generated disturbance propagation in the solar wind from the Sun to beyond Earth's orbit. These models should strive to simulate the CME structure itself as well as the perturbation caused by the CME in the ambient interplanetary medium. They should ultimately be able to describe disturbance initiation and propagation from the base of the corona to 1 AU using realistic initial conditions for the ambient wind and realistic boundary conditions for the CME disturbance itself.
- 3D models of particle acceleration by CME-driven interplanetary shocks. These models should be capable of predicting the intensity and time history of the CME-associated energetic particle events at 1 AU given a realistic model of the disturbance propagation as in the above bullet.
- Models of the CME-driven shock related radio emission process. These models are needed to optimize the use of radio noise as a remote sensing device and as a diagnostic of approaching CMEs.

Observations

- Soft X-ray imagers for use in understanding and predicting solar wind disturbances such as provided by Yohkoh and SOHO, in anticipation of the SXI X-ray monitoring spacecraft series.
- Radio facilities for tracking solar wind disturbances in interplanetary space from the Sun to the Earth, using both radio bursts and the interplanetary scintillation technique.
- Coronagraphs for studying the behavior of CMEs as a function of radial distance.

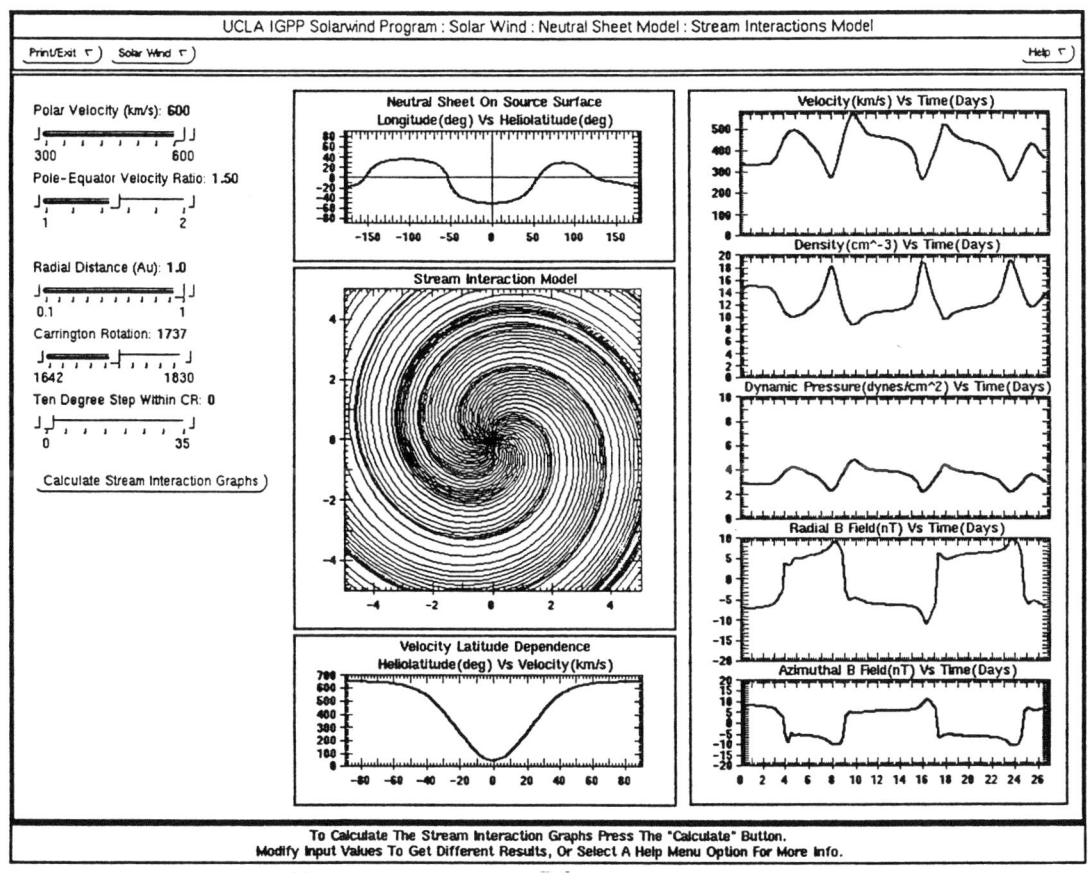

Figure 10. Display from a potential tool for solar wind stream structure forecasting from full-disk magnetogram-derived neutral lines.

- Ground-based coronagraphs to provide a measure of global solar CME activity levels, and to increase the data base for investigations of solar cycle variations of CMEs.
- Solar wind monitors placed near Venus or Mercury orbit for predicting solar wind disturbances near Earth.
- L1 or equivalent upstream monitors for carrying out the above investigations and for providing at least a one-hour forecast for major geomagnetic storms.
- EUV magnetographs for measuring coronal magnetic fields.

Because this list will constantly evolve as new knowledge is incorporated into the National Space Weather Program, it should be viewed only as a current perspective that serves as a "benchmark" at the program's beginning.

6. WHAT WE CAN DO NOW

These challenges and future efforts notwithstanding, there are things that we can do today, with current knowledge, to improve the state of the art of CME forecasting. For example, some recent investigations have concentrated on identifying signatures on the Sun that can be used as indicators of a coronal ejection event [e.g., see papers by *Martin and McAllister, Hudson and Webb, Bothmer and Rust*, among others in this volume]. The association of CMEs with disappearing filaments seen in H-alpha is well-known, while the long duration soft X-ray events, probably arising from the glowing coronal arcades observed in Yohkoh images, seem to follow low coronal disruptions (and sometimes filament disappearances). The newest possibility relates to dimmings of the low corona in the soft X-ray Yohkoh images [see *Hudson and Webb*, this volume]. Of course, one of the problems with the reliance on the filaments is that they are not always present (or seen), while the physical reasons for the soft X-ray arcades and low coronal dimmings are poorly understood. Moreover, the keys to geoeffectiveness, high velocity and substantial southward magnetic field, are not clearly predictable from these "smoking guns" (although *Bothmer and Rust*, [this volume], are optimistic about predicting the magnetic field orientation when an associated filament is seen). Nevertheless, schemes based on physical models may be available for testing soon.

For example, *Mikic and Linker* [this volume] are able to simulate eruption of realistic coronal helmet streamers subjected to photospheric shearing motions at the footpoints, while *Feynman and Martin* and *Wu et al.* [both this volume] are looking into the role of solar flux emergence in producing CMEs from observational and modeling perspectives. In the meantime, we can make better use of the abundance of new observations of the Sun to gain insight. In particular, the SOHO spacecraft instruments [e.g., see *Howard et al.*, this volume] provide a look at CMEs to several 10s of solar radii from the Sun, with the potential of helping us to understand when and why acceleration of a CME sometimes seems to occur quite far from the Sun. SOHO also provides images in many wavelengths of coronal interest, including full-disk magnetograms every 96 minutes that can be used to analyze the flux emergence concepts. Yohkoh continues to produce more data on the low coronal soft X-ray phenomena it discovered including coronal dimmings. The GONG network, in addition to possible helioseismological insights into the generation process, provides low resolution full disk magnetograms on a 20 minute time-scale.

In short, we have an unprecedented opportunity today to both test our present ideas about CME causes against observations and to use comprehensive solar observations to test new forecasting schemes based on these ideas. One can envision, for example, a system where high time resolution full disk magnetograms are used to observe flux emergence and also to update interplanetary stream structure conditions. The solar wind model might be used to produce a display that looked like Figure 10, for example. Then, if an injection location, width and speed was inferred, even approximately, another model could be used to propagate the transient through the solar wind, as illustrated by Figure 11, at least forecasting the disturbance pressure and velocity and time of arrival. Even the probability of a local shock-accelerated energetic particle event could be given by considering the orientation of the interplanetary field and the strength of the preceding shock [e.g., see *Reames*, this volume]. How we exploit the new observations and our current knowledge is up to us.

Starting next year, the construction phase of the International Space Station (ISS) is scheduled to begin. The ISS has a new high latitude orbit that takes it into the disturbed-time expanded auroral oval to which energetic interplanetary particles have access, and it requires a large number of EVAs to build. At the same time, large

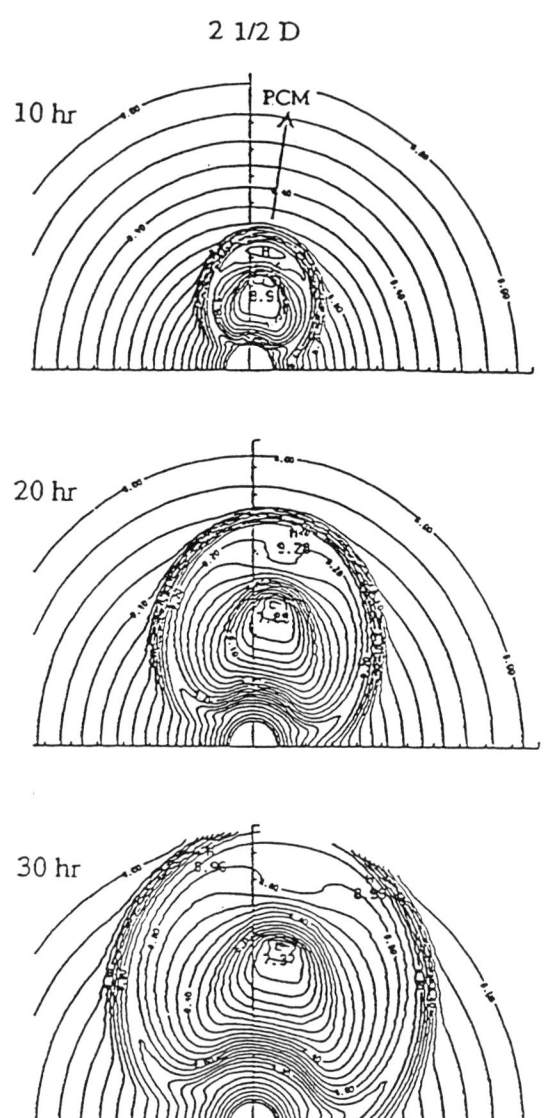

Figure 11. Display from a potential tool for forecasting the properties and timing of CME-driven interplanetary disturbances [from *Smith et al.*, 1992].

networks of navigation and communications satellites such as GPS, Iridium and Teledesic will become essential to our way of life. Solar activity, including the frequency of CMEs, will begin to increase as the 1999-2002 solar maximum is approached. Space weather will become more noticeable. It is a perfect time for us to unravel the physics behind CMEs.

Acknowledgments. The author thanks the conveners for organizing the excellent and timely Bozeman Chapman Conference on CMEs. Her work in this area is supported by NSF grant FD95-31741 from the Solar Terrestrial section of the Atmospheric Sciences Division.

REFERENCES

Blake, J. B., et al., Injection of electrons and protons with energies of tens of MeV into L<3 on 24 March 1991, *Geophys. Res. Lett.*, 19, 821, 1992.

Bothmer, V., and D. M. Rust, The field configuration of magnetic clouds and the solar cycle, this volume.

Farrugia, C.J., M.P. Freeman, L.F. Burlaga, R.P. Lepping, and K. Takahashi, The Earth's magnetosphere under continued forcing: Substorm activity during the passage of an interplanetary magnetic cloud, *J. Geophys. Res.*, 98, 7657, 1993.

Feynman, J., Evolving magnetic structures and their relation to coronal mass ejections, this volume.

Gonzalez, W.D., J.A. Joselyn, Y. Kamide, H.W. Kroehl, G. Rostoker, B.T. Tsurutani and V.M. Vasyliunas, What is a geomagnetic storm?, *J. Geophys. Res.*, 99, 5771, 1994.

Gosling, J.T., The solar flare myth, *J. Geophys. Res.*, 98, 18,937, 1993.

Gosling, J.T., D.J. McComas, J.L. Phillips, and S.J. Bame, Geomagnetic activity associated with earth passage of interplanetary shock disturbances, *J. Geophys. Res.*, 96, 7831, 1991.

Howard, R. A. et al., Observations of CMEs from SOHO/LASCO, this volume.

Hudson, H. S., and D. F. Webb, Ejections of coronal material seen in soft X-rays, this volume.

Hughes, W.J., The magnetopause, magnetotail, and magnetic reconnection, in *Introduction to Space Physics*, edited by M.G. Kivelson and C.T. Russell, Cambridge Univ. press, New York, 1995.

Joselyn, J.A., Geomagnetic activity forecasting: The state of the art, *Rev. Geophys.*, 33, 383, 1995.

Lindsay, G.M., C.T. Russell, J.G. Luhmann, and P. Gazis, On the sources of interplanetary shocks at 0.72 AU, *J. Geophys. Res.*, 99, 11, 1994.

Lindsay, G.M., C.T. Russell, and J.G. Luhmann, Coronal mass ejection and stream interaction region characteristics and their potential geomagnetic effectiveness, *J. Geophys. Res.*, 100, 16,999, 1995.

Martin, S., and A. H. McAllister, Predicting the sign of helicity in erupting filaments and coronal mass ejections, this volume.

Marubashi, K., Interplanetary flux ropes and solar filaments, this volume.

McAllister, A. H., and N. U. Crooker, Coronal mass ejections, corotating interaction regions, and geomagnetic storms, this volume.

McComas, D.J., J.T. Gosling, S.J. Bame, E.J. Smith, and H.V. Cane, A test of magnetic field draping induced Bz perturbations ahead of fast coronal mass ejecta, *J. Geophys. Res.*, 94, 1465, 1989.

McPherron, R.L., Magnetospheric Dynamics, in *Introduction to Space Physics*, edited by M.G. Kivelson and C.T. Russell, Cambridge University Press, New York, 1995.

Mikic, Z., and J. A. Linker, The initiation of coronal mass ejections by magnetic shear, this volume.

Pizzo, V., The evolution of corotating stream fronts near the ecliptic plane in the inner solar system 2., Three dimensional tilted dipole fronts, *J. Geophys. Res.*, 96, 5405, 1991.

Reames, D. V., Energetic particles and the structure of coronal mass ejections, this volume.

Scurry, L., and C.T. Russell, Proxy studies of energy transfer to the magnetosphere, *J. Geophys. Res.*, 96, 9541, 1991.

Smith, Z., T. R. Detman, and M. Dryer, Comparison of $2^{1/2}$D and 3D simulations of propagating interplanetary shocks, in *Solar Wind Seven*, edited by E. Marsch and R. Schwenn, p. 667, Pergamon Press, Oxford, 1992.

Tsurutani, B.T., W.D. Gonzalez, A.L.C. Gonzalez, F. Tang, J.K. Arballo, and M. Okada, Interplanetary origin of geomagnetic activity in the declining phase of the solar cycle, *J. Geophys. Res.* 100, 21,717, 1995.

Webb, D.F., and R.A. Howard, The solar cycle variation of coronal mass ejections and solar wind mass flux, *J. Geophys. Res.*, 99, 4201, 1994.

Wu, S. T., and W. P. Guo, A self-consistent numerical magnetohydrodynamic model of helmet streamer and flux-rope interactions: Initiation and propagation of coronal mass ejections, this volume.

J. G. Luhmann, Space Sciences Laboratory, University of California, Berkeley, CA 94720-7450.